H. F. Ebel, C. Bliefert, W. E. Russey

The Art of Scientific Writing

Hans F. Ebel, Claus Bliefert
William E. Russey

The Art of
Scientific Writing

From Student Reports to
Professional Publications in Chemistry
and Related Fields

Second, Completely Revised Edition

WILEY-VCH

WILEY-VCH Verlag GmbH & Co. KGaA

Dr. rer. nat. habil. Hans F. Ebel
Im Kantelacker 15
64646 Heppenheim
Germany

Prof. Dr. Claus Bliefert
Meisenstraße 60
48624 Schöppingen
Germany

Dr. William E. Russey
3508 Cold Springs Road
Huntingdon, PA 16652
USA

1st Edition 1987
2nd Edition 2004
   Corrected Reprint of the 2nd Edition 2005
   2nd Reprint 2008

Library of Congress Card No.: 2004555122

British Library Cataloging-in-Publication Data
A catalogue record for this book is available from the British Library

Bibliographic information published by
Die Deutsche Bibliothek
Die Deutsche Bibliothek lists this publication in the Deutsche Nationalbibliografie; detailed bibliographic data is available in the Internet at <http://dnb.ddb.de>.

© 2004 WILEY-VCH Verlag GmbH & Co. KGaA, Weinheim

Printed in the Federal Republic of Germany
Printed on acid-free paper

Cover design: Grafik-Design Schulz, Fußgönnheim
Printing: betz-druck GmbH, Darmstadt
Bookbinding: Litges & Dopf Buchbinderei GmbH, Heppenheim

ISBN: 978-3-527-29829-7

By three methods we may approach wisdom:

First, by reflection, which is noblest;
second, by imitation, which is the easiest;
and third, by experience, which is the bitterest.

CONFUCIUS

Preface

The first edition of this book appeared in 1987. Many reviewers—representing a wide assortment of scientific–technical–medical disciplines worldwide—greeted this early product of our efforts with considerable enthusiasm, and much to our satisfaction it was also warmly accepted in the "marketplace". After only two years a first reprinting was required, and the book has continued to be available ever since. Excerpts of it have been published in Russian, and the entire volume was translated into Chinese and released by the respected Beijing scientific publisher Science Press.

At the time, the book gained a reputation for being highly modern, not only from a content standpoint but also because of the way it was produced. In the words of Robert SCHOENFELD (to whom many are indebted for *The Chemist's English*), published in *Chemistry in Australia*, "When I looked at the book production my eyes popped. The volume is impeccably printed by the latest in desktop publishing methods, and the text is entirely author-generated. That explains why all sections are spectacularly up-to-date, and it also explains the remarkably low price...". Today, 17 years later, that first edition in our eyes resembles a fossil out of the Mesolithic, but we feel confident that with this second edition we have given life to a worthy and viable "off-spring". The road from the one to the other turned out to be a long one, fraught with unforeseen delays and unavoidable (but nonetheless frustrating) obstacles. We are extremely appreciative of the indulgence of all those who have waited so patiently, most especially our publisher, who never lost faith in the project.

In the meantime, a German counterpart to *The Art of Scientific Writing* has been enthusiastically received (EBEL/BLIEFERT *Schreiben und Publizieren in den Naturwissenschaften*), the 4th edition of which (1998) was in fact adopted as a basis for this new English work, although the latter also breaks considerable new ground in its own right.

Perhaps no aspect of reality has been more heavily influenced by the "electronic revolution" than communication, including communication in the sciences, and the effects have been profound for both writing and publishing. The techniques of writing today are radically different from those of 20 years ago. Entirely new modes of information transfer have become central to everyday life, and all our assumptions about "publishing" in the traditional sense have been put to the test. That said, we have a great deal of confidence in the future of scientific publication, the foundation of which will continue to be writing, irrespective of precisely how information will be disseminated. Virtually every page of this book testifies to our faith in this regard.

Just as it was 17 years ago, our goal remains that of informing and instructing, perhaps also shining new light on what underlies scientific communication as we know it and at the same time providing guidance of a very practical sort. This second edition, even more than its predecessor, has assumed much of the character of a reference work. To help ensure that you, our readers, are able to profit to the fullest from its reference potential we have devoted substantial thought and effort to preparing an unusually comprehensive index, much of which might have been spared had we been able to make the material available to you in a database format. We strongly suspect that in the not too distant future that dream could become reality.

The three of us owe a major debt of gratitude to our physicist friend Walter GREU-LICH in Weinheim (Germany), who was recently named senior editor of the Brockhaus encyclopedia *Computer und Informationstechnologie* ("Computer and Information Technology")—someone who can truly be said to have been "bathed and baptized in all the waters of scientific publishing", meanwhile developing his own successful consulting firm serving the publishing industry. We are proud to acknowledge the extent to which this junior colleague has generously helped to ensure that this latest edition of our book can legitimately aspire again to be regarded as "up-to-date" in nearly every sense of that phrase. Work is already underway on the 5th edition of the German publication, in which Walter GREULICH will join us as a full coauthor. Thanks are also due to Dr. Ann RUSSEY CANNON, Assoc. Prof. of Statistics and Mathematics in the Mathematics Department at Cornell College, Mt. Vernon, Iowa, for her critical appraisal and guidance with respect to our comments regarding statistics, and to Dipl.-Chem. Dipl.-Ing. Frank ERDT and Florian BLIEFERT for their support during manuscript preparation.— Finally, we wish once again to express our thanks to our publisher, Wiley–VCH, and especially to the editor there chiefly responsible for this project, Dr. Frank WEINREICH, who has been steadfast in his support of our efforts and a paragon of understanding and encouragement. And anyone who has ever engaged in a project of this magnitude will appreciate how thankful we are for the faith and tolerance and sympathy and patience we have always been able to rely upon from our wonderful wives: Inge, Annie, and Lainie.

Heppenheim	H.F.E.
Schöppingen	C.B.
Huntingdon	W.E.R.

December 2003

Contents

I
Goals and Forms
in Scientific Writing

1 Reports

1.1 The Scientist as Writer

1.1.1 Communication in the Natural Sciences

Underlying all of natural science is a rather remarkable understanding, albeit one that attracts relatively little attention:

- Everything measured, detected, invented, or arrived at theoretically in the name of science must, as soon as possible, be made public—complete with all the details.

It is in fact absolutely essential that scientific results be shared openly, because without the constant exchange and dissemination of information there can *be* no science. In this sense, the humble scientific *report*, once published, is raised quite literally to the status of "indispensable end product of any piece of research". This observation serves as an excellent starting point for our examination of "the scientist as writer".

Every discovery in science is based to some extent on previous discoveries, frequently made by others, and each new piece of information constitutes but one further stage in an endless journey. Results and conclusions emerge from a laboratory, or from the theoretician's imagination, and sooner or later they stimulate or contribute to other investigations elsewhere, which subsequently produce their own results and conclusions. This is what in the aggregate we call "scientific progress". If the creative process is to unfold smoothly and efficiently it is obviously essential that new insights be shared widely—*communicated*—in a timely fashion, thus becoming readily accessible to others whose own (unpredictable!) future accomplishments might, under favorable circumstances, be heavily influenced by them.

The word *communication* has several distinct meanings. From a *linguist's* perspective, "communication" suggests a direct, substantive exchange (interaction), either oral or written, involving two or more distinct parties. In *information technology*, "communication" implies only that two or more parties have mutual *access* to a common store of information (Lat. *communicare*—to have something in common, to share with one another). *Oral* communication relies on speech and the spoken language, and may itself take many forms (dialogue, oration, debate, etc.), and the same diversity applies to *written* communication. Sometimes the latter is called upon to serve as an emergency substitute when speech fails. Thus, a Westerner traveling in China who lacked the ability to communicate orally with passersby—but was in a rush to honor an important engagement—might well resort to handing a taxi driver a business card or letterhead bearing the relevant address, spelled out in (for the traveler indecipherable) Chinese characters. A correspondence conducted by letter represents one obvious "low-tech" example of written communication facilitating meaningful "direct exchange" at a distance. The idea of "communication" needs to be expanded, however, if it is also to encompass information transfer via a textbook, because in this case the book will almost certainly have been prepared for a vaguely defined, anonymous readership. Its use as an educational tool is therefore unlikely to produce any "direct interaction". A similar situation arises in use of the term "scientific communication" in reference to a journal article, but it is communication in the latter sense to which we nevertheless direct our attention most often in this book.

Fine. It is a simple matter to establish the fact that communication plays a fundamental role in the sciences, but what form or forms should this communication take? The *spoken word* has under most circumstances been mankind's preferred medium for information transfer. Like other people, scientists do of course communicate orally, and they frequently share with one another in this way insights into their professional activities: informally in the office, over lunch, on the telephone, in the lecture hall, in the laboratory, or in the corridor outside a conference room. Occasionally they present (and attend) formal *lectures* that function at least in part as primary channels of communication.[1] Colleagues use vocal means to exchange ideas and scientific news in all kinds of settings.

But the spoken word is a fleeting thing, and even if it were not, it would rarely provide the precision and depth essential for official announcement of a profound scientific observation or deduction. Moreover, one person's voice is unlikely to have a lasting effect on more than a token fraction of the multitude of individuals worldwide actively engaged in the scientific enterprise. In other words, speech cannot be an adequate substitute in this context for *written communication*, with its unique capacity for conveying the most complex message, unambiguously and as often as necessary, via a sophisticated combination of tightly constructed verbal arguments, exhaustive (or perhaps only illustrative) compilations of relevant technical data, and pictorial representations for complex entities or concepts the description of which in words alone would at best be inefficient, not to mention far less thought provoking. Information in written form has the further advantage that it can be propagated with ease, for distribution as widely as necessary or desirable. A written message can also be encapsulated in ways that allow it to be preserved over indefinitely long periods of time, available to anyone with access to the appropriate deciphering tools. Writing in one form or another is the indispensable medium for serious communication within most academic disciplines, but especially in the sciences. Written communication plays such a central role in the scientist's professional life that most practitioners actually spend more time writing than engaging in any other activity.

In recent years the *process* of writing has undergone a remarkable series of transformations, acquiring new characteristics that have turned it into a vastly different activity from what was taken for granted only two decades ago. Where writing

---

[1] It is only fair to acknowledge, however, that a lecture also provides the speaker with a unique mechanism for achieving greater visibility!

was once associated in everyone's mind with pencils and pens, typewriters, manuscript paper, and formal documents produced professionally, every aspect of composition, editing, and printing in the modern world centers increasingly around electronic work-stations consisting of *computers, modems, display screens (monitors), keyboards, scanners, printers,* etc. With these facilities the average scientific professional is personally able to prepare documents, elaborately formatted, of every conceivable type (e.g., reports, journal-article manuscripts, grant proposals, product descriptions, safety sheets, etc.), and the finished product is often superior in quality—at least in a technical sense—to anything that could possibly have been generated in a home or office setting 15 years ago. Moreover, the same humble computer that constitutes the centerpiece of such document production also enables one to transmit a perfect copy of the new work virtually anywhere in the world—in *electronic* form—within a matter of seconds, thanks to almost universal access to the *Internet* (see below).

Millions of people around the globe, most of whom have no direct relationship whatsoever with the sciences, take advantage every day of these same communication possibilities, but scientists can perhaps be forgiven for feeling a special affinity with the new media because the scientific enterprise was so intimately involved in their development. Computer technology we now take for granted would be unthinkable were it not for a host of discoveries and inventions that can be traced directly to laboratories devoted primarily to basic research in chemistry and physics. Scientists were also among the first to foresee and work toward development of the highly interconnected "computerization" through which so much of contemporary society has been reshaped. For example, the information protocols that define the *World Wide Web* ("WWW"), responsible in turn for making the Internet such a vibrant part of everyday life, were first conceived and tested in 1989 in high-energy physics laboratories at the *Conseil Européen pour la Recherche Nucléaire* (CERN) in Geneva, Switzerland.[2]

The now-familiar expression *desktop publishing* (DTP) was coined nearly two decades ago[3] to underscore a valuable synergistic interaction of a suite of powerful tools, including both "hardware" and "software" components, each of which individually had made its own significant contribution to the information revolution. DTP is simply one manifestation of the information-processing potential inherent in the high-tech devices mentioned above in the context of the modern scientist's "electronic workstation". The original goal of DTP was to permit generation of high-quality hard-copy output, including both textual and graphic information, with versatile electronic means used to bypass unwieldy and costly equipment traditionally restricted

---

[2] The Internet itself is an outgrowth of an early United States Department of Defense computer network called *Arpanet*, which was developed at the start of the 1970s, long before most people could envision even the possibility that computers might soon become everyday communication devices for society in general.

[3] Probably by Paul BRAINERD of the Aldus Corporation, in conjunction with development and release (1985) of the pioneering PAGEMAKER page-layout software (see "Basic Considerations" in Sec. 5.4.1) for the then relatively new Apple Macintosh computer.

to full-time publishing professionals sequestered in dedicated printing establishments. Since its inception in the mid-1980s, the so-called *communication revolution*, which includes DTP, has been responsible for sweeping changes, long since eclipsing the original, rather modest goals.

The most recent phase of DTP has had a somewhat different aim, what has come to be called *electronic publishing*: the transmission, by electronic means and with the aid of various *digital storage media* and *electronic network facilities*, of "virtual" messages,[4] often highly sophisticated from a structural standpoint and capable of reaching large numbers of potential recipients ("readers") almost instantaneously.

We will return to specific aspects of these developments frequently in the sections and chapters that follow.

As suggested above, the venerable writing desk with its stock of pens, pencils, and paper has been effectively displaced from its long-held position as the essential point of origin of a scientist's[5] most important written communications. That crucial function has now been assumed by the electronic workstation (or *communication station*), with a desktop (personal) computer at its heart.[6] This versatile little facility—interpreted here as simply a "high-tech" substitute for pencil and paper—is in most cases directly connected with various other computers in the same complex (research facility, industrial site, or university), each unit constituting one particular *node* in a *computer network* (more precisely, a *local area network* or LAN). This in turn serves as a gateway to the broader Internet, linking millions of computers worldwide.[7] The scientist in

---

[4] The word *virtual* has taken on something of a mystical aura in the age of rapidly proliferating, interconnected computers; see, for example, Howard RHEINGOLD's book *The Virtual Community* (2000). The term has its origins in the Latin *virtus*—"virtue", in the sense of "a potentially available power or possibility". The adjective "virtual" first came to be associated with the notion "thought of" or "conceived" in the world of medieval philosophy, finding its way eventually into physics through the principle of "virtual advance", which in turn became the basis for the "golden rule of mechanics".

[5] Often in this book, points we make explicitly with respect to "scientists" apply equally well to those engaged in developmental work, for example, even though according to a strict interpretation the latter would probably not be considered practitioners of ("pure") science. The same holds true with regard to a host of other technically oriented areas, the educational training for which is likely to come from some institution other than a "traditional" university. Despite infrequent direct mention here of technicians, engineers, and the like, these people were very much on our minds as we formulated our arguments, and we certainly would not wish such respected colleagues to feel neglected! Our examples are biased as well, most being drawn from the "classic" sciences, especially chemistry, but only because that is the territory with which we ourselves are most familiar.

[6] Even before the advent of powerful personal computers, pioneering attempts were made to exploit computing devices—mostly massive mainframe behemoths of the type ubiquitous in those days at universities—in the production of high-quality printed material. Legacies from that era are still occasionally encountered, like the following observation from a still-valuable monograph: "...These commands, which presuppose LaTeX-based processing, are often system-dependent in nature, and thus subject to important local variation or modification. For further information you should consult with experienced personnel at your institution's computer center" (KOPKA, 1st ed., 1991). Warnings like this were exceedingly common only a few years ago, but they now reflect a stage of technology long since superceded.

[7] One sometimes speaks in an institutional context of a local *intranet*, defined as an internal network based on the same transmission technologies (TCP/IP) and services (e.g., e-mail, FTP) common through-

→

charge of a networked computer will also have been assigned a unique *e-mail address* (*e-mail*: electronic mail), opening the way to one-on-one communication with countless other individuals and institutions throughout the world. The personal computer thus assumes responsibility for routine distribution of a high proportion of everyday business correspondence—and may simultaneously act as home base for a personal *Web site* as well. Our scientist is now in a position to generate with ease documents of every conceivable type, and to subject them to previously unimaginable levels of aesthetic refinement (a consideration to which we return in Chapter 5), ultimately dispatching the documents personally to wherever they belong.

Electronic mail differs from conventional mail in several important respects, including the fact that in most cases the sender assumes there will be almost instantaneous delivery—and no need for direct human involvement anywhere along the way. The same of course applies to the scientific manuscript transmitted today over telephone lines to the editor of a journal, usually in the form of an e-mail *attachment*. The composition (or reading, for that matter) of almost any sort of "electronic" document can be pursued at the most basic work station, all the creative initiatives involved being monitored on a video display screen with a diagonal dimension unlikely to exceed 17 inches (ca. 43 cm). The concentration in one facility of so diverse a set of activities is actually quite remarkable. The developments that led to these new manifestations of communication *technology* have of course also produced new forms of communicative *behavior*, along with a reward system that reflects the centrality of skills different from any one would have associated in the past with either reading or writing.

Writing in the broadest sense implies communication conducted on the basis of agreed-upon sets of *visual symbols*. Typing text into a computer is a novel alternative to communicative "writing" in the classical sense, but one also subject to unique limitations that depend, for example, upon what facilities happen to be available to the targeted recipient of the message, and to some extent as well upon the particular path the message is destined to follow in the course of its journey.[8] At the receiving point, electronic signals representing the message can of course be transformed into an ordinary printed document, for leisurely reading in the traditional manner, but this presupposes the recipient is so inclined. Electronic messages in fact arrive initially as mere *titles* displayed on a video screen, which might be glanced at only in a cursory way before being consigned unceremoniously to an "electronic wastebasket" residing unobtrusively in a corner of the user's electronic "desktop".

● The content of an electronic message becomes intelligible only after it undergoes "translation" into a conventional visual format with the aid of, among other things, appropriate *type fonts*.

---

out the Internet, making it especially easy for an intranet to forge two-way links that can span the globe.
[8] Interestingly, the Internet is so structured that various words making up a document typically reach the recipient over a multitude of *different* pathways. The complete text is then efficiently reconstituted as if by magic once all of it has reached its destination.

This translation process is an indispensable but often overlooked step the sending party should actually take carefully into account to increase the probability that a message will in fact reach its intended reader in an understandable form. Modern computers typically offer their users a wide assortment of type fonts, but only a few of these over the years have come to be regarded as "standard" (e.g., Times, Courier, Arial, Verdana, Symbol, etc.). If the sender in cavalier fashion specifies that a particular text should be rendered in some exotic font not recognized by the recipient's computer, the message's *content* may become hopelessly garbled, leaving its meaning utterly obscure, at least until an appropriate "fix" has been devised.

Even though communication *techniques* have changed radically in the last few years, basic *arts*, *skills*, and *practices* long associated with good writing—and addressed in innumerable guides, handbooks, treatises, seminars, and even entire curricula—must continue to represent a high priority for anyone aspiring to develop into a truly productive, successful professional, held in high esteem by others. If anything, arguments in favor of learning to write and present one's thoughts effectively in visible form should now be more compelling than ever, because in their role as "communicators" today scientists find themselves under increasing pressure to function as "desktop publishers". They are thus expected to assume added responsibility for ensuring the overall *quality* of their communicative efforts, including aesthetic and technical considerations previously attended to—routinely, albeit rather discretely— by professional editors (or "designers"; cf. KOPKA and DALY 1991). This is a matter to which we will necessarily return frequently in the chapters that follow (see especially the discussion of electronic publishing in Sec. 3.1.2).

1.1.2 The Maintenance of Quality in Science

We have already noted that active communication—*written* communication—between practicing scientists scattered throughout the world is the powerful engine that propels virtually all scientific progress. This observation provides one important justification for the assertion with which we began: that written documentation is a mandatory adjunct to every legitimate scientific concept or finding. But the role of documentation in science extends well beyond communication. Information in written form, unlike speech, is distinctive in that the medium supports intense *critical evaluation*.

Academics and the scientific community as a whole have long appreciated the need for rigorous *quality control* with respect to scientific activity, an attribute that has most effectively been provided at the level of the scientific *literature*. There has thus developed, over time, what might be called a powerful system of *values management* in scientific communication, which in turn exerts a palpable influence on scientific attitudes and behavior generally. Editors play an especially important role in the enforcement of standards governing communication, with indispensable support from a host of dedicated and conscientious *reviewers* (see Sec. 3.5.1). Nearly as crucial, however, is the high set of expectations of an avid and critical *readership* with respect to the quality of information circulated in print. The most dramatic assessment of published results comes in the form of (largely unforeseeable) attempts on the part of other interested scientists to duplicate proclaimed achievements, efforts undertaken quite naturally in the course of pursuing their own research interests. This independent testing is an essentially random process, one that not infrequently leads to critical follow-up publications. In a sense, the pages of a venerable periodical like the *Journal of the American Chemical Society* assume characteristics of an ongoing "reader's forum", a capacity in which *J. Am. Chem. Soc.* has served the scientific community in exemplary fashion for over 125 years.[9]

In the early days of modern scientific publication—toward the middle of the 19th century, say—it was widely recognized that scientific controversies could be highly emotional affairs, and virulent arguments often played themselves out publicly in the pages of technical journals. Salvos launched in these "debates" were sometimes incredibly bitter and personal (one might even say vicious), and essentially no linguistic holds were barred. From our vantage point the remarkable journal articles that testify so eloquently to the ferocity of these encounters may seem quite amusing, but for the participants such intellectual duels assumed an almost life-or-death character. Times change, and criticism and controversy today tend to be far more subdued and decorous, but it is sobering nonetheless to reflect on the dire consequences that could flow from a demise of this priceless "values management" system: if, for example, prestigious refereed journals were stripped of their quality-control responsibility and replaced by some loose system of "private publication". Would that cause scientific debate to be channeled exclusively into *virtual* arenas, by their very nature unsuited to producing a lasting record? Might serious critical evaluation as we know it gradually come to be regarded as impractical and obsolete in view of the fact that unfounded assertions or interpretations were no longer susceptible to effective challenge from every conceivable direction and over an unlimited period of time?

Modern science is heavily dependent upon the notion that reported results will always be *verifiable*. This fundamental premise in turn places strict limits on what one can legitimately submit for publication.

- Publication in a reputable journal implies that reported findings and observations are capable of surviving in every way the potential test of *replication* by one's peers within the discipline.

Furthermore,

---

[9] Service records are even more impressive in the case of a number of distinguished European journals; the record-holder is in fact a serial known as the *Philosophical Transactions of the Royal Society of London*, first published in England on May 6, 1665, the direct descendant of which is still a highly respected publication.

● Only results that have actually *been published* are accorded an enduring place in science, and only the scientist with a *personal* record of publication is regarded as having made meaningful contributions to his or her field.

As straightforward and innocuous as these assertions may seem, some of the corollaries appear not to be widely appreciated. How else could it be that universities so blatantly avoid devoting significant attention and resources to training students in the skills and habits essential for effective scientific communication: authorship, oral presentation, and publication? How can academe possibly justify its almost total neglect in this respect of the emerging generation of scholars?[10] Perhaps one factor contributing to the crisis is unconscious arrogance on the part of our educators: senior scientists under the illusion that research itself—exemplified, of course, by their own masterfully conceived investigative programs—is the hallmark of true scientific achievement, not effective *dissemination* of one's results and ideas. Another factor could be honest, widespread skepticism about whether it is in fact possible, especially within a research context, to significantly enhance a student's communication skills. Or perhaps the educational community has simply failed somehow to recognize the singular importance of communicating effectively. Whatever the explanation, the quality of the communicative aspect of science is currently in jeopardy, with a potential waste of untold amounts of time and energy—translating into money!—in the face of ineffective and inefficient dialogue.

There may be some truth in the notion that acute sensitivity to language is a gift from above, and that linguistic skill is "teachable" only to a limited extent. It could also be that stylistic feel is to some degree inherited, or transferable only at a very early age. Nevertheless, much of the secret of communicating well lies in clear, systematic analysis of the specific ideas one wishes to convey, and success depends heavily on taking the trouble—consistently and conscientiously—to implement well-established strategies and to exploit proven techniques for transforming ideas convincingly into words. At this level it is surely possible for senior scientists, with their advantage of experience, to render much needed assistance to the budding junior colleague short on communicative ability. One obvious tonic to administer would be frequent assignment of concrete information-transfer challenges of various sorts, coupled with penetrating follow-up analysis and critiquing of the results. A strong conviction that communication skills *can* be enhanced is what has motivated us to prepare the book you hold in your hands.

It is encouraging to be able to report a few early signs that some in the educational establishment are beginning to recognize widespread failings in the current teaching of writing skills. We are aware, for example, of an increasing number of technical schools that now offer interdisciplinary courses emphasizing scientific and technical

---

[10] Until relatively recently it was assumed that mastery of basic communication skills would be a certain outcome of the *general education* of anyone contemplating graduate study, but this is clearly no longer the case, as studies in many countries have confirmed.

writing. Enrollment in such courses has even become compulsory in a few scientific curricula, though the hurdles themselves are sometimes dressed up with more seductive names (e.g., "Research Seminar"), and associated with a broader agenda that also addresses traits and skills useful in the search for satisfying long-term employment, or in public speaking. The sundry arts and crafts underlying effective scientific communication—whether in presentation of a first-rate dissertation, consistent publication of praiseworthy journal articles, or occasional delivery of the outstandingly successful lecture—surely these merit concentrated attention as part of the ongoing education of every aspiring or practicing scientist.[11]

● Many of the skills and much of the insight required for composing effective scientific prose—and for maximizing the probability that one's efforts will be found acceptable for publication—can in fact be acquired, and extra attention paid to this objective can pay rich dividends.

Our principal goals as we set out to prepare this book were (1) to help raise the typical scientist's level of consciousness with respect to the nature and importance of effective professional communication, (2) to pass along tools of the trade that over the years we ourselves have come to value, drawing on our experiences as both writers and editors, and (3) to assemble an authoritative source of technical information and suggestions applicable under a wide range of circumstances.

The pursuit of research in the sciences might be compared with organizing and conducting a series of exploratory missions into unknown territory. No one can be sure at the outset where a particular initiative will lead. Nevertheless, the consequences stand a better chance of being positive ones if sound, conservative preparations are made in advance. No expedition into the wilderness should begin until one has at hand all the equipment likely to be required, or before careful consideration has been given to dangers that may lie in wait. Trusting the protection of one's feet in a jungle setting to nothing more sturdy than a pair of house slippers would obviously be foolhardy, and the same can be said of consigning the first news regarding one's latest exciting research findings to a haphazardly prepared manuscript dispatched on the spur of the moment to the editors of a randomly selected scientific journal, especially doing so under the illusion that the manuscript will be enthusiastically accepted and rapidly transformed into a respected part of the public record.

---

[11] Two of the authors of this book have themselves engaged in the design and presentation of practical classroom instruction of the type suggested: in the form of a one-week course devoted to "The Preparation of Theses, Reports, and Publications" under the auspices of the Department of Chemical Engineering of the Institute for Technology and Design in Mannheim, Germany (1995, 1996), as well as a two-day seminar entitled "Lecturing and Writing for the Natural Sciences and Engineering" offered annually since 1990 by the Scientific Technology Department of the East Frisian College of Technology in Emden, Germany. Basic principles of effective communication were for many years also a central theme in an "Introduction to Chemical Research" course developed by the American member of the team as a one-semester required precursor to senior undergraduate (bachelor-level) research in the chemistry department at Juniata College in Huntingdon, Pennsylvania.

1.2 The Purpose and Significance of a Scientific "Report"

At every stage in your career you will almost certainly find yourself called upon repeatedly to prepare documents that fall in the general category of the "report": *student laboratory reports*, *interim* and *final progress reports* (typically solicited in the context of research), *project proposals*, *grant applications*, *public relations documents* (in both the academic and corporate worlds), *descriptions* applicable to products or services—the list is endless. We obviously have chosen to address in this chapter a very diverse set of materials from the standpoint of both form and magnitude.[12] *Patent applications*, *dissertations*, and even *monographs* also display many of the characteristics of "reports". One of us encountered the following interesting observation in a memorandum distributed some years ago to all the scientific personnel at a major research center:

● "Scientific text [i.e., a report] translates and organizes into a presentation format results derived in the course of the research process, with a special emphasis on consistency and coherence".

We will later address various niceties peculiar to more formal scientific writing, as well as aspects of specific types of specialized reports, restricting our attention for now to those small, unassuming—but absolutely essential!—reports that everywhere constitute a part of "business as usual", documents one associates more often with a practitioner's workbench than a scholar's study. Papers of the sort we have in mind seldom aspire to the recognition accorded the publications whose indispensable function were our focus earlier in the chapter, but they quite often serve as a *basis* for publications at a later date. For our purposes it is irrelevant, however, whether or not information in a particular report will ever be published, because the features that distinguish a report and influence how it should be prepared have little to do with public availability.

We will regard here as a *scientific report* any document that presents, for whatever purpose, a systematic account of selected scientific or technical subject matter—whether it be results from a recent experiment, a summary of one's current understanding of a controversial research issue, or a proposal outlining projected future activities.

● The most fundamental characteristic of a report is that it constitutes an enduring, independent, self-standing piece of *documentation*.

With respect to report *style*, brevity is nearly always regarded as a cardinal virtue. The rationale for this bias from the writer's standpoint, apart from efficiency, is a desire

---

[12] Valuable normative documents dealing with reports include ANSI Z39.16-1979 (*Preparation of Scientific Papers*), ANSI/NISO Z39.18-1995 (*Scientific and Technical Reports: Elements, Organization, and Design*); BS-4811:1972 (*Presentation of Research and Development Reports*); and DIN 1422-4 (*Publications in Sciences, Technology, Economy, and Administration: Presentation of Scientific and Technical Reports*, 1986, in German).

that the cynical reader not suspect one of attempting to conceal a lack of information by embellishing the trivial with an overabundance of verbiage. At the same time, however, an account that is too terse risks failing to convey the true depth, significance, or extent of a problem, concept, or investigation.

● A report is prepared not only for a specific *purpose*, but also for a specific *recipient* (or *audience*).

The intended recipient may be someone as close to the reported activity as the director of one's own research group, but reports are sometimes directed toward remote and impersonal entities like outside granting agencies. The distinction is a crucial one for the writer to take into account, since it can heavily influence where primary attention should be focused: probably on experimental details in the former case, but in the latter situation broad conclusions would be more appropriate to emphasize. Subtle shadings in this respect can profoundly impact a report's success, and only an author possesses the insight and background needed to establish an optimum balance. Above all, be sure you give close attention to the *informational needs* of your most important prospective reader or readers.[13]

● A report is often the subject of rather strict guidelines with respect to form.

Form in this case refers to *appearance* and *organization*, but also to *style* and *language*, and it can play nearly as important a role as *content*—an assertion that would certainly be seconded by the publication professionals called upon regularly to evaluate scientific manuscripts (see Sec. 3.5.3). In this regard, try not to forget that a report functions almost like a "personal representative" of its author (the person seeking to *communicate*). Surely you would want your personal representative always to make the best possible impression!

There is another good reason for setting high standards for yourself whenever you are called upon to write a report: an unclear, poorly focused, jumbled presentation may raise serious questions in the reader's mind about whether the thought underlying the reported work—or even the work itself—was also pursued in a superficial or careless way.

● The reader should quickly receive a clear sense of the *context* within which a particular report has been prepared; i.e., what has preceded it.

As noted previously, virtually all scientific progress, including that which is to be the subject of an impending report, relies to a great extent on past results, most of them probably announced and elaborated upon publicly in some setting. Every report writer is morally obligated to establish a clear connection between what is being reported now for the first time and information available when the reported activity began. Rel-

---

[13] Note that we have begun, on occasion, directly to address *you*, the reader. Actually, it is high time we began to do so, since we prefer to envision ourselves as engaged in a rather informal way in a "virtual dialogue" with real partners—namely our peers!

evant past work merits explicit acknowledgement in even the briefest report—in the form of clear-cut references to particular "sources" (i.e., *documents*) upon which the new findings are based, or at least key aspects of them. Meeting this demand obviously entails that the writer have a firm command of the relevant scientific or technical *literature* as well as an understanding of accepted ways of citing it, subjects to which we return in Sec. 2.2.11 and especially Sec. 9.3 (with further information available in Appendix A).

● Finally, a good report offers insight into not only the *nature* and *significance* of what is being reported, but also the level of *effort* entailed in achieving the reported results.

Once a report has been formulated and carefully "massaged" to the point that you are completely satisfied with its content, there remains a need to *present* the finished product in such a way that it will be accorded the stature and degree of respect you believe it deserves. Surprisingly, one of the most basic considerations here is too often overlooked:

● Be sure your report is readily *identifiable*, and that it has been *labeled* in a straightforward but at the same time distinctive way.

One important element in the "identification label" accompanying any report is the *date* the document was prepared or submitted. In the case of a particular type of "report" we discuss at length in the section that follows—the laboratory notebook entry—this dating requirement can be an especially stringent one: in most cases it is *mandatory* that the assigned date reveal precisely when the described activity was carried out (see Sec. 1.3.1).

● Other anticipated features of a report label include a carefully selected *title*, prominent and unambiguous identification of the *author*(s), and indication of the *place of origin* of the work (an institution, but perhaps also a specific department or research group within that institution).

Official reports often require some even more exacting designation, such as a formal *report number*, assigned either by the author or the sponsoring institution, or a *project number*, especially in the case of work funded by an external source. British standard BS-4811:1972 is especially comprehensive in its treatment of requirements typically applicable to "official" research reports, also providing interesting observations on the subject of security classification.

● Even though a report is almost by definition a document directed toward someone else, *you*—the author—can expect to benefit from its preparation as well.

Being confronted with a requirement that you regularly submit detailed summaries of your recent research activities—at six-month intervals, for example—would seldom be perceived as grounds for celebration, but reports of this sort often turn out to be surprisingly valuable to the person obligated to prepare them. In the first place, the very act of carefully reviewing and analyzing one's work of the past few weeks, and trying to cast it in the best possible light, brings into sharper focus what it is that actually has been accomplished and what still remains to be done, making it easier as well to establish priorities for the time immediately ahead. Moreover, if you *fail* to reflect occasionally upon your work, and to commit the resulting insights to writing, the consequences can be nearly as detrimental to future progress as loss of a set of vital computer files, because much that seems obvious today will have been completely forgotten by next week (e.g., Why did I carry out this particular experiment? What did it really prove?). The availability of a series of carefully prepared periodic reports assures you will have long-term access to an accurate, comprehensive, orderly record of knowledge you have painstakingly assembled, together with contemporaneous interpretation of its significance. In other words, there is every reason for claiming that

● One of the very most important recipients of any report is its author!

Analogous to the oft-repeated maxim that all data entered into a computer should be "saved" in a timely and systematic fashion, and that complete *backup files* should, on a regular basis, be set aside in a secure place, it is just as crucial that you retain *at least two copies* of every report you write, to be kept permanently in safe locations. This is the best insurance there can be against accidental and perhaps irretrievable loss of precious time and effort already invested in your work.

1.3 The Laboratory Notebook

1.3.1 The Role and Form of a Scientist's Notebook

A *laboratory notebook* serves, in effect, as an experimental scientist's diary—perhaps not his or her only one, but certainly the most important from a professional standpoint. In contrast to other diaries, a notebook usually cannot be regarded as one's personal property, or something to be shielded jealously from the scrutiny of others. On the contrary: scientific work described in a notebook is usually work carried out at someone else's expense, and that person or institution obviously deserves full access to any acquired information.

The director of a research group may or may not have a monetary interest in projects the group has under investigation, but he or she will most certainly have an *intellectual* stake in the content of every notebook stemming from that group—as a consequence of being the person who has sanctioned the research activity, established its themes, and presumably participated directly as a source of ongoing advice and counsel. Corporations and research institutes usually regard staff notebooks as *their* property, not the property of individual scientists.

It is equally self-evident, however, that a laboratory notebook is not intended for just *anyone* to view, most especially a notebook maintained within a corporate setting, where "trade secrets" tend to be a sensitive issue.

● The laboratory notebook might be described as the "germ cell" of the scientific literature.

The high esteem in which notebooks are universally held is a direct consequence of their *authenticity* and *irreplaceability*, but also their *purview* or *scope*. The conscientious researcher—by definition a meticulous observer—makes a point of recording in his or her laboratory notebook everything that transpires during an experiment, including fleeting phenomena and the seemingly inexplicable. At the precise moment when a particular observation is made it might seem unimportant, but sometime later that could prove to be the most significant aspect of the entire exercise.[14] For this reason, whenever the occasion arises to *repeat* a previously conducted experiment, for example, it is important once again to prepare a complete and precise (but also *concise*) description of the entire experience, including applicable conditions, every modification no matter how trivial, and all observations.[15]

In addition to facts, a notebook should also reflect spontaneous flashes of insight, and spur-of-the-moment attempts at explanation. These are deemed worth preserving for posterity because they, too, might well prove of considerable interest at a later date.

Much of what one enters into a notebook is likely never to see the light of day in an official report or publication (cf. Sec. 1.4), and for this reason alone a notebook acquires unique documentary value. At the same time, raw notebook data is not infrequently called upon to provide crucial evidence in judicial proceedings of various sorts.

The contemporary German writer Johannes Mario SIMMEL once remarked during an interview that an important factor in the success of his novels is his obsession with *precision*. For example, SIMMEL would never be content to describe one of his heroes

---

[14] There exists an interesting account, attributed to Otto BAYER, entitled *The Role of Chance in Organic Chemistry*. It is derived from a lecture BAYER apparently delivered many years ago while director of research at the German chemical colossus Bayer AG. (That his own name also happened to be BAYER is, by the way, sheer coincidence.) We understand that the complete original text of the address unfortunately no longer exists. What is known, however, is that BAYER used the occasion to present a number of remarkable examples—which he implied could, from his own experience, easily be supplemented by a great many others—illustrating the fact that entire classes of compounds and new synthetic principles came into being quite fortuitously in the Bayer research laboratories "simply" as a consequence of seemingly unscientific misadventures: a result of the fact, for example, that a mercury thermometer was accidentally broken in the course of an experiment, or that—for no particular reason— a condenser made of iron was one day substituted for the copper one usually employed. BAYER aptly cited in the context of one of his illustrations an insightful remark from Friedrich SCHILLER's *Don Carlos*: "Chance provides opportunity, but it is man who must transform the latter into something useful". Implications for the scientist and his or her notebook are obvious!

[15] The account of a replication might conveniently be assigned space on a blank page (properly dated!) adjacent to that bearing the original entry.

as boarding "a train" in the "late afternoon". Instead, he would carefully research every aspect of the supposed event, taking careful notes that would later allow him to craft a persuasive account in which "Martin's train, Intercity 234, left platform 7 as usual at precisely 4:58 PM". SIMMEL allegedly once spent $150000 on background preparations for a single novel—a stunning example of "meticulous scientific rigor" from a context usually perceived as infinitely remote from science! Should we as scientists or engineers allow ourselves to be trumped by a novelist when it comes to attention to detail?

In this latest version of our book on writing we decided to digress here a bit and explore briefly the *concept* of the "laboratory notebook" from a historical perspective. To begin with, we note that the *word* "laboratory" itself (from the Latin *laboratorium*) was first employed by chemists—more precisely by the medieval *alchemists*. Laboratories as we know them today in a broader scientific sense evolved as an amalgamation of the workplace associated with the typical alchemical practitioner and shops maintained by various skilled craftsmen, including dyers and goldsmiths. The first laboratories we would probably recognize as such were set up early in the 17th Century by Andreas LIBAVIUS, as recounted in the *Book of Famous Chemists (Buch der grossen Chemiker*, BUGGE 1929, 1965; Vol. 1, p. 119).[16]

In many branches of science, important orderly tendencies in nature are detected and explored through the careful examination of existing forms and relationships: that is, by observation and description of that with which mankind is naturally confronted. Only the scientist who chooses in a systematic way to *alter* selected *variables*, thereby creating conditions *not* present in nature, practices *experimentation* in the sense we associate with a "laboratory". Thus, a "laboratory notebook" would not, strictly speaking, be applicable in every area of scientific endeavor.

● In the more descriptive sciences a counterpart to the laboratory notebook might be a notebook for recording *field observations*.

This could be designated, for example, for documenting the discoveries of a wildlife biologist, or a geologist. It would not be surprising, incidentally, if a journal of this type were found to be liberally salted with statistical analyses (cf. Sec. 1.3.3).

There will be other occasions on which we will find it necessary to draw distinctions between various scientific disciplines regarding aspects of professional writing. Most of what we have to say about the laboratory notebook, however, will in fact be equally

---

[16] LIBAVIUS is known to have been active around 1600 as a school teacher in the picturesque village of Rothenburg ob der Tauber (Bavaria). The first laboratory officially connected with a *university* was apparently that of Franciscus SYLVIUS DE LE BOË (1614–1672), a professor of medicine at Leyden (PARTINGTON 1957), but the first university laboratory devoted to *educational* activities was not established until 1811, in Jena by the German chemist DÖBEREINER with encouragement and support from the literary giant Johann Wolfgang VON GOETHE, who was one of DÖBEREINER's students (BUGGE 1929, 1965; Vol. 2, p. 23). Soon thereafter, in Giessen (1825), Justus VON LIEBIG laid the foundations for modern chemical education by assigning to students practical laboratory exercises he devised in conjunction with his university lectures.

applicable to scientific records of the more descriptive type as well. Conceptually, the two might even be merged under the broader category of "scientific logbook".[17]

● Notebook entries, if they are to be widely accepted as authoritative, must be recorded *immediately* after the corresponding observations take place, while the experiment in question is still in progress.

Experimental information should always be committed to writing *directly* in an official notebook, not consigned temporarily (for later transcription) to scratch paper or a handy chalkboard. The risk of loss or erroneous transfer with informal jottings of this sort is too great to be tolerated.

● Under no circumstances should one trust observations to *memory*, even for brief periods, with the idea that this will permit preparation of a "cleaner" notebook entry at a more convenient time—unless for some reason there is absolutely no alternative.

Things "remembered", regardless of the time span, far too often turn out to be remembered only vaguely, or perhaps incorrectly!

There are additional important reasons, however, for recording experimental information in a permanent notebook immediately—in "real time". If one were instead to wait until later, it might be, for example, that some instrument, a source of crucial data (which on reflection you realize must be supplemented!), will in the meantime have been turned off, so you could no longer safely assume it will still function precisely as it did earlier.

Coupled with our remarks at the beginning of this section regarding completeness, the preceding injunction regarding timely transcription rounds out our general overview of what a laboratory notebook should contain and how it should be maintained. The fact that a notebook will obviously evolve into a spontaneous, *handwritten*, "unpolished" document in no sense diminishes its value, and there is no justification whatsoever for scornfully dismissing this apparently crude account as an obsolete "fossil", an artifact more in tune with an earlier, technologically inferior scientific tradition. On the contrary! We earlier characterized the notebook as "the germ cell of the scientific literature", so it quite properly displays features reminiscent of a more "primitive form": individualistic penmanship, for example. Fig. 1–1 reproduces a page from a typical modern-day scientific (in this case chemical) notebook. The sample shown is actually a memento of important investigations carried out by Sir Harold W. Kʀᴏᴛᴏ, work for which he (together with Robert F. Cᴜʀʟ, Jr., and Richard E. Sᴍᴀʟʟᴇʏ) received the 1996 Nobel Prize for chemistry. In contrast to data that might reside in some transient

---

[17] A mariner understands a "logbook" to be an official ship's record, a formal repository of all the important technical information relevant to a particular sea voyage, including position data, weather conditions, routine observations, unusual occurrences, and certain essential calculations. A similar account is required in conjunction with every trip a licensed truck driver completes. A rough parallel from the aeronautical world might be the "black boxes" one so often hears described as objects of an intensive search—sometimes even on the sea bottom—in the aftermath of an airplane disaster.

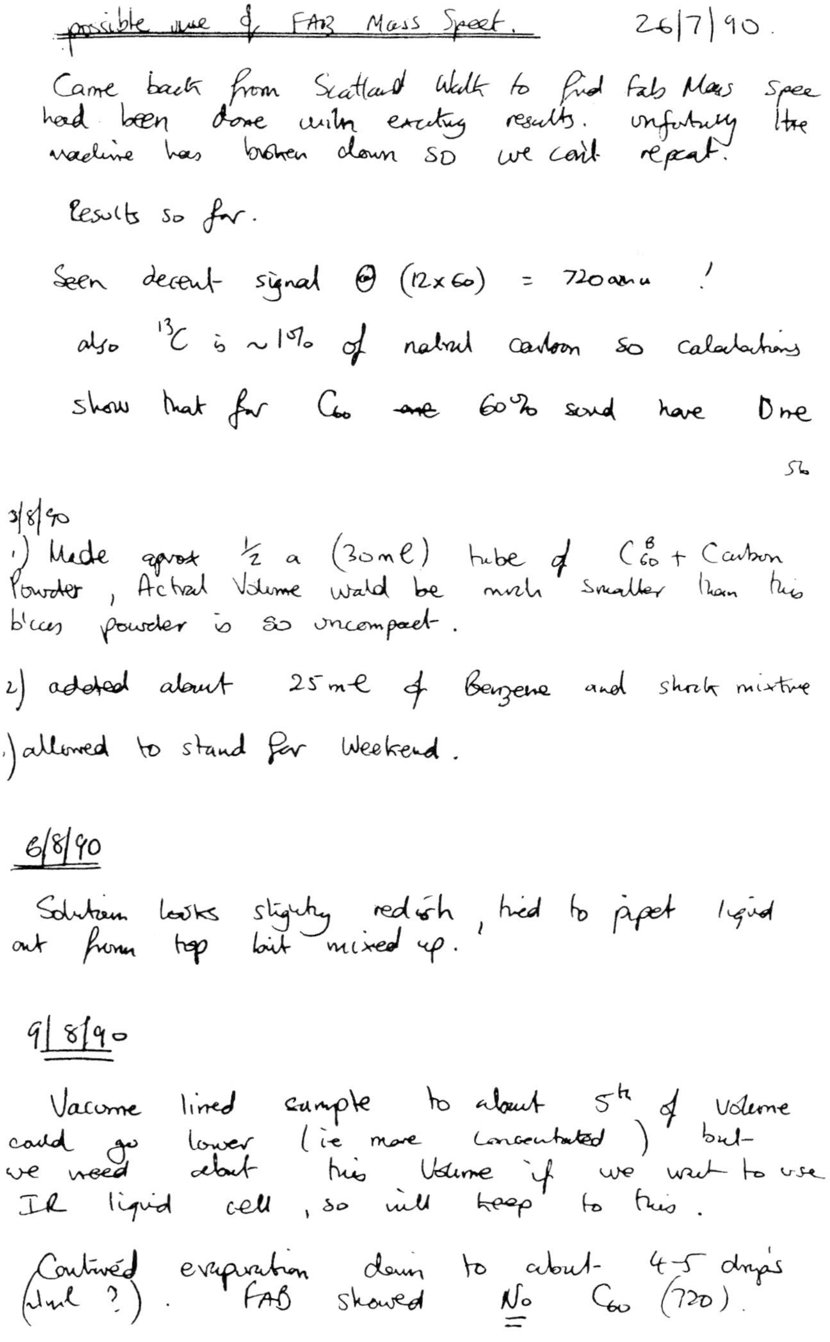

possible use of FAB Mass Spect.   26/7/90.

Came back from Scotland Walk to find Fab Mass Spec
had been done with exciting results. Unfortuly the
machine has broken down so we can't repeat.

Results so far.

Seen decent signal @ (12×60) = 720 amu !

also ^{13}C is ~1% of natrul carbon so calalations

show that for C_{60} are 60% sould have One

s_6

3/8/90
1.) Made aprox ½ a (30ml) tube of C_{60}^B + Carbon
Powder, Actual Volume wald be much smaller than this
b'cus powder is so uncompact.

2.) added about 25 ml of Benzene and shook mixtre

3.) allowed to stand for Weekend.

6/8/90

Solution looks slightly redish, tried to pipet liquid
out from top but mixed up.

9/8/90

Vacume lined sample to about 5th of volume
could go lower (ie more concentrated) but
we need about this Volume if we want to use
IR liquid cell, so will keep to this.

Continued evaporation down to about 4-5 drops
(1ml?). FAB showed No C_{60} (720).

Fig. 1–1. Historical laboratory notebook entries documenting man's first encounter with the compounds known as fullerenes (KROTO HW. 1992. *Angew. Chem.* 104: 113–115, p. 127; reproduced with permission of the author).

electronic "memory", or a printed page called up from a static, impersonal computer file, this unique record preserved in a laboratory notebook represents a "genuine original", and it deserves to be the object of considerable respect.

● From the standpoint of *form*, a laboratory notebook should be a sturdy volume with a *permanent* binding, and its pages should reflect paper of high-quality.

Information should always be entered into one's notebook using a *ballpoint pen* to ensure permanence and also to testify to the document's authenticity.

● All pages within a notebook should display prominent, sequential *numbers.*

Every discrete experimental record should begin on a new numbered page—although not necessarily the next available blank page. All pages should also be *signed* and *dated.* Unused portions of a page should be "canceled" with a heavy diagonal line, introduced concurrently with data entry. This will help in reassuring a skeptical reader that nothing has been deleted or added "after the fact". (Physicians are generally urged to take similar precautions with unused portions of their prescription forms as one way of discouraging forgery.) Companies tend to impose even more stringent rules regarding authentication of official laboratory records as insurance against frivolous legal challenges—with respect to patent rights, for example. Thus, it is common for a firm to insist that each experimental record be *certified* in timely fashion through a testimonial signature applied by an authorized official or coworker. Many organizations outfit their researchers with custom-designed notebooks with provision for a (mandatory!) table of contents and formal validation of all entries.

The preceding observations with respect to proper form and management of a laboratory notebook are analogous to similar advice we gave in the first edition (1987) of our book. The essence of these suggestions gained unanticipated endorsement and widespread attention a few years ago in the wake of the disconcerting fraud episode in the United States still known as the "BALTIMORE affair" (a reference to the principal investigator involved). Interested readers—non-chemists included—might wish to look at an essay in the March, 1992, issue of the official news organ of the German Chemical Society (ROTH 1992), which includes the following summary observation:

> One's records should be kept not as 'a pile of loose papers' ... but in a permanently bound laboratory notebook. Had this regimen been followed [here] it would have made it much more difficult [for them] to back-date [falsified] experiments and engage in other improper manipulation, including the convenient omission of offending pieces of data.

1.3.2 Content

Heading and Introduction

We turn our attention now to a more detailed consideration of how a typical entry in a laboratory notebook should actually look. To begin with, sets of related operations and observations should obviously be grouped together in the form of discrete "experiments".[18]

- The account of every experiment should begin with a *title*, followed by a brief *introduction*.

The formal (concise) title actually should be supplemented with additional information to produce what might better be termed a "heading"; this could include, for example,

– the *numerical* designation assigned to that experiment;
– the *date* on which the experiment began, perhaps expanded to include the *time* and *place*;
– one or more relevant *literature* sources, or cross-references to other key notebook pages.

The subsequent introduction would logically commence with a brief summary of the overall *goal* of the exercise. To the extent that unusual materials or apparatus may have been required, this should be noted here and elaborated as appropriate. An introduction can be augmented with other background information such as:

– a list of special *conditions* that may have been associated with certain measurements;
– a sketch of relevant *special apparatus* or unique experimental facilities;
– a list of the commercial *consumables* employed (e.g., chemicals, solvents), including important specifications (e.g., purity, grade);
– the source and nature of important *noncommercial materials* utilized, again with relevant specifications;
– in the event that *animals* happen to play a role, detailed characterizations, including genus, species, strain, holding conditions, and, where appropriate, information regarding permits that may have been issued in conjunction with the work (e.g., by an ethics committee);
– unusual *preparatory steps* that may have been required (e.g., purification of starting materials);
– potentially relevant *environmental factors*, such as temperature, atmospheric pressure, or humidity.

The introductory portion of a notebook entry is also a good place to sketch out longer-range plans you may have with respect to the experiment. In the case of synthetic

---

[18] We postpone to the next subsection (Sec. 1.3.3) a closer look at what is actually *meant* by "an experiment".

work, for example, careful preliminary thought could influence your judgement regarding optimal amounts of starting materials (taking into account what you know or can reasonably anticipate about stoichiometry and probable yield). Needless to say, *speculative* observations of this nature must be clearly distinguished from the *actual* observations and experiences that mark the course of the experiment itself.

In the case of analytical work it would be important to document in the introduction the "life history" and source of samples under investigation:

- What sampling procedures were employed?
- Where, when, and under what conditions were samples collected?
- Who collected them?
- To what on-site treatments (if any) were samples subjected?
- Who actually submitted the samples, and how were they transported and stored?
- When did the samples arrive at the laboratory?

In a modern analytical laboratory, work often revolves around highly automated operations, with routine automatic documentation at each step. Where this is true, the status and function of the classic laboratory notebook might appear to diminish to little more than a convenient data repository, but for a variety of reasons we strongly resist the notion of an automated database ever making the traditional notebook completely obsolete, even though we are well aware of commercial software marketed to serve as the backbone for a potential "electronic laboratory notebook". The senior officer in an organization—a *person*—is always charged with ultimate record-keeping responsibility, in the interest of maximizing the reliability and integrity of officially acquired information. This person is also expected to oversee matters related to data security, access privileges, etc. Many of the associated tasks are of course delegated down through a chain of command, eventually reaching to the individual rsearcher, but no such chain of responsibility can pass through a *machine*! Should it ever become necessary for authoritative laboratory records to be presented as evidence in the course of a legal proceeding, for example, there is a real risk of an "electronic transcript" being treated at best as the equivalent of an unauthenticated laboratory protocol.[18a]

- The inexorable trend toward automation and utilization of sophisticated laboratory robots makes it more and more difficult to assign personal responsibility for the maintenance of strict, uniform quality standards with respect to laboratory data and the oversight of record-keeping.

In light of these developments, the personal, conscientiously maintained records of the individual scientist will actually have an increasingly important part to play in sustaining the integrity of the scientific enterprise.

Automation and related phenomena have led not only to the need for attention to organizational and logistic matters, but to new ethical issues as well. Thus, the past

---

[18a] To be truthful we must acknowledge that migration to electronic record-keeping—involving elaborate safeguards to evolve over time—is virtually inevitable. The reader interested in related issues and current „state-of-the-art“ should visit the Website www.CENSA.org for a glimpse of what may lie ahead.

provides no precedent for analyzing broader implications associated with the "virtual laboratory".

The Experimental Section

The introduction to a notebook entry should be prepared before the experiment itself begins, but everything that follows must represent a complete, contemporaneous account of the actual experiment, reflecting precisely how every step is carried out. Certain narrative devices and conventions, if observed consistently, can help ensure a clear distinction between expectation and incontrovertible fact. Thus, whereas an introduction might include conditional phrases like

> "... it is believed that ...";
> "... the procedure should show ..."

an experimental section is more properly built around concise passive constructions that unambiguously point backward in time, such as:

> "... [was] stirred 3 h at (25.0 ± 0.2) °C;
> "... found ..., calculated ..."

Anyone with publishing experience will almost automatically formulate the corresponding text in ways stylistically analogous to the prose typically found in a journal article, an expedient that considerably simplifies the later task of revising a notebook entry to make it suitable for incorporation into a more formal document (cf. Sec. 1.4).

When an experiment involves a series of distinct phases or steps you will find it advantageous to introduce into the written account a few *subheads*, and perhaps even *dividing lines*, as a way of structuring the information and making it more readily accessible at a quick glance.

Absolutely no harm is done if a notebook is allowed to contain a certain amount of "laboratory jargon" clearly inappropriate for a publication. Colloquial phrases are after all easier to formulate than more elegant formal prose, and getting the facts down *quickly* has a high priority! Brevity is also desirable, so long as nothing important is omitted and all ambiguity is avoided, and jargon can contribute positively toward this attribute as well. There is thus nothing wrong with referring in a notebook simply to a "buchner" when what you obviously mean is a "Büchner *funnel*", and every chemist would understand that the word "erlenmeyer" was meant to be synonymous with "Erlenmeyer flask". "Ether" would also suffice so long as it was clear from context that "diethyl ether" is in fact the solvent in question.

Don't be shy either about occasional use of a conveniently invented verb, like "to rotovap", as shorthand for the much wordier (albeit more formal) description of "concentrating a solution with the aid of a rotary evaporator", and in a laboratory notebook slipshod phrases like "taking an IR" or "getting the lines to separate" should not be provocation for raised eyebrows, even though they would certainly give a case of heartburn to even the most lenient journal editor. What is most important is ensuring that you and others like yourself (i.e., people equally at home with common "lab slang") would come away from your notebook with a solid (and absolutely correct!) understanding of all that transpired in the course of a reported experiment, and be in a strong position to repeat the process, in every detail.

● Always record in your notebook *raw* data and observations, not interpretations.

An example of "raw data" in this sense would be initial and final burette readings taken in the course of a titration, or weights determined for a crucible before and after an important drying step.[19] Such information may of course later be transformed into *processed data* more suitable for inclusion in a report, such as a volume of titrant consumed, mass loss upon desiccation, or (better) "amount-of-substance" values (quantities expressed in terms of the unit "mole"; cf. Sec. 6.3.1). The latter are bits of insight with more scientific significance, but experience has shown that mistakes are disconcertingly common in the performance of even so trivial an operation as subtracting a tare weight in order to ascertain an amount of added starting material, just as mistakes inevitably creep into data copied from one place to another (one of the several reasons we insist you enter *all* your notes *directly* into the notebook!). Spurious conclusions based on processing errors of this kind can sometimes be located and eradicated after the fact, leading to "sanitized" results with which you can "salvage" an experiment, but only if the notebook contains *every* applicable primary observation, and sufficient information is also present to establish which specific pieces of apparatus were used, for example, in the course of making precision measurements.[20]

In some cases you will probably find yourself taking advantage of a pre-programmed laboratory computer when the time comes to perform certain routine computations. If so, try to incorporate into the anticipated printed output both entered values and the corresponding results—clearly labeled (and in the ideal case automatically dated as well)—in a form that will allow you to paste the entire printout directly into your lab notebook. The "leitmotiv" here is "full documentation". Situations frequently arise that call for a complete record of primary data, as well as unrestricted access to it. A non-laboratory example would be your desire at the end of the month to consult a comprehensive, detailed statement of your recent financial transactions as you attempt to reconcile a checkbook balance with the "authoritative" cash balance reported by your bank.[21]

---

[19] The examples we present here may seem trivial—indeed archaic—but they illustrate the point, and the concept lends itself easily to straightforward extension to more "high-tech" situations.

[20] Correction for instrumental errors is of course facilitated if up-to-date calibration records for critical measuring devices are also readily accessible.

[21] Real-world analogies like this are sometimes surprisingly informative—and persuasive! Consider another example: While a politician might ordinarily be perfectly satisfied with reading a "joint statement" released by the participants at the end of a complex series of negotiations, such an account would be of no use whatsoever to the diplomat charged with explaining *how* particular terms in fact came to be agreed upon in the course of formulating a newly announced treaty.

- Under no circumstances is a "results summary" an adequate substitute for a complete contemporaneous account of an experimental procedure—including any steps leading to *interpreted* results.

If you are able to anticipate that certain collected information might later be required in a different form (after unit conversions, for example), carrying out the appropriate computations in the course of documenting the experiment would certainly be advantageous. Nevertheless, all raw numerical data subjected in this way to further processing, as well as relevant conversion factors, should be incorporated into the notebook record and labeled to facilitate subsequent review and confirmation of the accuracy of your work should that ever become necessary.

Also appropriate for inclusion in a notebook entry is an account of *quality-control* activities that may have accompanied an experiment. These would include, for example, significant *adjustments* to which analytical instruments were subjected, *calibration factors* applied to measurements, and unusual *repair* or *maintenance procedures* that may have been followed. "Every such modification should be documented in the laboratory notebook, together with when it occurred and the identity of all other personnel who may have been involved" (translated from a respected treatise on quality control in analytical chemistry: FUNK, DAMMANN, and DONNEVERT 1992, p. 104). Routine attention to such minutiae now plays a decisive role, incidentally, in determining whether or not a particular analytical laboratory is entitled to "Good Laboratory Practice" (GLP) certification.[22] GLP has "as its sole purpose to discourage falsification", with reference "exclusively to the documentation and archiving of test results" (GÜNTHER 1993). Objectives of the GLP initiative are summarized in familiar terms in EN 45 001 (1989), *General Criteria for the Operation of Testing Laboratories*: "All original observations, calculations, and derived data are to be preserved for a reasonable period of time, together with descriptions of all calibrations and final test reports. The account of every test must contain sufficient information to permit a repeat of that test".

A Scientist's Ethical Responsibilities

It is perhaps appropriate at this point for us to address briefly one awkward question certain at least occasionally to arise for every active experimentalist:

- Almost by definition, a notebook will contain *more* information than would ever be incorporated into a report. But how *much* more?

---

[22] GLP designation is a sign of compliance with an internationally accepted set of standards primarily addressing administrative matters, including "conditions under which laboratory tests should be planned and carried out" as well as standard procedure in the "documentation and reporting of results" (quoted phrases are from the German Chemicals Law of 1990). The GLP protocol presents detailed guidelines for such things as the organization of laboratory activities, report preparation, and the systematic archival of data. One of the long-range goals is increased acceptance of analytical results on an international level, and appropriate standards have been incorporated into numerous statutes and legislative actions in countries throughout the world (cf. the German Chemicals Law, as noted above).

Our repeated emphasis on "completeness" in this context is certainly proper, although it should not be misinterpreted as encouraging excessive verbosity or wasteful elaboration. An abundance of detail of course makes an enormous contribution to ensuring reproducibility, but beyond this there is a second consideration:

● Probable exclusion of certain data or experimental details from a subsequent report cannot be allowed to serve as an excuse for subtle *manipulation* of a set of results.

Selectivity, properly exercised, is undoubtedly a virtue, but it is quickly transformed into a deadly sin if applied such that "unwanted" results are deliberately obscured. The falsification of results, or feigning of accomplishments that represent only wishful thinking, constitutes unethical behavior of the worst kind—and in some cases it can even be criminal. Perpetration of fraud in this way can destroy a promising scientist's career if the facts ever become known. Germany was the scene of a disturbing parallel to the reprehensible "BALTIMORE affair" (mentioned earlier), with consequences that ultimately proved more devastating and far-reaching even than those suffered earlier in the United States. The context in this case was a cancer research scandal—leading to the serious tainting of a heralded project that had enjoyed high-level government sponsorship.

The most irritating dilemma commonly encountered in conjunction with "omission" of information is actually more subtle in nature. Documenting all the details of an experiment that *fails* (perhaps because a thermostat was inadvertently left unplugged) would arguably be a pointless exercise—unless the fiasco happened to be one from which unexpected, interesting observations emerged. But where should one draw the line? What may legitimately go undocumented? Obviously no general rule can be expected to provide the answer. In case of doubt, too much information is always better than too little. One thing is certain, however: *selective* (or even *arbitrary*) omission of *one* of a *series* of analytical results would be problematic at best. Once obtained, every such result—even an embarrassing one—must be documented. In a subsequent report the awkward data point can perhaps be brushed aside as an inexplicable "outlier", but in the ideal case it will ultimately lend itself to explanation, or at least plausible rationalization.

● Science's commitment to authenticity and lack of ambiguity presupposes absolute, unadulterated honesty.

A few hardened critics are firmly convinced that scientific "achievements" are the source of as many negative consequences as positive. It is certainly their right to hold that view, but no scientist should ever place in jeopardy acceptance of the fundamental *validity* of scientific results (as distinct from certainty that they have been *interpreted* correctly, or *applied* wisely). True science cannot be manipulated and still claim to be science. It is bad enough that some people suspect there are "scientific" things they need to know which have been *withheld* from them, but no one should ever find reason to doubt that what science *does* report is honestly regarded as the unvarnished truth.

This indirectly brings up another important matter. Scientists in general find it very difficult to break free of an "insider's" professional jargon when called upon to communicate and express themselves in a language the public might be expected to understand. Many at the same time underestimate how urgent it is for society to be kept abreast of science and the effects of the scientific enterprise on everyday life. Public understanding and active, rational public engagement are crucial to the retention of a fundamental trust in science itself, especially in view of the enormous impact science now has on almost every aspect of our world. Important communication problems that have arisen in this context cannot of course be laid *solely* at the scientist's door; a certain amount of blame must also be ascribed to the politics of publication and forces that society has come to rely upon for the dissemination of information, including the press and electronic media.

Space considerations prevent us from launching additional forays into aspects of science with profound societal and ethical implications, one of which is of course the role test animals (including humans!) are called upon to play in scientific research. We will at least add, however, that with respect to *pharmacological research*, routine use of animals has sharply declined in recent years, in large measure due to remarkable developments in *medicinal chemistry*, especially the vastly expanded application of sophisticated *molecular modeling* methods, permitting many aspects of drug development to be pursued in theoretical ways, and with far greater selectivity than in the past.

1.3.3 Organizational Matters

What is an Experiment?

We conclude this section of the chapter by looking at a few issues not obviously central to the themes "writing and publishing", but ones we nevertheless regard as highly germane to our overall objective. Attempting to deal with any subject in a comprehensive way almost always shows it to be in fact only a bit of foreground, closely tied to some higher organizational scheme. We have so far taken it more or less for granted, for example, that laboratory or field work in science can be conveniently analyzed and treated in terms of a limitless series of individual *experiments* (or, alternatively, an endless train of descriptive activities). This is indeed largely the case, but sometimes establishing boundary lines between so-called components is not so straightforward.

● Maintaining an organized laboratory notebook forces one to make conscious decisions about where one experiment will be said to end and the next to begin.

There clearly is no such thing as a "best general solution" to ambiguity in experiment definition; each researcher is left to his or her own judgement, and the decisive factors will necessarily differ from one experiment or project to the next. Take, for example, a chemical synthesis, work-up of which leads to a dozen product fractions. Further examination, purification, and characterization of each of the fractions might, in principle, be regarded as an experiment in its own right; alternatively, the elevated status of "experiment" could be reserved for work associated with the crucial third fraction—or perhaps *all* follow-up efforts should simply be treated as continuations of the synthesis itself. Several different management approaches can thus appear to have merit.

The most important considerations in resolving an issue of this sort should be efficiency and transparency: in other words, how long and involved are corresponding notebook entries likely to be, and how difficult would they be to organize? What would make things easiest for someone seeking information from the ensuing write-up at a later date? Another factor perhaps worth taking into account is whether all the planned subsequent steps will be carried out immediately, perhaps over the course of a day or two, or if for now the process has no foreseeable end.

Depending upon the nature of an investigation, as well as one's personal inclinations, one attractive expedient is establishing as a rule of thumb that a given independent (numbered) notebook entry—an *experiment*—should never be allowed to extend over more than one or two pages. Even so, observations contributing to a single entry could still occur over a long period of time, requiring various *parts* of the corresponding entry to be dated accordingly (cf. Fig. 1–1).

Defining precisely what is meant by an "experiment" tends to be especially troublesome in the case of the more *descriptive* sciences, and the optimal approach to documentation correspondingly less obvious. Despite the challenges, however, every biologist or geologist manages somehow to erect workable sets of "fences" around his or her activities. The most extreme situation of all is perhaps posed by investigations involving *behavior* (e.g., in the context of *ethnological* research), but even here researchers traditionally devise ways of interpreting their studies in terms of discrete "experiments".

Work in the latter categories is commonly subject to one further organizational factor that only rarely influences something like a chemical investigation in a significant way:

● With certain types of projects, the *sampling techniques* adopted, and decisions regarding the optimal *number* of discrete observations can prove to be crucial.

The role of multiple experiments and data accumulation becomes most apparent when the time comes to evaluate and interpret one's results. Some critics (and disciplines!) are notorious for insisting that a set of conclusions is meaningful and reliable only if it has been subjected to a thorough and truly valid *statistical analysis* (cf. Sec. 6.4). Apparent trends, especially counterintuitive ones, will be taken seriously and judged *significant* only if, long before any actual observations are recorded or measurements taken, the researcher has sketched out and then carefully perfected an elaborate "statistical plan of action". Simply put (LAMPRECHT 1992, p. 73):

- "Think first, then measure!"

It is beyond our scope to dwell on this point, but we urge especially those with relatively little research experience who anticipate engaging in work subject to such considerations to study the organizational suggestions in *Design and Analysis of Experiments* (COBB 1998) as well as the analytical guidance provided by *Applied Linear Statistical Models* (NETER, KUTNER, NACHTSHEIM, and WASSERMAN 1996). Beginners contemplating serious research in the life sciences would be especially appreciative of LAMPRECHT 1992 (available—unfortunately—only in German).

Experiment Numbers

In Sec. 1.3.2 under "Heading and Introduction" we implied that

- Every experiment should be assigned its own unique identification number, sequential in character.

The person entering information into a laboratory notebook will also have accepted responsibility for carrying out the reported experiments, and thus establishing the order in which they are conducted (and described), their *sequence*, which in turn will determine numbers assigned to them. We recommend you incorporate into each of your formal experiment numbers the *page number* of the corresponding notebook entry—more specifically, the page on which that account *begins*. In some cases a single page will probably contain information on more than one experiment (cf. Fig. 1–1), and it is also likely that some experiments will extend over several pages, so page numbers alone are not a satisfactory substitute for experiment numbers. A properly maintained notebook also rapidly becomes full, so over the course of time you will probably find yourself the author of several notebooks. These can conveniently be distinguished from one another by roman numerals, which should then also become part of your experiment numbers. Preceding all this with your initials results in an official designation for each experiment confirming that it indeed represents your work rather than that of a coworker in the same laboratory.

Combining these suggestions, Sam Schmidt's experiment 307, which commences, say, on page 117 of his third notebook, would be designated "SS-307-III–117". If this notebook entry represented a synthetic operation in chemistry, a specific product fraction it generated (the third one, for example) could be called "SS-307-III–117–3". That material (so described!) might in turn constitute the starting point for a subsequent experiment, such as "SS-308-III–119". Notebook-related designations for specific materials prepared in a laboratory environment are handy for definitive labeling of such associated things as sample vials, spectra, chromatograms, etc., and they also prove valuable in conjunction with paperwork for follow-up activities pursued elsewhere (elemental analyses, for example). Descriptors like these are valuable temporary "names", especially for experimental materials of uncertain character to which reference needs to be made: in an interim report, perhaps, or elsewhere in a notebook.

● Systematic experiment numbers can even facilitate operation of the laboratory itself.

A straightforward "information management" scheme like the one we propose can contribute in important ways to the establishment of order, especially in a laboratory serving as home to several scientists. Large research groups find it helpful, for example, to maintain a centralized database with records of all materials under investigation, based on their locally assigned identifiers—from the time these are issued—and providing at least a limited amount of descriptive information for each.

● Productivity in the natural sciences demands close attention to organization, and a set of well-maintained notebooks is an important prerequisite.

Our discussion of laboratory notebooks has obviously taken us rather far afield when we find ourselves discussing laboratory organization in general! There are other detours we might profitably pursue as well: for example the fact that some scientists entrust to their notebooks the additional responsibility of serving as a "personal planning assistant", with the notebook literally becoming a "professional diary". For many, it is thus common practice regularly to jot down in the lab notebook miscellaneous work-related insights or ideas, stimulated by a phone conversation, perhaps, as well as vague, general bits of idle speculation (possibly with exciting long-range implications!) from reading an interesting journal article. One consequence of ascribing a role this broad to the laboratory notebook is a temptation always to carry the record around with you when you are professionally "on duty", especially on trips to the library—despite the inconvenience, since in order to fulfill even its normal functions a notebook must be relatively large. (A bound volume consisting of 200 or more sturdy, letter-sized pages is typical.) Many scientists nonetheless find the effort worthwhile, although the recent introduction (and enthusiastic reception) of "PDAs" ("personal digital assistants": the PALM PILOT®, for example) is changing behavior in this regard. If you in fact decide to assign to your notebook a variety of functions, we recommend you at least organize the pages with *dividing lines* or other clear markings to prevent the contents from becoming too convoluted.

This leads us to issue a last cautionary note: the prospect of *losing* one's notebook, or seeing it become the victim of serious physical abuse, should send chills to the very core of the conscientious researcher's soul. As noted earlier, a laboratory notebook is literally irreplaceable! Be sure that one prominent feature of your notebook is a reliable, up-to-date list of addresses and telephone numbers through which you can always be reached, whether at work or at home: just in case. Make it a point also, on a regular basis, perhaps every weekend, to take the further precaution of photocopying all entries made since the last copying round, subsequently storing the copies in a secure location apart from the laboratory—someplace where they can easily be retrieved if it ever becomes necessary.

1.4 From Laboratory Notebook Entries to a Formal Report

1.4.1 Describing an Experiment

Transformation of an experimental description as recorded in a *notebook* into a form suitable for a *report* is illustrated here with two excerpts. Fig. 1–2a reproduces a page from a typical notebook, a page describing the synthesis of an organic compound; more precisely, the entry of interest documents purification and characterization of the newly synthesized substance. (The synthesis itself, as the opening observation implies, would be on an earlier notebook page.) The account is quite *detailed*, but the information itself is presented in characteristically compact fashion, with what might be described as a high level of "information density". The record consists primarily of isolated words and cryptic abbreviations, supplemented by numerical data. Notice that the researcher reports with almost no elaboration a melting *range* for the substance after recrystallization, a range whose breadth strongly suggests material that is not (yet) pure. A result is shown for one simple computation (fraction of material recovered after recrystallization), based on weights—explicitly in grams—determined before and after the purification step. Fig. 1–2b shows how the corresponding information might subsequently be treated in the context of a *typed report*.

● Information in the report is presented even *more* concisely, and in conformance with certain formal conventions.

In contrast to the notebook entry, everything in the report is expressed in complete sentences. Certain bits of raw data have been replaced by more meaningful *derived* values. Thus, the percent recovery from a particular recrystallization is no longer shown, having been taken into account instead as one factor in an overall-yield calculation reflecting what might theoretically have been anticipated for this synthesis, given the amounts of starting materials introduced and *molar masses* of the substances involved (see Sec. 6.3.1). Furthermore, a pH value from a test solution of the compound has been interpreted to give an estimate of the more fundamental *acidity constant* (pK_a). According to the notebook, the composition for the recrystallization solvent was established on the basis of exploratory tests. This solvent mixture, described in the notebook in terms of a CH_3OH/H_2O volume ratio ("ca. 1 : 3"), is expressed more formally in the report as a "volume fraction of methanol" value ("$\varphi = 25\%$"). The preliminary solubility studies were evidently felt not to be sufficiently important to warrant mentioning in the summary report. Similarly, no attempt has been made, at least in this part of the report, to explain or justify physicochemical assumptions involved in approximating a pK_a on the basis of a measured pH. Finally, the observed melting range is accompanied in the report by a melting point value from the published literature.

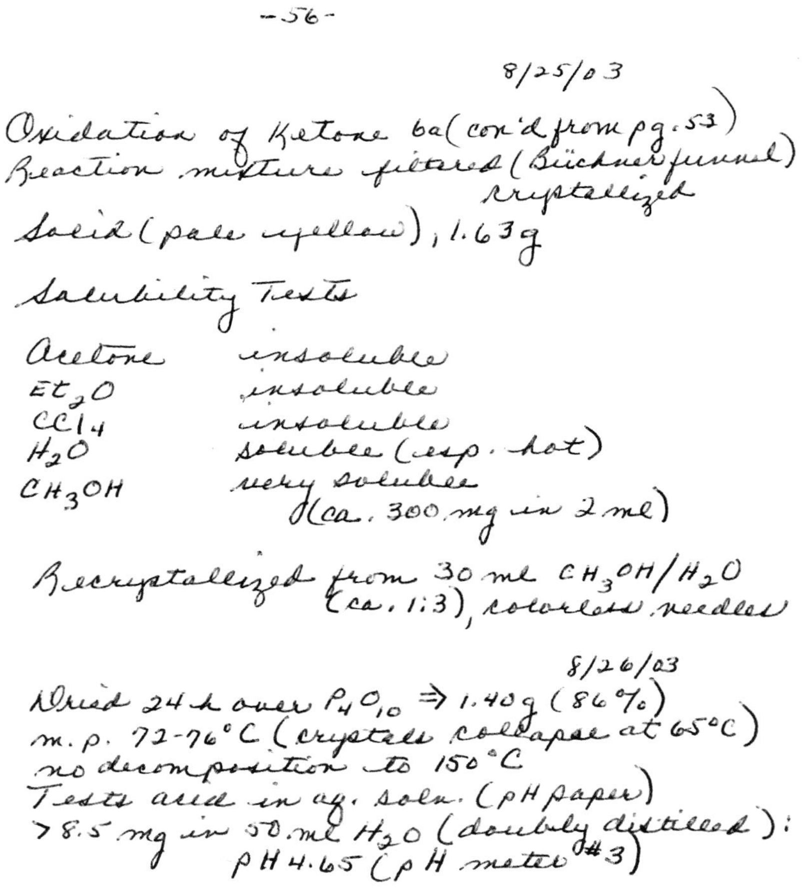

– 56 –

8/25/03

Oxidation of Ketone 6a (con'd from pg. 53)
Reaction mixture filtered (Büchner funnel)
                              crystallized
Solid (pale yellow), 1.63 g

Solubility Tests

Acetone   insoluble
Et₂O   insoluble
CCl₄   insoluble
H₂O   soluble (esp. hot)
CH₃OH   very soluble
                 (ca. 300 mg in 2 ml)

Recrystallized from 30 ml CH₃OH/H₂O
                   (ca. 1:3), colorless needles

8/26/03

Dried 24 h over P₄O₁₀ ⇒ 1.40 g (86%)
m.p. 72-76 °C (crystals collapse at 65°C)
no decomposition to 150 °C
Tests acid in aq. soln. (pH paper)
> 8.5 mg in 50 ml H₂O (doubly distilled):
        pH 4.65 (pH meter #3)

b

... The crude product (pale yellow) was isolated from the reaction mixture in crystalline form. <u>3a</u> was obtained by recrystallization from a water-methanol mixture (φ = 25%) as colorless needles (m.p. 72-76 °C; Jones 1980: 78 °C) in a yield of 75.8%. An aqueous solution of the compound tested acidic (pK_a ≈ 9).

Fig. 1–2. Example of an experimental procedure – **a** as described in a laboratory notebook; **b** recast as part of a typed report.

Of further interest for our purposes is the fact that the particular notebook page illustrated bears two separate dates, but no title. As previously noted, the account of the experiment in question actually begins on an earlier page, so that is where a title should be sought. The opening notation immediately under the date 25 August 2003 points the way. The crystals mentioned here were not available in dry form until the following day, but this was not considered problematic; such a minor delay is no reason for declaring the start of a new experiment. In other words, the calendar need not have as much influence on a laboratory notebook as it usually would on a personal diary.

From the introductory comment one can deduce that the two preceding notebook pages (pp. 54 and 55, a double-page spread: i.e., a left-hand page and the corresponding right-hand page) were used for documentation of at least one *other* experiment proceeding simultaneously. Note also that the substance referred to in the notebook as "ketone 6a" bears the identification "3a" in the related report, so the two accounts must be organized somewhat differently.

The fact that most reports lack important details regarding the way an experiment was actually conducted—information that of course *has* been conscientiously recorded in a notebook—is a consequence of the almost universal pressure to limit the content of reports (especially *published* ones) to the barest essentials. Most readers would in any case not be willing to devote the necessary time to reading an account that was *too* complete, nor would there be reason for them to do so. The extent to which details are omitted will naturally vary with the situation.

● The greater the separation—conceptually—between the author of a report and its recipient, or the *wider the distribution* of the document, the more sparing will probably be the content.

The impulse to condense, first manifest when a laboratory notebook entry must be reformulated for an interim research report, persists into the preparation of theses and dissertations, and beyond that to the various higher-level forms of publication (research papers, review articles, monographs, etc.). The farther up the chain of reports one encounters information, the more impractical it becomes to use that source as a guide to repeating an experiment. For this reason it is not uncommon for a researcher to breathe a sigh of relief upon discovering that it is actually possible to gain access to a key dissertation or interim research report—perhaps even to a long-neglected laboratory notebook gathering dust somewhere on another researcher's shelves. This is an important incentive (apart from nostalgia) for the conscientious research director to archive laboratory notebooks accumulated over the course of decades, and it also accounts for our admonition earlier in the chapter that laboratory notebooks should be not only comprehensive, but also orderly, and legible. In other words:

● A laboratory notebook should be so structured that *others* would also be able to find in it information they need, not only now, but for the foreseeable future.

We conclude this subsection by considering briefly an excerpt from the "Experimental" section of a typical *journal article* (Fig. 1–3). Here one would no longer expect to find any indication, for example, that it may have proven impossible to repeat one important measurement because of a sudden loss of vacuum (apparently the case when fullerene was discovered; see Fig. 1–1). Application of a wide variety of analytical methods, however, is painstakingly described, even to the point of identifying specific instrument models and conditions under which they were employed, but the information has been compressed into a few lines full of technical jargon and abbreviations.[23]

1.4.2 The Preparation of a Formal Report

Proposed Subdivisions: An Outline

Our objective in this section of the chapter is first to describe a typical interim or final research report summarizing results obtained in an experimental (laboratory-based) project, and then to offer suggestions on how a report like this might best be prepared. Much of what we say may seem self-evident or trivial—and yet it is remarkable how rarely even the most straightforward and obvious recommendations are taken really

Schmp. (unkorr.): Gallenkamp. –$[\alpha]_D^{20}$: Polarimeter 141 (Perkin-Elmer). – NMR: Varian XL 300 (TMS), ^{1}H: 300 MHz, ^{13}C: 75 MHz. – MS: Kratos MS 50, Einlaßsystem Hot-Box, Quellentemp. 180 °C: 70 eV. – GC/MS: Finnigan/MAT 1020 B; Datensystem: INCOS 1. Kapillarsäule: DB-S (J & W Scientific Inc. Rancho Cordova. USA), Fused-Silica, 30 m × 0.32 mm i.D., 0.25 μm Filmdicke. Trägergas: Helium, Vordruck 26 psi. μ = 60 cm/s, Gerstel-Split-Injektor. Split 1 : 20, direkte Kopplung. Quellentemp.: 180 °C, Ionisierungsenergie: 70 eV, Multiplierspannung: 2000 V. Analyt. DC: Fertigplatten (Merck), Kieselgel 60F-254, 0.25 mm.- Präp. DC: Kieselgel 60 PF-254 (Merck), 0.75 mm. – SC: Kieselgel 60, 0.063-0.200 mm (70-230 mesh ASTM) (Merck). – Trimethylsilyl-Derivate (TMS-Derivate): 1 mg Substanz wurde in 100 μl *N*-Methyl-*N*-trimethylsilyl-trifluoracetamid (MSTFA) gelöst und 1 min im Ultraschallbad behandelt.

Fig. 1–3. Excerpt from the experimental section of a journal article [taken from *Arch. Pharm. (Weinheim). 1992. 325: 47–53; p. 49*] .

---

[23] The literature example shown is taken from a 1992 issue of the journal *Archiv der Pharmazie*, first published in 1822. Note, incidentally, the close relationship this example suggests between laboratory routine and specific "brand names" associated with high-quality instrumentation and reputable, internationally distributed reagents and supplies. Careful inspection of the example also reveals that a surprising amount of information can be derived (or at least inferred) from an experimental record in an unfamiliar language—here German. The latter point has been eloquently and persuasively elaborated in a remarkable book entitled *Chemistry Through the Language Barrier* published some years ago (1970) by the American chemist E. Emmet REID—who was in his late 90s at the time!

to heart. We suspect even the veteran researcher will encounter here an idea or two worth remembering the next time a report is required.

First and foremost:

- Never start writing before taking the time to formulate—*in writing*—a thoughtfully organized *tentative outline*!

The first version of an outline for a report may be little more than a loosely arranged collection of relevant *keywords*, perhaps implying a rough general *sequence* for presenting the necessary ideas. It nevertheless constitutes a first step toward creating an overview of what needs to be reported. A mere *sketch* will suffice, a crude framework suggesting *contours* for the document that ultimately will evolve. Making a serious effort to consider the optimal structure for a report at the outset of the writing process (and not after it is already well advanced!) greatly increases the chance you will put the essential information on paper efficiently and in a logical order, incorporating all the required elements the first time around and sparing yourself much tedious reorganization and frustrating elaboration (and/or pruning) at the last minute, when time will probably be short.

A distinction is sometimes made between a *topic outline*—a structured list of headings heavily reliant on keywords—and a *sentence outline* composed of pithy introductory sentences, each of which will later be expanded into an informative paragraph.

- The larger and more complex the report, the more thoughtfully one should develop the corresponding outline, and the more attention should be directed toward *structure*.

A typical report actually has a surprising amount in common with a *dissertation*, even though the latter obviously represents documentation at a much higher level (as discussed in the chapter that follows, especially Sec. 2.2). The organizational principles are in fact similar for reports of all types, and the same principles are applicable to the *journal article* designed to familiarize the *public* with a piece of research for the first time, through the pages of a professional periodical (cf. Chap. 3). The almost universally accepted pattern features five major elements (cf. Sec. 2.2.1):

- *Introduction — Experimental — Results — Discussion — Conclusions*

The *Experimental* section is sometimes placed nearer the end of a report than this list would indicate: immediately ahead of (or even after!) the concluding summary (see, e.g., the standard BS-4811:1972).

Depending in part on the overall length of a report, some of its sections may differ significantly from others in relative "weight", and the extent of further subdivision will vary correspondingly. In an extreme case the "Introduction" might be reduced to a single sentence, though it should never be omitted entirely. Sometimes two of the suggested sections are combined—results and discussion, for example, or discussion and conclusions—especially if there is a concern that too much fragmentation would cause a brief report to look "padded".

The longer the report the more likely it is that the original outline will undergo significant modification during the writing process. This is the main reason we referred earlier to a *tentative* outline, whose structure might well differ considerably from the subdivision pattern ultimately adopted for the report itself.

An example of an outline or preliminary subdivision scheme proposed by a student preparing a report on a synthetic chemistry project, generally consistent with the principles described above, is shown in Fig. 1–4. The numbering system applied to the various subheadings is described later, in Sec. 2.2.5.

Drafting the Text

Once a tentative structure is available, work should begin on a *rough draft* of the entire document. Ideas will no longer be presented as mere keywords, but rather in complete sentences, although at this point you should pay little or no attention to the impact specific *words* might have.

● What is important to ensure now is that each sentence follows naturally from the one before it, and that sentences fit comfortably together into cohesive paragraphs.

Paragraphs serve the purpose of organizing text into blocks of closely related thoughts, but at a level below that of formal subheadings. If you encounter trouble at this stage, the fault probably lies with shortcomings in your outline.

The immediate challenge is to come up with a logical sequence of self-contained paragraphs encompassing related observations, which taken together produce a smooth flow of ideas in a channel that not only avoids confusing, turbulent stretches resembling rapids but also discontinuities perhaps more nearly analogous to falls or cataracts. At the transition from one paragraph to the next, marking a conceptual break, printed text will spring abruptly to a new line—which may also be indented. Such shifts can be further emphasized by inserting extra, blank lines between the paragraphs, producing a visual effect comparable to that produced by the mortar in a brick wall. The average reader will not be especially conscious of these subtle indicators of structure, but *would* be inclined to protest if confronted with a document consisting of a single, uninterrupted stream of words.

Thanks to tools available in word-processing software, paragraph signals can easily be subjected to subtle "tweaking". For example, there is no rule that paragraphs must be separated from one another by precisely *one* blank line.[24] An author's software typically permits a gap of any desired width to be placed between the last line of one paragraph and the first line of the next. Incidentally: if one *does* introduce space at every paragraph break there is no point in also *indenting* first lines (though uniform

---

[24] Indeed, it is usually considered preferable that the space inserted *not* be exactly equivalent to a line of text.

Grignard Reaction of 2-Methylbutylmagnesium Bromide with 3-Pentanone

1	Introduction
1.1	Historical Background
1.1.1	General Overview of Grignard Reactions of Ketones
1.1.2	Chemical Equation, Mechanism, Likely Byproducts
1.1.3	Beilstein: Properties and Alternative Syntheses of 3-Ethyl-5-Methyl-3-heptanol (1)
1.1.4	Recent Literature on the Synthesis, Applications, and Physiological Effects of 1
1.2	Proposed Synthetic Path and Reasons for Choosing It
2	Experimental
2.1	Materials
2.1.1	Chemicals
2.1.2	Equipment
2.2	Description of the Work
2.2.1	The Reaction Itself
2.2.2	Workup and Isolation of the Product
2.3	Characterization of the Reaction Product
2.3.1	Chemical Tests
2.3.2	Physical Properties (Melting Point, n_D^{20}, Density)
2.3.3	GC Studies
2.3.4	Spectroscopic Data (IR, NMR, MS)
3	Results
3.1	Yield, Basis for its Estimation
3.2	Purity (Interpretation of Gas Chromatograms and Physical Data)
3.3	Structural Evidence (Interpretation of Spectra, Comparisons with the Literature)
4	Discussion and Conclusions
4.1	Comments on Yield and Purity, Estimates of Experimental Error
4.2	Interesting and Unexpected Observations; Attempts at Interpretation
4.3	Recommendations Regarding this Experiment
5	Literature

Fig. 1–4. Formal outline for a report on a laboratory exercise in preparative organic chemistry.—The main subdivisions correspond to sections found in many types of reports, although in this case there is no separate section labeled "Conclusions".

indentation by any amount one wishes is easily achieved). One signal is enough! Refraining from superfluous indentation is actually better from an aesthetic standpoint, and it saves a bit of precious space. In this way indentation can be reserved for signaling the presence of some other feature, such as a *list* or a *free-standing formula* (see Chap. 6).

● Especially serious thought should be given to the *first sentence* of every paragraph.

An opening sentence (sometimes called a *topic sentence*) should offer the reader a preview of the ensuing paragraph's content and purpose. The *last* sentence in a paragraph plays a special role as well: it provides closure for the current thought and at the same establishes a conceptual link to the *next* paragraph.

● A common rule of thumb holds that the optimum *length* for a typical paragraph is somewhere between four and eight sentences.

With letter-sized typewritten or computer-generated sheets, this recommendation leads to about two or three paragraphs on the average page. This is not to say that a paragraph consisting of only one or two lines is necessarily a bad thing. On the contrary: a short paragraph can be a valuable device for focusing the reader's attention, in that it will be perceived almost as a "free-standing line". It in fact acquires something of the character of a *key sentence*. In this book we draw the reader's attention toward such "key sentences" even more blatantly by starting each with a bold, black dot (in printing terminology: a *bullet*).

First Refinements: Perfecting the Language

Once your report has been roughed out in its entirety, with everything now expressed in complete sentences, it is time to begin the long, tedious process of reworking the document in an effort to improve the way its content is expressed. A point we want to emphasize is that you should perceive a clear distinction between a *rough draft* and the *first revision*. One can of course always find ways to improve a document at any stage, but this transition is special. Until now all attention has been directed toward completeness, continuity, and the logical flow of ideas, but throughout the upcoming revision process your effort should instead center on expressive qualities and the *impact* you achieve with each word and sentence.

● Document refinement should be considered *fine-tuning* of your side of what you want to become a kind of dialogue.

Above all, pay close attention to clarity, since unlike the situation in an ordinary face-to-face conversation, the dialogue you are striving for here is one in which you would have no immediate way of sensing, let alone responding to, the other party's reactions, nor will you be able directly to answer unexpected questions that may be raised.

In place of the "eye contact" every accomplished orator depends upon for guidance as well as a sense of one-on-one communication, you must try to establish a degree of "intellectual contact" with the other party, attempting to *imagine* how your text will be received by a first-time reader. Does every sentence *really* say what you hoped it would? Could there be any misinterpretation? What conclusions might the reader be likely to draw at various points?

It is a remarkable fact that many scientists—despite an obligatory apprenticeship in analytical thinking—produce written arguments unparalleled for their obscurity. The scientist–writer commonly fails to express in writing, clearly and unambiguously, the message he or she actually wishes to convey. For example, *words* and *verbal images* that create slightly misleading impressions can cause a great deal of damage, and yet they crop up surprisingly often. Awkward or incorrectly utilized *sentence components* also confuse more readers than you probably realize. Vague *references*, or nonexistent connections that nevertheless *seem* to be implied are other common sources of knitted eyebrows, frustration, and awkward unanswered questions. Something as simple as an inappropriate *conjunction* (e.g., one suggesting causality in the wrong direction, as through a mix-up in the pair "because/therefore") can sabotage the logic of an entire paragraph, and a sentence that ends in a way inconsistent with what the opening led the reader to expect can be a major obstacle, potentially extinguishing even the determined scholar's interest.

We could not resist adding a few miscellaneous observations related to language and the stylistic aspects of writing.[25]

Suggestions Related to Writing Style

● Even when discussing a highly technical scientific subject, resist the temptation to rely too heavily on *nouns*.

We all fall victim on occasion to this shortcoming—but at least *try* not to let yourself succumb too often! *Activities* should be represented by lively *verbs*, not stodgy nouns forced to accept an assignment in a boring, all-purpose verb-like formulation. "Bring to one's attention" as a substitute for "point out" sounds bureaucratic, and "the necessary calculation was accomplished" rather than "was calculated" is just as technocratic. If actions *must* be expressed in a nominative way, then at least consider the possibility that the infinitive form of a verb might be preferable to tacking on a tired "ing" (e.g., "To err is human" vs. "Erring is human"). Issue your reader an active injunction like "*draw* a diagram" when the *activity* is what you want to emphasize, and reserve "drawing" (a noun) for a situation where the *consequence* (e.g., "Drawing manually with India ink on vellum has, at least in science, become obsolete during

---

[25] It certainly is also not amiss for us to urge you to consult what at least a few *others* have had to say about writing. In particular, every aspiring writer should spend an hour or two with one tiny classic, almost 70 years old but still as fresh and pithy as the day it was written: STRUNK and WHITE's *The Elements of Style* (2000, 4th edition). Those who take special delight in deliciously zany and bizarre examples are encouraged to seek out the books by Karen Elizabeth GORDON, especially *The Well-Tempered Sentence* and *The Transitive Vampire* (both 1993; described, respectively, as "a handbook of punctuation" and "a handbook of grammar for the innocent, the eager, and the doomed"). Especially those who are chemists are likely to enjoy reading Robert SCHOENFELD's *The Chemist's English*. Finally, no serious writer should be without access to the authoritative *Chicago Manual of Style* (2003; "The Essential Guide for Writers, Editors, and Publishers") with its treasure chest of highly organized, singularly prescriptive "rules" for virtually everything (originally compiled for use at the University of Chicago Press, and now in its 15th edition).

the past twenty years …") or the *outcome* of the action (e.g., "a line-drawing of the apparatus") is of primary concern.

● Events should be carefully set in the proper grammatical *time frame*.

That which *was*, but is no longer—e.g., an occurrence, a one-time phenomenon, an observation, some former understanding or rationalization—is correctly expressed in the *past* or *past perfect* tense (e.g., "was decanted", "had thus been excluded"), whereas something *valid now*—e.g., a characteristic, a relationship, a logical sequence—belongs in the *present* (e.g., "absorbs at *xy* nm"). If something began in the past and continues into the present it can legitimately be addressed in the *present perfect* tense (e.g., "has therefore proved to be …").

● Avoid superfluous "filler" words and empty superlatives—anything that makes a passage sound overblown, "mushy", or bombastic.

This description often applies to words like

"now", "certainly", "also", "too", "above all", "quite", "especially"

among a host of others, even though they may occasionally be truly necessary, and their *sparing* (optional) introduction can add life to an otherwise dull piece of text. On the other hand, unnecessarily raising the stakes by proclaiming an example of "extremely high" or "extraordinarily great" purity is absolutely unwarranted in a scientific context. Be content with a plain but instructive factual expression like "was obtained with a purity > 99.3%".

● Avoid words, phrases, and images that are—perhaps subtly—redundant.

Pointless repetition (*pleonasm*, from the Greek for "more") is commoner than you might imagine. For example, attaching the modifier "so-called" to a term set in *quotation marks*—as in "the so-called 'Murphy effect'"—is an example of needless redundancy, because quotation marks and the extra "so-called" indicate precisely the same thing. Choose either "the so-called Murphy effect" or simply "the 'Murphy effect'".

Several sins could be said to be coexisting (awkwardly) in an expression like "the very best": unwarranted exaggeration, wasted space, and tiresome repetition. Characterizing someone as "the very best grandpa" *might* be excusable in paying tribute to your favorite patriarch in a birthday card, but there is no justification for claiming a field experiment to have been conducted under "the very best conditions possible". First of all, "best" *always* means "*the* best" relative to anything comparable, so the word "very" adds nothing substantive whatever. Even alleging that these *were* "the best conditions" almost certainly overstates the case. Some other possible observation period—a few years ago, or perhaps next year—might well correspond to even *better* conditions, in which case you are clearly misrepresenting the facts.

Words like "self-evident" and "naturally" or "of course" should also be used with caution, and only when your intent is to emphasize strongly the fact that something is *especially* self-evident, and for this reason merits no further discussion. On the other hand, if it really is *that* obvious, should you be bothering to point it out at all?

Even a meager knowledge of Latin can be helpful in avoiding certain erroneous formulations, like "most optimal", where the origin of "optimum" (from *optimus*, best) makes it apparent that nothing could possibly be better. Attempting as in this case to elevate a superlative form even higher makes a mockery of the language. Note, too, that many common adjectives are immune to intensification, quite apart from such obvious examples as "dead" and "pregnant". Whole *classes* of adjectives of a technical nature fall into this category, including words ending in "-less" and "-free" ("frictionless", "error-free").

It is probably safe to say that even *your* writing sometimes suffers from lapses like the ones we have been spotlighting; certainly ours always has—and we fear it always will, though hopefully to a gradually diminishing extent.[26]

● It is often useful to pause and consider in detail the words you propose to employ, especially ones with which you are only marginally comfortable.

Take the time to consider word *derivations*, for example, which most good dictionaries provide, and also look to see how the dictionary *defines* the word in question. Close attention to a particular word can lead you beyond a literal meaning and origin, offering insight as well into natural relationships between that word and specific *contexts*. Many words, you will find, automatically conjure up distinctive images in the sensitive reader's mind, suggesting scenarios into which certain other words fit very comfortably, but some wouldn't fit at all. The nature of such a scenario sometimes becomes apparent through thinking about original sources of the various fragments from which a word is constructed.[27] This piece of advice should actually encourage you to engage frequently in something all of us should do more often: thinking analytically about what we say and how we say it (especially in print), an exercise that at the same time sharpens and strengthens vocabularies.

The English language is infinitely rich in metaphors, which in turn evoke powerful images. But too much prose reads as though a linguistic paint pot had overflowed while the text was being assembled. A wit once remarked, for example, that something or other was analogous to an "apple cart in a china shop", a marvelous juxtaposition of the famous "upset apple cart" with the clumsy bull careening toward a display of porcelain. The sensitive reader often finds occasions to chuckle over passages like this, with expressions that present a kaleidoscopic panorama in which a number of incompatible images overlap. Usually the author responsible had no intention what-

---

[26] We don't mean this as a challenge for you to examine our work with a magnifying glass! In fact, we are happy to point to one nearby spot where we quite obviously resorted to redundancy: a few sentences ago as we cautioned against the use of "superfluous 'filler words'". A rigorous definition of "filler words" would certainly *include* reference to verbiage that is "superfluous".

[27] Did you notice here the deliberate pleonasm? A "source" is by *definition* original ("source" and "origin" are virtually synonymous). This intentional bit of repetition was meant to supply emphasis; in its absence the message could have been overlooked too easily.

soever of being funny; the humor is an accident, and a sign of carelessness—but also a distraction, as well as a lost opportunity to confer upon the work a deeper dimension of meaning.

- Always pay attention to *imagery* associated with language you use, and be sure images are grouped in such a way that they interact harmoniously.

Yet another challenge:

- Be vigilant in trying to assure that relationships you imply are all intentional and unambiguous.

A relative clause beginning with "which", in company with a main clause containing multiple nouns, almost always rests on an insecure foundation; *which* of the nouns is the antecedent of the subordinate "which" clause? We surely need not provide specific examples in this case. Open almost any newspaper (or even a scientific journal!), look closely, and you will find yourself confronted with a rich supply of references that are at least technically ambiguous.

- Avoid sentences that are too complex!

We have in mind here the sentence with multiple, interwoven clauses, one that might be represented

$$A_1\text{-}[A_2\text{-}[A_3\ldots[A_n]\ldots A_{3'}]\text{-}A_{2'}]\text{-}A_{1'}$$

with a level of complexity that is a function of n. A scientist should have no trouble interpreting our notation, or with constructing an illustrative (probably amusing) example. Try analyzing in this way the next frightfully tangled sentence you come across in your daily reading ("The soldier, becoming, as a result of an intense wave of feeling, from which he could find no way to free himself, deeply depressed, jumped back into the trench"; $n = 4$?). From our experience, high n values are especially common in *philosophy*, a discipline followed closely in this respect by *theology*. (Scientists are rarely the worst offenders.) By the way: If you are interested in increasing the likelihood that your papers will be intelligible even for *non-native* speakers of English[28] you will try to be sure that n rarely exceeds two in manuscripts you prepare.

- You may be surprised to find how much you can learn from reading freshly written text *aloud*, thereby becoming sensitive to each individual word and also experiencing how the whole actually "sounds".

Hearing your words makes it easier to ascertain how far removed your academic prose is from (formal) spoken language—which is what you should actually be trying to simulate—and how complicated your formulations may have become. As suggested above, long, involved sentences are among the most common obstacles to fluent interpretation. For this reason:

---

[28] Too few authors consider the possibility that this description may in fact apply to a *majority* of their potential readers, many of whom are influential within the discipline.

● Most sentences in modern scientific text should be kept reasonably short.

It has become increasingly fashionable in recent decades to encourage efficiency at the expense of "elegance". The stately prose of the Victorians is today definitely "out", and while some of us may still admire its rich vocabulary and vivid metaphors, and in some ways very much regret its demise, the younger generation finds this sort of writing extremely difficult to comprehend. An inevitable consequence of "modernity"—or perhaps "post-modernity"? It is not our place to pontificate or even speculate about such lofty matters, but we still note with sorrow that much marvelous scientific literature from the 19th Century is now dismissed (or ignored) as "dull", "boring", "obsolete", and too hard to read (and thus presumably irrelevant).

At the same time, however, we would not be able to associate a precise numerical value with the concept "optimal sentence length". For the sake of your less patient readers, though, you would be wise to consider dividing most of the exceptionally long sentences you find yourself writing into two or more shorter ones. But the reverse strategy has a place too: merging two very short sentences sometimes markedly improves the overall flow of ideas. Remember also that *variety* has its benefits: *variatio delectat*! What may look like a casual, unplanned interplay between long and short sentences may in fact be rescuing an otherwise monotonous piece of text by arresting the reader's attention now and then and thereby rekindling flagging interest. This observation is more important than may at first be apparent. Whenever you find a brief declarative sentence immediately followed by several more of the same type, take note of the fact, and then create a little syncopation by deliberately introducing a sentence containing a lengthy dependent clause.

● Variety in one's *choice of words*, often regarded as a characteristic of good writing, is generally taboo when dealing with technical terms.

For a sportscaster describing a baseball game, the person standing near the center of the diamond throwing the ball can be referred to interchangeably as the "pitcher", the "hurler", the "man on the mound", "this season's strikeout king", or a host of other descriptive appellations (not all complimentary!), and no harm whatsoever will be done. Every fan in the stadium or watching the game on TV will after all know who is meant—and can probably even *see* that person—so there is no serious potential for confusion. The situation is entirely different in a scientific treatise, however. Never tinker arbitrarily with *technical terminology*. It is categorically wrong in scientific text to try to achieve variety with respect to technical terms in the interest of avoiding monotony. Once a concept has been labeled, if you mention it again later, apply to it precisely the same descriptive name, despite the risk that your prose may end up sounding dull.[29]

---

[29] *Terminology* has been defined as an "organized collection of concepts from within a particular subject area, together with associated concept symbols" (FELBER and BUDIN 1989, p. 5) A *technical term* always serves a defining function, and its content and scope are constrained by fairly rigid boundaries established formally or informally by authoritative practitioners of the discipline—although these boundaries may

→

Consider also the fact that at some point an editor may want to make your work accessible by way of an *index*, thus making it possible to locate something quickly, and from several vantage points. Effective indexing is almost impossible in the absence of a standardized *vocabulary*. In other words, if there is some defined quantity that you (and others) routinely represent by the Greek letter ξ, don't refer to it at one point in your text as the "coherence length" and somewhere else as a "coherence distance". The especially alert and conscientious reader would very likely struggle to discern a "distinction" here that you in fact have no intention of suggesting.

No one expects a technical report to be a paragon of literary style. Especially with an in-house report, for example, an experimental description can still be perfectly acceptable even if it contains a certain amount of loose laboratory jargon and technical slang inappropriate for a journal article. This is particularly true if such informality facilitates comprehension and interpretation. The coined verb "to freeze-dry", for example, is not a very elegant substitute for "to remove water by subjecting a frozen sample to high vacuum", but few practitioners would be offended by it. More important, the low-key formulation is clear, concise, and to the point, and therefore advantageous (at least so long as there is little risk that an uninitiated outsider—or notorious pedant— will ever be confronted with the passage).

● One who anticipates writing much should also read much.

A role reversal we suggested earlier (with the writer attempting to occupy the seat of a typical member of the target audience) can help one discover—in the same way a reader might—text that "works" as opposed to that which does not.

Thoughtful reading, perhaps with the goal of self-improvement, makes it abundantly clear that not every respected "scientist–author" is equally proficient at communicating. By all means devote serious effort to seeking ways to improve your writing—and make it clear you are open to suggestions from others. An author's poor form drastically reduces the value of a document because it prevents the content from being fully appreciated. You will never reach the point where it no longer is worthwhile to work on your writing style. The scientist preparing to submit a manuscript to a professional journal, and certainly the aspiring book author, will unavoidably encounter one important *quality control inspector*: an *editor* likely to suggest (or insist upon) changes in the document being presented for publication (cf. Sec. 1.1.2). Your first reaction to such criticism will probably betray a certain amount of indignation, but *study* the proposed changes anyway, and learn all you can from them, even if only for the purpose of discovering how your precious words were *perceived* by someone deeply committed to excellence in communication.

---

shift over time. Moreover, certain "things"—specific chemical elements or compounds, plants, microorganisms, etc.—are officially known by very definite, "nonnegotiable" *names*. The sum of all such names conferred within a given discipline is known as a *nomenclature*, and this in turn constitutes one part of the terminology of that field.

Subsequent Drafts

Once a first draft has been closely examined and extensively revised, what results is a *second draft*—which will require equally serious attention. The extent of the burden at each stage in the editing process has been considerably reduced in recent years by computers, with their "Cut", "Copy", and "Paste" commands, and tools for moving pieces of text cleanly and instantaneously from one place to another by "dragging" with a mouse. Nevertheless, don't let yourself be tempted to do all or most of your revising from a computer screen. Constraining your field of vision to a miniscule "text window" robs you of perspective, and it makes it almost impossible for you to maintain a balanced overview of your work. There is also a high probability of the technology seducing you into thinking that editing is really a simple matter, quickly disposed of. *You* may rashly conclude after a first hurried pass that you've done enough, but if you quit too early, your readers will be seriously short-changed, and *they* are the ones who count!

Serious editing should always be accomplished on a *printed* copy of the most recent draft—with pencil, so that hasty, misguided moves are easily retracted. Once a manuscript becomes fairly cluttered, make all the proposed changes in the electronically stored version and print a fresh copy—for *further* intensive editing. (Only clean copy can be processed easily—and swiftly—with each word exerting its full impact.)

● Depending upon your overall skills as an author, and on the complexity of the document, *several* printed versions will almost certainly be required before you should even consider producing a "definitive edition".

If you wish evidence in support of this contention, consider the fact that each page of *this* book was subjected to an average of at least *six* revisions—by authors able to point to a fair amount of writing experience.

Finished Copy

Eventually, it *will* be time for you to prepare a *finished* (or *final*) *copy*: a document truly satisfying your highest aspirations and revealing your full capabilities with respect to content, organization, style, language, and form.

● Once you are in fact dealing with final copy, attempting to introduce even minor changes or corrections will have consequences that are both unsightly and distracting.

Whether at this point a few last-minute essential changes should be dealt with by printing a fresh version of the entire document, or instead through careful revision of a few individual pages, will depend to some extent on the nature and intended purpose of what you are preparing—and thus on standards established by (or appropriate to) the prospective recipient. Another factor may be the availability of support personnel who can execute the changes for you, but since nearly everyone now has personal access to a word-processing system and a high-quality printer, the latter consideration has almost become obsolete. There is today scarcely any excuse for submitting a disfigured, hand-corrected document—not even one carefully altered with the "white-out" correction fluid so prized during the typewriter era.

The pages that constitute final copy should of course carry *page numbers*. One published standard, DIN 1422–4 (1986) is exceptionally prescriptive in this regard, asserting categorically that

● Pages should be numbered consecutively from beginning to end, with the *title page* acquiring the number "1".

Not everyone would agree. More common is to begin regular pagination instead with the introduction. Text identified as a *preface*, *foreword*, or *table of contents* is traditionally paginated separately, as in books, with roman numerals (i, ii, iii, iv, …), thus forming a sort of prologue (technically: the *front matter*). Note also that an explicit page number need not appear on *every* page. For example, imposing a number on a carefully designed title page, or on a set of acknowledgements, or even a preface—can be distracting and unsightly. The same holds true for a full page of illustrations, or the blank left-hand page that often precedes the start of a new chapter.[30] Such a page must of course be accounted for in the *pagination* (or *page count*), but a formal printed number is optional.

DIN 1422–4 fails to prescribe a specific *location* for the page number. DIN 5008 closes this apparent loophole quite dogmatically, however, insisting unambiguously that the number be centered above the text; more precisely, that its place is on the equivalent of the fifth potential line (counting from the top of the page), flanked by a pair of dashes. A corresponding British standard (BS-4811:1972) sees matters rather differently: this source claims numbers should always be centered at the *bottom* of the page (i.e., *below* the text), adding the observation that each page of an *official* report should also bear the appropriate *report number*.

We would encourage you in the case of most reports (as well as with job applications, minutes of meetings, etc.) to supply your page numbers in the form "3/5" (read "3 of 5"), where 5 is the total number of pages in the document (the page count) and 3 is the number assigned to the current page. With a "loose" (unbound) report this allows the reader to be sure that every page is in fact present, including ones that fall at the end.

Assuming a document is prepared with a word-processing system, page numbers are most easily attended to in a *header* or a *footer*, in which case counting and marking occurs automatically each time a new copy is printed. This of course has the advantage that editorial changes will not upset some previously assigned (fixed) pagination. Page

---

[30] The beginning of *Chapter* 1 is almost always defined, at least in a book, as *page* 1, a right-hand page in the case of *two-sided print*. Subsequent chapters traditionally start on right-hand (and thus odd-numbered) pages as well. The beginning of a new *section* can be announced more dramatically by introducing an extra sheet that is blank on both sides, which is to say that here both a left- and a right-hand page will go "unused".

numbers in a header or footer can be specified as centered or adjacent to either the left or right margin. Other placements can be achieved through use of the *tab* feature. Additional useful items to consider including in a header or footer are the *date* (perhaps even the *time!*) of printing, as well as chapter or section titles. Word processors usually also offer the means for creating a special "title page", and facilities for paginating different sections of a document in different ways—with either roman or arabic numerals, for example. (As noted previously, front matter is commonly treated as a discrete section, with roman numbering.)

A comprehensive, up-to-date outline of your document can conveniently serve as at least a crude *table of contents* so long as it is supplemented with page numbers.

Body text in a report usually reflects a line-space setting of one-and-one-half (cf. "Manuscript Style and Markup" in Sec. 5.5.1). With letter-size paper, text blocks should be centered, surrounded by at least one-inch (25 mm) margins. If a normal "typewriter-size" font is employed (e.g., "12-point" type), copies will remain clearly legible even if reduced somewhat in size in the course of reproduction.

More detailed suggestions applicable to the preparation of final copy are provided in Chapter 5. "Special features" required by some reports—formulas, equations, tables, and illustrations, for example—are discussed in various chapters in Part II.

1.5 Various Types of Reports

1.5.1 The Academic Environment: Laboratory Reports, Grant Proposals, and the Like

In preceding sections we have considered at length preparation of a report in general, but without really examining the detailed nature of the resulting document. There is nevertheless merit in pointing out differences characteristic of specific *types* of report—what a purist might call various "text forms"—and this is the subject we have chosen for closing the chapter. In particular, not every "report" the scientist is called upon to write is based on a prior laboratory experience. Indeed, the close relationship we have frequently stressed between a report and one's laboratory notebook can in fact sometimes be quite tenuous. Even so, most of what we have already discussed will continue to be applicable, at least in principle, as we extend our horizon.

● The *Laboratory Report:* A Place to Begin

Academic laboratory experiences in the natural sciences and engineering typically require students to submit regular written accounts of their experimental activities, with straightforward *description* to be accompanied by more or less comprehensive *discussion* of what was done, including why the work was carried out in a particular way and what principles it was expected to illustrate. The exercise itself may have been a preparative experience in organic chemistry, or measurements associated with some natural phenomenon analyzed in an undergraduate physics laboratory, or even a training session involving a complex instrument, like an electron microscope. Whatever the circumstances, there is a certain clearly defined body of information to be dealt with—within the confines of a few pages—for critical review by a laboratory assistant or possibly a professor, who will typically assign a grade to the effort. That grade may also reflect the outcome of the experiment or activity itself. Sometimes reports of this type will be collected over an extended period of time in a loose-leaf binder to serve as documentation for the work of an entire course rather than a single afternoon.

Reports in this category share certain characteristics with both laboratory notebooks and the more formal professional reports all too familiar to researchers. Unlike most other report types, student reports may occasionally be submitted in handwritten form.

Whatever the nature of the assignment, nothing should be committed to paper until you have carefully examined all available guidelines for clues about the recipient's expectations with respect to both content and style. (Actually, the same advice applies to virtually every writing assignment you will ever face.) Preparing a laboratory report of this kind may be distasteful, and even seem rather pointless in the larger scheme of things, but it can be a valuable learning experience, introducing the fledgling scientist to skills, habits, and ways of thinking that are the prerequisites for a successful future.

● The *Seminar Report*: Where Learning and Teaching Overlap

In this case the assignment involves exploring, partly in the interest of others (your cohorts in a research group, for example), some specialized aspect of your discipline: through independent study of relevant library resources, perhaps, or information from documents prepared by your peers. Often this type of report will never be submitted to anyone in written form, but instead communicated *orally*—preferably from notes rather than complete text. The chief penalty attending a poor presentation is the embarrassment of making a fool of yourself in front of your colleagues—incentive enough to induce you to come up with a minor masterpiece: a comprehensive, intelligible, instructive account the others will find both interesting and informative.

The challenge here obviously goes far beyond arranging your own previous experiences and thoughts in some logical pattern. Basic skills of *literature searching* (Sec. 9.2) and *oral communication* (EBEL and BLIEFERT 1994, for the reader familiar with German) come into play as well. It may be you will want to dress up and enliven your presentation by including at least a few *graphic* elements (schematic diagrams, structural formulas, etc.), possibly projected in the form of transparencies or slides, or else shared as a computer-based package (e.g., a Microsoft POWERPOINT presentation). You should find various chapters in Part II to be of considerable assistance in this regard.

● *Research Reports*: Getting Down to Serious Business

Documents in the "research report" category are typically requested (better: *required*) by one's supervisor under the heading of *progress* (or *interim*) *reports*. Examples include reports expected at regular intervals from graduate students summarizing their recent investigative efforts, which are expected to culminate eventually in a thesis or dissertation, or the overviews postdoctoral fellows typically prepare in the course of (or upon completing) fixed-term research appointments. Most companies also require staff members engaged in research and development to submit regular *quarterly reports* of their activities, or even to report at more frequent intervals.

The importance attached in commerce generally to the ubiquitous "report" can be inferred from terminology employed on an everyday basis in an automatic, almost unthinking way throughout the world of business: "Y *reports* to X" has become a standard way of indicating that "X" occupies a higher position than "Y". Similarly, it is taken as a compliment when someone is identified as "an official spokesperson" (i.e., *reporter*) for a company or agency. The leader of a corporate research team is almost always obligated to *report* regularly to management, in part to account for resources expended but also to document progress in achieving specific corporate goals.

The most important thing to try to achieve in a research report is the organization of what is almost certainly a complex set of detailed facts within some broader, more meaningful context. At the same time one also wants to underscore *positive accomplishments*, making it clear precisely what it has been possible (and to a lesser extent not possible) to achieve. There may of course also be a need to explain (rationalize?) obvious failures, or justify why some assignment was carried out in an unexpected manner.

A good research report concludes with at least a brief look into the future, singling out promising work still in progress and making a case for why certain avenues of exploration may be worth special attention.

Official research reports are sometimes prepared for readers well outside the confines of one's own institution. A report might be solicited, for example, by an *external agency* involved in the sponsorship of one's research, perhaps a *government entity*. If so, the report will probably be expected to conform closely to a prescriptive set of guidelines. Widespread interest in models for such guidelines has been the impetus behind development of a number of national and international *standards* (e.g., ISO 5966-1982; BS-4811:1972; DIN 1422-2, 1984), a source of some of the ideas and recommendations we present here and elsewhere (e.g., in Sec. 2.2 and 3.3, among others places). Guidelines from a number of such sources reflect a consensus that the typical research report should incorporate the following elements, preferably in the order shown (items in parentheses are often considered optional):

Front cover • Title page • Summary • Table of Contents • (Glossary)
• (Preface) • Introduction • Body • Conclusions • (Recommendations/

Future prospects) • (Acknowledgements) • References • (Appendices) •
Documentation page • (Distribution list) • Back cover

The "Body" is commonly subdivided into "Results" and "Discussion" sections (cf.
"Proposed Subdivisions" in Sec. 1.4.2). Only rarely is a separate "Experimental" section
necessary, particularly if the report is to be directed beyond the research-group level.

Prominent on the cover and title page should be a formal *title* descriptive of the
work, together with the name of the *author* and the *date of submission*. In many cases
a recipient also expects to see here an identification number (*project number* or *report
number*; cf. Sec. 1.2) and explicit acknowledgment of the sponsoring entity, although
sometimes the identification "number" includes unambiguous indication of the latter.

The reason for making sure this information is especially apparent is to facilitate
rapid recognition of the report, and retrieval of the work from a bank of archives.
Should a report also happen to be assigned an official ISSN or an ISBN designation
(see Sec. 3.1.1 and 4.5.2), then the document automatically rises above the murky
realm known pejoratively as the "gray literature", gaining the status of a *publication*,
which means it would at least theoretically be available to anyone interested
in reading it. Information regarding a report's "identity" might also become part of
some public database, further ensuring its widespread availability in the future.

It should go without saying that all the pages in a research report should be
numbered, and that the integrity of the document should be secured by a suitable
binding.

● The *Grant Proposal*: One Key to a Scientist's Survival

The chief difference between a research report and a grant proposal is the fact that the
former is largely retrospective in character, whereas the latter is a request for
sponsorship of concrete ideas or open-ended suggestions related to *future* experiments
or scientific endeavors in general. In other words, activities shaping a grant proposal
have yet to occur. This sort of "report" necessarily serves as a forerunner of the work
discussed, because its principal task is securing the most important prerequisite:
adequate funding.

Almost all university-based research today is funded by third parties, which means
grant proposals assume enormous importance for the scientist in academe—including
from the standpoint of the time and intellectual attention they consume. In almost no
other arena is the interaction between research and writing so intense.

Every grant proposal shares a common goal—convincing a potential donor or
financier of the validity of certain premises:

– that results likely to be achieved in the course of the proposed research will be
  valuable,
– that completion of the project as described is *feasible* within prescribed temporal
  and fiscal limits, and

– that the applicant has the *abilities* required to accomplish the envisioned ends.

A grant proposal is usually accompanied by a selection of representative reprints illustrating the applicant's previously published work, together with an inventory of *all* such publications, in part as testimony to the applicant's abilities.

It is also assumed in most cases that one will include a detailed description of essential personnel and facilities already accounted for with respect to the proposed project. Thus, if the success of a particular investigation will depend upon results from a high-resolution mass spectrometer, it is important that a request for funds demonstrate convincingly that such a device will in fact be available, or at least at the principal investigator's disposal as required. Failure to provide such evidence could in itself be sufficient grounds for one's application being rejected (although in some cases, of course, acquisition of a key instrument constitutes a *part* of the request, as will be clearly apparent from both the narrative and the projected budget).

Grant proposals generally require that requested funds be specifically allocated within a small set of fairly narrow categories (e.g., personnel, apparatus, supplies, overhead, etc.). Be sure you don't overlook such peripheral expenses as the cost of research-related *travel* and *publication*.

Granting agencies typically provide special forms for the submission of grant applications.[31] The purpose of the form is to expedite the application process, but it also ensures a measure of "uni-*form*-ity" from request to request. Especially with respect to budget categories a standard form has the additional advantage that it facilitates the decision-making process, since the true dimensions of each request are clearly revealed. Standard application forms help guarantee as well that all essential information is actually provided in a clear-cut way (the applicant's full name, professional title, place of employment, address, telephone and telefax numbers, e-mail contact address, etc.), since in the absence of explicit guidance important items are easily overlooked.[32] Grant applications are discussed in somewhat greater detail in the earlier editions of *The Art of Scientific Writing* (EBEL, BLIEFERT, and RUSSEY 1987, 1990).

● Other (Personal) Report-Like Documents

The last step up the educational ladder for a scientist almost always requires preparation of an especially comprehensive "report" called either a *thesis* or a *dissertation*. This is such an important and complex document that we have dedicated an entire chapter (Chap. 2) to its characteristics and development.

---

[31] There is now a trend toward encouraging submission of *electronic proposals* via the Internet, typically as PDF files (cf. Sec. 5.4.1). In fact, the U.S. National Science Foundation recently began *requiring* that proposals be submitted in electronic form as a part of the new "fastlane" management process (which is also designed to provide up-to-date status reports electronically).

[32] Should you be among those grant applicants who tend to become buried in paperwork, you may find it useful to scan a paper-based form into your computer and then treat the scanned image as a background over which the desired information is simply "typed", with the aid of an appropriate graphics application, for example.

Finally, we should mention very briefly the nature and preparation of one other critical professional document that might be compared to a "report": the *resumé* (*curriculum vitae*) and/or *job application*. In addition to being a "report" this might even be looked upon as a "proposal"—but this time the focus is not on a scientific investigation in which you have a stake, but rather on *you yourself*! One could certainly be forgiven for regarding this type of "report" as the most important a person will ever write. The fact that we find ourselves in a world characterized by a limited job market and ever-increasing emphasis on flexibility and mobility means that the employment scene everywhere is very fluid (and probably will remain that way). Resumés may therefore need to be prepared repeatedly, each time almost from scratch as the years go by and one's career evolves.

The subject of resumés is not one we are in a position to explore at length in the present volume, but *you* should give it a great deal of attention, and from a variety of perspectives: technical and aesthetic, tactical and strategic. Entire books have been written on the subject (e.g., ROSENBERG and HIZER 1996), and the acquisition of one or more of them could turn out to be among the most rewarding investments you ever make.

1.5.2 The Corporate Environment: Technical Documentation

More and more of the products we rely on in our increasingly high-tech world—laboratory instrumentation, of course, but also shop tools, household appliances, entertainment devices, computer software, etc., etc., etc.—are virtually useless in the absence of intelligible, user-friendly, comprehensive, and reliable written explication. For this reason, *technical documentation* and *technical reports* are of increasing concern. Producing material of this sort is often the responsibility of someone with a job title like *technical editor*; common alternatives include *technical author* and *product editor* (an especially charming variant we once encountered is "descriptive engineer").

Technical information for the consumer has been the subject of widespread attention only within recent decades. One professional organization specifically committed to quality information transfer in this context is the Society for Technical Communication (STC), with roots in the United States reaching back as far as 1953, though its present name and form date only from 1971. The Society currently serves 25 000 members in 150 chapters, a few of them abroad. In its own words, the STC is dedicated to "encouraging the development of better-educated professionals whose jobs are to make complicated information usable by many". International efforts related to the transfer of technical information are a special concern of the International Council for Technical Communication (INTECOM), to which the STC belongs. Another somewhat relevant and interesting organization—but with a much broader and loftier mission—is the National Association of Science Writers (NASW), incorporated in 1955 to "foster the dissemination of accurate information regarding science through all the media normally devoted to informing the public". It describes itself as a body which "above all, fights for the free flow of science news". All three organizations (STC, INTECOM, and NASW) sponsor informative Web sites worth visiting.

The major purpose of technical documentation is to "forge a link between a product (producer) and its application (user)" (Sheet 1 of the German guideline VDI 4500, 1995, *Technical Documentation — User Information*); this particular characterization is actually taken from a discussion dealing with a relatively limited document category that includes *operating instructions*, *maintenance manuals*, *user's guides*, and *installation information*, but the principle is applicable as well to *educational aids*, *data sheets*, *product brochures*, and *catalogues*: i.e., virtually any piece of literature designed to be supportive of a commercial product. Other challenges within the purview of the technical editor might include surveys, magazine articles, and especially *Web sites*: the conveyors of on-line information. Indeed, much of the field of "technical communication" as defined in the academic world—often as a part of or closely associated with a department of information technology, IT—concerns itself with electronic messages distributed via the World Wide Web.[33]

Technical documents in printed form can sometimes swell to massive proportions: operating instructions accompanying a major piece of instrumentation, or in support of a production facility, may well comprise several thick volumes. The "instructions" compiled for one type of submarine are reported to weigh more than 18 tons in printed form, and Airbus Industries claims the technical description of one of its jetliners requires 400 000 pages!

In the long run, all technology is dependent on clear articulation of its potential applications. Erroneous and unintelligible user information can be an amazingly costly problem. It has been estimated, for example, that poor documentation is responsible for loss and damages on the order of $500 million a year in the German economy alone. There is probably no consumer anywhere who has not at some time become extremely frustrated because of poorly prepared product information, finding it perhaps too superficial, or characterized by far too much technical jargon, as if intended for a professional fluent in the specialized vocabulary of a particular industry or service. Another common source of complaint is the document with too much (confusing) text and too few illustrations, leading one to speculate that it actually applies to something other than the device before one's eyes. In principle, no consumer should ever be severely challenged by a set of operating instructions—and interpretation of a manual certainly should not presuppose technical training! Fortunately, evidence is accumulating that industry has slowly become aware of the seriousness of the problem and sincerely committed to resolving it.

---

[33] Degree programs in technical communication are now quite common. The Association of Teachers of Technical Writing (ATTW) in the United States provides links on its Web site to 51 such programs at the bachelor's level, 36 master's programs, and 12 that even confer a Ph.D.

Technical communication extends in other directions as well, including toward the public relations arena. Consider the matter of environmental fears, which have manifested themselves in the "NIMBY" ("not in my back yard") attitude. What effect might it have if a group of sincere but concerned citizens were invited to pay a leisurely visit to a modern waste-disposal incinerator or a nuclear generating plant in the company of a credible, well-informed guide who was also an effective communicator? The participants could well find themselves drawn into long-term constructive dialogue with their hosts simply on the strength of having acquired a better appreciation for the way management and operators of the plant are already determined to take the environment seriously. But the guide/communicators who could be the key to such harmonious interaction are rare, and they are made, not born. Their educational background would need to include a solid combination of technical knowledge, communicative abilities, *and* people skills.

Apart from descriptive and explanatory assignments, technical documentation specialists also have an essential part to play in developing effective and reliable *safety information* for alerting consumers to hazards potentially associated with a product, or risks posed by its inappropriate use. Virtually all technical documentation is now accompanied by explicit safety instructions. This is in part because of legal dimensions product safety has acquired in the face of severe *product liability* laws. Manufacturers today find themselves shouldering very broad responsibilities with respect to damage and health claims occasioned not only by product defects but also by lack of attention to safety issues, even on the part of suppliers.

Technical documentation was long an undervalued commodity, but it has clearly come to enjoy greater respect. User information is increasingly treated as "one of the inherent constituents of a product and/or its delivery" (DIN V 8418, 1988). It is thus becoming commonplace to find reference in marketing campaigns to "quality management" (cf. DIN ISO 9000-1, 9001, and 9004), and many manufacturers and/or distributors have taken to heart the adage that "a product is only as good as its description", at the same time being forced to acknowledge that attention to product documentation can have a significant influence on market share.

1.5.3 Commissioned Reports

On several occasions we have been asked to add at least a few words on the subject of "book reviews", and this is perhaps the appropriate place (for further comment see O'CONNOR 1991, pp. 185–187). In a sense, a *book review* might be regarded as one more example of a "product description", albeit one ordinarily prepared by a "consumer". Since the product in this case—a book—is also a cultural object, it should come as no surprise that some book reviews resemble more closely the diatribe of a theater critic than the subdued pronouncement of a staid scientist—but in a sense that is as it should be! A review that passes no judgements with respect to whether the book in question would be of value to the target readership, and displays no passion, is of marginal value.

Someday you may find yourself, quite unexpectedly, one of the select few in your field privileged to review a newly published book. If so, remember that the review you prepare will have an impact not only on the book's success, but also on your own reputation.

Our first piece of advice is never put off completing such an assignment, and by no means treat it as of low priority. Otherwise you do your colleagues a great disservice and impede the flow of valuable communication. If you conclude that some aspect of the work deserves to be severely criticized, then by all means criticize it—but in a way that is *fair*. It would be wrong (or should we say cruel?), for example, to condemn a book and severely impair its chances of success simply because you take offense at the absence of what you regard as an important literature citation (i.e., one connected with *your* work?).

A good way to begin formulating a review is to ask yourself a series of penetrating questions. What does this book actually have to offer? What makes it novel relative to other books on the same subject? What aspects are handled especially well, and where has the author perhaps missed the mark, at least in your opinion? For whom might this book prove particularly valuable, and in what ways? Has the subject matter been treated at an appropriate level, given the target audience? The answers to these and related questions should serve as a solid foundation for your evaluation.

It is always a good idea, by the way, to familiarize yourself at the outset with other book reviews that have appeared recently in the publication soliciting your input.[34] This will help you gauge the depth to which you should take your analysis, and perhaps suggest something about an appropriate writing style.

Once you have a complete first draft, check to see if you in fact answered all the questions you posed earlier—competently but also in a balanced way. Be sure also that your review contains all the essential bibliographic information, because someone might wish to order a copy after reading your comments. Then proceed to polish the text as extensively as you would any other important piece of writing, which almost certainly means working your way through several more drafts.

The preparation of a book review actually has much in common with the "peer review" to which journal articles are subjected (cf. Sec. 3.5.3). In a sense, if you were to write a *monograph* or a *textbook* (Chap. 4) you would also be engaging in a reviewing process, in that countless judgemental decisions would be called for. Preparation of such a volume is in fact largely an exercise in selectivity, necessarily conducted with discretion and tact. Certain sources or points of view are invited to join into the discussion, while others are ignored or explicitly rejected. The author thus ends up articulating a set of opinions related to what he or she considers important.

---

[34] Note the assumption that you probably won't have decided on our *own* to become a reviewer!

Every reviewer temporarily assumes a seat on the bench of justice. Keep this in mind as you put the finishing touches on the next review you write—and also when you feel inclined to *refuse* to write a requested review.

This latest detour has carried us into the realm of criticism in general. Written critiques and evaluative recommendations of every sort will occupy an increasingly large fraction of your attention the farther your reputation spreads within your discipline and into the "scientific community" at large. Writing recommendations is something one only rarely engages in as a matter of *choice*: you will almost always find yourself acting upon a *request*. In higher education, for example, recommendations or evaluations of one's colleagues are routinely solicited in conjunction with promotion and tenure reviews, and students are obliged to submit recommendations from their instructors when they seek to gain admission to graduate school or an attractive postdoctoral appointment. Recommendations sometimes play an important part as well in the distribution of research grants. Constructing a recommendation that is both just and informative can be an agonizing and time-consuming task, but it must be confronted. Useful guidance is available from many granting agencies, including the National Science Foundation.

2 Dissertations

2.1 Nature and Purpose

In the English-speaking world, advanced training in science beyond the bachelor's degree (whether a bachelor of science, B.S., or a bachelor of arts, B.A.) generally entails first earning a degree at the master's level (M.S. or M.A.), typically to be followed by the "terminal" degree known (for historical reasons) as a "Doctorate of Philosophy" (Ph.D.). In some cases the master's degree is a mere formality awarded without ceremony at the completion of a certain amount of coursework early in a doctoral program. On the other hand, schools that offer *only* a master's degree (and not a Ph.D.) usually expect each candidate to submit for rigorous evaluation the fruits of a substantial piece of independent research: an exhaustive description of the person's work in the form of a lengthy written report known as a "thesis".

The Ph.D. degree *always* requires a major research effort, culminating in most cases in an even more formidable document: a "dissertation".[1] A final examination (or *defense*), usually oral, plays a significant part as well, but the written description of a major exercise in independent discovery is regarded as the true hallmark of a candidate's achievement. In other words, one's last days as a student constitute an occasion upon which the art of writing plays a truly decisive role in the scientist's (or *aspiring* scientist's) professional life. The present chapter was conceived primarily in terms of preparation of a doctoral dissertation, but master's theses and even the *undergraduate* theses required (or strongly encouraged) at many baccalaureate institutions differ from dissertations primarily in that they tend not to be as long, and they are based upon less formidable pieces of research.[2] Thus, the questions posed at the beginning of a study directed toward a degree at the master's level are more limited in scope, consistent

---

[1] A few institutions have in recent years begun to offer the option of substituting for a dissertation one or two major *journal articles* (see Chap. 3) published in the refereed professional literature, but this alternative remains the exception. Many would in fact argue that one of the most important aspects of graduate work is acquisition of the skills needed for preparing a comprehensive, formal report of a complex and successful research project. This includes achieving a solid, profound understanding of the background and significance of the project, without which preparation of an acceptable dissertation would be impossible.

[2] In fairness it should be pointed out that the words "thesis" and "dissertation" are often used interchangeably. Our chapter could just as well have been entitled "Theses and Dissertations", but we elected to stress dissertations because we expected an especially large number of our readers to be doctoral candidates. We hope in the near future to publish another, less exhaustive treatment of much of the same material, this time directed explicitly toward an undergraduate readership [cf. the German volume *Diplom- und Doktorarbeit: Anleitungen für den naturwissenschaftlich-technischen Nachwuchs* (EBEL and BLIEFERT 2003), which was similarly designed to be a sequel and companion to *Schreiben und Publizieren in den Naturwissenschaften* (EBEL and BLIEFERT 1998)].

with a realistic expectation that meaningful results can be achieved within a relatively short period of time.

Some academic traditions, especially in Europe, erect yet another research-and-writing hurdle beyond the Ph.D.: a capstone experience that confers eligibility for a professorial appointment at a university. Successfully meeting this ultimate test brings as its reward the *venia legendi*: authorization to teach others. In the German-speaking world the corresponding document is called a *Habilitationsschrift*, successful presentation of which is acknowledged as a very major accomplishment.[3]

The true significance of a thesis or dissertation is not that it testifies to the author's right to a particular title or position, however. Just as with other reports, the primary goal is dissemination of information, in this case a rather large body of newly acquired insight into a narrow, closely defined aspect of nature. This insight is in turn to be supportive of a novel *premise*, an idea put forth for the first time by the author and called the author's *thesis* (not coincidentally one of the very words also applied to the associated *document*; cf. the standards BS-4812:1972[4] and ISO 7144-1986).

Circumstances surrounding much graduate (and postdoctoral) work have in recent years been the subject of considerable debate—and criticism. There have long been pockets of uneasiness, for example, about whether extensive fundamental research leading to the usual type of dissertation should in fact continue to be a prerequisite—often the *sole* prerequisite!—for teaching at the college or university level.[5] This controversy notwithstanding, students everywhere engaged in graduate research can take pride in knowing that a substantial fraction of all the achievements of modern-day science are an outcome of student efforts, one of the reasons we might use to justify our devoting an entire chapter of this book to the subject of dissertations.[6]

At least in the natural sciences, the basic questions driving a graduate student's research usually originate with a faculty member in the student's academic department—who in turn becomes the degree candidate's official *advisor*. This is in sharp contrast to an age-old tradition in the humanities, in which a "thesis"—the underlying idea—is expected to be the product of a student's *own* imagination. It has also become

---

[3] Questions have recently been raised, however, about whether Habilitation should continue to be an absolutely *essential* criterion for appointment to a professorship in Germany, at the same time deploring the often difficult conditions under which work of this sort takes place (see, for example, ERKER 2000).

[4] This particular standard contains the following definition of "thesis" in the sense of a document: "A statement of investigation or research presenting the author's findings and any conclusions reached, submitted by the author in support of his candidature for a higher degree, professional qualification or other award."

[5] Even graduate schools are now raising this question and exploring its implications. Graduate departments at a number of institutions have begun to sponsor research into factors specifically associated with successful *teaching* of their respective disciplines, going so far as to establish professorships with this inquiry as their chief focus. Illustrative of the trend is the University of New Hampshire (USA), where several graduate departments—including chemistry—have shown extraordinary interest in improving the training of potential college and university faculty.

[6] Not surprisingly, numerous *books* have also been written on the subject; a list obtained from Amazon.com in fact identifies 20! A classic, currently in its sixth edition, is TURABIAN 1996.

common in science for graduate research topics to relate at least indirectly to commercial interests of some sort, even though work on the project generally takes place in a university setting. Irrespective of the source of ideas, however, or where graduate research is conducted, awarding of a degree is the exclusive prerogative of an accredited university or technical institute.

Quite apart from being a teaching credential, the Ph.D. degree is also the most widely recognized ticket of admission into the circle of *professionals* in a discipline. Successful completion of a doctorate in science is taken as strong evidence that a young person is capable of functioning *independently* in the search for answers to scientific questions. Beyond this, a doctorate signifies that the recipient has the skills required not only to tackle scientific problems, but also to communicate their nature and import—together with the evidence attending their solution—in a competent and methodical way to a knowledgeable and receptive audience through the medium of a dissertation (cf. *lat.* dissertare, to expound).

● Preparing a dissertation is a demanding exercise, presenting challenges of many sorts. It would be a serious mistake to attempt it "on the side", or under the pressure of rigid time constraints, or when you suspect you may find yourself subject to major distractions.

What is entailed is not merely formal submission to a review body of a detailed account of how you went about collecting a set of results—results you (and/or your advisor) regard as sufficiently numerous, persuasive, and significant to satisfy university expectations. It also involves presenting those accomplishments in a specific *way*, where once again there are high expectations to be met. Your dissertation could well assume dimensions comparable to those of a monograph! In other words, a doctoral degree candidate is expected quite literally to write an entire "book", and then *publish* that book—without benefit of meaningful instruction, and despite having little or no prior experience in either writing *or* publishing. It should thus come as no surprise that dissertations vary widely in quality with respect to both content and form. Our suggestions in this chapter are directed exclusively to issues of presentation.

A brief word is perhaps in order here with regard to one important peripheral issue:

● At what point is it appropriate to begin writing a dissertation?

There is obviously no way for us to provide a satisfactory general answer, but we can at least address the spirit of the question. In the first place, there is nothing to be gained from starting to write until *all* essential experiments, calculations, and interpretations associated with the project are satisfactorily attended to. Serious writing demands single-minded concentration, and you dare not run the risk of being diverted by a nagging awareness that a few important loose ends are still dangling somewhere. Moreover, tying up those "loose ends" could well shed important new light on the entire project!

Once writing begins, steer clear of engaging in any further activity of an experimental or theoretical nature. As you write, you *might* nonetheless suddenly become aware of something important you had previously overlooked, making it necessary to return at least briefly to the workbench.

● For your own protection, try to retain a right of access to your research space for as long as possible, and avoid dismantling crucial apparatus or losing track of key materials until the writing process is essentially complete.

There is no easy way to establish categorically that "enough" information has in fact been accumulated and interpreted to warrant preparation of a dissertation. The decision should in any case be made in close consultation with your research advisor. The extent of your accomplishments relative to what you perceive to be typical for a doctoral project in your field—and at your university—can perhaps serve as a rough guide, but what is most important is reaching the point where you can draw a reasonable number of novel, informative conclusions that coalesce comfortably into a convincing, comprehensive, largely self-contained intellectual whole. Once these conditions are met, one should strenuously resist—*you* should resist—succumbing to the temptation to probe even more deeply into unexplored territory, or permitting someone to seduce (or unreasonably coerce!) you into doing so.

2.2 The Components of a Dissertation

2.2.1 Overview

We turn now to a detailed consideration of the structure and composition of a typical dissertation. Our focus will be on a document proudly announcing the fruits of a piece of experimental work, the type of dissertation most common in the natural sciences. The majority of what follows is equally applicable to doctoral studies of a more theoretical nature, however, as well as to the preparation of a *thesis*, whether at the undergraduate or the master's level. Indeed, it will serve to characterize a major scientific report of almost any type (see also "Proposed Subdivisions" in Sec. 1.4.2).

The following "standard menu" of possible components is broadly representative. It lists in parentheses a few components that would not appear in every dissertation, but their inclusion should at least be considered (cf. Sec. 1.5.1).

● Typical dissertation elements, arranged in a commonly recommended sequence:

  • Title page • (Signatory page) • Abstract • Table of Contents
  • (List of Figures and Tables) • (List of Symbols) • (Dedication) • (Preface)
  • (Acknowledgements) • Introduction • Results • Discussion • (Conclusions)
  • Experimental • References • (Appendices) • (Remarks)
  • (Biographical information) • (Declaration)

A full-fledged *book* always begins with what is known as "front matter" (cf. Sec. 4.5.2), usually represented in a dissertation only by a *Title Page*, an *Abstract*, and a *Table of Contents*. The front-matter segment of a dissertation could be extended to include a page of *Acknowledgements*, a *List of Figures and Tables*, a *List of Abbreviations* (or *List of Symbols*), a *Glossary*, and possibly even an *Index*, although the latter (if present at all) would more likely be placed at the very end.

Certain of the headings associated with the body of the document itself, such as "Results" and "Discussion", are sometimes combined. An "Introduction" occasionally gives way to a section called "Nature of the Problem"—a title offering the advantage that it provides a bit more insight into the role the associated text plays. A dissertation in medicine often will be found to contain as a substitute for an "Experimental" section (or "Description of the Experiments") a section labeled "Methodology", whereas in biology the comparable information might be presented under a heading "Methods and Materials", or in some cases "Field Work", analogous to a similar section appropriate to the geosciences. With a more theoretical discipline (e.g., theoretical chemistry, mathematics) such a section might be replaced—or supplemented—by one simply entitled "Theory",[7] whereas a better heading for certain physics dissertations would be "Experimental Design and Measurement Techniques". Actually, the theoretical premises, experiments, evaluation, results, and proposed models associated with a physics project are sometimes so intertwined and complex that one would be served better by a different structural scheme altogether.

Just as all sonnets share a distinctive (14-line) canonical framework, a certain amount of structural uniformity characterizes most science-based dissertations as well. This verse-like architecture in fact shapes scientific reports of almost all types, regardless of their discipline of origin. The reason is that in nearly every case the fundamental objective is the same: formal presentation of a question addressed to nature, postulation of one or more potential solutions, and subsequent development of a case for the particular alternative that is most plausible.

Devising the right question to pose in a scientific investigation is an art in its own right, as well as the key to successful research. A suitable question must first of all be potentially answerable. Indeed, one should from the outset usually have an inkling of what a plausible answer could at least *look* like. A "first guess" of this sort—formally known as a *working hypothesis*—will be productive, however, only if it is likely to be *testable* within a reasonable span of time given the resources available. Broad questions like "how does an egg transform itself into a chicken" continue today to defy anything resembling a complete answer despite 100 years or so of intensive, careful research into almost every aspect of developmental biology. It would obviously be folly to select for a dissertation topic a sweeping inquiry with that kind of scope!

---

[7] The word "theory" has a somewhat different connotation in a dissertation based on experimental work: here the combination of Introduction, Results, and Discussion/Conclusions is sometimes distinguished collectively from experimental material by describing it as the "Theoretical Part" of the document (cf. the discussion of "parts" under "Structure and Form; Decimal Classification" in Sec. 2.2.5).

But you presumably dealt with strategic issues like this long ago, when work leading to the dissertation first began—ideally, in fact, before you first gave serious consideration to a particular theme. Interestingly, however, it is now time for you to try recalling as clearly as you can the thoughts running through your head when you selected your research topic, probably several years ago, and to review in a systematic way all that has happened since; i.e.,

- The topic—What actually was the question I (we?) set out in the beginning to study?
- Experiment/theory—How did we originally propose to tackle the problem in search of answers?
- Results—What specifically transpired over the course of the search? Did the nature of the question itself undergo any change?
- Discussion—What answer(s) did we in fact find to the question in its final form? What conclusions can we draw, and how significant are they?

We urge you to take our advice here seriously, because if you do you will soon find yourself ready to sketch out almost the entire dissertation. Ponder the questions above, and then begin organizing your odyssey the way you would if you were to tell it as a story—which might be treated as a rather dramatic one if you choose to see it that way (see Sec. 2.3.1 for concrete suggestions). The story should eventually unfold as a broad, continuous arc stretching from a specific starting point to what hopefully constitutes a satisfying, definitive ending (although you are of course aware that your own achievements represent just part of a continuing saga, one in which you may or may not elect to be involved in the future).

As noted previously, there is often good reason for deviating significantly from what has come to be regarded as the "standard" outline for a scientific report; some alternative may in your case offer major advantages. Methodological approaches associated with the various scientific disciplines are too diverse for reports to conform to a single mold. Nevertheless, the broad organizational scheme outlined above has stood the test of time and served as a basis for countless contributions to the scientific literature (cf. Chap. 3). Still, if you see a sound reason for altering the pattern, or discarding it completely, then by all means do so (unless considerations related to the point raised in the next bulleted paragraph make this impossible!). One of the most common modifications, incidentally, involves moving the "Experimental" section so it falls immediately after the introduction. We will later (in Sec. 2.2.7) look briefly at why this might be done. Before we proceed, however, one point must be stressed most emphatically:

● *Universities* (or departments) sometimes prescribe a very specific order in which the various parts making up a dissertation must be presented, and they may even dictate organizational terminology to be employed—not to mention such additional details as one particular (mandatory!) pagination scheme, strict rules regarding page layout, types of paper that are acceptable, and countless others.

Formal requirements of this sort will in general be spelled out in official university and departmental documents (subject to change over time!) that apply to every dissertation submitted. We cannot urge too strongly that before you even think about beginning the writing process you should acquire the *latest versions* of all relevant directives and recommendations, and then become intimately familiar with their content, bearing in mind that some of the requirements you encounter may directly contradict suggestions in our book—and the *official* guidelines must of course take precedence.

All the dissertation "sections" we discuss in what follows have one formal characteristic in common: each begins on a new page, specifically a *right-hand page* in the case of double-sided copy.

2.2.2 Title and Title Page

The *title* of any scholarly work should be crafted in such a way that it accurately and meaningfully describes the content, but in as few words as possible.

● An ideal title is a concise (but informative!) summary of the document to which it applies.

The hallmark of research is that it breaks new ground. That being the case, one could not possibly assign a fitting title to a piece of research before the work itself was complete, at least not with any assurance that the words selected would be truly apt, because some of the adventure has yet to transpire. Instead, one settles initially for what is known as a *working title*. The "topic", in the form you and your advisor originally conceived it, may continue to provide a working title for your project for several years, until the time finally arrives to begin writing. Once that point is reached, however, and you start to organize and evaluate your results, you will almost certainly recognize that the generic characterization you have relied on for so long is no longer adequate. Actually, a *permanent* title should never be formulated until the very end, after the bulk of the dissertation has actually been *written*. Only then will you be in a position to summarize accurately and in a few choice words the essence of what you have spent a hundred or more pages of text describing.

● The definitive title must be one upon which you and your research advisor agree.

The advisor is generally the one who suggested the topic in the first place, and who provided (or at least offered) considerable guidance during the course of the work, and it is likely that he or she will soon be writing recommendations for you. For these reasons alone your advisor should be intimately involved in the title-selection process.

● A good title contains at most ten words.

If ten words are not enough for a satisfactory description, then extra detail should be provided by a *subtitle* appended to the *main title*. Thus, instead of

> An Investigation into Influences of Temperature, Solvent, and
> Metallic Salts on the Degradation of Phenols by Yeast

a better choice would be:

> The Degradation of Phenols by Yeast:
> Influence of Temperature, Solvent, and Metallic Salts

Besides yielding a shorter main title (5 words instead of 17), the change has accomplished a second valuable objective: the most important words are now at the beginning (i.e., in the main title), and less important ones have taken on a more subordinate role. Titles are often incorporated into collective *title lists* with considerable bibliographic significance, and the change suggested here would go a long way toward ensuring that a potentially interested reader would someday encounter your work while examining such a list. Formulations structured around phrases like "An Investigation into ..." (as in the example above) or "A New Method for ..." are littered with useless "filler" words, reason enough to regard them as inferior. The second variant suggested is especially poor, since one always *assumes* that what is being reported is *new*!

Insofar as possible, a title should be based on terminology that would be understood across a broad spectrum of scholarly interests. At the same time, however, make an effort to build in as much *specificity* as you can. In place of the dull and vague formulation

> "Influence of ... on the Transport Velocity of ..."

consider the benefits of approaching the same thought from a different direction:

> "Increasing the Transport Velocity of ... by ..."

The latter would be much more likely to pique a reader's interest (see also Sec. 3.3.1).

● Abbreviations—apart from ones widely recognized across disciplinary boundaries, such as "IR" or "DNA"—should be rigorously excluded from titles. Special symbols are also taboo.

Discipline-specific abbreviations are likely to prove unintelligible to at least a few people who should in principle be interested in your work. A serious problem in the case of *symbols*—including letters of the Greek alphabet—is their lack of availability in standard type fonts, which in turn places severe limits on where a title containing them could be properly cited. (Web pages are a good case in point.) Symbols also stand in the way of computerized information searches.

We turn now from the title itself to the dissertation's *title page*, which usually includes (besides the title) the author's full name, the name of the institution where the work was carried out, and a submission date. University guidelines may dictate the presence of additional information as well. Give careful thought to optimum *placement* of the title on your title page, although you may again find that a particular location is specified by institutional rules. As a matter of fact, the entire title page is often subject to especially close scrutiny by "the authorities", extending even to such minutiae as the spacing of words—or letters!

● Unless you receive instructions to the contrary, *center* all text on the title page.

"Centering" a line means placing it such that it symmetrically spans the *optical* center of the page, which is in turn defined by the relative widths of the two vertical margins. If, for example, the left margin is 30 mm wide, while that on the right occupies only 20 mm, then the "optical center" is shifted to a point 5 mm to the right of the central axis of the page as a whole. All word-processing programs support a "centering" function for achieving the requisite placement.

2.2.3 Abstract

The title page is usually followed by a summary—more precisely: an *abstract*—of the work as a whole.[8]

● The *Abstract* should be limited to a single page, perhaps even a single paragraph. It should state explicitly the *goal* of the work, mention any unusual *methods* that may have been utilized, enumerate the most important *results*, and identify major *conclusions* that can be drawn.

A person reading your abstract should come away with a good general understanding of the nature of your accomplishments and also be in a position to assess their significance with respect not only to the discipline but also to the reader's own professional interests. The abstract must of course be clear, and it should be capable of standing entirely on its own, in the complete absence of the document upon which it is based. Despite the strict injunction that an abstract be brief, the text still must consist of grammatically sound, complete sentences. By convention, references to oneself (i.e., "first person" constructions, such as those that include the pronoun "I") are strictly avoided.

Finding the best way to compress a hundred or more pages of information into a few sentences is a real challenge. No one should expect it to be easy. Every word must be carefully weighed (Is this word really necessary? Is there a better, more concise way of expressing the chief conclusion?), and individual sentences should be combined with the greatest of care.

At least two standards [DIN 1426 (1988) and ANSI/ISO Z39.18-1995, *Scientific and Technical Reports: Elements, Organization, and Design*] distinguish between two *types* of abstract: the *indicative* abstract and the *informative* abstract. The former avoids serious treatment of what the master document contains, focusing entirely on its

---

[8] Some institutions require that, between the title page and the abstract, there be an additional page bearing the signatures of all members of the candidate's evaluating committee to provide official evidence that the work has been judged satisfactory.

purpose. An extreme example is the following typical entry in the professional abstracting literature (e.g., *Chemical Abstracts*) serving as the abstract for a review article: "Review with 134 references". The abstract in a dissertation will of course be of the "informative" type—or at least it *should* be.

The *format* applied to the abstract (type size, line spacing, etc.) usually leads to more compact presentation than would be appropriate for the rest of a dissertation: perhaps spacing based on one-and-one-half lines per line of text rather than the double-spacing more common for body text. Formatting of the abstract is one more thing that may well be specified by university guidelines. Tight spacing, perhaps coupled with the use of unusually small type, can help ensure that an abstract will fit comfortably on a single page, a definite advantage for documentation purposes.

The abstract is occasionally placed at the *end* of a dissertation instead of near the front. *Preparation* of the abstract should in any case be delayed until the very end—for the same reason that a title should acquire its final form only after the corresponding document is otherwise complete (cf. Sec. 2.2.2).

It is often recommended that, above the abstract itself on an abstract page, there should appear the *word* "Abstract", then the full title of the dissertation, in turn followed on the next line by the complete name of the author. An abstract page containing all this information can almost be viewed as an "ultra-abbreviated version" of the complete dissertation.

2.2.4 Preface

The *Preface* of a dissertation should be brief and characterized by restraint. In contrast to a typical book preface, a dissertation preface should not address the nature of the contents, nor should it comment on the significance of reported results. Such a preliminary "evaluation" from the perspective of the author would be out of place, especially because the work described is still subject to the critical assessment of a graduate committee. The preface in a dissertation is intended only as a place for calling attention to special circumstances that may have affected how the project was carried out, and—perhaps most important—expressing appreciation in the form of a few *acknowledgements* (though the latter might instead be incorporated elsewhere: see below). Gratitude should be acknowledged in the first place for the contributions of your advisor, after which you will wish to thank others who have helped you in some important way (by offering advice in the construction of a key piece of apparatus, for example, or assisting with a measurement or an important aspect of data analysis). A few words might also be directed toward an institution, granting agency, or corporation that provided support: financial backing, or assistance in some other form (e.g., the donation or underwriting of special equipment or supplies).

The Preface would typically conclude with a "dateline" like "Boston, October 2001", followed by the author's full name, although here, too, you should be alert to institutional guidelines, and perhaps consult recent dissertations submitted by others.

Alternatively, a simple "Acknowledgements" page might take the *place* of a preface. In some dissertations, such expressions of appreciation are consigned to a place near the very end—following the "Conclusions", for example. The Preface might also be preceded by a brief "Dedication" (cf. Sec. 2.2.11).

2.2.5 *Table of Contents; Section Headings*

Basic Considerations

The *Table of Contents* (or *Contents* page for short) is always placed at or very near the front of a lengthy, complex document, even though it, too, is actually prepared *after* the text itself has been composed and given its final form. Earlier creation of a meaningful table of contents would in fact be impossible. It must be based on a printed copy of the final document, because *page numbers* are required for the various entries, and these can only be known after the entire document is printed.[9] Furthermore, the *headings* under which a document is organized—which shape the Table of Contents (see below)—are often subject to last-minute change, though most will probably continue to resemble elements presented in the *outline* prepared very early in the writing process (cf. Sections 1.4.2 and 2.3.1).

A "Contents" page is a directory to help a reader locate specific parts of the text, but it should also be a place where one can gain a clear sense of the document's *structure*.

● The table of contents for a dissertation is a list of all the *headings* in the document, accompanied by corresponding page locations; in *form* it closely resembles a typical outline.

Every listed heading must reflect precisely the wording actually present in the document. This may seem self-evident and hardly worth stressing, but mistakes in this regard are exceedingly common, often as a result of impulsive, late editorial changes. For the author equipped with a good word processor, preparing at least an accurate *draft* of a table of contents can be a very simple matter: let the program do most of the work! Taking advantage of this handy software feature requires only that headings in the text be assigned appropriate (pre-defined) *heading formats* (or *heading styles*).[10] Once this has been done, issuing a special "Insert Table of Contents" command

---

[9] Trusting the *screen image* of a digital document for accuracy in this respect is dangerous, because what actually emerges from one's printer not infrequently has at least some line or even page breaks in slightly different places.

[10] One would normally make use of several *different* "heading formats" in a document, since this is an easy way to produce consistent typographic distinctions among headings at various "levels" (see below).

will instantly produce a list of the sort desired, consistent in every way with what appears in the document.[11]

Headings and Hierarchical Structure

The *headings* prepared for individual sections of the text should be subjected to many of the same considerations as the document's title: each should describe ensuing text in as precise and meaningful a way as possible, but with a minimum number of words.

- Headings in a large document like a dissertation will vary in their relative importance.

A complex document typically requires headings to introduce text passages operating at various hierarchically related "levels". Some headings will be *chapter* titles, some will be titles of the *sections* comprising these chapters, and some will apply to *sub*sections, which themselves may function at multiple levels. These organizational elements reflect a document's *internal structure* and prevent it from overwhelming the reader by looking like one monotonous continuum. To further clarify a scholarly document's hierarchical structure, and to help orient the reader, it is useful to attach to the headings a set of systematic *numbers*, so arranged that their complexity increases as one moves *down* the hierarchical heading scale (see below). Numbers of this sort precede the corresponding heading text and become an integral part of the heading as a whole.

As already noted, these headings eventually become the entries for a table of contents, where they will be *arranged* in such a way that the reader can see at a glance how the document is structured.

By tradition, the hierarchical status of a particular heading within a book is underscored by a set of typographical characteristics peculiar to headings at that specific level. For example, the most important headings, chapter titles, are usually set in the largest and/or boldest type. In the days of typewritten dissertations, subtle typographic indicators of hierarchical level like this were unavailable to the average person preparing a dissertation, but that changed dramatically with the advent of computer-controlled word processors. It is now a matter of routine for anyone to produce elaborate documents of "near-typeset" quality, which among other things means that every aspect of typography—type face, type style, type size, spacing—becomes subject to the author's wishes.[12]

---

[11] In Microsoft WORD, for example, there is an "Index and Tables" function in the "Insert" menu that provides a wide variety of stylistic options ("classic", "fancy", "simple", etc.) for an automatically constructed table of contents. Once created, the latter becomes an integral part of the document, situated wherever the screen cursor happened to be at the time table construction was requested.

[12] In Europe, typewritten dissertations were never common. Until relatively recently, standard practice there was for every dissertation to be typeset by professionals and then "published" by a commercial printer, in part so that enough high-quality copies could easily be prepared for distribution to university libraries all over the continent.

If the need arises for multiple formal copies of a document, a single carefully prepared "final version" of the manuscript, generated with a high-quality laser or ink-jet printer, can be dropped off at a neighborhood copy shop where efficient and relatively inexpensive reproduction techniques make possible the rapid generation of any desired number of duplicates for subsequent assembly and "binding" in any of several ways.[13]

Structure and Form; Decimal Classification

As mentioned above, it is now common practice to indicate hierarchical organization in an academic document—whether a dissertation or a monograph—not only typographically, but more directly as well, with the help of a so-called *decimal classification system* for assigning numbers to the various headings. Recall, too, that these numbers are understood to constitute an integral part of the headings themselves.[14]

- "Decimal classification" as we use the term here leads to numerical identifiers consisting of *sets* of digits, individual elements in the sets being separated by periods.

In fact, referring to this as a "decimal" system is misleading, since the resulting identifiers sometimes include double-digit elements. The number "10" (*lat.* decim, ten) thus plays no special role whatsoever, and the word "decimal" reflects only the fact that periods—which in *other* numerical contexts serve as "decimal points"—are present as separators.

Irrespective of what it is called, application of the system is quite straightforward. First, the most important sections of the text, typically referred to as *chapters*, are numbered sequentially, starting with 1. Chapter headings thus acquire "level one" identifiers (1, 2, 3, ...). Headings at the next level in the hierarchy—corresponding to major *sections* within chapters—are given sequential numbers as well, this time restarting with 1 at the beginning of each new chapter. *Complete* section numbers are more complex than chapter numbers, because they consist of one of the newly assigned lower-level numbers together with—more accurately, *preceded* by—the appropriate chapter number, the two being separated by a period (e.g., 1.1, 1.2, 1.3, ...; 2.1, 2.2, ...). In other words, subdivisions at the second classification level are distinguished by "two-element" numerical identifiers. The same procedure is followed, level by level, until all subheads in a document have been assigned numbers.

---

[13] In fact, "binding" in this context most likely involves a "gluing" process, not the stitching with thread that a reputable book publisher would prefer. If no convenient copy shop offered binding services of the quality required, it would be necessary to avail oneself, as in the past, of the skills of a professional bookbinder. Note that it is *never* appropriate with a dissertation to resort to an informal "binding" technique based on holes punched in the pages—nor would it be permitted, since this would not afford the durability required for library usage.

[14] We refrain from describing alternative systems based on combinations of roman and arabic numerals together with upper- and lower-case letters. This admittedly can lead to unambiguous hierarchical distinctions among a document's headings, but not in a way that is at all intuitive.

What follows is an excerpt from the table of contents of a hypothetical document structured according to this system:

Notice that no period is present after the *last* element in a set, and that the entries above have been aligned such that the status of each section is easily recognizable in terms of the complexity of its numerical identifier.[15]

As indicated, the term *chapter* is usually associated with sections at the highest level of numerical classification, text blocks at all other levels being regarded as *sections* or *subsections*. In principle one might reserve the designation "subsection" for segments with "three-part numbers", calling the ones below these *sub-subsections*, but this degree of linguistic "precision" seems pointless.

It is unwise to carry the decimal numbering system to extremes. Some readers regard even three- or four-element numbers as "too mechanical".

● Section numbering that reflects more than four hierarchical levels becomes cumbersome, and it is also unsightly.

Our recommendations in this regard are essentially identical to those in the norm DIN 1421 (1983) *Gliederung und Benummerung in Texten – Abschnitte, Absätze, Aufzählungen* (cf. also BS-4811:1972 and ISO 2145:1978, *Numbering of Divisions and Subdivisions in Written Documents*). This source does not explicitly refer to a "decimal classification system", but no other scheme is even mentioned, so there was apparently felt to be no need to give the system a formal name. DIN 1421 is actually more restrictive than what we are suggesting with respect to levels of complexity, as is BS-4811:1972. Both indicate that

● Subdivision should cease with the *third* level in order to ensure that section numbers are easily scanned, readily interpreted, and conveniently articulated.

You will notice that we have chosen to adhere to this limit in our book, although our publisher, Wiley–VCH, sanctions as many as five levels of subdivision with especially complex treatises.

Certain "tricks" allow the author to subdivide text a bit further without resorting to more complicated numbers, as our book also illustrates. Thus, subdivisions at the last *numbered* level can be broken down into segments with *unnumbered* headings. The

---

[15] The table of contents created automatically with a word-processing program will probably require a certain amount of manual editing before it conforms to these layout criteria.

latter can be given a characteristic appearance by setting them in type of an unusual style and/or size. Structure at this more detailed level is ordinarily *not* reflected in a table of contents.

One can actually create several additional levels of structure in this way, although that would normally be necessary only with a book considerably more complex than the average dissertation.

Structure *above* the level of the chapter can also be achieved: by declaring groups of adjacent chapters to be *parts*.

● "Parts" in this sense are often assigned sequential *roman* numbers (Part I, Part II, …, Part IV, etc.).

Note that even if a book is divided into parts, for the sake of clarity the *chapter numbers* should continue to run sequentially throughout. For example, our own book is divided into two "parts". Part I contains four chapters, so Part II commences with Chap. 5.

There may be reason to organize a dissertation such that some components from the list in Sec. 2.2.1 are treated as "parts" rather than chapters: Further subdivision of the sort described here might then be applied to the *parts*, the latter being limited perhaps to "Results", "Discussion", and "Experimental", possibly together with an "Introduction".

These rather inclusive "parts" could of course also be regarded as the document's *chapters*, and therefore be assigned *arabic* numerals, in which case the lower-level subdivisions would need to be interpreted and labeled accordingly. In our earlier example, *Chapter* 3 ("Reactions in Nonpolar Media") might have been the *first* chapter in a "*Part* II: Experimental". Alternatively, the same text might have been called *Section* 3 in a "Chapter 2 Experimental" which, with four-level numbering, could result in the following table-of-contents excerpt:

A dissertation devoted to underwater sonic studies, in which the results have been collected in a chapter near the front, might include as opening lines above the text of that chapter the collection of headings

2 Results
2.1 Measurements of Sound Velocity
2.1.2 Deep-Water Measurements

As noted earlier, there is sometimes good reason for not adhering to the "classical" organizational scheme "Results–Discussion–Experimental". This is actually rather common with dissertations in physics, a discipline less subject than chemistry to investigations based on an arsenal of "standard experimental methods".[16] An outline perhaps better adapted to a hypothetical physics project is illustrated in Fig. 2–1, close examination of which might serve as a stimulus for you to direct extra creative attention to the unique structure characterizing your own dissertation work.

I	**Introduction**
II	**Conceptual Approach and Measurement Techniques**
1	Merged Beams
2	Experimental Setup
3	Ions in High-Frequency Fields
4	The Laser System
5	Procedure for Establishing Flight-Time Distributions
III	**Jet Streams**
6	Effusive Streams and Jet Streams
7	Pulsed Jet Streams
8	Experimental Studies Associated with the Piezoelectric Valve
9	Cooling Experiments
10	Relationship Between Effective Jet Diameter and Velocity
IV	**Achieving an Ionic State**
11	Overview of Multiphoton Ionization
12	State-Selective Preparative Techniques
13	Resonant Multiphoton Ionization of the Hydrogen Molecule
14	Fragmentation and Autoionization Studies
15	Discussion
V	**Rotational Dependence of Ion–Molecule Reactions**
16	The Hydrogen Reaction System
17	Data Collection
18	Discussion
VI	**Summary**
VII	**Appendices**
19	Jet-Stream Theory
20	Valve Specifications and Operating Parameters
VIII	**References**

Fig. 2–1. Example of a "customized" dissertation outline.

---

[16] Standardization plays such a dominant role in chemistry that certain research groups (teams specializing in organic synthesis, for example) organize all their experimental efforts around *standardized data forms*. Every team member is thus expected to record in a uniform way, and in prescribed places, specific pieces of information collected in the course of preparing and examining each (new) substance. The resulting "data sheets" become a convenient source of brief characterizations of the sundry compounds produced, serving simultaneously as checklists for ensuring that all mandatory routine studies have in fact been carried out. At the risk of repetition (cf. Sec. 1.3) we feel it worth stressing again, however, that summary data sheets of this sort can *never* be allowed to replace conscientiously prepared notebook entries!

It will be apparent from the foregoing examples that a table of contents alone can provide the interested reader with considerable insight into how a dissertation is structured and what types of information it contains. To this end, the various entries should be positioned on the page in such a way that one can recognize at a glance from the nature of the numerical identifiers which headings apply to the most important subdivisions of the book in contrast to others that are more secondary.

As implied in a sample presented without comment earlier, however, within the document itself wording that confers substance on a heading is always set directly after the numerical identifier, separated from it by no more than two blank spaces:

> 3 Reactions in Nonpolar Media
> 3.1 Reactions in the Condensed Phase
> 3.1.1 Photolysis Reactions with Cyclohexane as Solvent
> (start of the text for Chapter 3)

A "chapter" typically opens with a series (or "cluster") of headings like this, and nothing is gained by adopting the more structured alignment that clearly has advantages in a table of contents.

Adhering rigorously to the orderliness at the heart of the numerical (decimal) subdivision system requires attention to a few additional details. For example, consider the following:

● Creating a section at a new level implies that at least one *additional* section at that same level will follow.

Were this *not* the case, the newly introduced subdivision would serve no useful purpose. In other words, a numerical sequence of the type 1, 1.1, 2, … or 4.2, 4.2.1, 4.3, … is never justified. This is a technicality frequently ignored by authors as they prepare their manuscripts, and even an editor will sometimes carelessly permit a transgression of this sort to slip into a published work.

Another affront to proper form—more common still—is creating a piece of text that does not relate logically to *any* subheading. Consider, for example, the "cluster" of headings depicted above. Suppose now that there were a paragraph of introductory text immediately after the heading "3 Reactions in Nonpolar Media"—before the next heading, which bears the number 3.1. This is the type of trespass to which we are referring, because there is no logical, systematic way of expressing the location of the "orphaned" paragraph. In other words,

● Text must never be left to float in a classificational "no-man's land".

One exception is sometimes permitted in textbooks: chapters are occasionally allowed to commence with brief paragraphs of "introductory remarks". Even so, it is best to set opening material of this sort in distinctive type, or assign to it some other unique formatting characteristic so that it is clearly differentiated visually from ordinary body text.

What should one do if an opening, lonesome paragraph of isolated text seems unavoidable? One simple solution is almost always available: assign it a heading of its own, at the same level as whatever heading comes next. In other words, confer upon the waif a nondescript title like "X.1 General Remarks" (where "X" is the number of the corresponding higher-level text element). What follows would then be "X.2". Other similarly vague wordings serve equally well, like "Background" or "Preliminary Observations".

● DIN 1421 (1983) suggests utilizing the number "0" as a way of dealing with this problem, as in "X.0 General Remarks".

The zero offers an especially convenient "work-around" with an isolated paragraph discovered after the fact, since it eliminates the need to change subsequent numbers that may already have been assigned.

DIN 1421 (1983) is explicit in pointing out one additional rule that we have so far only mentioned in passing:

● Periods ("decimal points") are used only to *separate* adjacent integers in a heading with a decimal-system identifier; *no* period should appear at the *end*—not even after a single-element identifier (e.g., "3 Results", *not* "3. Results").

Violations of this rule are again common, but we urge you not to become a culprit.

We conclude this rather extensive discourse on the mechanical aspects of subdivision and headings with a rule of thumb regarding the optimal *length* of subsections. In our judgement,

● Eight pages between headings should represent an absolute upper limit for "typed" (or computer-generated) text, correspondingly less in the case of a printed work.

If at the end of eight pages you still find no convenient opportunity for introducing a break, it is quite likely your argument is poorly structured and should therefore be reconsidered. Keep in mind that headings have the valuable secondary function of enlivening a piece of text—and for this reason alone readers appreciate their presence. On the other hand, too many section breaks make a work seem choppy, causing too much attention to be directed toward structure at the expense of content (which is clearly more important).

2.2.6 Introduction

A reader should experience the *Introduction* (sometimes referred to as the *Background* or *Problem Definition* section) of a dissertation as a kind of panorama, depicting the landscape that confronted the author when the reported work began. How and when did this particular research topic first suggest itself? What made it clear that there were interesting questions awaiting answers? What facts and relationships were already apparent then, and what key information was perceived to be missing? If as a writer you succeed in arousing the reader's curiosity about what the previously known terrain was concealing, then your Introduction will have gone a long way toward accomplishing its primary purpose.

- The introduction to a typical piece of experimental work begins by posing one or more specific questions associated with some once obscure characteristic or unexplained phenomenon in the natural world.

Usually the Introduction also describes a proposed *methodological approach* to solving the problem of interest, perhaps illustrated by specific examples of that approach applied in other circumstances. If there is a great deal of material you wish to treat here it might be worth dividing the Introduction into subsections.

- Be conscious as you are writing that you must distinguish clearly between *your* work and past contributions of others (cf. also Sec. 2.2.8).

No other section of a dissertation depends so much upon *literature research* as the Introduction.

Before you even began working on your project you undoubtedly read several key background papers recommended to you by your advisor, but you have probably not looked at them again recently. Now you absolutely *must* not only reread these, but also delve deeply into the references they provide as a way of becoming intimately familiar with all the crucial sources of scene-setting information. This will of course also put you in a much stronger position to evaluate past work critically. As you proceed, your breadth of vision and overall mastery of the field will gradually expand in ever-widening circles.

A broad background search must also be conducted for relevant material so far unfamiliar to you. This is often facilitated by concentrating on certain *keywords*. Occurrences of these definitive words are sought in the literature by examining various comprehensive *databases* as a starting point. Each must then be checked for its pertinence. Today such a quest is almost always pursued in "online" mode, taking advantage of data services like STN International, Datastar, MEDLINE, and Dialog, which in turn serve as gateways to a wide variety of specialized databases (see Sec. 9.1). Access to the services is by subscription, which usually means working through an intermediary such as a library. It is quite likely that, in the interest of efficiency, at least some parts of the search will be conducted—under your guidance—by an experienced member of the library staff, since time spent connected to a database can be quite expensive. Computer-based access to the literature is expanding at a rapid rate, with ever more resources becoming available in full-text electronic versions.

Many libraries maintain their own *internal* ("in-house") databases, accessed via CD-ROM or a local server. Be sure to check with your librarian at frequent intervals so you remain abreast of the current situation at your particular institution.

● It is your obligation to personally read, understand, and internalize every single literature source with a significant bearing on your project, and later formally acknowledge (cite) and reference in your dissertation each of the important ones.

This is not a requirement that can be met with abstracts. You need to familiarize yourself with original sources, since only in that way can you acquire a comprehensive appreciation for how the "facts" of interest were established.

In all likelihood you will find yourself becoming discouraged at some point as you begin to realize that the reservoir of relevant publications is infinitely deep—and the more generously you draw the boundaries the greater will be your harvest. The process obviously must be cut off somewhere, but it will do no harm for you to read more papers than absolutely necessary—and certainly more than you are likely to cite.[17] The more intimately a (re)searcher becomes acquainted with his or her surroundings, the more confidently can the landscape be presented to others.

Nevertheless, don't fall into the trap of trying to master literally *all* the peripherally relevant material. This can only lead to frustration, and probably to questioning unnecessarily your competence—quite apart from having a negative impact on your enjoyment of the work. A reasonable upper limit for background reading might be about 50 or so original papers and a few reviews, within a time frame of one or two months. Additional key literature may turn up as you begin to write, but at some point you simply must say "Enough!", because getting on with preparing your *own* document is after all what now counts the most.

● If, as you pursued the experimental phase of your work you took the time to study, carefully evaluate, and take notes on every important publication you encountered, you will have saved yourself a considerable amount of effort at the writing stage.

Systematic review of the literature during the early weeks of a research project—and at frequent intervals thereafter—also minimizes the likelihood of your later being confronted with some unpleasant surprise ("If I had only known, I could have accomplished such-and-such far more effectively, or more speedily, or with so much less effort …!"). The worst possible surprise, of course, is the catastrophic discovery that your latest hard-won results have already been reported by someone else, and are thus no longer "novel".

Returning to the Introduction, the extent to which you see this as a place to drop hints about important outcomes of your work, or reveal some of the insights you have gained, is really up to you. Some graduate advisors caution against exposing too much too early, using as an analogy the premature unmasking of an archvillain in the first chapter of a mystery novel. Still, good tales have been written in which one *does* learn early on who committed the evil deed, suspense thereafter emanating from an intriguing account of how the culprit was identified and apprehended!

---

[17] Interestingly, Albert EINSTEIN in his dissertation entitled *A New Determination of Molecular Dimensions* (University of Zürich, 1905)—which was only 21 pages long, by the way!—saw no need to cite a single piece of work by anyone else.

2.2.7 *Results*

The author's own scientific contributions are usually first dealt with in a section entitled *Results*, assuming the more detailed *Experimental* section is saved for the end. (The chief rationale for delaying the presentation of experimental details is that this material is rarely of interest to the general reader, who is anxious to find out as soon as possible—in *broad* terms—what it is you have accomplished.) On the other hand, institutional guidelines sometimes insist that—just as with certain research reports (see "Proposed Subdivisions" in Sec. 1.4.2)— description of all the experiments should *precede* serious discussion of results, or of conclusions that might be drawn.[18] Once again, be sure to check university or departmental rules closely, and also ask about standard practice in your particular research group.

● Whereas an Introduction acts rather like a *conceptual* primer for the reader, what needs to be provided under "Results" is a clear account of the author's discoveries, albeit without for the moment going into experimental details, or the wider significance of the findings.

A scientific context has already been established, so there is no need once again to justify having undertaken the various experiments, either individually or collectively.

● Limit experimental details in this section to what would be absolutely necessary for an outsider to understand the results.

For example, you might want to explain the precise nature of some piece of equipment, or the unusual role played by a particular chemical, but do so only if the reader would otherwise have difficulty imagining how one of your results could have been achieved. A meticulous survey of all the various devices you happened to use should be reserved for the "Experimental" section. Decisions regarding what bits of information belong where can be among the thorniest problems associated with preparing a dissertation.

● Results that are interrelated, or likely to be the subject of comparison, are advantageously presented in *tables* (see Chap. 8).

On the other hand, exhaustive records cataloguing a multitude of very similar operations, with all the associated data, are often reserved for an *appendix* (Sec. 2.2.11), the account in the Results section being limited to a summary, perhaps illustrated by one or two examples. This has the advantage that the reader is spared from being so overwhelmed with detail that it becomes impossible to appreciate what is actually significant in the dissertation.

● Functional relationships lend themselves to effective and succinct presentation in the form of *diagrams* or *graphs*.

---

[18] Remarkably, British standard BS-4812:1972 fails to discuss how the heart of a "thesis" should be organized.

Graphs can take many forms, but in general they are line drawings designed to facilitate the depiction and recognition of numerical relationships and quantitative trends. If possible, choose a format for your graph that permits you to indicate probable error in the associated data. Chapter 7 provides more specific information about diagrams and graphs of various types, as well as how they are prepared.

- Announcement of your results in this section should be accompanied by very little *interpretation*.

This is not the place for you to comment on the meaning or significance of your findings; reserve that for the subsequent "Discussion" section. We concede, however, that this injunction is not always easy to observe in a rigorous way.

- One factor to consider in deciding whether something does or does not belong in the Results section is what the potential reader would think, keeping in mind likely ranges of interests and background.

By "reader" we mean not only members of the committee charged with evaluating your dissertation, but also others who might someday have reason to seek out this particular "publication". You can reasonably assume that any reader would want first of all to know what is actually being offered: the scope of your investigation, the amount of care and imagination that went into it, and its eventual outcomes. Only then might he or she want to learn—from the Discussion section following your account of the results—how you, the author, interpreted your results in the context of what was known at the time you wrote your dissertation. For this reason we feel that the smooth flow of information is actually *impeded* if a dissertation is so structured that Results and Discussion are combined into a single section. This combination not only makes the document less convenient to read, but it also complicates the writing, because there then needs to be even more attention devoted to careful and repeated distinction between fact and interpretation, and between new information and what was previously known.

- A good Results section requires very few literature citations—perhaps none at all.

The virtual absence of citations follows from the stated purpose: revelation of that which is *new*. One possible exception might be a reference related to methodology; e.g.:

> "Miller's procedure [100] produced somewhat more scatter than the alternative approach suggested by Maier [101]".

What *should* be present in the Results section, however—and not relegated to obscure, scattered treatment in a "Methods" or "Experimental" section—is explicit discussion of the reliability of reported data, and ranges over which suggested correlations are thought to be valid. You should also acknowledge here procedural shortcomings that may have impacted negatively upon your work or its evaluation. Camouflaging a problem, or acknowledging it only in the detailed account of an experiment (where it is least likely to be noticed!), or burying in an appendix a weak link you recently discovered in one of your chains of logic—all these trespasses are too suggestive of someone trying to be deceptive. This represents another situation, however, in which the "right" course of action may not always be self-evident.

● Most Results sections profit from being broken judiciously into *subsections*, again to prevent the reader from feeling overwhelmed.

Logical subtopics in the case of a project in synthetic chemistry, for example, might be (a) the preparation of starting materials, (b) synthesis of key intermediates, and (c) characterization and structural analysis of products—perhaps together with a "catch-all" section for miscellaneous (peripheral) results. Whether *methodological* or *structural* considerations should play the decisive role in organization here is something that will depend on the nature of your investigation.

Results should be introduced in straightforward, lucid sentences, preferably (in the interest of clarity) rather short ones. Just as later in the Experimental section, standard practice is to write in the past tense, and with passive voice:

> " It was thus demonstrated …"
> "… somewhat exceeded …"
> "No obvious relationship could be established …"
> "… therefore had no influence on …"

We believe in the value of a long tradition (which some deplore) arguing that it is inappropriate for the author of a scientific document to refer to himself or herself directly, in the first person ("I added …", "I discovered …", "After I realized …"). It is not that we fail to appreciate the importance of the creative individual—in science no less than in other settings—but the "Results" section of a scientific report is supposed to be a repository exclusively of *facts*, accompanied by at most a few noncontroversial interpretations the author can safely assume will be universally shared. There is no place for the subjectivity implicit in personal intrusion on the part of the one who conducted the research—especially since the section is explicitly labeled "Results". Furthermore, the reader has no reason to be interested in the personal dismay conveyed by an observation that "unfortunately" or "regrettably" some experiment or other, devised by the author, proved to be a failure, or at least was not a source of useful information. Things just worked out that way! If first-person pronouns are appropriate anywhere in a dissertation, it would be in the *Discussion* section (the one that comes next), because different people might indeed draw different *inferences* from a given set of facts—even in the natural sciences!

Generally speaking, the Results section should not be all that difficult for you to put together, especially if as the underlying work progressed you regularly took the time to prepare thorough *interim reports*. In a favorable case, most of this section could come directly from these earlier summaries.

2.2.8 Discussion

In many respects, the *Discussion* section represents the capstone of a dissertation. Here (and again in the "Conclusions" section, if there is one) the reader can expect to find answers to the question or questions raised in the Introduction, and to obtain reassurance that these answers have been effectively defended and shown to be consistent with the author's stated "thesis". This is also where the reader will find out to what extent you as author have learned to think contextually.

● The goals of the Discussion are to present a careful *analysis* of your results, to put your findings properly in the context of what was previously known, and to offer an honest appraisal of your work's significance.

Obviously every important conclusion presented in the Results section must be dealt with again here. A certain amount of repetition is unavoidable, even though the more *interpretive* presentation of the facts in the Discussion should cast new light on things relative to the *descriptive* treatment in the Results section.

Working from a fresh vantage point you must now try to merge what was previously known with that which you yourself have recently discovered. The newly enriched landscape you have necessarily learned to know so intimately should now be illuminated for your reader. Think in terms of a mosaic, where your own contribution represents a group of special tiles whose distinctive colors have had a unique impact on the larger, ever-evolving pattern.

● Other people's results, to whatever extent you need to introduce them here, must come with unambiguous references to *retrievable literature sources*, delineated so clearly that anything undocumented in the Discussion may safely be interpreted as based on new work you are making public for the first time.

Be especially vigilant with respect to this latter point, because an omitted literature citation would be legitimate grounds for a charge of *plagiarism*, an accusation that you are making an unjust claim to originality. If you cannot point to a specific literature reference that documents some key bit of background you may need to resort to a phrase like "in contrast to conventional wisdom, the present results show that …".

● Whatever has (or perhaps—unexpectedly—has *not*) been achieved through your work should be spelled out very clearly, and also related explicitly to goals established at the outset of the project.

In formulating the Discussion, make a special effort to distance yourself from the true *chronological evolution* of your experimental findings. This may seem odd advice, but it relates to the fact that one can never "program" the path of scientific progress. Every experimental investigation is a trip into the unknown, and uninteresting, irrelevant—usually irritating—detours and dead-end streets are an inevitable part of the journey. A dissertation is not intended to be a detailed and systematic report of a *process*, which was probably inefficient and on occasion fraught with frustration. It is instead an announcement of new discoveries and insights, or, as suggested at the beginning, the proclamation and defense of a novel *thesis*. The precise *way* the results were achieved may play a role in how a research advisor judges the candidate's work, but beyond that it is of interest only if aspects of the convoluted process were somehow crucial to achieving the reported results.

As noted earlier, this section, the Discussion, is the place in which it is most likely that as author you will choose to discard your anonymity and openly express *opinions*. Statements of personal opinion can contribute significantly to the liveliness of a presentation. Nevertheless, we recommend that even here you exercise a certain amount of reserve. Within the scientific context, a "discussion" is viewed not so much as a forum for the expression or evolution of individual beliefs, but rather as a colloquy for promoting the development or solidification of a consensus or *collective* judgement. For this reason a guarded, impersonal phrase like "… almost certainly means that" still seems to us preferable to "I therefore believe that …".

As you near the end of a document so much colored by the notion of change—what most would also regard as *progress*—it is natural for you to want to speculate a bit about "how things will go from here". The desire is a healthy one, and the Discussion is the logical place for it to be fulfilled.

● So long as it is restrained and not pursued to the point that it overwhelms or obscures your actual accomplishments, speculation—prediction of future revelations, perhaps also with your perception of important steps waiting to be taken—will be welcomed in the Discussion.

At the same time, as a doctoral candidate you must not forget the expectation that, by definition, you have already succeeded in fully establishing the validity of a particular thesis. Visionary projections ("Further studies are likely to show that …") run the risk of eliciting from the doctoral committee a response like "So why haven't you gone ahead and proven it?" The best way to avoid such a trap is to emphasize ways in which results *already* achieved meet the goals you articulated in your Introduction.

2.2.9 Conclusions

Not every dissertation contains as a separate entity a *Conclusions* section. Nevertheless, we see real merit in providing yourself an opportunity to underscore the most important outcomes of your work in a special section dedicated to that purpose, situated immediately following the Discussion. In many respects a Conclusions section can be viewed as a concise summary of the Discussion. Here the casual reader can hope rapidly to acquire a general sense of the spirit, content, and import of the dissertation as a whole.

If near the end of your dissertation you do introduce a Conclusions section, then the Abstract (Sec. 2.2.4) should by all means be placed at the front, with its focus on goals, methods, and results. Concise comments regarding significance are then largely confined to the Conclusion.

● A "Conclusions" section serves to put the finishing touch on the main body of a dissertation, drawing special attention to its evaluative character.

While this section is intended mainly to provide an overview, it may also offer hints of what you think the future might bring.

2.2.10 Experimental Section

In contrast to the intellectually and creatively demanding effort required in the composition of all that precedes it, you should find assembly of the *Experimental* section to be largely a matter of routine. Still, the responsibility must not be taken lightly, especially since this is the part of the dissertation likely to prove most enduring. In fact, the main reason why a scientist in the future might show an interest in this, your *magnum opus*, is that it represents a source of experimental details not readily available elsewhere. Take special pains therefore to be sure that *every* novel activity that played a role in your achieving the reported outcome (observations, measurements, data workup, equipment design, preliminary studies, etc.) is well documented. One permissible shortcut is treating experiments of a serial nature in terms of a single prototype so long as you also report all deviations from this standard procedure. Try not to omit anything that another scientist would need to know in order to repeat your work. We might claim as a special reason for your devoting extra attention to the quality of the Experimental section an ergonomic factor: one would expect you to demonstrate exceptional proficiency in the section most reflective of your investigative expertise.

Incidentally, if you take the route of consigning experimental details to a section with the more descriptive title "Methods and Materials", then be sure *both* keywords are adequately addressed, perhaps by assigning them to separate subsections. A "Materials" section in a chemistry dissertation would presumably focus on reagents and other consumables, and it would include detailed information on sources and levels of purity (including manufacturer's specifications), as already recorded in the introductory parts of experimental descriptions in your laboratory notebook (see Sec. 1.3.2). If biological materials are involved, identify all species and strains as precisely as you can, adding in the case of animals, perhaps, information about age and sex, weight, nutritional state, relevant environmental or conditioning factors, and so forth. In addition, with animal experiments of all types it is important to document ways in which ethical issues were taken into account.

● Every experiment should be described in sufficient detail that another trained scientist could carry out the work exactly as you did.

The priority assigned to "reproducibility", and some of its wider implications, is discussed in Chap. 1. Any special factor or consideration that arguably played a significant role in the success of one of your experiments clearly deserves to be noted, even if (or perhaps especially if!) that information would be unlikely to find its way into a more widely distributed publication (e.g., a professional journal).

Apart from interim reports, your most important source of information for the Experimental section is of course your set of laboratory notebooks. The ease with which you are able to mold the essential facts into a satisfying shape, consistent with the quality of your dissertation as a whole, will be a direct function of the organization and thoroughness that characterizes your overall effort as an experimentalist.

2.2.11 References Section and Miscellaneous Components

The *References* section is where you must identify in an unambiguous way all the sources you consulted as you carried out your work and prepared the formal account of it. References are expected to conform precisely to a rigid set of specifications, though several acceptable models exist (see Chap. 9). As we have already stressed several times, all ideas and information attributable to others and utilized in any way in your dissertation must be explicitly acknowledged. This is accomplished by declaring exactly where in "the literature" the corresponding information appears, and the References section is the place for such declarations.

Details regarding the various ways different types of literature can be cited are examined in depth later (in Chap. 9). For the moment what needs to be stressed is the broad point that

● A literature citation has the important task of providing the reader with all the essential facts about who deserves credit for having entered a given piece of information into the public record, when that occurred, and precisely where the account is located, making it relatively easy to consult the original publication itself.

The word "citation" is used by scientists in two distinct ways in referring to the literature. On one hand the word may refer to a *passage* taken directly from another document, but it can also be used to point to *the other document itself*.[19] It is in the latter sense that we will here be speaking of "citations" or "literature citations".

Unpublished results are seldom subject to formal acknowledgement—indeed, it would be pointless, since you have no systematic way of directing your reader to an original source. Casual mention of a lecture you happened to hear somewhere, or the assertion that you were privy to a "personal communication" about something, is unlikely to be taken very seriously by your colleagues. This provides a good opportunity

---

[19] One's intent would be clearer if the word "quotation" were employed in the former case and "reference" in the latter.

for a reminder of something we said in Sec. 1.1.2: "Only results that have actually been *published* are accorded an enduring place in science".

The important point we are trying to make is that information which *has* been published in an orderly way, whether in a journal or in a book, *can* (in principle) be retrieved: by locating the public document (publication) in which it appears and examining its contents. It is your responsibility as author of a dissertation to take this access potential very seriously and provide your readers—in the form of *bibliographic data*—with all the necessary means for proceeding directly to each definitive source of information underlying your work. Unusual sources you may have encountered could of course present special problems, and no all-purpose guidelines can provide all the answers. It is also not yet clear, by the way, what value you should attach to sources like "info@camsci.com" or "http://www.feuchter.organik.uni-erlangen.de", especially since their long-term accessibility remains questionable.[20]

One sometimes wishes to call the reader's attention to sources of information other than those cited explicitly in the text. The author so inclined should establish two *separate* reference categories—e.g., "cited literature" and "bibliography" (sometimes called "additional sources")—and provide lists for each.

● Any of several other components might be incorporated into a dissertation after the References section.

For example, material supplementary to that in the body of the text can be accommodated in a section called an *appendix*—perhaps even in several *appendices*. Be certain, however, that you don't consign to an appendix truly *essential* information inappropriately excluded from the main body of the dissertation. Appendices are seldom consulted, and the average reader will expect them to contain only details tangential to work adequately reported elsewhere.

On the other hand, an appendix would be a good place for documenting a long series of essentially repetitive experiments, for example, or for sequestering spectra, flow charts, mathematical derivations, computer programs, equipment specifications, and similar *backup material*. Above all an appendix should be considered as a potential site for parenthetical information that would unnecessarily disrupt the reader's train of thought were it to be in the main body of the text.

An appendix is also a useful venue for the pursuit—in a suitably restrained way—of interesting findings not really germane to the principal thrust of one's thesis: material omitted from the body of the document because it is not a good fit conceptually.

In the event several appendices are required, standard practice is to label them "A", "B", "C", … (as in this book), perhaps subdivided into "A.1", "A.2", …

---

[20] This issue is currently a subject of intense study by librarians and other information specialists. Various proposals have been developed and circulated in an arduous search for the best means of ensuring long-term availability of Web-based sources. We return to this point briefly at the end of Sec. 9.5.2.

- Another section to consider including near the end of a dissertation is a set of "Remarks".

This would be one place to house supplementary observations, explanations, tips, and the like that might conceivably be of interest to the specialist, but clearly fall outside the scope of the dissertation itself. "Remarks" of this sort are actually rather rare in scientific documents, however, being more often associated with the humanities. The few parenthetical comments a scientist might wish to interject are more likely to be found in *footnotes*.

- One further element commonly found near the end of a dissertation is a brief biography of the author, typically in tabular form.

The appropriate information here is essentially what you would expect to provide (or be *asked* to furnish) in the course of applying for a job. Important stages in your education would of course be listed, as would professional positions you have held, significant honors bestowed upon you, original publications you have authored, professional organizations to which you belong, and perhaps experience in the military or a service organization. Circumstances that may have led to your transferring from one university (or research group or scientific discipline) to another would also be of interest, as might an account of lengthy experiences abroad, scholarships or fellowships, and sundry professional relevant activities.

- Occasionally it is expected that a dissertation will conclude with a *formal declaration* of some sort.

A "declaration" in this sense could be a statement affirming that the work presented is in fact your own, that all special assistance you received has been duly acknowledged, and that all published sources of information you may have utilized are properly identified. You might also be asked to attest to the fact that the work described has not previously been submitted in the context of a dissertation to any other academic institution.

Finally, a brief word regarding an element we mentioned in passing at the end of Sec. 2.2.4: a doctoral dissertation often includes a *dedication*, which might appear near the front, but could also come at the very end. Like every other "section", this would be assigned its own right-hand page. You should limit any dedication to a very few words, such as "Dedicated with gratitude to my parents". It would *not* be appropriate, incidentally, to dedicate the dissertation to your graduate advisor; your doing so might even be scorned as suggesting condescension. (The document is, after all, still subject to the withering scrutiny of every member of your doctoral committee!)

2.3 Preparing the Dissertation

2.3.1 From Outline to Final Draft

Developing a Concept

We have already addressed the nature and purpose of an *outline* in the context of reports (Sec. 1.4.2). It will have become apparent earlier in this chapter (in Sec. 2.2.5) how complex the outline for a more extensive piece of written work can become. This seems a good place for a few more suggestions on how to approach the organizational problems inherent in building a framework you can develop smoothly into a profitable outline.

● The biggest challenge is coming up with the right "pattern". Devising the pattern requires you to be creative, and creativity is generally facilitated by *free association*.

Every writer brings to the project at hand a wealth of valuable ideas, though initially they will probably be somewhat vague. The first priority is to work your ideas into a compatible set of mutual *relationships*: to "associate" them. Start by trying to acquire a sense of detachment, coupled with the freedom of thought which alone makes it possible to articulate your various ideas independently of one another. The term "brainstorming" was coined primarily to describe free exchange of thoughts within a *group* of people, but in this case the goal is to see what happens when you try brainstorming with yourself! The next challenge will be to open yourself to the widest imaginable set of possibilities for *associating* your ideas, though you should avoid concentrating too early on any specific relationships.

● One frequently productive approach entails the use of *idea cards*.

To apply this technique you need to have before you a stack of blank file cards. A separate card will be used for expressing each "idea" that comes into your mind (hence "idea cards"): a piece of background information, perhaps, or some experimental finding, or one of the several conclusions to which your research has led. Only after you have what you perceive to be a truly comprehensive set of cards—when you have literally run out of ideas and have filled out a card representing every relevant thought you have had—only then should you begin to *organize* your cards: that is, to relate the various ideas one to another. Gradually, the stack will begin to reflect a tentative sort of structure. Groups of related ideas begin to coalesce, and individual cards within each group little by little settle into what you recognize to be a plausible sequence. Finally, you will need to link the various groups together into a rational whole.

As the process unfolds you will probably find some cards would fit better if moved from one group to another, and while at first various arrangements might appear to work, some will turn out to work more effectively (or efficiently!) than others as the structures become more complex. New cards will occasionally need to be created as new ideas occur to you, and you will undoubtedly conclude that some of the original cards are superfluous or redundant. Finally, after one last meticulous examination of the entire stack you will no longer be able to find serious shortcomings or awkward transitions, and the organizational challenge will have been met: the stack of cards in its current state is a model for precisely the outline you need![21]

● A related technique involves writing down *keywords* in random order, just as they occur to you, on a blank sheet of paper.

Groups of ideas or concepts that are somehow related typically come to mind almost simultaneously. Equate each idea that arises with a distinctive keyword, jot the word down somewhere on an initially blank sheet of paper, and in all probability the resulting collection of words will be seen to form a loose set of arrays, more or less mirroring the way ideas materialized out of the subconscious reaches of your brain. Some of your keywords will represent concepts relatively high on an organizational pyramid, under which others will naturally group themselves. Words that intuitively seem most closely related should be moved as necessary so they will be near each other. Eventually, more explicit connections can be established by enclosing small groups of obviously associated words within rough circular boundaries. What will ultimately emerge is a surface littered with *keyword clusters*. In a subsequent step this more or less unconsciously created assembly is examined more closely and elaborated in a methodical way, working from what might be called a *critical–analytical* perspective. The developing structure is improved little by little, and *tie-lines* are gradually added to connect the various clusters, until the whole thing is ready to be recast as a crude outline ripe for refinement.

● Yet another approach entails picturing your project as a kind of *tree*, one which you will attempt to sketch in terms of a growth pattern.

Trees seem to know intuitively what twigs are supposed to develop on which branches: buds mysteriously appear, and the buds quickly develop into vigorous sprouts that in turn spawn buds of their own. Growth tends to be most spirited in places where light is able to reach the branches in abundance. As a "budding" author, try your luck at working your way through a developmental pattern for your dissertation inspired by the example of a living tree!

We have chosen to address the fundamental problem of organization from a variety of perspectives because the novice author almost always experiences a more or less serious panic attack upon first being confronted with the stack of empty sheets of paper that must somehow be transformed into a polished manuscript. The challenge is equivalent to overcoming an "activation barrier", and it must not be allowed to develop into paralysis. It is obvious that you are capable of achieving a promising start, but to reach this point you will need to expend some suitably targeted effort, preferably along

---

[21] Instead of file cards you could of course have used slips of paper tacked to a bulletin board, or flipcharts.

a path in harmony with your own peculiar instincts and talents. After that, the entire project should proceed quite naturally to a satisfying conclusion. The trick is to try several initial approaches until you find the one that suits you best.

There also exist computer programs designed to help in the organization of ideas and information (a process sometimes described as *mind mapping*). *Outlining software* encourages you to type your various thoughts in what again amounts to a rather random way, analogous to creation of the "idea cards" described previously.[22] In this case every keyboard entry assumes the form of a line of text, which you attempt to input into the computer as rapidly as possible. The resulting collection of lines, visible on the screen, is then ready to be processed, taking advantage of the fact that individual lines can be

- freely moved about,
- combined in any way desired,
- related hierarchically,
- deleted,
- expanded,
- annotated (with notes that can be either constantly displayed or temporarily hidden), and
- automatically assigned numbers to reflect the current state of a developing hierarchy.

Once a promising structure begins to emerge, the computer-based line collection should be printed, with wide margins and generous spacing, to serve as a hard-copy starting point for further analysis and development.

Full-featured word-processing programs (e.g., Microsoft WORD) now typically include tools to facilitate precisely this sort of "framework construction". For example, the "outline view" in WORD (accessed via the "View" menu) permits one effortlessly to examine and also systematically modify the structural backbone of a document. Changes introduced at this "outline level" are faithfully and instantly applied to the "normal text" mode as well, which is to say that any modification of the outline simultaneously results in a restructuring of the master document. For example, if a heading—regardless of its level—is displaced within the existing structural hierarchy it will promptly be reformatted so that it conforms in every way to the revised scheme, even acquiring a numerical identifier consistent with other headings in the document (cf. Sec. 2.2.5). You can also shuffle the sequence of paragraphs *within* a subsection by taking advantage of an *expanded outline* view that presents you with not only all headings but also the first few words from every paragraph. If a footnote happens to be affected by a text displacement, it will automatically move in conjunction with the

---

[22] Software of this type (e.g., THINKTANK, MORE, etc.), enjoyed rather great popularity a decade or so ago, especially with the MACINTOSH computer due to its powerful graphics capabilities, but interest quickly died down, perhaps in part because of (somewhat more limited) outlining capabilities introduced into word processors (see below). Interestingly, some of the "obsolete" software from the past has once again become available for free downloading from the Web.

text passage to which it is related, and all notes in the document will be renumbered as appropriate to reflect the new arrangement.

Similar features are available to the author who prefers working with a more specialized document-design system, like TEX or LaTEX (cf. Sec. 6.6). These amenities can be invaluable when you find yourself preparing a very long, complex document—such as a dissertation. Advanced outlining and word-processing tools provide the author with an astonishing breadth of flexibility, helping transform writing into a remarkably relaxed, efficient, and productive activity—provided one has taken the time and trouble to master the many features there to be enjoyed.

● Outlining software is a powerful stimulus for loosening fetters that otherwise inhibit the free exercise of one's imagination, constraints that inevitably cause agony in premature attempts to articulate one's thoughts. The same tools also encourage you to capitalize most effectively on fresh insights that strike you *during* the outlining process.

We see no reason for beginning work on your dissertation anywhere *but* at the outline (and/or framework) stage. As you proceed, you will probably be tempted to try out already at this juncture specific wording that seems appropriate for at least some of the formal headings you expect eventually to devise. Don't hesitate to take advantage of such impulses. You have at your disposal all the space you could possibly want for creative experimentation, and absolutely nothing you write now is in any sense cast in stone.

One big advantage of a computer-based outline—often overlooked—is the way it can help you work conveniently anywhere you choose, whenever you choose, within an emerging document. If an idea suddenly occurs to you, you can capture it immediately and incorporate it roughly where it belongs, for refinement later.[23] Don't hesitate to skip back and forth between "outline" and "normal" views as you work. Remember: every modification you make within either mode takes immediate effect in both, since the two display modes are based on one and the same data file.

It can be very convenient to set yourself up as you write with a "split-window" screen. In this way you can have simultaneous access to *both* the full-text version of your document *and* the corresponding outline. Among other things this makes it easier to maintain a balanced perspective on the project as a whole, at the same time facilitating activity in whichever viewing mode seems to promise the greatest efficiency. In effect you will find yourself using your outline in precisely the same way a reader will later take advantage of the dissertation's table of contents: as a vehicle for gaining rapid access to a specific piece of text within the document. Remember, too, the benefits of the "expanded outline" display feature mentioned earlier, since that can make it

---

[23] In the past, writers and compositors often jotted down ideas on their cuffs. Today all sorts of computer-like alternatives are available for the purpose, including the PDAs mentioned in Sec. 1.3.3. You may even have remote access, at home or on the road, to the personal computer in which your manuscript is stored.

even easier to locate and then jump directly to a particular passage you wish to examine or alter.

Regardless of *how* you ultimately proceed to write, the most important thing is *that* you proceed, and that you allow yourself to be guided by a thoughtfully prepared outline. The fear so common at the outset of a project that threatens to consume an enormous amount of time is terribly counterproductive, and it can be overcome most effectively by starting work on some "easy" aspect of the job, one certain to produce immediate results: in this case, creating a solid, informative, carefully organized backbone. Equally important, if you compose your thoughts around a definitive outline the chances are greatly reduced that you will waste huge amounts of time on a misguided draft that will later end up in the wastebasket.

Writing Techniques

The question of how *physically* a major document should be prepared was itself once a matter of heated controversy. Many argued that the first draft should always be *handwritten*. "Typing" was said to present too many "obstacles to free creativity". Attention lavished on a mechanical keyboard—and on the heavily flawed copy it produced, flawed as a result of careless keystrokes delivered by unskilled fingers— was seen as attention *diverted* from the vital creative act. On the other hand, who could reasonably be called upon to do a reliable job of transforming a nearly illegible, hastily composed pencil manuscript—full of obscure technical terms—into a clean typescript? And through how many typing stages would the manuscript ultimately need to pass? What would all that cost in terms of both time and money?

Today, the questions seem rather quaint. Anyone with a recent college degree almost certainly takes it for granted that writing—at least demanding writing—is an activity undertaken by the author exclusively at a computer. It would in fact be folly in the 21st Century to attempt composition of a dissertation "manuscript" without a word-processing system![24] The document's author is the logical person to prepare every version, from first draft to final copy. It is of little consequence that the writer probably cannot type as rapidly or as accurately as a secretary. No lasting harm comes from "typing" something incorrectly: every error will later be swiftly and completely expunged—perhaps with welcome help from reliable, built-in proofing routines.[25]

One thing has not changed, however. Major revision and stylistic polishing should today as in the past proceed exclusively from a hard-copy version of your text. Most people tend to be less vigilant and less creative when they try instead to engage in on-

---

[24] Note that the very *word* "manuscript" has lost much of its significance: it comes, after all, from *manu*, derived from the Latin for "hand".

[25] This is of course somewhat of an oversimplification. Modern "spell-check" software is very powerful, but the best spell-checker cannot detect the "to" you inadvertently and mistakenly entered as "too" or "two". In our judgement, so-called grammar-checkers, widely acclaimed by the purveyors of word-processing software, are worse than useless, wasting the writer's time by balking at constructions that are perfectly correct, while often failing to flag truly hideous formulations.

screen correction, which can also lead to serious eye strain. Even large numbers of hand-entered changes can be later transferred to the master electronic text file, fairly quickly and with surprisingly little effort. "Writing" (composing) and producing attractive, clean copy has become *so* easy in our time that a new threat has arisen: the likelihood that the job will be dashed off *too* quickly, and with too little reflection, resulting in a document that is inappropriately long, inelegantly formulated, and poorly organized. In the past, when so much depended on skillful use of a cumbersome typewriter, there was considerably more incentive to write sparingly and thoughtfully. Today's "writing machines" demand a high level of self-discipline on the part of an author.

Thanks to technological advances, first drafts now mature smoothly, in an almost continuous way, into final copy. Indeed,

● Computer-based composition essentially eliminates visible distinctions between a rough draft and a finished manuscript.

It is always advisable to keep (electronic) copies representing multiple *stages* of manuscript development. In other words, there should be *several* evolutionary versions of the dissertation on your hard disk at any time; this simple expedient could prove well worth the modest cost of the required storage space. Early versions can be an invaluable resource, for example, if a need ever arises to reconstruct how you arrived at some statement or conclusion. Also, a change that earlier seemed quite reasonable occasionally results in text you later recognize to be cumbersome, or even misleading, and you may wish you could remember exactly how you handled that particular passage originally.

Label each stored version according to some simple, unmistakable system (e.g., "vers. 3" or "D5" for "version 3" or "Draft 5", respectively) and append the appropriate date (i.e., when that particular revision was completed). It is wise even to keep *backup copies* of the various versions—especially the most recent one—in some other location (on a diskette at the laboratory, for example) so that in the event of a real catastrophe "only" work from the past few days will be lost, not the fruit of weeks of labor.

Eventually, your dissertation will assume a shape that seems satisfactory in every respect: it reads smoothly, it says precisely what you want it to, and there are no more glaring errors.

● This "final draft" should be shared with one or two sensitive colleagues, and of course also with your advisor, for reactions and critical comment.

If you are so fortunate at this stage as to enlist the aid of cooperative and perceptive critics, the euphoria and relief you recently experienced is likely soon to give way to a more sober reality. Additional effort will almost certainly be called for, addressed to both structural and linguistic issues. Someone may even point out to you—convincingly!—that whole sections of your manuscript completely miss the mark. In other words, thanks to your self-selected critics, this "final draft" will have betrayed its true character as only a *draft* of the final draft. Once again it is time for a major overhaul. Fortunately, thanks to your word processor, the newly discovered problems can be dealt with expeditiously with relatively little retyping.

● At long last you will finally find yourself in possession of a *true* final draft.

Assuming you have at your disposal a high-quality printer, you should now be able to prepare a copy of the dissertation that meets even the most demanding typographic and aesthetic specifications. Indeed, a modern laser or inkjet printer is capable of turning out a document that looks almost as though it had been produced professionally, especially if you install a fresh toner or ink cartridge prior to preparing final copy. Technical considerations responsible for this revolutionary state of affairs—and a few matters pertaining to your role—will occupy our attention later (in Sec. 5.2). By the way: our use here of the word "revolutionary" may seem like hyperbole, but the fact is that the way professionals are expected to respond to writing assignments has changed drastically in the past few years, and recipients of documents, for their part, also have much higher expectations than did their predecessors as little as a decade ago.

One serious shortcoming of early word processors—especially inexpensive systems—was a lack of sophisticated *graphics capability*. By a *graphic element* in a document we mean anything you would not associate with a standard set of typographic characters (letters of the alphabet, numbers, punctuation marks, etc.). This broad category of course encompasses *illustrations* of every type, but also almost all the *structural formulas* so prevalent in a chemical manuscript. With only a primitive word processor—especially in the absence of supplementary graphics software—it was generally necessary to create these and other "special elements" manually (painstakingly with pen and ink, for example) and literally paste them into the document. The same was true for complex *tables*, and especially mathematical *equations*, which even now can almost never be prepared using ordinary word-processing software in such a way that they meet professional standards, at least not without an unreasonable amount of effort.

● If alternative software is used for creating special elements in your manuscript— the results ultimately being "pasted in" *electronically*—be sure to retain copies of the elements in native form as separate data files, since changes of some sort may later be required.

The fact that certain word processors (WORD, for example) are "bundled" with their own graphics applications is not an unmixed blessing. An "all-purpose" word processor tends to become unwieldy because it is trying to do *too* much, but beyond that the advertised "extended capabilities" often prove awkward in practice and of more limited value than you were led to anticipate.

Authors who need to work extensively with mathematical formulas—especially physicists and mathematicians—in some cases still find it worthwhile to prepare documents with the less "user-friendly" software systems TEX or LaTEX, designed from the outset to support professional-quality formula typesetting, and thus exceptionally powerful and robust in this respect (cf. Sec. 6.6).

● Be sure to become fully acquainted with the capabilities (and limitations!) of your word-processing system very early in manuscript preparation, especially in terms of *formulas*, *tables*, and *graphics*, and plan accordingly how your needs can best be met.

One last remark on the subject of "final drafts" and "final copy". Generally speaking you will not actually submit to the university the final, highest-quality copy of your dissertation you are able yourself to produce. Instead, that will serve as a basis for *photocopies*, and one of the latter will become the "official document". In other words, *your* final print should be kept exceptionally clean and wrinkle-free in loose-leaf form, never to be subjected to binding. At this point there is no way for you to know whether at some later date you might want to make more copies—in the context, for example, of a job application.

2.3.2 The Final Product: A First-Rate Doctoral Dissertation

We have spent a great deal of time talking about how a dissertation should be *developed*: what remains is to add a few words regarding the physical form it should take.

● Details associated with the appearance of a dissertation, as well as the quality of all materials used in its production, tend to be subject to very strict guidelines unique to the university where the document will be submitted.

We have already referred to university guidelines on innumerable occasions, but we do so here yet again simply to underscore further their importance. Ignoring the rules at any point could result in your dissertation being summarily rejected for purely formal reasons, and that could in turn seriously interfere with timely receipt of your degree. Similar considerations apply to other academic documents (e.g., theses), but rules are taken especially seriously in the case of doctoral dissertations.

At least one copy of your dissertation will eventually be housed in the *university library* where it will remain indefinitely as a semipublic document available for consultation by others.

● Issues related to *physical preservation*, *microfilming*, potential requests for *interlibrary loan*, and the possibility that a *facsimile* of the dissertation will be solicited for archival purposes elsewhere are responsible for the fact that universities develop and actively police such stringent rules.

As a doctoral candidate it is your responsibility to become familiar as early as possible with all applicable regulations and standards.

● Be sure also to ascertain the minimum number of *copies* of your dissertation you will need, and whether all are subject to the same requirements.

Most universities require that you provide several *hardbound* copies of your dissertation. One is of course for the library (perhaps *more* than one). Another may be designated for the academic department where the work was carried out. At least one copy should be presented to your graduate advisor, irrespective of whether the university has a rule to that effect. Also, you will certainly want copies for yourself, and perhaps a few extras for parents, friends, fellow students, and colleagues, though not all the latter need to be hardbound.

Speaking of binding, you are also responsible for determining in a timely fashion the best place to have your dissertation duplicated and bound, and for seeing that deadlines are met.

● Special paper may be required at least for the copy destined for the university library.

Inferior paper, especially in combination with the wrong ink, deteriorates surprisingly rapidly; it can literally be "eaten away" by the print. Incidentally: while you are inquiring into paper and ink requirements it would also be a good idea to make sure there are no problematic rules regarding permissible reproduction techniques.

We suggested earlier that university guidelines tend to be especially detailed and inflexible with respect to such things as page format, margins, line-spacings, and the placement of page numbers, headings, and footnotes. Pay careful attention to all such matters—collectively referred to in the publishing world as elements of *design* and *layout* (discussed more fully in Chap. 5). Should you happen to entrust the preparation of final copy to someone else, the obligation remains yours to ensure that all rules are followed to the letter. You would be well advised under such circumstances to insist on receiving sample pages from your "production team" at a very early date, and these should at once be compared directly with pages from other recently submitted dissertations available in the university library.

2.3.3 The "Electronic Dissertation"

Everything we have said so far has presupposed a "classical" dissertation in the form of a *printed* document destined to be shelved in a conventional library. Someone wishing to consult such a publication would normally be forced to travel to its "home", or arrange to have a local library borrow it briefly via interlibrary loan. Things began to change radically during the 1990s, however, as academic institutions started experimenting tentatively with the idea of accepting dissertations in the form of *data files*: what would soon come to be known as *electronic dissertations*. Several potential benefits of this innovative notion were recognized immediately:

– Electronic data files take up far less precious space than printed documents;
– Dissertations in electronic form are, in principle, much more easily disseminated: they might even be made accessible 24 hours a day from anywhere in the world via the Internet;

- Complex and demanding graphic material is often more readily incorporated into a data file than reproduced in hardcopy form;
- Exciting new possibilities open up, such as the inclusion in one's dissertation of audio material, video clips, animations, and above all hypertext links (cf. Sec. 3.1.3);
- The act of preparing a dissertation of this type can provide the graduate student with experiences likely to be valuable in a future career.

The first high-level, intense discussion of electronic theses and dissertations (ETDs) occurred in 1987 under the auspices of University Microfilms, Inc. (UMI), which had since 1938 acted as a repository for dissertations from all over the world (and a supplier of duplicate copies to all who were interested).[26] Among the most enthusiastically supportive of the academic institutions in these early days was the Virginia Polytechnic Institute. The seminal 1987 meeting became the basis for what is now known as the Networked Digital Library of Theses and Dissertations (NDLTD), to which well over 100 universities worldwide belong.

Apart from establishing as a premise that all ETDs should take the form of either SGML (Standard Generalized Markup Language, Sec. 3.1.3) or PDF (Portable Document Format, Sec. 5.4) files, no attempt has been made to infringe upon an individual university's right to establish guidelines for what a dissertation should contain or how it should be structured. In general, universities that accept ETDs assume they will be prepared in precisely the same way as any other dissertation (i.e., with the help of a conventional word processor). The only difference is that what the candidate ultimately submits to the institution is an electronic data file (or perhaps several files) rather than a bound document. Cooperating universities offer special support services to assist authors in preparing files of the proper type.

You should by all means inquire as early as possible about your own university's policies with respect to electronic dissertations. If it is permissible, we suggest you make every effort to submit your work in this form—which clearly represents the wave of the future.

2.3.4 Last Steps on the Road Toward Acquiring Your Degree

Several copies of the formal dissertation are normally placed ceremoniously in the hands of a designated administrative officer of either the university or the candidate's academic department. At most universities in the United States every member of the candidate's *doctoral committee* will then independently examine the dissertation, after which the committee may immediately request certain revisions. (Presumably you will

---

[26] UMI, now a division of the ProQuest Company, also offers doctoral candidates assistance with respect to copyright matters (cf. Sec 3.1.3 under the subhead "Authors and Authorship in the Twenty-First Century"). It even distributes modest royalties to authors of dissertations for which reprint requests are received. Active in the ETD movement since the beginning, UMI has expressed an intent to make available online all electronic dissertations submitted after 1996.

already have attended to all suggestions from your graduate advisor.) The next step in the process is typically an *oral examination*, at which members of the committee will exercise their right to raise any questions they wish regarding the work itself, its potential significance, and even peripheral matters you might regard as irrelevant. Most of the questions will represent honest attempts to gain additional insight into your work, but some will certainly have as their objective eliciting an impression of your general academic skills, interests, and agility when called upon to deal with the unexpected. This oral encounter presents one further opportunity for you to discover that a few more minor changes are required in your dissertation before it will be permanently archived, though requests for major changes are rare this late in the game. It is also unusual for an oral "examination" to result in a candidate's petition for a degree being rejected, but a bitter outcome of this sort is unfortunately not entirely without precedent!

● Clearly it is in your best interest to be *exceptionally* well versed in everything in your dissertation when the long-awaited day arrives for your oral examination.

Once this last hurdle has been cleared, all that remains is for you to wait patiently for your university's next official *commencement exercise*, after which you will finally be able to gaze with satisfaction into a mirror and savor the image of a scholar who has just withstood the most rigorous challenge the world of academe imposes.

3 Journal Articles

3.1 The Scholarly Journal as a Medium of Communication

3.1.1 "Publication" as a Concept

"Science originates in dialogue" (HEISENBERG 1973, p. 7)—one-on-one contact, conversation within a research group, perhaps the spirited discussion that sometimes follows a lecture. The scientific process only reaches its culmination, however, when a carefully verified set of discoveries is made openly accessible to everyone: i.e., the work is *published*.

● The most important medium for publishing, and thus communicating scientific results widely, is the *professional journal*, a type of *periodical* with the dissemination of new scientific information as its primary mission.

The designation "periodical" is applicable to any print medium issued on a regular basis (at least quarterly). Newspapers, for example, are periodicals that typically operate on a daily or weekly publication schedule and concentrate largely on current events ("news"), whereas successive issues of a *journal* are normally released less frequently (though still with predictable regularity). The effective life span of the latter is vastly greater at the same time that the scope is significantly narrower. A given journal usually tries to appeal to a rather limited circle of professionals, all typically active in the same general field, many perhaps with a professional organization in common.

The "public" toward which a journal is directed is thus not necessarily *the* public in the everyday sense, but since—at least in principle—pages of a journal are freely accessible to everyone, anything contained in a journal classes as published. The converse is not true, however: there obviously exist legitimate modes of publication *other* than periodicals, including *books*, *patent documents*, and, more recently, *electronic* venues (see the section that follows). The specific role played by scientific books is addressed in the chapter after this one.

Scientific journals of the sort we will mainly be discussing—those based on a printed (paper) format—have existed for roughly 200 years, during most of which time they have essentially defined the concept of announcing new scientific information.[1] Scientific journals in this sense constitute one subset of a broader category of publications known as "professional periodicals". To be absolutely precise one should

---

[1] Our publisher, Wiley-VCH, maintains responsibility for over 100 different journals. The oldest of these, *Liebigs Annalen der Chemie*, has appeared continuously since 1822, albeit since 1998 under a new name: *European Journal of Organic Chemistry*.

probably add a further qualification and use the expression *disciplinary professional periodicals* to characterize scientific journals and closely related scholarly publications. Statistical summaries compiled by publishers, for example, recognize also a broad category of "*other* professional periodicals". What we are calling scientific journals usually confine themselves to material relevant to practitioners and students of specific traditional academic disciplines, while publications in the "other periodicals" category tend to address topics of interest to collections of people defined somewhat differently: all those professionals allied to a specific industry, trade, or administrative function, for example, or scientists generally, irrespective of specialty. Journals are maintained primarily for the spread of *verbal* information, and their pages are generally dominated by closely spaced text blocks, accompanied by the occasional technical diagram or collection of formulas. Professional periodicals in the "other" category, by contrast, often reflect a desire to attract attention through imaginative use of typography or eye-catching illustrations. Categorization at this level is bound to be somewhat arbitrary, however, and it sheds little light on issues with which we will be concerned.

In the discussion that follows, occasional use of the term "scholarly journal" is meant to conjure up a wide range of periodicals, including those directed toward scientists in the narrow sense of the word, but also publications for physicians, lawyers, linguists—indeed, scholars of every conceivable stripe, with each group laying claim to its own space in a common edifice dedicated to increased dissemination of truth and knowledge, and all sharing a common commitment to opening their doors widely to anyone wishing to visit.

Periodicals most closely resembling "scientific journals" in the narrow sense would be the "technical journals" and "medical journals". Similarities in content, form, and objective that link these three groups of publications have led to formation of a worldwide organization known as the "International Association of Scientific, Technical & Medical Publishers" (logo: STM), and the acronym STM is now widely understood as a shorthand way of embracing these three branches of scholarship. People in the publishing trade, for example, routinely make reference to "stm publishing", confident that conversation partners will know what they mean. The publisher of this particular book has adopted "STM" as the official designation for one of its largest and most successful organizational units.

Incidentally: the "periodical" nature of journal publication warrants a bit more comment. Scholars, like everyone else, tend to be creatures of habit. Among other things, this means they are accustomed to receiving informational updates in their field on a *regular* basis. They also expect some assurance that information they themselves wish to communicate will be made available to others in timely fashion. It goes almost without saying that advertisers rely on issues of publications they support being released according to a fixed schedule.

● The scientist in possession of information ripe for dissemination—i.e., research results begging to be "published"—requires only a very general awareness of technical factors that might distinguish one publication from another; more important is the ability to adapt to the particular style and mode of presentation appropriate to the medium of choice (see also Sections 3.1.4 and 3.2.3).

The word "publication" is actually used in two quite different ways. In some contexts it means the *process* of *bringing* certain information to the public, but it can also refer to something or other that has already *been* brought before the public. A scientist thus speaks of "publication" of his or her work, but also of a *particular* "publication" (journal, issue of a journal, or copy of an article) that could be lying open at the moment on the desk.

- One would regard as *published* in the traditional sense only information that has been *set in print*, *reproduced* in substantial quantity, and then *distributed* reasonably openly.

The production and distribution functions are normally attended to by a *publishing house*, despite the fact that, in principle, anyone or any institution could undertake the printing and circulation of large quantities of a particular document. *Effectively* doing so, however, presupposes the availability of significant financial, technical, organizational, and staff resources. As it happens, recent advances in methods for producing (or *re*producing) text and graphics have pushed the potential for publication more nearly within the reach of almost everyone, as suggested by the currently fashionable expression "personal publishing". But the real challenge in scholarly publishing has much more to do with *acquisition, refinement*, and *distribution* of intellectual material than with more mundane physical operations, one of the important reasons why we regard it as quite unlikely that the publishing responsibilities vested in scientific societies and profit-making corporations will disappear in the foreseeable future.

- A second key aspect of publication relates to the fact that a document is recognized as "published" only if it is *unambiguously identifiable* and also *readily retrievable*.

This second criterion is more far-reaching than the first, and its importance surely will not diminish in an era that has witnessed the birth of *publishing on demand* (more accurately: *printing* on demand). In principle, current technology would allow an author to "publish" a piece of work simply by depositing in an archive somewhere a single "original" of a document describing the corresponding findings: *provided* the managers of the repository would (better: *could*) guarantee permanent availability of means by which *copies* of that document—perhaps *digital* copies—could be readily obtained by anyone so inclined. The information in question would then clearly class as "public", even though this mode of "publishing" does not involve preparing a "substantial number" of physical entities. In other words, the fundamental issue is "retrievability". Unfortunately, uncertainty with respect to long-term retrievability is one of the most vexing problems casting a shadow over the current impetus toward electronic communication. So far, perpetual retrievability seems much more difficult to guarantee with electronically stored information than with its counterpart in the "paper world", the realm in which we have so long been accustomed to dealing. This stumbling block is a cause of widespread concern, and it has engaged the creative efforts of some of those most actively and enthusiastically promoting the future of electronic interchange. The sheer volume of information in circulation today of course exacerbates the problem, as does the diversity of portals for potentially accessing it.

Among the most reliable tools for achieving "long-term-accessibility" with respect to conventional publications—what might be described as the "classic" solution—is the *International Standard Serial Number* (ISSN), a descriptor publishers depend upon for uniquely characterizing a particular journal or other periodical, and the *International Standard Book Number* (ISBN), which plays a similar role in the world of books. The average reader rarely has reason to be particularly conscious of either, but both are of immense importance to librarians and the distributors of publications. ISSNs will be introduced briefly here, but detailed consideration of the ISBN is deferred to Sec. 4.5.2. To these two identification systems have been added others in recent years, in particular the printed *bar codes* upon which book stores now rely heavily.

The ISSN had its origin in a joint initiative of the American National Standards Institute (ANSI) and the International Organization for Standardization (ISO). ISSNs have been utilized worldwide since 1971 as an instrument for unambiguous *identification* of periodicals and other serial publications. Each such venue is assigned a unique ISSN, in the form of an eight-digit number, by a centrally coordinated network created specifically for that purpose and managed by the ISSN International Centre in Paris. The system operates as a service assisting the world's publishers and those whom they supply. An ISSN always consists of exactly eight digits,[2] and once a given set of digits has been assigned it remains permanently and exclusively associated with that particular serial publication. A registered serial is expected to display the appropriate eight digits—arranged as two groups of four, joined by a hyphen—in a prominent location in every issue, preceded by the initials ISSN. The numbers themselves have no inherent meaning; they are simply serial-specific.

● A particular ISSN can only be assigned once; the corresponding number is permanently withdrawn from the system if the publication with which it has been associated ceases to exist, or even changes its name.

The ISSN International Centre maintains records of all the essential information required for characterizing any listed serial, as well as the corresponding publisher. These records are in turn at the disposal of a range of interested parties, including various international serials directories (e.g., *CASSI*, the *Chemical Abstracts Service Source Index*, one of the most convenient places to access publication information related to *chemical* serials for purposes such as citation or acquisition).

---

[2] The last character in an ISSN is a "check digit", which sometimes takes the form of an "X" to represent 10; cf. the more detailed discussion of check digits in conjunction with ISBNs in Sec. 4.5.2.

3.1.2 "Electronic Publication": Its Nature as Defined by Early Manifestations

The Beginnings

We have noted briefly that scientists today have open to them an important new communication option: dissemination of research results to colleagues throughout the world *electronically*, especially via the Internet. Widespread awareness of the potential inherent in this new vehicle for information transfer first developed in the early 1990s. The promise of electronic communication seemed awesome, especially since it was expected soon to become available almost everywhere: through personal computers, networked in such a way that they could truly be regarded as "windows on the world". At first, publishers and editors interested in exploiting the power of computer-based communication focused their attention primarily on generating and then circulating electronic *synopses* of traditional publications, most commonly in the form of periodic updates, distributed initially via diskette but increasingly through the internal networks that began to flourish within universities and industrial complexes. Individual scientists in many places soon found themselves in a position on a regular basis to peruse title and author lists—sometimes even abstracts—derived from a wide variety of important publications, and to do so in the comfort of their own offices, on their personal computer screens. Not surprisingly, those who sampled and developed a taste for the new services began to ask about the possibility of developing a similar gateway to complete publications, including graphic elements that might be present. Within a few short years, "full electronic access" was shown to be an eminently viable concept, and today it is virtually the norm in cutting-edge publishing, despite a few very fundamental issues that still remain to be resolved (see especially *Professional vs. Private Publishing* in Sec. 3.1.3).

In principle, communication of scientific results via the Internet can be remarkably direct, essentially free of cost, and—again, in principle—unfettered by external controls or censorship, not to mention virtually instantaneous! Some have even speculated that someday a system of "electronic bulletin boards", on which anyone would be free to post messages of any type, will meet most if not all the communication needs of the active scientific community. (Later—in Sec. 3.1.3—we consider reasons why this is in fact unlikely.) More or less complex "Web sites"[3] already play an important role in

---

[3] A *Web site* is effectively a "node" somewhere along the World Wide Web. Each such site bears a unique address, and serves as a place where specific information—packaged in the form of Web *pages*—can be sought and solicited (viewed somewhat differently: where it can be posted for the use of others) at virtually any time, and from any "networked" location. A Web site might be compared to a hotel playing host to a scientific conference, where many people meet to exchange thoughts and the latest professional news. Pursuing this analogy, the lobby of the hotel would be the equivalent of a *homepage*. Here the visitor can learn precisely what meetings are occurring at the moment and what sorts of things are being discussed or presented, as well as where. Web addresses have rapidly become nearly indispensable components of advertisements and printed material of all types. The fact that an intriguing address one has somehow acquired frequently proves to be incorrect or obsolete is simply one of the unavoidable

→

the international exchange of technical information, with valuable "pages" of material maintained by a host of scientific societies, organizations, and institutes, as well as academic departments, research groups, and even individual scientists. One beneficial consequence is that laboratories and other research facilities throughout the world have been brought into much closer contact than could previously have been imagined. A few scientific research organizations like CERN actually had decisive parts to play in the organization and early development of this new vehicle for professional communication (see Sec. 1.1.1).

But we have strayed a bit too far from our primary focus: scientific journals. Early on, some visionaries argued that *electronic journals* (now often referred to as *e-journals*, a shortcut we also will take) were destined to constitute a valid and perhaps even preferable alternative to conventional print journals, despite the fact that the latter seem so firmly entrenched. That the envisioned potential might in fact be real became increasingly apparent, especially in the late 1990s. Many were moved to begin speculating about the more poignant question of the extent to which e-journals would leave any room whatsoever for traditional "p-journals"! Could print journals count on retaining a meaningful foothold on the strength of their historic record alone, or was mortal combat between the two alternative media unavoidable, with the upstart inevitably the victor? (Jumping ahead a bit in our story, the odds at the moment—in the year 2003, as this chapter is being completed—would seem to favor persistence of some form of peaceful coexistence.)

Publishers and other purveyors of information gradually became enthusiastic participants in the fascinating evolution of electronic communication. Actually, most were quick to recognize they had little choice but to become engaged if they wished to assure their own survival. It is probably fair to say that almost no one would today seriously wish to impede the progress of the electronic communications revolution. It would in any case be virtually impossible to turn back the clock. That is of course not the same as to suggest that every optimistic vision regarding the future of electronic information transfer will become reality. Some promising, innovative efforts will certainly fail. Ideas that initially seemed very attractive have on occasion already shown themselves, when actually implemented, to be more trouble than they are worth. At the same time, a number of the traditional patterns of information transfer will almost certainly endure, even if in substantially modified form.

It is remarkable how many newspapers and magazines in countries throughout the world are currently prepared and distributed simultaneously in *two* modes—"p" and "e"—usually by a single corporate entity, but sometimes by separate companies cooperating in a close "mother–daughter" relationship. These "alternative modes" of dissemination exist peacefully side by side, with no hint of a potential for "cannibalism"

---

shortcomings—albeit an annoying one—associated with this newest of all approaches to knowledge and experience.

someday threatening their friendly coexistence. "Web-based" newspapers have in fact been found to serve a function somewhat different from that of their print counterparts.

In the course of arriving at the present equilibrium some publishing houses have managed actually to increase circulation, permitting them to take on additional staff to meet new challenges and demands. Irrespective of why this peculiar synergy works economically, the fact is that it is now possible for almost anyone to examine—on screen and in "real-time"—many or indeed most of the articles published in the print version of his or her favorite general periodical, and in some cases to acquire supplementary material in the same way. Information can often be requested either in an abbreviated form, for survey purposes, or as full text. Publishers exercise considerable care to ensure that complete articles displayed on the screen are essentially equivalent to the corresponding "print versions". Pages of interest can even be printed, or downloaded to a hard disk, in most cases without cost. A reader (or "Web surfer") thereby acquires an interesting article gratis, and a paid subscription to the print edition of the publication is not a prerequisite. The "system" (some would say *lack* of system!) so far seems to be working astonishingly well, providing incontrovertible evidence that the dire predictions of many "futurologists" have at least until now completely missed the mark. Newspapers and "newsscreens" are turning out to be complementary rather than competitive media, with a relationship not unlike that between the print media and broadcasting/telecasting.[4]

In the early part of this chapter our comments were directed largely to the scientific journal in its "classic" guise, but recent developments make this exclusivity inappropriate. Even the very *word* "journal" now has a meaning much broader than in the past: it can no longer be assumed to refer solely to information-transfer based on print. A great many scientific journals, especially those produced and/or distributed by publishing houses, now flourish in both print *and* "electronic" versions; more accurately, in both *analog* and *digital* form. Recent developments in information technology have clearly changed not only the scientists who read journals, but also the publishers who prepare them. e-Journals have become a very significant force in the "information market", and most publishers—sometimes with the help of partners of various kinds—are engaged in wide-ranging activities directed toward the introduction, circulation, and continuing evolution of electronic precursors to what may well represent very different dominant modes of scientific communication in the future. The most progressive publishers are confident they have demonstrated a level of "electronic" performance, expertise, and service that assures their competitiveness for the foreseeable future.

---

[4] This outcome is incidentally reminiscent of what has become of another famous prognosis of some years ago: that the "paperless office" was the wave of the future. In fact, however, more paper is being consumed by bureaucracy, research, and other "official" activities now than before the "electronic revolution"!

The Prototype E-Journal: Current Clinical Trials

As the developments cited above were unfolding, an entirely new type of stm journal made its debut: a journal existing *only* in electronic form. To the best of our knowledge, the first peer-reviewed, full-text journal in this category (in particular, the first one that also featured extensive graphics) was an "electronic publication" called the *Online Journal of Current Clinical Trials* (*OJCCT*). This pioneer actually came on the scene very early: in 1992, two years *before* (!) the World Wide Web began its triumphal march across the landscape, and its birth was considered at the time to be a harbinger of an important phenomenon. It is instructive to examine a few of the characteristics and stated goals that distinguished this particular innovative information service.[5] The chief objective of the new "journal" was dissemination of important information *rapidly*. A strong emphasis on timeliness would have been expected in any case for the first electronic journal, but it was especially apt here since breaking news from the hospital front can have dimensions that are literally life-saving.

With a traditional journal, printing and physical delivery are important factors contributing to the significant delay between submission of a manuscript and arrival of the corresponding publication in the hands of interested readers. These are areas in which an electronic alternative would be expected to offer major advantages.[6] It was thus hoped, for example, that physicians at hospitals and research institutions worldwide would soon be able "to read a 'paper' almost the moment it [was] accepted"—an ambitious but seemingly reasonable objective at the dawn of the *telecommunication* age. Notice, by the way, the quotation marks surrounding the word "paper" in this passage (from the previously cited article in *European Science Editing*); they were meant to call attention to use of a word that, while clearly derived from a particular *medium* (one whose day may be passing!), is now understood to be associated with the entire *concept* of scientific publication.

At the same time, those responsible for developing the new journal firmly believed in the importance of strenuously resisting shortcuts in the *manuscript reviewing* process (see Sec. 3.5.3). This is why the word "accepted" appears in the preceding quote, not "received"—to avoid any suggestion that papers would automatically be passed along to the public once they had been submitted. The point is made more clear as the passage continues: "Acceptance will only follow full peer review". A solemn promise was also given that contributions accepted into the system would be "permanently" archived in static form, immunized against any risk of corruption or revision.

OJCCT, as the first of the truly electronic journals, owed its debut and early evolution not to a random group of publishing amateurs, but to efforts of experienced professionals. The launch came also not as some grudging reaction to the inevitable,

---

[5] Cf. the news report "New electronic journal" in *Eur. Sci. Ed.* 45 (1992): 26–27.

[6] A distribution system providing an edge of only a few hours in delivery time has long been recognized as a valuable asset with respect to conventional publication. For example, scientists in California literally *demand* that air-freight be used to supply them with journals from Europe so that copies reach them the same day their East-coast colleagues are served.

but rather as an innovative initiative marked by a great deal of enthusiasm: a potential model for the future, crafted quite deliberately and creatively. Edward J. HUTH was recruited to be "editorial director" of the new medium. HUTH was a respected scientist with extensive experience as editor of the journal *Annals of Internal Medicine*, one who was also for many years chairman of the Council of Biology Editors (CBE). He was widely known in addition as the author of several important books on communication, including *How to Write and Publish Papers in the Medical Sciences* (1982, 1990), *Medical Style and Format: An International Manual for Authors, Editors, and Publishers* (1987), and *Writing and Publishing in Medicine* (3rd ed 1998). The driving forces behind the brash electronic journal were thus people who, despite having acquired their skills and experience in the context of traditional scientific publishing houses, clearly possessed the imagination to explore new avenues.

The initial development of *OJCCT* occurred in close cooperation with the American Association for the Advancement of Science (AAAS): a cleverly conceived partnership because it helped confer on the venture an air of "legitimacy" from the very beginning. In other ways as well there was ample evidence of creative impulses at work behind this attempt at successfully launching a publication of a boldly new type. Thus, an arrangement was made quite early to include abstracts of articles from *OJCCT* on a regular basis in the distinguished British medical journal *The Lancet*. Moreover, plans were developed to release certain urgent contributions immediately upon receipt—*without* review—under the special heading "works in progress".

Other innovations conceived for *OJCCT* were announced through periodic press releases. Thus, each subscriber to the electronic journal service was to be offered the opportunity to direct *comments* to the authors of incorporated articles. An editorial staff would play a role in transforming such submissions into an ongoing dialogue, in that all comments were to be collected, sorted, and publicly archived alongside the articles themselves. As a result, it was hoped that the new "journal" might develop into a *forum for discussion*, with opinions and observations from readers made available for public consumption immediately upon their arrival. One could almost envision the birth of a gigantic, internationally accessible Hyde Park Corner dedicated specifically to scientists!

Finally, the infant electronic journal was designed to exploit in an exciting way one of the most powerful tools associated with Internet communication: wide-ranging, reliable, and instantaneous "navigation" on the basis of *hyperlinks* (described below). Thus, for the reader with appropriate access privileges, "clicking" on a bibliographic reference of interest within an article was expected to open a direct pathway (via a network "log on"[7]) to the massive professional database MEDLINE, yielding an abstract of the cited article, information regarding libraries with subscriptions to the

---

[7] This fashionable new-found noun, together with the corresponding verb form "to log on" (or "log in"), is derived ultimately from the Greek *logos* (λογοσ) meaning "word"—the source as well for such loosely related terms as "logbook" and "logistics". Curiously, the English word "log" also conjures up the quite different image of a piece of wood cut directly from a tree.

journal containing it, and guidance in expeditiously acquiring a printed copy of the original document itself. The creators of the *Online Journal of Current Clinical Trials* were obviously determined to break important new ground with respect to the entire concept of scientific communication![8]

Further Ramifications of the Digital Revolution

We have so far only loosely (and then hesitantly) made use of the term "electronic publishing", an expression that is in any case increasingly giving way in professional circles (for reasons that will become more obvious in Sections 3.1.3 and 5.2.2—in the context of "SGML") to a more accurate characterization: "*digital* publishing". How and within what limits digital technologies will prove most effective for scientific communication over the long term remains to be seen. One thing is certain, however: the electronic delivery of information almost certainly will *not* distinguish itself from alternatives by virtue of being a *cost-free* phenomenon (some of the reasons why this is true are presented in Sec. 3.1.3).

One of the most far-reaching respects in which electronic documents reveal advantages over conventional printed documents is their ability to support *links* or *hyperlinks*. The average person first encounters the phenomenon of the hyperlink through Web pages. A link is a tool that permits a Web-page author to provide interested readers with the means of instantly acquiring additional relevant information from some source other than the one currently displayed. The alternative source must of course be specifically identified and targeted by the author in advance. A link could be used to transport the reader to another page on the current Web site, to a different document on the same server, or indeed to any document on any connected server anywhere in the world.[9] The presence of links is often signaled by an *underlining* of appropriate words or passages, but sometimes links operate through discrete page elements (fields or "buttons"). Activating a link by clicking in the appropriate spot directs one's browser immediately to open a connection to the Internet address corresponding to the linked document. (A "back" button lets one return at least as rapidly to the source document.) In other words, the current screen display disappears, to be replaced by a different one, drawn from the referenced location. Many e-journals now take advantage of this potential as a way of providing readers with access to such supplemental material as *spectra*, for example, or other experimental data that would ordinarily—for space reasons—not be included in a typical article from a traditional journal. The trend toward supplying "extra services" in this way is a growing one.

---

[8] Unfortunately, pursuing the story requires acknowledging that, for all its promise, *OJCCT* failed to win over a sufficient number of readers to survive. In 1995, AAAS elected to withdraw financial support. Rights to the journal passed into the hands of the commercial publisher Chapman & Hall, but a decision was soon made there to allow the title to die.

[9] Links are also used to facilitate navigation *within* especially long documents, since "jumping about" can otherwise prove awkward due to the relatively small "window" provided by a typical video screen.

"Electronic journals", just like "electronic books" (see Sec. 4.1.3), also offer other advantages over their print counterparts. For example, the electronic version of a journal can put at an author's disposal all the expressive possibilities conjured up by the term "multimedia": astonishingly complex graphics (including facilities granting a reader the power to simulate three-dimensional manipulation of an image—a cathedral, perhaps, or a molecular model), as well as animated components, film clips, sound—the range of opportunities is staggering.[10]

Archiving is also more convenient with an electronic (digital) source relative to its print equivalent, and in several senses. For example, with an ordinary journal a library normally feels obliged to gather up the most recent issues every few months and send them away for binding as a way of better preserving them, despite the fact that even in a bound state pages remain susceptible to wear and tear. There is obviously no such physical processing required with an e-journal. Traditional journals must also be allotted an enormous amount of library shelf space. By conservative estimate, switching to electronic storage reduces a library's space requirements by a factor of at least 1000. Moreover, production of an e-journal consumes absolutely no paper (although interested readers may well decide to print certain articles themselves), an advantage conceded even by those with little overt sympathy for radical environmentalists.

Finding a particular piece of information within the electronic data universe can pose problems, but the challenge has led to a number of innovative and powerful strategies and tools. Thus, approached in the right way, electronically processed data actually turn out to be extraordinarily well-suited to sophisticated forms of searching. Under ideal circumstances, an electronic journal might thus be viewed as a completely searchable, full-text database, a characterization that ideally encompasses even such "special features" as *tables* and *figure captions*. Specific content, contributions from particular authors, or even isolated passages containing keywords of interest can be rapidly located and scrutinized, irrespective of where they occur within the journal. Moreover, every search can be relied upon to be truly exhaustive. Imagine the difficulty of trying to unearth, from somewhere in several years' worth of a traditional journal, one particular article that you're sure you once read—even though at the moment all you recall is that it contained a table with information concerning substance X or concept Y. Such a venture would be prohibitively time-consuming in the "paper world". In a worst case you would be forced to examine in detail every page from a huge stack of issues. With an electronic data search the comparable feat can be accomplished (much more reliably!) in a matter of seconds, or at most a few minutes.[11]

---

[10] So far as we know, no one has yet offered olfactory support to the author wishing to publish an "illustrated" research account in a hypothetical *International Journal of Perfumery*!

[11] In undertaking a search it is nevertheless important that one be conscious of limitations that may handicap the available "tools", since characteristics of one's tools will dictate the optimum *search strategy*. For example, any search dependent upon an *index* will inevitably suffer from shortcomings inherent in the indexing concept itself—regardless of whether the information source is electronic or print-based. Thus, an index is essentially an organized collection of *keywords*. These keywords may in turn be restricted by having been drawn exclusively from words that actually appear in the original text

→

One special problem associated with searching an electronic database is the fact that too *many* hits may turn up, in the form of pointers to more potentially relevant sites than one has the time or inclination to investigate. When this happens the only recourse is to "narrow" one's search criteria and try again. The problem becomes more acute the larger the database; some repositories offer information from literally millions of sources.[12]

We revisit the subject of data searches in the subsection of Sec. 3.1.3 entitled "Information Acquisition Today: Search Capabilities".

Questions of "Permanence"

We noted earlier that the founders of the *Online Journal of Current Clinical Trials* were anxious that their journal, despite its "electronic" nature, be taken seriously as a *permanent* part of the STM literature. But how "permanent" electronic (or digital) media will ultimately prove to be remains a troublesome open question. Researchers have for centuries relied on the fact that it is always possible unambiguously to *identify* and *date* source materials obtained from a library or similar repository—even a handwritten manuscript from the library of an obscure monastery—and this in turn provides a firm foundation for claims of *authenticity*. But who will be able to guarantee, for example, that *electronically* stored documents, especially ones readily accessible to a great many people, have never been subjected to modification, surreptitious or otherwise?

(though obviously not *every* word will have been indexed; e.g., "and", "the", etc.). Alternatively, they might have been supplemented by inclusion of related (perhaps even *standardized*) terms assigned manually and in a descriptive way by the author or some professional indexer, but there will still be limits, set by that person's proficiency and conscientiousness. From an intellectual standpoint, creating a good index is a frightfully demanding assignment. No index should ever be regarded as truly comprehensive: omissions and errors are unavoidable. Indexes furnished with many print journals are especially constrained due to the fact that they are constructed solely on the basis of *abstracts* of published articles. The subject of indexes, obviously peripheral to the present chapter, is treated at greater length elsewhere (e.g., Sec. 4.5.1 of this book).

[12] The largest electronic collection devoted to literature anywhere in the world is maintained by the Online Computer Library Center (OCLC) in Dublin, Ohio (USA). This noncommercial supplier of library support, with contributions from over 8650 member libraries, is currently (2002) relied upon by roughly 41 000 libraries in the United States and 81 other countries. OCLC and its *Online Union Catalog* (referred to on the Web as *WorldCat*) encompassed nearly 47 million titles in mid-2002, with over 852 million listings of document locations. It could fairly be described as the general directory to a hypothetical "world library". Most of the titles represent books, but 9000 full-text electronic journals are also included. Other organizations have accomplished heroic bibliographic feats of their own. Thus, the search engine of the German *Zeitschriftendatenbank* (periodicals database; ZDB, maintained by the Deutsches Bibliotheksinstitut/Staatsbibliothek Berlin) provides information about roughly one million distinct publications housed in 4300 German libraries, with a total of nearly 6 million discrete references. Another noteworthy German enterprise is the Electronic Journals Library ("EZB"), maintained in the city of Regensburg. As its name implies, the EZB deals only with e-journals, but its files numbered 12 129 from 189 member libraries in 2002! Included were 1256 purely online journals (i.e., journals for which no print counterpart exists), and 2986 journals of the "full-text" variety. We can't help but wonder how St. BONIFACE would react to all this. It was he who first established a bishopric at Regensburg in the 8th Century, thereby laying the groundwork for the future imperial city on the Danube which is now home to the EZB.

Uncertainty in this regard must necessarily be a serious concern for the sponsors—but also the patrons!—of every electronic journal.[13]

Furthermore, how long is it in fact possible to archive a digital document with absolute assurance that it will not have *deteriorated*? Too little experience is so far available for anyone to answer such questions in a definitive way, or even confidently offer trustworthy technical advice. *Magnetic hard disks*, the primary storage media for most computers, are known to be of questionable reliability after as little as two years in service. *Compact disks* (CDs) are alleged to have a lifetime of at least 50 years, but at this point the assertion must be taken somewhat on faith, since no CD was ever used for text archival until the late 1980s. Perhaps even more worrisome, who is to say with complete assurance that appropriate *play-back devices* for today's media will still be widespread 50 years in the future? Indeed, experience from the past few decades provides strong circumstantial evidence to the contrary: consider the rapid disappearance of 8-track tape players and the turntables required for enjoying classic vinyl recordings.

Recent events raise another specter as well. Extensive "magnetization" of the sources of primary information might tempt an unscrupulous "enemy of information" to plot a demagnetizing campaign directed against sensitive records. Could an infallible shield be developed to prevent such an act of sabotage? At a more mundane level, what limits exist on the practical lifetime of *any* electronically stored information? The questions are complex and delicate—but also urgent. Only a fool would today claim to be in possession of all the answers.

Librarians have come increasingly to rely on *microfilming* as their conservation measure of last resort (for print media), but even this approach to preserving information must class as unproven over the *truly* long term. How far removed this all is from the demonstrated resilience of the Laws of HAMMURABI, "documents" chiseled originally in stone and still clearly legible after nearly 4000 years!

The Road Ahead

Pressure on publishers, librarians, and individual scientists to respond to change and make the most of the various new media is having a multitude of far-reaching effects. The apparently unstoppable, ever-accelerating cost/price spiral plaguing the publishing

---

[13] Aspects of this particular issue and various others of a technical, organizational, and legal nature have been carefully explored in a book we highly recommend—to readers conversant with German—for its currency, scope, and thoughtful preparation: ZIMMER 2000, *Die Bibliothek der Zukunft* (*The Library of the Future*). Other (somewhat less up-to-date) standard and relevant sources include PAGE, CAMPBELL, and MEADOWS 1997, and BUTTERWORTH 1998. We are not particularly reluctant, incidentally, to cite outstanding sources we have encountered that happen not to be in English. Regardless of where you may be living and working it should be possible today for you to obtain any source in whatever language it is written, and you can perhaps call upon a friend or associate outside your "tent" to assist you with interpreting a source in an "exotic" tongue. Indeed, we hope you will come to recognize that the confines of one's own tent can be unduly restrictive!

world would in any event have forced fundamental changes upon the traditional "information-transfer industry" within the near future.

With respect to journals it is actually rather surprising how little serious interest there was in electronic information transfer on the part of individual scientists until traditional publishing houses themselves took the initiative and began to offer parallel electronic and print access to existing publications. Perhaps researchers were at first reluctant to risk the possibility of their precious communications ending up buried in an electronic data-graveyard few would ever visit, where even the grave markers might lapse into a state of decay within a short time. Nevertheless, it appears that analog and digital forms of publication have now come to enjoy nearly equivalent status in the information marketplace, although the durability of the present equilibrium remains to be seen. One reason for skepticism in this regard is the fact that journal publishers currently feel obligated to market in two separate formats what from a content perspective is virtually a single product—hardly an efficient state of affairs, and not one conducive to holding down costs.

By the way: we still hesitate to apply very freely the label "publishing" in this new electronic context. The major problem as we see it—apart from permanence—is *acceptance,* not technical feasibility. Consider for a moment the ways in which, given a choice, scientists tend to gather their information. What habits do they easily fall into with respect to maintaining a state of currency, both in their specialties and more generally across the wider landscape of science? What sorts of activities appear most compatible with prevalent ingrained patterns and what serious negative factors need to be taken into account?

Effective and Efficient Acquisition of Information

Picture, if you will, a typical conscientious scientist forced by circumstances to spend several hours traveling by air, time that threatens to class as "unproductive". Would the prospect of passing some of that time squinting at a laptop computer really act as a serious inducement to ingesting along the way large blocks of digitally stored information? Given the option, our hypothetical traveler would almost certainly much prefer pulling out of his or her briefcase the latest issue of a favorite print journal, settling into a comfortable position, and perusing at leisure the few articles that look most interesting. Most people would probably concur that, while a computer can certainly be a convenient place to "look something up", it is not particularly appealing as a dispenser of large doses of conceptually demanding ideas requiring serious study. No computer yet introduced lends itself, for example, to relaxed use as an "information agent" by the overworked professional stretched out on a sofa over a quiet weekend— in sharp contrast to printed matter in almost any form.

Comfortable conditions, congenial surroundings, and information packaged in an inviting format might almost be considered prerequisite to absorbing and effectively "processing" complex technical subject matter on a routine basis. The traditional print version of a scientific journal—with its absence of secondary demands (e.g., power cords, batteries)—still seems to us to offer the greatest potential in this respect. It can be called into play almost anywhere, and it caters to the reader wishing to sample its content in an informal way—paging forward or back as impulse dictates, skimming rapidly over passages of minimal interest but probing more deeply whenever that seems appropriate, processing the available material very selectively. Incidentally, activity of this sort is associated with a surprisingly high level of intellectual effort. The serious (casual!) reader must stay constantly open to new insights of unexpected kinds. Whatever information is retained will have been strictly self-selected, priority always attaching to that which appears most relevant to one's own activities and needs. We find it hard to imagine it ever really becoming preferable for one's "reading" to be confined to an electronic version of a journal.[14]

The Quest for Knowledge

Knowing is of course not the same as *possessing wisdom*, and simply being *informed* falls short even of knowing. The best scientists have always felt a sense of urgency about staying broadly current in their disciplines—knowledgeable, but insofar as possible also wise—and we sincerely hope this drive will persist as the era of electronic communication continues to unfold. There is at least one serious cause for concern, however: specialized libraries all over the world are every year canceling large numbers of subscriptions to technical journals, strictly for economic reasons. Tens of thousands of titles are being dropped, and there is no end in sight. The situation is reminiscent of the extermination of animals or plants from a rainforest or some other carefully balanced biotope. Only a few scientific "environmentalists" are available to oppose unrestricted development, and their influence is limited due to the fact that the number of people deeply interested in this particular "habitat" is relatively small. Even the few who do cherish the "flavor" of their favorite part of the forest often seem unaware of changes that threaten its future.

Abandoning analogy and returning directly to the issue of depleted libraries, individual scientists everywhere will *necessarily* find it increasingly difficult to shore up their knowledge base in traditional ways. Was it an illusion to imagine that the "information age" would only make our lives richer? Probably the most effective intellectual "bandaid" available to the conscientious seeker of wisdom, at least for now, is subscribing to a wide variety of e-mail title lists, selected with care so as to ensure ample breadth, and then conscientiously and consistently downloading, from whatever source is most convenient, all the individual articles that appear promising. Even more important will be taking the further time and trouble necessary to *print* the articles once they arrive. As pages accumulate these can be collected in stacks, which

---

[14] A dogmatic stance on this point should be treated with caution, however, at least until the unveiling of the promised next generation of electronic reading devices and "electronic books"; see also the comments in Sec. 4.1.3.

might be regarded as roughly analogous to single issues of one's very own, highly individualistic, "personalized journal". The most recent stack should then be kept conveniently at hand for relaxed (priority) reading whenever circumstances are favorable.

3.1.3 More from the World of Tomorrow (or Today!)

Professional vs. Private Publication

In the late fall of 1997, tens of thousands of students, notably in Germany, launched vigorous protests decrying among other things the slashing of library budgets, an act they characterized as a prime example of the state-sponsored propagation of ignorance—perhaps with a measure of justification. Among the most thoughtful responses to the protestors' questions and demands were those offered by publishing houses, though they were not issued directly within the protest forum or in the streets. For example: "Especially in view of … increasing digitalization of our products, worldwide shortfalls in funding from public sources, and the resulting cuts in library budgets, etc., we must be clear about the fact that we have a responsibility to be something special" asserted the publisher of this particular book (Wiley–VCH) at the onset of 1998 as a prelude to unveiling a new electronic system for dispensing journal information.

The objectives of the enhanced Wiley–VCH information system were described as provision of an inclusive

> search potential based on tables of contents and abstracts"—for almost 400 different journals!—and introduction of the novel concept of a "personal *homepage* [our italics] … providing every user with the ability to activate an individualized and specific search procedure. Users are thus able to store their own lists of preferred journals, topics, and search criteria. Along with this comes the ability to download articles to a local computer station and then print them, complete with all illustrations, tables, and miscellaneous graphics. Moreover, each of the [publisher's] journals has also been assigned a unique homepage, with extensive information related to that particular publication. In many cases … contact is enhanced by the availability of hyperlinks to related products, organizations, and external Internet sites.

Other major scientific publishing houses—including Elsevier in the Netherlands[15] and Academic Press in the United States—as well as a host of smaller firms. have of course also contributed significantly to the development of e-journals. The list is too long for us here to acknowledge them all.

---

[15] Especially memorable is this publisher's contribution to project TULIP (1991–1995), with the objective of providing electronic journal access to universities with subscriptions to standard versions of Elsevier journals. Elsevier has its headquarters in Amsterdam.

Professionals in the realm of information delivery clearly believe their services will continue to be in demand. At the same time it seems to us likely that, working together with publishers, "middlemen" ("information brokers") in the information and data market, such as CompuServe and STN,[16] will find themselves playing an increasingly important role in the transfer of technical information; indeed, the trend is already apparent.

No matter how attractive a more informal substitute might seem, we are convinced it is unrealistic to expect the most trusted sources of information ever to be of a type that operates on a casual no-fee basis. Instinct alone warns us that "whatever is free is likely also to be worthless". Many well-intentioned dreams to the contrary, attributable to computer revolutionaries and Internet pioneers (cf. RHEINGOLD 2000)—"Boundless freedom above the clouds"—are in our judgement destined to remain unrealized and impractical utopian fantasies. Completely independent electronic "publication" of the individual scientist's original discoveries, or "free" dissemination of definitive results painstakingly acquired by small groups of professionals, has shown no real sign of flourishing, and we are convinced this was a dim prospect from the outset. Unrestrained, uncensored, *unreviewed* "private publishing" on the part of whoever seizes the initiative, and for the benefit of all, is an extremely limited phenomenon, and that will surely remain the case.[17] Generally speaking, "traditional" publishing houses with their well-established procedures and quality controls are likely to continue to dominate the "information-transfer scene".[18]

Why do we not expect the work of the professional *information industry* soon to be displaced by extensive *private publication*, with distribution over the so-called *information superhighway*? Strong arguments against such a development actually follow from premises that have played an essential role in the shaping and development of the Internet: an absence of oversight, minimal organization, and an abhorrence of "tolls". Thus, the scientist who, through private publication, wished to circumvent payment for professional efforts devoted to nurturing and facilitating the flow of information would also be obliged to forego, among other things, the benefits that derive from a reputable publisher's assurances regarding the reliability of that which

---

[16] STN itself, with headquarters in Falls Church, Virginia, USA, has for some reason always seemed reluctant to divulge what its tantalizing initials stand for: Scientific and Technical Network. In any case, they are easily contacted through their homepage at http://www.stn.com. (If this book were already "online" you would need only to activate the foregoing underlined hyperlink to reach the Website with a mouse click.)

[17] At the same time, distribution of *preprints* of work yet to be published has become extremely popular in certain disciplines. That this applies especially to physics (see the second footnote in Sec. 3.2.2) may be related to the fact that much physics research today is conducted under the auspices of large, prestigious research institutions, especially in the case of high-energy physics. Often the research teams involved are so all-encompassing that any results emerging have in a sense already been "refereed"—internally—thereby automatically conforming at least to the intent of the reviewing process so central to conventional journal publication [cf. EBEL H.F., "Travelling the information highway: CERN World-Wide Web days", *Eur. Sci. Ed.* 55 (1995): 9–11].

[18] Incidentally, we also find it impossible at this juncture to imagine a "virtual library" that would ever completely displace "bricks and mortar" libraries as we have always known them.

is communicated. In the absence of compensation there is no meaningful incentive for an intermediary to monitor the quality of information dispensed. (It is significant, by the way, that publishing houses today actually assume a *legal* responsibility for their products.) Viewed somewhat differently, refusal to pay a fee for valuable services conscientiously rendered by another undermines the latter's economic base, in effect placing the entire concept of professionalism at risk. We return to the important subject of "quality control" later, especially in Sec. 3.5.

One often overlooked beneficial aspect of professional publication is worth interjecting here, however. A good journal is expected to provide a respectable *index*, one supportive of meaningful searches on the basis of "rational" terminology. This implies, for example, that a search should not be constrained by particular modes of expression an author may have chosen to employ. With a quality journal the author therefore must bend to terminological whims of a senior colleague, in this case the editor, who will probably have selected a set of "official" terms for preferential use in that journal. How does this particular set of in-house "rules" compare with those promulgated by other journals? The best way to find out is to glance through indexes derived from issues of the previous year.

Providing an index is hardly a trivial consideration, by the way, because creating it requires that all information published by the journal be *processed* in a systematic way, at least from a terminological standpoint. The best journals go a significant step farther in manuscript processing: they also undertake *editorial polishing* of submitted work in hopes of making that which eventually appears in print easier for the reader to comprehend and interpret.

These supplementary benefits, and the extent to which they are applauded, reinforce our conviction that professional involvement in the publication process is unlikely ever to disappear. Still, perpetuation indefinitely of the status quo is far from inevitable. Some of the forces lobbying for change are extremely vocal, and proponents of various radical schemes have shown themselves to be surprisingly tenacious; academics in particular seem to thrive on lofty promises of what the future should bring. There is little doubt that, over time, the information landscape as we know it will be reshaped in fundamental—but not yet predictable—ways.

The Scene Today

The preceding observations largely reflect the state of our thinking in 1998 as the fourth German edition of this book was being prepared. Given the pace of change, 1998 classes as "a long time ago", so we felt it important to reflect a bit more on the situation as we assess it now. One reservoir of insight we found particularly helpful in this regard has its roots in Germany, but it was the American member of our team who encountered it quite by chance on the Web. One might therefore describe it as a source especially in tune with our transatlantic impulses and sympathies. The Web address (*URL, Uniform Resource Locator*) in question is http://www.harrassowitz.de/ms/ejresguide.html

Otto Harrassowitz KG has been an active participant on the information scene for over 125 years. Situated now in Wiesbaden (Germany),[19] Harrassowitz is a major international purveyor and distributor of scholarly literature—books, and especially journals—with the motto: "from Europe to the USA". Many publishers rely heavily on the expertise and experience of this venerable (but still lively!) firm as they deal with complex problems related to subscriptions, transport, delivery, etc. The abbreviation "ejresguide" in the URL above stands for a remarkable repository of e-journal information, freely available on the Harrassowitz Web server, with the full title "Electronic Journals: A Selected Resource Guide". When we first perused this guide (in June, 2000) we were impressed to note that it had been thoroughly updated as recently as six weeks previously. In printed form it provided us with 27 pages of information containing roughly 300 references, most of them accessible through hyperlinks. In other words, the website itself constitutes a dynamic piece of literature—a document—in the form of an organized and annotated directory to a wealth of other resources, most of them digital (or at least available in digital form via the Internet). This timely treasure trove was for us a godsend! As we studied it, however, we came to realize that there have actually been relatively few earthshaking developments on the journals front in the last five years. Various earlier expectations and goals have been refined, but in general the "electronic journal" has simply continued to come of age in a conceptual sense, gradually earning the status of a standard commodity of "e-commerce".

The Harrassowitz document did introduce us to several journals dedicated exclusively to matters of electronic publication, and we found it a valuable guide to a remarkable array of books, articles, and conference reports dealing with topics related to developments on the journal front. Evidence from several of the cited sources suggests, by the way, that staid, traditional publications like *Nature* (*London*) and *Science* (*Washington DC*) have contributed just as much to evolution of the literature and its documentation as more narrowly focused organs like *Scholarly Publishing*. We were struck, too, by the fact that in the course of intense debates concerning vital publishing issues, a great deal of thoughtful input has come not only from librarians, editors, and publishers, but also from individual scientists, leading us to surmise that to a far greater extent than in the past there now exists a lively cooperative interchange between professionals in the areas of archiving and documentation on one hand and both publishers and scholars on the other.

A few journals, like the *Journal of Electronic Publishing* and *D-Lib Magazine,* have apparently evolved into publications one would expect to find on the shelves of a good library. Professional societies such as the Society for Scholarly Publishing, and organizations like the Association of Learned and Professional Society Publishers (ALPSP) have been at the forefront of much that has transpired. Widely circulated

---

[19] Its original home was Leipzig, in what after World War II became the "German Democratic Republic". When its assets were seized by the authoritarian government the firm was reestablished in the West.

trade periodicals, including *European Science Editing* (on whose editorial board one of us actively served during some of the early decisive years), have diligently kept their readers abreast of important trends, with the result that editorial staffs everywhere have been able to adapt responsibly and in a timely fashion to changing circumstances.

As indicated above, electronic journals have matured considerably since their dramatic debut. Traditional publishing houses are now the source of most e-journals, many of which are actually alternative editions of familiar print publications. At the same time publishers have been forced to perfect fundamentally new forms of contracts, licenses, and billing systems. An array of options like "individual licenses", "single-use network licenses", and "multi-use network licenses" is now essential if the diverse needs of a wide range of clients (subscribers) are to be met. Virtually all publishers are supportive at least as an option of the goal of widespread distribution of scientific reports in full-text electronic form.

While considerable attention has necessarily been devoted to implementing and perfecting progressive modes of information transfer, publishers have by no means abandoned their traditional commitment to *quality* as this is understood by the professional community. Submission alone no more guarantees today that a contribution will be published than it did in the past. Documents destined for public consumption—whether in print or through the Internet—are still challenged to meet stringent criteria with respect to both form and content. Given the recent emphasis on "audits" as an essential tool in *quality management* generally, it would be unconscionable for safeguards serving a similar purpose in the scientific community suddenly to be discarded (see the discussion of manuscript review in Sec. 3.5.3 for more thoughts on this subject). Is it indeed desirable that everyone be encouraged to present their ideas on an equal basis in the scientific information marketplace, without evaluation, censure, or restriction? Surely the answer must be "no" if quality scholarship is the goal. Everyone? In principle, perhaps yes, but the position is one we are extremely hesitant to adopt. The analogy may be far from perfect, but most would be reluctant to take off in a jumbo jet if there were uncertainty about whether the captain was in fact licensed to fly the machine.

Generally speaking, the "business" of scientific journal publication has actually been flourishing of late. For example, our own publisher, Wiley–VCH, recently introduced several new scientific and technical journals, and so far all would be described as highly successful. Many existing journals have witnessed a steady increase in the number of qualified manuscripts submitted for publication, and in annual page count,[20] and it is not uncommon to find cases in which the total number of subscribers is growing as well. As an example consider the following information from the 1999 Annual Report of the German Chemical Society with respect to the publication

---

[20] Another measure of "size" that leads to the same conclusion is the number of bytes of electronic data processed annually.

Chemistry—A European Journal (in which the Society has both intellectual and financial interests).[21]

> The unexpectedly large number (694) of manuscripts received in 1999 (22% more than in the preceding year) meant that in this fifth year [of the journal's existence] 350 more pages appeared than were originally planned. The 3750 pages actually published represented 24 review articles and 356 full papers. Total content thus increased by 42% relative to 1998 ...

What follows is a set of statistics related to "international acceptance" as reflected by number (and geographic distribution) of manuscripts submitted from abroad, which—to the delight of the sponsors—increased 26% over the preceding year. The report concludes with the observation that, judging from the *impact-factor* value of 5.15 achieved by the journal, worldwide reception was impressive indeed.[22] It is interesting to note that this particular journal has been available since its inception in 1995 in both print and electronic versions. Due to demand, the publication schedule was recently advanced from monthly to twice-monthly.

The health or survival of an "established" publishing house appear to be less a function of technological expertise (which all claim to have) than of varying degrees of success in developing viable marketing schemes for innovative products.[23] Most have proceeded from the premise that in a world accustomed to (and tolerant of) several versions of "pay TV" there would also be a willingness eventually to accept subscription charges for electronic journals. An example of the apparent validity of this assumption is the fact that the fee-based electronic document dispensing system at www. interscience.wiley.com processed during the single month of October, 1999, reader requests for half a million pages of text.

The dream of every scientist with a thirst for information is being able to look in one place, under a specific set of keywords, with the assurance that everything known about the corresponding subject will quickly be unearthed, irrespective of how deeply the facts might be buried. To put the matter differently, it would be a godsend if one could pursue a reliable search of all the potentially fruitful databases simultaneously. Not the least of the obstacles in the way of realizing such a vision is the need for uniformity to be achieved in the way keywords are assigned in different resources.[24] As it happens, the National Institutes of Health in Washington, D.C., is already engaged in the development of just such a database ("PubMed"), one in which *all* biomedical

---

[21] The report itself appears in *Nachrichten aus der Chemie 7/8*, 2000, p. 42.

[22] The "impact factor" is essentially a complex measure of the extent to which papers published in a journal have recently been *cited* in *other* sources, which is believed to correlate with the esteem in which the journal is held by scholars generally (cf. Sec. 3.2.4).

[23] Publishers unwilling to participate at all in the "electronic revolution" have in most cases simply ceased to exist.

[24] This is not the only major hurdle, however. Who could be induced to take on the awesome responsibility for keyword assignment? And where would the qualified indexers required for such an undertaking be trained? These questions alone could serve as focal points for a whole series of high-level, multidisciplinary conferences!

articles published everywhere in the world are to be represented. Astonishing as it may seem, gigantic projects on this scale may actually be feasible, at least technically. Perhaps the most worrisome issues relate to the best approach to managing what will surely be a veritable flood of inquiries once such a "superdatabase" becomes fully operational.

Just a few short years ago—in 1990, say—the most one would have been likely to foresee in the way of a worldwide electronic information system was a record of titles and authors, possibly expanded to include abstracts. In the space of a decade this has evolved into unhindered electronic access to almost every major serial publication, in its entirety and regardless of source. E-publishing has become a matter of routine. The pages of important journals can typically be examined either on paper or via the Internet. Publishing houses, dismissed at one time by cultural curmudgeons as the bane of any "true" information system, have reasserted their legitimacy by offering traditional services in dramatically new garb, leading to the ironic consequence that such ostensibly archaic terminology as "article", "contribution", and "page count"—even "paper" in the sense of a publication—continues to be widely employed.

At the same time, recent developments have brought with them numerous unforeseen advantages for the consumers of information. Access to journals that budgetary constraints have caused to disappear from library shelves has for many suddenly been restored thanks to provisions for selective electronic acquisition—an outcome few predicted. Whether a particular journal is available *only* in digital form or *also* in print is becoming increasingly irrelevant. So-called core journals, those held in the very highest regard in a given discipline, will probably continue to be treasured by librarians in printed form, if only because these lend themselves better to informal browsing, and readers appreciate their convenience. As a matter of fact, most professionals *enjoy* the occasional opportunity to glance at the advertisements, announcements, etc., that are integral adjuncts almost exclusively to print journals, and they welcome a chance now and then to thumb through *all* the articles in a given issue, an activity far less pleasurable—and practical—when one's "field of view" is limited to a video screen.

The Current State of the Electronic "Environment"

The very name associated with the World Wide Web provides an indication of its founders' ambitions: implementation of a pervasive, seamless *information network* on a global scale. Before it attracted the attention and fascination of technicians and scientists at the European Nuclear Research Center (CERN), the structure of the precursor to the Internet was far too cumbersome and disorganized to be amenable to occasional (or even everyday) use by the layman, and data flow through the system was painfully slow. Thanks to an innovative new data-transfer protocol ("http": "hypertext transfer protocol"—an acronym now widely familiar in the context of *web addresses*), developed by CERN staff working in Geneva, those with information to offer and those seeking it are now in a position of being able to enter easily into direct "conversation".[25] Any multimedia document prepared according to http standards can be sent or received instantaneously from anywhere on the Web irrespective of the nature of a participant's computer or operating system, and usually without complications. Furthermore, any computer connected to the Internet is potentially able to function not only as a *client*, but also as a *server*.

The term "server" in this context refers to a dedicated computer located in most cases at the hub of a computer network. It may be configured to serve any of several purposes, such as database management, traffic control, or even the provision of convenient and flexible storage space. Assuming it is connected to the Internet and sufficiently conversant with http, it is suited to function as a *Web server*, thereby becoming one of the innumerable computers constituting "nodes" and thus helping define the complex "web" we call the Internet. A server's task is to act upon requests relayed from "clients"—other computers that may or may not also be capable of functioning as servers, but are at least *connected* via servers to the network. Most clients are in fact workstations assigned to individual users. Active participants in this worldwide information system make themselves available to others through individualized portals that take the form of *homepages*.

An indispensable secondary role in the resulting communication system is played by sophisticated *user software* for extracting data from the Web and placing it at the disposal of the individual wanting to take advantage of a networked workstation. The corresponding programs are known as *browsers* (cf. also Sec. 9.2.1).[26]

Browser-based communication occurs in more or less complex "windows" defined on a computer screen, often incorporating features that permit direct interaction on the part of the user. The typical window offers a variety of navigational tools and various controls affecting things like the display mode. Advanced browser technology encourages multimedia spectaculars reaching far beyond anything that information suppliers could create on paper. The variety of available illustrative techniques has in scientific applications been especially heavily exploited—often quite imaginatively— by innovative biologists and chemists. Much current attention is focused on achieving the full potential of Web-based "streaming video" presentations.

Acquiring a rapid preliminary overview of a complex Web page is facilitated if the user temporarily dispenses with graphic elements, replacing them with simple *placeholders* as a way of minimizing data-transfer demands. Nevertheless, display of a full

---

[25] The fundamental ideas underlying "http" are usually credited to Tim BERNERS-LEE, a technician from the UK working at the time at CERN. Few now recognize the name, and we find that odd. In our judgement he deserves a prominent place in the annals of information technology (IT).

[26] Browsers manage the acquisition and display not only of text, but also *graphics*, *videoclips*, and audio presentations—often with help from *plug-ins* (subsidiary programs that extend a browser's capabilities). The most widely used browsers at the present time are INTERNET EXPLORER (from Microsoft, Inc.) and NETSCAPE (from Netscape Communications Corp.), though others have their advocates as well. These giants have been engaged over a long period of time in a bitter battle for market domination, with both technical and legal implications, a struggle that has frequently created waves in the business pages of the popular press.

set of illustrations (and even capturing them in print) in most cases requires only a few additional seconds—especially for the user with access to a high-speed *ISDN-*connection (ISDN, Integrated Services Digital Network) and a fast computer with extensive storage capacity. By now we feel it safe to assume that virtually every practicing scientist or student in a scientific discipline is well acquainted with routine utilization of the "net", with a knowledge of the subject probably greater than would be appropriate for us to attempt to enhance.

Chemistry has long been recognized as an especially fruitful and challenging testing ground for innovations in *information technology,*[27] largely because since the 19th Century it has represented a highly *systematic* study focusing on clearly defined entities (molecules, structures, reactions, etc.) and, to a lesser extent, concepts. Some of the observations in the paragraphs that follow stem from insights shared at a gathering of chemists in the fall of 1996 at the Chemistry Computer Center of the University of Erlangen–Nürnberg.

Much of the interest in communication today arises not out of further development of traditional modes of publication in response to new technologies, but rather from the appearance of new forms and concepts in scientific communication in the broadest sense. The keyword is "networking": linking isolated computers in such a way that information can be summoned instantly to the screen of one computer from another located at a great distance, or facilitating the intimate interaction of several computers— perhaps even encouraging the computers to establish their *own* productive contacts. The starting point is the local network (LAN, Local Area Network), also referred to as an *intranet*, serving to interconnect multiple data-processing devices within a single institution—e.g., a corporation or a university. These local networks are in turn joined together through the Internet, permitting efficient communication between computers at far-flung corners of the globe. In this environment, information can be exchanged essentially instantaneously—at least in principle—a promise with truly revolutionary implications.

● "The network is the computer!"[28]

---

[27] This has been the case since long before the invention of that fashionable expression. For over a century chemists have been the envy of their academic colleagues because of the power inherent in such discipline-specific "information tools" as *Chemical Abstracts* and the exhaustive—and exquisitely organized—*Beilstein* and *Gmelin* encyclopedic compendia.

[28] This phrase was adopted as a slogan for the California computer company Sun Microsystems, perhaps best known for its development of the programming language "Java". It is of course a play on words, evoking a famous pronouncement by Marshall MacLuhan in the 1960s: "The medium is the message". The point here is that the Internet is far more than simply a source of stored *data*: it can also provide the means for *managing* that data in the form of networked *programs*. For years, even students in many places have been in a position, for example, to carry out complex structural calculations with resources supplied by a powerful computer somewhere else, which they have never even seen. The only prerequisite to exploiting this power is sufficient knowledge of how to gain appropriate access via the Internet. Scientists may be famous for their inventiveness, but maximizing the potential inherent in creative navigation within the new universe of electronic information presents a formidable challenge to even the experienced scientist.

As one example, consider the data emerging from modern analytical equipment, much of which is ideally suited to graphic display, as with *spectra*. It should come as no surprise that spectrometers are now perceived by the technically-minded as data-generating devices based on highly refined sensors in conjunction with sophisticated computing modules that serve as intelligent controlling links between the analyst and his or her sample. Spectra of course provide scientists with a great deal of information, but in the past they were rarely available in complete form beyond the limits of a single research group, largely for economic reasons: their reproduction (including publication) was too costly. Much valuable data therefore remained unshared, or accessible only in special archives or obscure databases, often in the form of COM microfiches (COM, Computer Output on Microfilm, an early application of computerization to microfilming). In principle, the situation has now changed drastically. Every researcher is potentially in a position to share with others vast amounts of information—almost without limit.[29]

Another illustrative example is the amount of attention being directed toward upgrading tele*communication* into tele*operation*: creating the means whereby the course of a procedure in some distant laboratory might be influenced directly through interconnected computers linked to *laboratory robots*.[30] In principle, all the computerized instruments in *your* laboratory could then be made available for instantaneous access from an unlimited number of locations, producing the impression that the equipment actually resides in the remote user's immediate vicinity. Cutting-edge physician's practices and hospitals are ideal places for witnessing effects of these developments. "Patient XY? Right. Here are your most recent blood-test results, this is an electrocardiogram we just recorded and had automatically interpreted in Stockholm, and ..."

Virtual laboratory? It would be inappropriate for us to expand further on scientific implications of technological advances. Suffice it to say that real-time acquisition of a spectrum via a networked computer, or obtaining a definitive piece of personal identification (perhaps a genetic fingerprint) from another continent, is becoming a matter of routine in today's world. Even the (accelerating!) *rate* of progress in some areas is truly awesome—for good or ill ...

Again we are straying too far from our avowed focus on electronic publication, an assignment which is complex enough! Realizing the full potential of digital commerce in research findings exacts a variety of demands on the participants. Scientists today

---

[29] Obstacles of course remain. For example, especially at "rush hour" the "information superhighways" making up today's Internet are subject to major traffic jams, and moving large amounts of data from one place to another can be both time-consuming and, in at least some cases, expensive. Resourceful Internet communicators have long since taken advantage of the fact that rush hour occurs at different times in different parts of the world, which means considering geography as one factor in data-sharing activity. Telecommunication companies everywhere are competing in a high-stakes race to increase rates of data flow through their systems.

[30] Remote control of certain complex surgical procedures is a dramatic case in point. This has already been achieved by teams of physicians—several years ago in Europe and in mid-2000 for the first time in the United States.

must think and behave in ways quite different from the past in order to become full partners in "electronic communication". This is an aspect of the revolution too often overlooked. The pair of reading glasses tucked so handily in one's pocket no longer suffices for "reading" an electronic journal. Efficient perusal of the literature dictates that that one have ready access to a fairly sophisticated personal computer system—with a fast processor, ample memory,[31] software as up-to-date as possible (in order to take advantage of the latest "bells and whistles" authors so much enjoy showcasing), high-quality video output, and—ideally—a rapid, high-resolution printer able to generate color copy with reasonable fidelity. A system in this category is most assuredly not compatible with a budget line conceived originally in the context of acquiring nothing more exotic than a few "journal reprints" in the traditional sense. Budget considerations can in fact loom large in meeting the legitimate informational needs of today's serious scientists and their students.

One of the biggest problems still to be resolved in the area of modern communication is not technological at all: it has instead to do with the law. The keyword here—which at this point we will simply mention in passing—is *copyright*. This is a subject to which we return below under the heading "Authors and Authorship in the Twenty-First Century". Once again, ZIMMER (2000) provides valuable and up-to-date background and guidance in his chapter dealing with copyrights and the Internet ("Urheberre©ht im Internet").

Information Acquisition Today: Search Capabilities

Scientists find especially fascinating the ease with which electronic data can be *searched*: in the records of electronic publications, for example. With electronic information it is possible to pose—and quickly find answers to—complex questions of almost any type. It is a relatively trivial matter to look for something to which this *and* that *and* the other as well all pertain, for instance, or for a place in a spectrum or graph of the *x,y* variety where a particular number is displayed. Even *tolerances* can be specified: e.g., requiring that, to be of interest, a number must fall between a particular pair of limiting values, or an article must have been published more recently than 30 June 1998, say. The most elaborate systems supply not only a list of "hits", but *ratings* of the various results, or estimates of plausibility. The user might even be alerted to possible typing mistakes in a question posed, or the system itself could be eager to suggest possible improvements in a set of search criteria, thereby opening the way to a somewhat broader (or narrower!) range of results than that produced by the originally conceived boundaries. In other words, a cutting-edge search system performs almost like a faithful trained assistant!

---

[31] Desktop systems with at least 256 MB of RAM, for example, are rapidly becoming commonplace—astonishing when one considers that just a few short years ago the first edition of this book was written, edited, and fully typeset in a rather professional way using nothing more than a proud little 512 KB (!) MACINTOSH computer (what was known in those days as the "fat Mac" because of its "prodigious" memory).

● A search addressed to an electronic journal can take advantage of every piece of information in the database.

In a traditional "search" based on a print medium, success is highly dependent upon selecting the proper set of keywords—ones that are recognized by the publication's *index*—and also upon the thoroughness with which the index was prepared. With an electronic journal what is being examined instead is the full text of every contribution. Nonetheless,

● The ability of a literature database to respond to queries is a function of how carefully the stored documents have been processed prior to incorporation.

State-of-the-art searching is greatly facilitated if documents destined for electronic archival have been subjected to a rigorous *markup* process. Among other things, this introduces (invisible) indications of whether a given passage is to be regarded as body text, for example—or as a header, a footnote, a literature citation, a figure caption, a table heading, a cell within a table, or virtually any other clearly distinguishable element. The most comprehensive and widespread system for marking documents consisting primarily of text is known as the *Standard Generalized Markup Language* (SGML; see also Sec. 5.2.2).

"Markup" of an Electronic Document

"Markup" as we will be using the term here is valuable for a number of purposes, most especially for conferring upon a document, in a flexible way, design characteristics that can be interpreted in terms of a variety of different media (analogous to formatting a document manually with a word processor or page-layout program). We are interested for the moment in the nature and significance of markup in a conceptual sense. Markup begins with classifying the overall text object at hand in terms of its *document type*, which in turn implies a particular set of *structural characteristics* to be anticipated. A convenient acronym has once again been invented to describe the process: DTD for *Document Type Definition*. "Languages" other than SGML also make use of DTDs. A familiar example—at least in terms of its acronym—is HTML, for *Hypertext Markup Language*, frequently encountered in the context of document exchange via the World Wide Web. To be more precise, HTML should probably be regarded as a limited subset of SGML, or as a small SGML application. The HTML system has undergone extensive revision and enhancement (e.g., "version 3.2") over the course of time, and in 1997 a new, more convenient DTD framework for Web pages was introduced: "XML" (*eXtensible Markup Language*).[31a]

With respect to the marking ("coding") of text in general, much interest is currently focused on what is called the *TEI* (*Text Encoding Initiative*), a development that also has its roots in SGML. This particular system recognizes roughly 150 kinds of

---

[31a] A good introduction to the subject of DTDs is MALER and ANDALOUSSI (1996); numerous examples are provided by JELLIFFE (1998).

"elements" potentially present within a document, each of them associated with its own unique *tag*. The presence of such tags in an electronic manuscript can be a tremendous aid to software charged with arranging for the display (or reproduction) of a document, and in many cases it is also able to increase the efficiency of a search routine. Unfortunately, no computer exists with the ability automatically to *introduce* the appropriate tags (<p> for "paragraph" and <q> for "quotation" are typical examples). This still requires much human effort, which unfortunately means the process is expensive. But expenses are unavoidable in any case: costs incurred in the preparation of a database, or in searching a database that has been less than optimally constructed, or in coming to terms with the fact that one must regrettably forego altogether some of the benefits of new technology for lack of database processing. This is just one more example of the age-old struggle over equitable distribution of scarce resources in the face of potential advantages clearly worth securing—as in parceling out the meager flow of water carried by the River Jordan.

A generalized resource like TEI, no matter how sophisticated, will never be capable of meeting the specific needs of everyone: chemists, for example. Intense efforts are therefore underway to perfect a markup language specifically tailored to this smaller market: *CML, Chemical Markup Language*. Among the goals in this case are making it possible to establish associations in a situation where there are multiple references to a given research group, for example, or to link words or numbers in a text block with information present in specific graphic entities (e.g., spectra, chromatograms) or formulas.

As indicated above, text that has been extensively "preconditioned" can be an especially fertile field for complex data searches. Most publishers have been applying SGML coding to documents for several years, not only in order to take advantage of currently available applications, but also in anticipation of (envisioned) future applications. Many important reference works have been subjected to retroactive assignment of SGML codings despite the enormous effort involved. Readers of such documents are usually as oblivious to the efforts expended as are the original authors themselves. A remarkable range of products bearing SGML tags already exists in the marketplace, the most important off-line example perhaps being documents distributed on CD-ROM. This medium is very attractive, incidentally, since a single unpretentious disk offers the potential for storing as much as 650 MB of information.[32]

Interesting and fruitful as this brief excursion into contemporary aspects of primary and secondary publication might be, we must now return to the more conventional and traditional—or is that also no longer what it once was?

---

[32] As one striking illustration of how immense the storage potential with this medium really is, our publisher, Wiley–VCH (Weinheim), several years ago (at the 1997 Frankfurt International Book Fair) introduced onto the market a CD-ROM version of the complete *Ullmann's Encyclopedia of Industrial Chemistry*—a work previously distributed as 34 (thick!) printed volumes, but now condensed onto one thin, flat, data-storage device only ca. 12 cm in diameter. In the meantime, this electronic version is even accessible via the Internet, so that an authorized "visitor" can actually page through the entire encyclopedia online—with nothing whatsoever cluttering up the desk!

Authors and Authorship in the Twenty-First Century

Scientific journals perform a dual function: they *take* and they also *give*—collection and distribution. They *take* from a scientist that which he or she wishes to make public, and then *give* (better: *offer*, in exchange for a fee) the corresponding information to other scientists who hope to make use of it. Every researcher at various times will be involved in both aspects of the transaction. For over 200 years, journal publishing houses have played an indispensable role in ensuring that the transactions take place smoothly and efficiently.

● The scientist who authorizes publication of a manuscript automatically acquires the status of *author*.

The author (*originator*) of a contribution is entitled, in most cases by law, to a certain set of rights. These are the *copyrights*, and law both national and international recognizes claims to these rights asserted by every creative originator or "author",[33] irrespective of whether the created "work" in question is a scientific article or a novel—or even a painting, sculpture, or other *objet d'art*. A creator retains control over all uses of his or her creations, as specified in detail under the terms of *copyright law*. In most countries only *natural persons* are entitled to be designated as creators of literary or artistic works, thereby qualifying to exercise the claims of an original copyright-holder. Through sale, licensing, or contractual agreement, however, it is possible for various "rights of usage" to be transferred in whole or in part to a *publisher*—or to a radio or television firm, a database, or almost any other "legal entity".

● A copyright guarantees rights of ownership (of intellectual matter) in both an economic and a moral sense.

For example, an author is legally entitled to receive a reasonable share of any income produced by the sale of his or her work. This is the basis for an *author's royalty* (or *honorarium*), although in the case of scientific publication such payments would usually be associated only with a book, not a journal article. Beyond this, personal rights provisions stipulate the permanent, integral relationship between an author and his or her work, thereby offering protection against plagiarism or unauthorized alteration.

In the case of a literary work the right of usage is primarily an issue with respect to the granting of permission to manufacture *copies* of the creation—i.e., to reproduce it in *print*—and then market those copies. (This incidentally explains the origin of the English word *copy-right*, although today the term obviously has a meaning much broader than its derivation implies.) Rights other than those of *reproduction* may be granted as well, including rights of distribution and display. In the case of a book (see Chap. 4) a unique and formal contract is commonly drawn up and concluded between every author (or translator, editor, etc.) and some publishing house. Among other things, such a contract defines the extent of any agreed-upon copyright transfer, and it often

---

[33] Note the significant link between the words "author" and "authorization"!

contains clauses related to the possibility of licensing authorized foreign-language editions, for example—as well as the nature and extent of a negotiated honorarium and terms according to which this will be disbursed.

Unfortunately, there are still many open questions with respect to protection of an author's rights and the limits of usage rights in the case of *electronically published* works, and these form the backdrop for considerable heated debate. The very fact that an electronic document lacks physical substance is enough to initiate the confusion. What does it mean to "copy" something in the virtual world of "cyberspace"? An urgent need has become apparent for reconsideration and revision of copyright laws at all levels to reflect a changed world. One example of the problem is provided by the music industry, which is presently in a state of trauma because of the vagueness of current law. Adaptation of existing laws and promulgation of new ones will almost certainly continue to lag far behind events. "Globalization", nowhere more evident than in the realm of telecommunication, has created pressures that underscore the vast international aspects of this problem.

Most parties involved have so far been willing to abide by the wisdom of allowing "common sense" and "fair play" to govern events for the time being. Numerous legal disputes are wending their way through the courts, however, in some cases at the very highest levels, but fundamental legislative clarification has still not been forthcoming. Deep concerns related to "technological progress" surfaced four decades ago as xerographic copiers began to proliferate in libraries throughout the world, transforming the latter in effect into self-service duplication stations at the disposal of everyone. What is (or should be) the extent of copyright protection with respect to preparing photocopies? Adding the further complication that fax devices now permit photocopies of almost anything to be *transmitted* at any time, anywhere, has led to mobilization of a host of trade organizations representing the publishing industry. In Germany, for example, an important legal battle was enjoined when publishers and book dealers challenged practices of the Technical Information Library (TIB, Technische Informationsbibliothek) in Hannover. In this particular case the library won, but the court nevertheless advised the industry that it should attempt to arrange as quickly as possible for some system of compensation from "information disseminators" as a way of legitimizing the practice of photocopying. All the problems connected with these issues have by no means been resolved, just as the Pentagon apparently has not succeeded in completely solving potential espionage problems posed by "high-tech hackers" who manage to tap into sensitive computer files.

It has become standard practice to share documents of all types electronically. The World Intellectual Property Organization (WIPO) in Geneva, Switzerland, has been charged with transforming the copyright system originally devised by the Berne Convention of 1886 into a vehicle meeting the needs of the present and at least the near future. WIPO functions as a subsidiary of the United Nations and currently operates in the name of 170 countries. Many hopes are pinned on its success in soon proposing and effectively promoting a viable set of comprehensive standards.

● As in the case of books, professional *periodicals* also depend upon contract-like relationships between publishers and authors, which take effect (usually with little fanfare) when a submitted manuscript is accepted for publication.

In this case the "contract" usually takes the form of a "boiler-plate" letter drafted by the publisher, similar to that shown in Fig. 3–1. (The words chosen to express certain key ideas will probably strike the reader as amusingly archaic).

Honoraria are rarely provided for such "minor works" as journal articles describing original scientific research, primarily because of extra financial strain such payments would impose on the existing information-transfer mechanism.[34] (Indeed, many journals—especially in the United States—instead confront authors with a *fee*, known as a *page charge*, for agreeing to publish their manuscripts!) Nevertheless, the author of a journal article is entitled to fundamental rights very much like those granted to the author of a book. Taken together, copyrights and various related rights applicable to publishers (to which must also be added journalistic rights), create a mixture of the material and the immaterial that merges into an extraordinarily complex mélange. For our purposes the brief summary above will necessarily suffice.

What are the *motives* that induce a scientist to publish? Apart from the inherent challenge in virtually every walk of life for one to become better known, scientists—along with other professional scholars—are subject to an even more compelling motivation:

● Publications are usually taken to be the surest sign of *productivity*.

The well-known short form of this dictum is "Publish or perish!" The merits of the "rule" may be questionable, but the fact remains that in most cases it accurately characterizes the real world of scholarship, both now and in the past. Suspicion regarding the underlying wisdom of the situation centers around a temptation to equate *quantity* with *quality*. It is nevertheless undeniable that a scientist's *list of publications* plays a crucial role in success with respect to both appointments and promotions. How this may ultimately change with the advent of electronic publishing is difficult to predict, but the short history to date of e-journals is consistent with a tendency to cling to tradition and the basic urge to seek proper recognition: the new medium experienced a breakthrough only after publications already well established began to become available in *both* printed and electronic form.

---

[34] In a test case pursued successfully by the American publishing house Academic Press against the Texaco oil company, which dealt with unauthorized reproduction of journal articles, an author's non-existent honorarium actually played a significant role. Citing from the court's decision: "… it is misleading to characterize the authors as unpaid. Although it is true that … publishers do not pay authors money to publish their articles, the authors derive benefit from the publication of their work far more important than any small royalty the traffic might bear". In the final analysis, publication is an important factor in enhancing an author's professional reputation, resulting in a net "growth in prestige".

·   ·   **Bestätigung des**
                                                    Manuskripteingangs

                                                    Confirmation of
                                                    Receipt of Manuscript

                                         ·   Datum:
·   Date: _____

Betr.: Ihr Manuskript   Re: Your manuscript _____

_____   _____

_____   _____

für die Zeitschrift _____   for publication in _____

Sehr geehrter Autor,   Dear Author,

wir bestätigen mit Dank den Eingang Ihres Manu-   we herewith confirm receipt of your manuscript, which,
skriptes. Sollten Sie von uns keine weitere Nachricht   if you do not hear to the contrary, may be considered
erhalten, geht das Manuskript nach redaktioneller Über-   as accepted for publication and will be edited and set
arbeitung zum Satz.   in type.

Für die Veröffentlichung gelten die folgenden Bedin-   The conditions of publication are as follows:
gungen:

1. Die Autoren besitzen das Urheberrecht und sind zur   1. The authors possess the copyright and are entitled
   Publikation des Manuskriptes berechtigt.   to submit the manuscript for publication.

2. Das Manuskript ist in der vorliegenden Form noch   2. The authors declare that the manuscript in its pre-
   nicht veröffentlicht worden, weder vollständig noch   sent form has not been published elsewhere, either
   auszugsweise. Es liegt in dieser Form auch keiner   as a whole or in part, that it has not been submitted
   anderen Zeitschrift zur Veröffentlichung vor und wird   in this form to another journal and that it will not be
   bei Annahme zur Veröffentlichung keiner anderen   submitted to another journal, if it is accepted for
   Zeitschrift angeboten werden.   publication.

3. Mit der Annahme des Manuskriptes zur Veröffentli-   3. With the acceptance for publication, VCH Verlags-
   chung erwirbt die VCH Verlagsgesellschaft das un-   gesellschaft acquire all publishing rights including
   eingeschränkte Verlagsrecht einschließlich des   those of reprinting, translating or reproducing and
   Rechtes, Nachdrucke und Übersetzungen anzuferti-   distributing the manuscript in other forms (e. g. as
   gen oder das Manuskript in anderer Form (zum Bei-   photocopy, microform or in machine-readable form).
   spiel als Photokopie, in Mikroform oder in einer für   Moreover, the provisions of the copyright law of the
   Datenverarbeitungsanlagen geeigneten Form) wie-   Federal Republic of Germany apply.
   derzugeben, zu vervielfältigen und zu verbreiten. Im
   übrigen gelten die Bestimmungen des Urheber-   4. The manuscript, originals of illustrations, and proofs
   rechtsgesetzes der Bundesrepublik Deutschland.   will be destroyed two months after publication of the
                                         manuscript, unless it is requested that this material
4. Manuskript, Bildvorlagen und Druckfahnen werden   be returned to the authors.
   zwei Monate nach der Veröffentlichung des Manu-
   skriptes vernichtet. Die Autoren haben die Möglich-
   keit, vorher die Herausgabe dieser Unterlagen zu
   verlangen.

                                         _____
                                                (Unterschrift/Signature)

Fig. 3–1. Sample of a typical letter in which a publisher acknowledges receipt of a manuscript submitted
for possible publication as an article in a scientific journal.

Publishing houses—more precisely, the *editors* of journals and *acquisition editors* responsible for book manuscripts—have inherited the responsibility of managing the day-to-day flow of scientific information. Editors are best known in their role of overseeing *quality* in publications by the fact that they all too often have the unenviable obligation of dispatching letters of rejection as the culmination of a reviewing process. We will return to this unpleasant subject later, and examine for now another often overlooked aspect of publication:

● Publication establishes an authors claim—which could have important legal implications—to *intellectual invention* and *priority*.

Quite apart from professional esteem or direct financial remuneration, most scientists are also anxious to receive personal credit for personal achievements. Almost every scientific accomplishment involves a certain amount of competition with other researchers working elsewhere, and it can in fact matter a great deal who is the one to first apprise the public of new findings: that person receives credit for the work, and in some cases a Nobel Prize could be at stake! In this sense science is just like football (American style, for the moment!). Everyone is interested in knowing who actually *caught* the game-winning touchdown pass; this is the athlete in demand for on-camera interviews after the game, not the one responsible for a key block without which the catch would have been impossible. "Winning" scientists also expect to be recognized for special achievements, even if that recognition rarely extends to the public at large.

● A scientific journal always documents when a given manuscript arrived for publication by publicly declaring a *date of receipt*.

The date is printed along with the contribution itself, and becomes the decisive factor if a question of priority ever arises. We await with interest the results of the first legal battle over this issue involving an *electronic* publication, or a hasty *prepublication* breaking the news of a stunning set of results.

Who will be acknowledged as a scientific author in the world of tomorrow? How will the definition of authorship change in the years ahead? What rights should authorship confer in an electronic age, and how will one be able to claim them? Only time will tell. One thing seems clear, however: "personal publishing" is not likely to influence the answer. The fact that it is strictly antithetical to the legitimacy (or value) of "authority" makes it incompatible with the informational needs of an enterprise—science—rooted in team spirit and the eternal quest for truth.

3.1.4 The Various Types of Journals

Most of the remarks that follow are just as applicable to electronic journals as to traditional publications distributed in hardcopy format. An *issue* of a scientific journal consists of a collection of individual *articles* (*contributions*) submitted rather randomly by many authors acting independently.

● Before you consider submitting an article for publication it is important that you be acquainted with the distinctive characteristics of various types of journals.

Every scientific, technical, and medical subdiscipline has given birth to a wide variety of journals, some of them designed to address very specific and unique communication demands. It is not unusual for several journals to be in direct competition. Most such publications fit comfortably into a relatively small set of *categories*, three of which we have selected for closer examination here.

● *Primary journals* have as their principal objective the dissemination of new information in the form of "original contributions" (*first reports, announcements*).

Journals in this category convey to the scientific public, for the first time, results derived from recent research. These are "journals" in the fullest sense of the word. A few statistics should provide a sense of the extent of the "journal world". The total number of professional journals ("titles", to use the language of the trade) published globally at the present time is thought to be in the range 200 000 to 300 000.[35] Considering a single discipline, about 15 000 journals are currently devoted to the science of *chemistry* and closely related fields, or at least feature chemical contributions on a reasonably regular basis, and the list of journals representing the *biological* and *medical sciences* is even longer. Most of these journals limit their coverage exclusively to original research, placing them unambiguously in the primary journals category. It is these to which most of the present chapter is devoted.

This astonishing number of primary journals—representing in turn a far larger number of published *articles* containing discipline-specific information (roughly half a million annually in the case of chemistry)—has long since made it impossible for the individual researcher to even consider dealing directly with the flood of information engulfing every field. The scientist is therefore dependent upon various tools—some of which are themselves journals of specialized types—designed to channel the incoming tide.

● *Abstracting journals* present concise *summaries* of research results reported in detail in the primary literature.

Since these serials attempt to encapsulate all the key findings described elsewhere, and because they are also a source of all the information required for accessing the original papers, they serve the researcher as an invaluable *orientational* aid. Everything they contain is in a sense "second-hand", however, so abstracting journals are assigned to the category known as *secondary literature* (note that primary journals account for most of the *primary literature*).

Abstracting journals achieve an information "concentration factor" on the order of 100 based on the space consumed by a typical article relative to the corresponding

---

[35] These in turn supply the public annually with something like 4 million individual articles. We refrain from presenting more precise data since detailed information would in any case be outdated by the time this book is read.

abstract. Obviously this expedient in itself would still be inadequate to permit the individual reader to gain a truly comprehensive overview of an entire discipline. As little as half a century ago it was still (barely) practical for a conscientious, determined scientist to be committed to literally *reading* one of the comprehensive abstracting journals in order to stay current, but those days are now only distant memories. The aim of abstracting journals today is rather to collect all the basic information entrusted to the primary literature, condense it, subject it to thorough indexing, and ultimately open it up to sophisticated *computer-based searches*. In other words, abstracting journals exist to facilitate *literature retrieval*.

The first abstracting journal dedicated to chemistry was *Chemisches Zentralblatt* (published starting in 1830 in Leipzig as *Pharmazeutisches Centralblatt*). In fact, this was the first abstracting journal devoted to *any* scientific discipline. Publication of *Chemisches Zentralblatt* ceased in 1969 as a consequence mainly of changes in the world political scene, and today the responsibility of maintaining an overview of activities in chemistry is almost entirely in the hands of *Chemical Abstracts*, a product of Chemical Abstracts Service (CAS), which is in turn an arm of the American Chemical Society. There exist several other abstracting services that reflect narrower interests, both national and disciplinary, but none begins to approach the scope and breadth of *Chemical Abstracts*.

Other disciplines have their own abstracting media, including *Physics Briefs*, *Physics Abstracts*, *Biological Abstracts*, *Index Medicus*, and *Engineering Index* to name but a few. Abstracting services and their associated *databases* are carefully tailored to facilitate targeted quests for information (*literature research*), typically with the support of some form of on-demand online access.

In addition to abstracting journals, researchers today also have at their disposal one other important orientational tool:

● *Review articles*, which present information drawn from a large number of (selected) original papers, organized and analyzed for the purpose of *educating* the reader regarding the current status of some narrow field of specialization.

Whereas in a primary journal the emphasis is on communicating results newly acquired through the efforts of individual researchers (or research teams), the goal here is to *assemble* and *critically evaluate* information already reported in the primary literature, thereby putting that which is known into perspective. The author of a review article provides a major service by agreeing to carry out a more or less comprehensive literature search from some particular self-selected vantage point. A good review accomplishes much more than merely collecting facts otherwise widely dispersed across a host of primary sources. The principal objective is to establish some degree of *order* among the facts. To the extent that this entails passing judgements, the reviewer actually becomes something of a "trend setter" with respect to future research. In an age suffering under increasing fragmentation of knowledge, reviews of this sort assume a more crucial role than ever, despite the fact that the "review literature" has been dealt an especially heavy blow in recent years by severe budgetary constraints everywhere.

Review serials represent a second cornerstone in the secondary literature. Such publications not infrequently carry titles that begin with

"Advances in ...", "Progress in ...", or "Fortschritte der ...."

While some (e.g., *Chemical Reviews*) resemble more conventional journals in outward appearance, many of the review publications take the form of sets ("series") of discrete *bound* volumes, and for this reason "serial publications" is a more satisfactory "umbrella designation" than "periodicals" or "journals" as those words are ordinarily employed.

● Review articles might be viewed as a transitional stage between articles in the primary literature on one hand and books on the other.

Whereas original papers in primary journals are *submitted* at the authors' own initiative, review articles are generally *solicited* by the publication in which they appear—that is to say, by the editors of the latter. An invitation to write a review article should be cherished as compelling evidence that one's expertise has been recognized by others. Young researchers especially should resist any temptation to turn down the opportunity of preparing a review, even if carving out the time to do so threatens to pose problems.

We postpone an introduction to other types of journals until Sec. 3.2.3.

Journals differ from one another not only in their membership in various categories, but also in such characteristics as style, frequency of appearance, and restrictions on the permissible length of published contributions. One further distinguishing feature is of the greatest importance, even though it is far from obvious how it should be described or even ascertained: the journal's *reputation* within the professional world. We will have more to say about this in Sec. 3.2.4.

3.2 Decisions Prior to Publication

3.2.1 Publish When?

Let us assume you have made the decision to publish your latest research results. The first thing you need to think about now is proper *timing*. There is no simple formula you can apply, so the best we can do is highlight a few relevant factors and perhaps raise your awareness level with respect to certain issues.

More important than anything else at this juncture is to guard against publishing data or information that will later prove to be wrong. A mistake of this type can cause just as much harm to one's reputation as drawing a patently wrong conclusion. Not that scientists are perfect; even famous scholars have been known to make serious mistakes without being destroyed in the process. Nonetheless, you should try very hard not to commit a blunder that suggests a lack of professionalism on your part.

● Overly *hasty* publication increases the likelihood of *faulty* publication.

A single published result that proves upon revisitation to be inaccurate or erroneous is all it takes to sap most of the joy out of publication—especially if the flaw ends up being discovered by someone else. Assigning a wrong structure to a new compound, when the mistake could have been avoided simply by careful examination of a ^{13}C-NMR spectrum, is the type of oversight at least some people are sure to hold against you.

● Newly collected data should always be scrupulously verified, and all chains of evidence fully secured, before anything is presented to the public.

The tried and true path to effective quality control in research is taking the time to test all your data carefully for *reproducibility*. In particular, an unexpected result that runs counter to established theory should always be viewed as suspect. Check it, and if necessary check it yet again, but if it persists in passing every test, then by all means publish. After all, theory is supposed to give way to facts, not the other way around.

Never let "time pressure" serve as an excuse for failing to carry out some experiment that a true professional in your discipline would regard as mandatory—testing for a placebo effect in a pharmacological investigation, for example.

● The earliest date for giving serious consideration to publishing is *after* conscientious execution of all the experiments a state-of-the-art study in your field would presuppose.

Nevertheless, even this may not represent the "right" time to publish. While the successful synthesis of a single representative of a new class of compounds might in some unusual situation justify going to press, claiming the discovery of a new synthetic *method* on the basis of one example would surely be inappropriate. Even if absolutely nothing is "wrong" with the heralded synthesis, your peers would quite legitimately expect you to demonstrate its effectiveness with *several* examples. A principle claimed on the basis of too little evidence is not likely to be taken seriously—nor should it be!

● The *optimal* moment for publication normally comes with the availability of a *set* of verified results, reflecting a reasonable amount of breadth.

The proper interpretation of "reasonable" here will depend to some extent on the *form* you expect your publication to take. Far less substance is required for a short communication serving only to announce experimental findings (see Sec. 3.2.3) than for a full paper.

There is one more consideration that provides a reason, at least occasionally, for not being overly hasty to publish. A person is not allowed to claim *patent rights* over something he or she has already published! Once information has been formally released—even if you were the one who released it—that material is considered to fall within the public domain and therefore represents "state of the art" knowledge, which means it is no longer patentable.

On the other hand, it is also a mistake in most cases to delay publication beyond what reasonable caution would warrant, because there is always the possibility that someone else will come forward in the meantime and present the same or closely related findings. Keep in mind that scientific meetings are good places to sense what topics in your specialty are currently "hot"—areas in which scientists other than yourself are almost certainly engaged in feverish experimentation, thereby becoming plausible candidates for "beating you into print" if you dawdle.

Even if the timing seems in every sense to be "right", sometimes circumstances can conspire to make it impossible to locate a journal willing to accept your cherished piece of work. This is especially likely to be a problem, for example, if you have already somehow released the information, perhaps in the form of a widely circulated preprint (irrespective of whose name was attached!). But then again, certain *physics* journals have taken it upon themselves to serve deliberately as vehicles for publishing information already communicated in preprint form (under their auspices, of course). Other cultures, other traditions! So many things are currently in a state of flux that it would be folly for us to be too dogmatic here about almost anything.

3.2.2 Publish What, and With Whom?

Just as filing a successful patent application requires that one meet strict criteria with respect to "novelty" and "ownership", a manuscript describing research must also display certain characteristics if it is to stand a chance of publication in a reputable scientific journal.

- A paper might report information of many different types—theoretical, experimental, descriptive, or simply illustrative of a new application of an established principle, for example—but whatever the nature of the report, it is expected to be based on material that is demonstrably *reliable*, *meaningful*, and *new*.

We have already addressed the issue of *reliability*: in part in the previous section in conjunction with our "When?" question. Put briefly, results should never be submitted for publication until they have been firmly established.

The criterion of *meaningfulness* (*significance*) is not such that it lends itself to measurement on something as simple as the time scale implied by "when". An investigation that from the outset is clearly a matter of routine (e.g., application of a standard synthetic method to a new substrate of a familiar type) will *never* lead to a "meaningful" publication no matter how long one waits. Lack of significance is widely understood to be a perfectly legitimate reason for an editor's refusing to publish a paper.

Perhaps most critical of all is the criterion of *novelty*. If the publishing industry is to avoid collapse, it is essential that repetition and double-publication be eliminated as completely as possible. Results already announced should never be submitted a second time for publication. This basic principle is broadly acknowledged as binding, but there can unfortunately be borderline situations. For example, the scientific community will tolerate it if a *few* of the results reported in a paper have already been described, provided *other* information present really is new and contributes in a significant way to a more satisfactory understanding.

Another legitimate exception is information released first in the form of a *short communication* (see Sec. 3.2.3) which is then presented again later, more formally, in a *full paper*—with amplification, extensive discussion, and especially enrichment from the standpoint of experimental details. Within bounds, this is perfectly acceptable. Nonetheless, you should expect to be severely criticized if you try repeatedly to bring forth essentially the same information, merely clothing it each time in different "wrappings", or perhaps changing the language of publication. Even worse is trying to take advantage of good will on the part of the editors of two or more journals. It is considered almost equally bad form, by the way, to take a set of closely related findings and deliberately chop it into small pieces for independent publication (sometimes described as exploitation of the "salami technique").

In recent years it has become rather popular in some circles for data or experimental results to be passed around in various *informal* scientific information networks prior to publication in a professional journal. Entire articles sometimes become available on computer screens everywhere under the guise of *previews* deliberately circulated before release of the hardcopy issues of a journal. Note, however, the ambiguity this could create with respect to establishing "priority", as discussed in the preceding section.[36]

All this of course presupposes one has established categorically that a particular result does *not* already appear somewhere within the international literature. Given the enormous volume of publication today, neither authors nor editors nor reviewers are capable of *absolutely* guaranteeing novelty. Duplication is especially difficult to detect and prevent in cases of nearly simultaneous publication, because abstracting services and other sources of documentation are unable to reflect the potential overlap before it is too late.

Now to the second question to be addressed in this subsection, which brings us to the subject of *coauthorship* ("With whom?"). It would of course be unrealistic to attribute the bulk of the original work reported in the scientific literature today to *individual* scientists working alone. In general, published information will have come from at least two scientists working together—possibly a professor and a student, or a senior researcher and an assistant—who (ideally) will also have collaborated in the

---

[36] Priority should not become problematic if the early release involves the electronic version of an article in a recognized journal, however, since a given article would bear a single submission date irrespective of the version.

drudgery of writing and submission. Every published study should bear the names of all parties who were heavily involved in its execution.

●   Anyone with a significant share in the responsibility for a reported set of findings, or even a major portion thereof, has earned acknowledgment in the form of "authorship" (cf. the discussion of authorship in general in Sec. 3.1.3).

On the other hand, a student, a technician, or a junior coworker who has done little more than record a few measurements (faithfully carrying out a set of instructions for which someone else was responsible)—a "participant" in this sense would *not* ordinarily be granted the status of coauthor. Such assistance does deserve to be *recognized*, but that would typically be accomplished in a note at the end of the corresponding article. It is an entirely different matter if, for example, some new method has been applied more or less independently by a subordinate, even if it is someone who has recently undergone a period of supervised learning. In this case the person *would* be entitled to a share in authorship of papers resulting from the work.

An article based on a dissertation is traditionally published under the authorship combination advisor–doctoral candidate. If a single publication represents work from several dissertations it follows that all the degree candidates involved should be treated as coauthors.

Consider another case: a publication in which reported results depend in a critical way on a statistical evaluation, and an outsider with a strong background in statistics has been called upon to assume an important share of the creative labor. Here a gratuitous "thank you" at the end of the article would *not* represent fair treatment; the statistician would have earned full-scale recognition in the form of coauthorship.

In practice, it is commonplace today for cooperation to reach even into other research groups, and dealing equitably with the implications becomes an important management issue in any complex scientific undertaking. Justice demands that such efforts from a second research group be explicitly credited by including in an author list at least one of the outside participants when results are finally made public. Those who shoulder an important part of the *responsibility* for a project deserve coauthorship in any publication spawned by the work.

●   The extent of the role played in seeing that a given piece of work actually comes to fruition should be the number one criterion in shaping the list of names to be permanently associated with the discoveries.

Given the extensive teamwork characterizing many research groups today, a balanced decision with respect to authorship can be difficult to reach. Further complicating the issue is the fact that the chairman of a department, or a chief of surgery, is likely to see matters rather differently from a junior staff person.

One special word of caution needs to be underscored: it is *never* appropriate to list someone as an author who has not explicitly agreed to such recognition. This also implies, incidentally, that every listed author should be given an opportunity at least to read carefully through and criticize the final version of a manuscript before it is submitted for publication.

There are no general ground rules regarding how the various tasks in manuscript preparation should be divided. Typically one individual assumes responsibility for drafting the entire paper. This might be the person who played an integrating role within a heterogeneous group of researchers, for example, or it might be the initiator of the project—the one responsible for the fundamental idea. However it is written, the preliminary draft should be circulated among all members of the project team in an attempt to elicit creative suggestions. This could well result in requests for major change: in the interest of making certain arguments more transparent, for example, or to achieve a more balanced presentation. Serious substantive input should be sought from all possible directions, but at the same time there is little to be gained from encouraging multiple people to recraft every sentence solely for purposes of cosmetic polishing.

A quite different approach to preparing an article would be for each member of an "authorship team" to volunteer from the beginning to compose one particular section. This would of course necessitate subsequently convening one or more joint sessions so that the various parts could be woven together into a single reasonably homogeneous document. The optimum writing strategy will be a function of the specific situation, one that takes fully into account the strengths and weaknesses characterizing the several parties involved.

As previously noted, there is at least one discipline in which advances have become extraordinarily dependent upon extensive teamwork: *high-energy physics*, with its singular reliance on exceptionally complex and costly tools (e.g., accelerators), which may exist in only one laboratory in the world. We have encountered an article in *Physics Letters*, for example, that acknowledges under an umbrella designation (e.g., "L3 Collaboration") the contributions of over 400 individual scientists and technicians from a dozen different research institutes! Under these circumstances our simplistic suggestions regarding the logistics of writing are obviously of little value.

By the way: this example also throws a different light on the notion of *peer review* (see Sec. 3.5.3). The members of such a large group of scientists could reasonably assert that they themselves had already brought to bear any necessary quality control; it is in fact quite likely that no outsider anywhere is in a position to improve upon a report these collaborators have agreed upon—for release, perhaps, in the form of a "preprint".[37] All the "experts" one might conceivably recruit for reviewing purposes are almost certainly part of the "collaboration"!

---

[37] In Los Alamos, the "atomic city" in the New Mexico desert famous for hosting the Los Alamos National Laboratory, there has existed since 1991 a computer—apparently subject to no particular limits with respect to capacity—which has served as an amazingly rich source of unreviewed preliminary versions of pending publications drafted by physicists and mathematicians all over the world (in the form of preprints posted on the Internet; address: http://eprints.lanl.gov, where "lanl" stands for Los Alamos National Laboratory). This electronic archive (known as the "e-Print arXiv") was conceived and later faithfully nurtured by the theoretical physicist Paul H. GINSPARG. It has been characterized as the "most rapidly growing journal in the world today", responsible at the turn of the millennium for

→

Once the issue of *membership* in a list of authors has been resolved, the next question that needs to be dealt with is the *order* in which the parties will be listed, especially since publications tend to be associated in a reader's mind only with the first-listed author—and this is no trivial consideration! It could once be taken for granted that the leader of a research group would automatically head an author list, but this is no longer self-evident. Not infrequently, sticky decisions dictated by any attempt at "placement by importance" are avoided simply by listing the names alphabetically. Usually, one name will then be singled out (and marked with an asterisk) as the so-called *corresponding author*, the person who has agreed to speak on behalf of the group if questions should arise.

In reaching a final decision regarding a sequence of names be sure to bear in mind that researchers typically take the matter of order among a set of coauthors quite seriously.

● The best practice is to seek consensus on the order of prospective authors' names at an early stage in the research itself rather than waiting until the time arrives to publish.

Following this advice can minimize last-minute friction and ill-will.

An important corollary to the considerations above also needs to be made explicit:

● Anyone listed as a coauthor automatically assumes at least partial responsibility for the work reported.

This notion of responsibility has ramifications that extend well beyond considerations of collegiality or undergirding a strategic alliance. If it should later turn out that observations or data reported in a paper with your name attached are flawed, *your* head will be one of those on the chopping block. It is also true, however, that the scientific community (and even the public at large, in the event that a widely reported controversy should develop) will presume that the lion's share of the responsibility for what has been reported falls to the leader of the research team that is the source of the report (see also Sec. 1.3.1). It could thus be a very costly mistake if a group leader were in a perfunctory way to allow his or her name to appear on a publication prepared entirely by a subordinate!

---

the release of ca. 3000 articles per month, a rate more than twice that registered five years earlier (KORWITZ 1995; see also STIX 1995). The resulting database, an outstanding collection supported by the American Physical Society, is freely accessible under the address http://arxiv.org, and it is said currently to log on the order of 120 000 visitors per day (during the work week). For comparison, the largest "bricks and mortar" library in the world, the Library of Congress in Washington (DC), experiences an internet demand little more than six times that great.

3.2.3 Publish In What Form?

The brief discussion of various types of journals presented earlier (in Sec. 3.1.1) requires a bit of further elaboration before we can take a closer look at the various *types* of contribution one might consider submitting. Primary journals typically accept one or both of two types of *original papers*: normal ("full") papers, and *short communications* (*notes*, *letters*). Publications in both categories play an important role in documenting scholarly activity, especially in the academic world.

Short communications can also be divided into two basic categories, with somewhat distinct goals. Those of one type, most often referred to as *notes*, differ from full papers mainly in extent. They are intended to facilitate routine reporting of small—but still relatively independent—units of original work. "Notes should be short, complete reports of investigations more narrow than is usual for an article" [*J. Org. Chem.* 1989, 54 (1): 10A]. Short papers of the kind described as *letters*, on the other hand, tend to be brief announcements (with few details), rapidly released, of especially important new developments—often in a high-profile subject area. One can legitimately expect to see that these accomplishments will be described in significantly greater depth at a later date. Not surprisingly, the line separating the two kinds of abbreviated communication is not always clear.

● The mode of publication one selects should reflect both the character of the reported work and the circumstances attending its planned announcement.

The nature of a proposed contribution can have significant consequences with respect to the mechanics of preparing the paper.

● Whereas communications published in traditional journals are always *typeset* (i.e., the manuscript is transformed by electronic or mechanical means into high-quality copy), journals devoted to short communications sometimes forego typesetting and instead reproduce submitted papers just as they are received (i.e., the manuscripts are subjected only to *photomechanical* processing).

The first journal to employ the latter model—and a highly successful pioneer at that!—was *Tetrahedron Letters*, a chemistry periodical introduced with considerable fanfare in 1959. The now common descriptive term "letter journal" is in fact derived from this source. With this type of publication the prospective author must understand from the start that a submitted manuscript (if accepted) will not be changed by the editor in any way, and this includes appearance. Guidelines provided by the journal must of course be followed meticulously, and final copy should be measured against the highest standards of quality, because direct reproduction would otherwise result in an unsightly publication. One not insignificant detail in this regard: a letter-journal manuscript should reflect a line spacing of one-and-a-half units, whereas other submissions (any that will be typeset) should be double-spaced (see Sec. 3.4.6). Over the years, authors gradually joined in the migration from typewriters to word-processing systems, and letter journals increasingly shed much of their distinctively amateurish, "typewritten look". Submitted "letters" are still processed for publication exclusively by reproductive means, but most of them now remarkably closely resemble "printed" (typeset) documents.

Another unusual type of publication, introduced over two decades ago, is exemplified by the *Journal of Chemical Research*. This periodical was launched as a joint venture of several European chemical societies in 1979, and in printed form it is limited almost entirely to *synopses* of research activities. Experimental details and other supplementary information entrusted to the editors is captured on microfilm ("microfiche") and distributed to interested readers in this or "miniprint" form only upon request, or as part of a special premium subscription arrangement. This concept also made publishing history, and led to the expression *synopsis journal*.[38] Again, potential authors must obviously take it into account in advance if a target journal is of this type, since this will have considerable bearing on the nature of the manuscript to be prepared.

● Letter and synopsis journals represent innovative responses on the part of publishers to critical issues confronting their industry.

Both were relatively early attempts to circulate word of scientific advances more *rapidly* and more *economically* than had traditionally been the case, utilizing the most advanced technical means available, but at the same time restricting coverage to the minimum amount of information that would still permit scientific progress to be maintained. Considering the profound technological developments we have witnessed in the last decade or so, there is every reason to anticipate an early demise for both "letter" and "synopsis" journals in their current form; indeed, the first fatalities have already been registered.

We defer largely to Chapter 4 (in Sec. 4.2.2) a discussion of editorial challenges introduced by the advent of digital manuscripts. These quickly became critical due to the enthusiasm with which authors embraced digital methodology. The reason we delay this discussion is that several of the most interesting issues were first dealt with in a constructive way in the context of *books* rather than journals—and for an important reason: close cooperation between author and publisher is essential if maximum advantage is to be taken of innovative methods, and the intense (and costly!) dialogue this entails is more easily justified in the case of the larger information package "book" relative to the many small contributions from a host of sources that make up each issue of a journal. We nevertheless note in passing that virtually *all* printing today involves digital preparation at some level, irrespective of how the final product actually

---

[38] Unfortunately, the forward-looking endeavor did not become very popular with the scientific community. As a result, the prototype project no longer has a sponsoring publisher in the traditional sense. The present distributor is Turpin Distribution Services Ltd. in Letchworth, UK; cf. www.turpin-distribution.com.

comes into being.[39] This applies not only to journals, but also to magazines, newspapers and virtually every other type of periodical.

A third type of publication that has evolved over the years deserves at least casual mention: an abbreviated version of a lecture or a poster presentation planned for (or already delivered at) a scientific meeting. The familiar word "abstract" has once again been called into service and adapted slightly to characterize this type of document. *Conference abstracts* typically appear in special publications assembled by the organization or society hosting a scientific meeting, and they are often distributed on site. Sometimes such a collection is treated as a special issue of a regular journal, but it more commonly takes the form of an independent publication, or of a volume within a series. Incidentally, a conference abstract is considered by most people to be a considerably less "serious" (and therefore influential) document than a paper published in a reputable journal. Space constraints prevent us from pursuing the subject further.

3.2.4 Publish Where?

As a prospective author it is important that you give careful consideration to *where* you will submit your work, primarily in the interest of maximizing the probability of your successfully addressing the most appropriate subset of readers within the wider scientific community. Given the wealth of potential publishing arenas, making the right choice is no easy matter.

An interesting service performed by the Institute for Scientific Information (ISI) in Philadelphia (now a division of the Thomson Corporation), and accessible through *Science Citation Index* (*SCI*), may assist you in your decision. *ISI* annually compiles statistics derived from about 2000 professional journals, including:

- a count of the number of papers each journal publishes per year;
- numerical "impact factors", revised annually, that reflect the extent to which, over the course of the most recent year, articles first appearing within the previous *two* years in a given journal were *cited* (ostensibly an indication of the importance attached to work appearing in that journal);[40]
- typical *publishing delays*; i.e., the amount of time that elapses, on average, between receipt of a manuscript by a particular journal and appearance of the corresponding article in print.

---

[39] The steps to which the observation refers would be described as "prepress operations". In contrast to the mechanical printing forms upon which "classical" print technology was based (e.g., offset plates in the flatbed process), the data carrier in a modern print shop has essentially been "disembodied". In principle—at least with a small print run—no printing form whatsoever is now required, because sophisticated arrangements make it possible to print directly from a data file, similar to the way a desktop printer can be controlled from an author's PC (cf. also Sec. 3.5.4).
[40] The "magic factor" itself is computed by dividing the total number of such citations recorded during the year in question by the total number of articles published in journal X during the two previous years [other approaches to assessing "impact" have also been suggested (GARFIELD 1972)].

From this information, anyone interested can gain insight into the physical extent and perceived "stature" of a journal, together with how quickly a journal's editors can be expected to process a manuscript. Interesting it certainly is, but the significance of such "scientific" data should not be overstated—one of the reasons we were reluctant to reproduce a sample "ratings list" in our book.

It is not our place to delve more deeply into criteria one might use for assessing the "quality" of a journal, or to comment on relative manuscript "acceptance rates" in the hands of editors of various publications. Suffice it to say that many people have expressed themselves with varying degrees of passion in both areas. Perhaps more to the point, you as a productive scientist have probably formed your own impressions over the years about journals in your field. We will content ourselves with examining here a few implications associated with such assessments.

● The more prestigious a journal, the higher will almost certainly be its *rejection rate* for unsolicited manuscripts.

A decision about whether or not you should *submit* a manuscript describing a piece of work to a particular journal rests entirely in your hands (perhaps subject to input from colleagues, including members of your research team). But whether that manuscript will ever actually be *published* is quite another matter—and in most cases today that decision will be made by the *editor* of the selected journal, usually in consultation with *reviewers* he or she selects. Considerations related to submission have already been addressed, and from various points of view. The selection process itself we examine more closely in Sections 3.5.2 and 3.5.3. At the moment we wish mainly to stress the *fact* that not every manuscript placed before an editor will be released in print. A significant number—often more than half of those received in the case of the most highly regarded journals—instead end up being *rejected*.

Among other things, this sobering reality means that once you develop a general sense of the relative "importance" of various journals in your field you must steel yourself to accept the consequences of some rigorous self-examination. The first question to ask yourself is this: How important *really* is my latest piece of work? Is it of the caliber one would associate with a premiere journal? The next question becomes: Which is more important, that I publish this work in the most highly regarded journal I can, or would it be better for me to do everything I can to ensure that my paper will be published *somewhere*? Each prospective author is faced with furnishing honest, personal answers to these most basic of questions. One obvious implication is that it can sometimes be counterproductive to "shoot too high" in selecting a target for a particular manuscript, but at the same time it is equally wrong to be unduly shy if the news to be communicated is truly important.

Apart from striving to identify a journal enjoying the appropriate measure of esteem, at least two other factors should be considered. Thus:

● The more prestigious a journal, the longer will almost certainly be the *publication delay*.

Under some circumstances, deferred publication can have serious negative consequences, and it will almost never work to your advantage.

At the same time:

● Highly regarded journals are also the ones most religiously and attentively read, and it usually *is* in an author's best interest to arrange for his or her work to achieve the widest possible recognition.

To some this may all be self-evident, but there are other things to bear in mind as well. For example, publishing something in timely fashion is a prerequisite to ensuring *priority*, and with some types of work that can be an overriding concern. In a different vein, publishing tantalizing tidbits in a widely read journal can be tantamount to throwing up a stimulus and challenge to competing research groups—which in itself almost certainly is *not* desirable.[41] In summary: you must try to strike a proper balance with respect to all the many issues involved if you wish to make the best possible publishing decisions.

One other matter technically subject to a decision on your part is the optimum *language* of publication (although—unfortunately—relatively few native speakers of English are ever tempted even to consider the question).

● In general, publications in *English* attract more attention worldwide than those in any other language.

Authors whose native language is not English often elect to publish their first papers in less widely read, national journals that rely on the local language, saving their *international* debut for a later occasion when the news to be spread warrants broader recognition. Actually, this option is becoming less viable with each passing year, as more and more journals worldwide adopt English as their primary language—largely in the hope of achieving greater visibility.

There are a number of other practical considerations one might choose to take into account—O'CONNOR (1991, p. 6) has in fact listed a total of 12—when deciding what journal to approach first, including, for example, whether or not a particular publication accepts photographs, or how intimidating a publication's page charges are, or whether you will be assured of an adequate supply of *reprints* (and, if so, at what cost). We refrain from pursuing the subject further, however. In point of fact, many scientists favor specific journals when it comes time to publish simply because of sentimental attachments. Or perhaps you have made the personal acquaintance of someone on the editorial staff at a particular journal, and for you as a prospective author that could be the most important factor. In the end, the choice is yours, to be made on the basis of whatever criteria you elect to apply.

---

[41] Some of the things we said early in Chapter 1 about the noble principles underlying scientific communication may cause the last observation to seem rather crass, but it is nonetheless accurate.

3.3 The Components of a Journal Article

3.3.1 General Observations; Title and Authorship

Before one begins to write any article it is important to read very carefully the target journal's "Instructions for Authors", printed in most cases at regular intervals in the journal itself or provided through the journal's web homepage. These instructions address formal aspects of the manuscript, and it is essential they be followed. Even if you think you know what would be an appropriate structure [such as one that conforms to recommendations in *Scientific Style and Format: The CBE Manual for Authors, Editors, and Publishers* from the COUNCIL OF BIOLOGY EDITORS (1994)], one ideally suited to the material at hand, if a manuscript fails to meet specifications adopted by the *editors* of the target journal it is likely to be returned to you before its content has even been looked at—with luck accompanied by nothing more biting than a stern admonishment to try again after first reading the guidelines.

Our first objective in this subsection is to point out consequences that can flow from your choice of a *title* for the proposed publication (cf. also Sec. 2.2.2).

● Many of the bibliographies associated with publications of various sorts include titles for cited journal articles. A paper with a poorly conceived title will obviously attract less attention in this context than one more carefully labeled.

A number of literature services, directories, and periodicals (e.g., *Current Contents*) offer title *lists* to assist the scientist engaged in a literature search or merely browsing. One of the most valuable and highly regarded publications in this category is *Science Citation Index* (SCI), which includes something called a "Permuterm Subject Index": an index that reproduces each title multiple times within a single alphabetical list, each time alphabetizing it under a different important word. Abstracting journals like *Chemical Abstracts* rely heavily on titles for *indexing* purposes, as does software designed for literature management (see Sec. 9.2.2), and the same could be said of many databases. Finally, title lists figure prominently in various Internet sites.

● Since you will want your paper to receive all the attention it deserves, and be encountered by readers who should know about it, it is vital that you pay close attention to words you choose for inclusion in the title.

Conceptually empty words such as "new" or "improved" should be studiously avoided in favor of descriptive words that actually characterize what you are reporting from a thematic standpoint. Also of marginal value is a vague formulation like "The Dependence of … on …". It requires only a little effort—and sensitivity!—to transform that construction into a more informative proclamation like "The Increase of … with …", and readers will be grateful you made the change.

Verbs are especially useful for injecting *life* into a title. Active verbs are remarkably rare in titles,[42] but they offer an easy approach to attracting attention and arousing a reader's curiosity. Consider, for example, the following title, from a recent issue of *Accounts of Chemical Research*:

> "Follow the Protons: A Low-Barrier Hydrogen Bond Unifies the Mechanisms of the Aspartate Proteases"

Surely this will have more impact than something bland like "The Role of Hydrogen Bonds in the Chemistry of Aspartate Proteases".

● In general you should strive to limit your title to about eight words, even though this obviously makes it impossible to touch upon every important idea expressed in a paper.

This is not meant to be regarded as a hard-and-fast rule, and sometimes there is good reason for ignoring it (note for example that the otherwise well-conceived title reproduced immediately above fails to comply—but we would not suggest shortening it!). The point is that you should force yourself to be as brief as possible.

A universal emphasis on brief titles is one reason many journals request authors to supply along with each manuscript a list of applicable *keywords*, although the keyword list will in most cases not become an integral part of the article itself. What it *will* do is contribute in an important way to compilation of the journal's *subject index*. In selecting keywords, concentrate first on the most important concepts addressed in your paper and search for terms that point toward these in as specific a way as possible. Many journals prefer not to give authors a free choice of keywords, asking that words instead be drawn exclusively from a master list developed by the editorial staff, a list consisting of standard terms that particular journal relies on for reference purposes (see Sec. 9.2.1).

The *bibliographic* part of a paper—i.e., that portion essential for identification—includes not only the title, but also the names of all authors. Questions related to who should be included as an author were addressed in Sec. 3.2.2. Here our concern is literally with the *names* associated with these authors, because authors' names play an important organizational role from a bibliographic standpoint.

Ideally, all papers that include you as an author should appear *together* in any directory in which contributions are arranged according to author. To help ensure that this will be the case,

● It is important that you always identify yourself in precisely the same way.

To minimize the risk of confusion it is recommended that you employ at least one given name in conjunction with your family name (do not rely exclusively on initials).

---

[42] Something you can easily confirm by examining the table of contents of almost any issue of any scientific journal!

Above all, be careful not to express your name differently on different papers. From the perspective of a computer attempting to compile a directory of names,

> Eva Schultz, Eva M. Schultz, Eva-M. Schultz, Eva-Maria Schultz, and Eva Maria Schultz

would be five different people!

Every formal list of authors should be accompanied by at least one *business address* containing the official name(s) of the institution(s) where the majority of the reported work was conducted. This information is also subject to indexing, as in the "Corporate Index" that forms a part of *Index Chemicus*.

By tradition, papers released by a single research group, often spanning a period of years or decades, have sometimes been treated as belonging to a discrete *series* of reports (a point emphasized by inclusion of an explanatory notation like: "The xth contribution in the series entitled ..."). It is not uncommon for various parts of such a series to be distributed over several journals, and not all will necessarily be ascribed to the same (first, principal) author. Unfortunately, the practice can be a source of major bibliographic headaches. For example, a title like

> Investigations with Oxidizing Bacteria.
> Part 6: α-Ketoglutaric Acid through Oxidative Fermentation

might well need to be truncated in a title list, leaving perhaps nothing more than the first four words ("Investigations with Oxidizing Bacteria"), a singularly uninformative residue. It would have been far better if a title had been fashioned from the proposed *subtitle* (everything following the colon: "α-Ketoglutaric Acid through Oxidative Fermentation"). If it is important to establish an explicit relationship between the new work and studies carried out previously this can just as well be accomplished within the body of the article, for example by judicious choice (and suitable emphasis) of key literature citations.

3.3.2 Abstract

We have already addressed at some length matters of optimal form and content with respect to an *abstract*: in the chapter on dissertations (Sec. 2.2.3). Important attributes of a good abstract can be identified as

> completeness – precision – objectivity – clarity – brevity

"Brief" in several abstract-related contexts has actually been defined for authors in the United States by an official *standard*. According to *Guidelines for Abstracts*, ANSI/ NISO Z39.14–1997, an abstract for a short communication should contain no more than 100 words, for an ordinary journal article no more than 250 words, and for a dissertation no more than 500 words. As a matter of fact, we strongly recommend that this particular standard be read carefully in its entirety by every publishing scientist;

four of the nine examples it offers of "informative abstracts" are taken directly from stm journals.

The abstract of an article usually precedes the article itself, and because it is often exploited for documentation purposes it is important that it convey as much information as possible.

● One should be able to read and fully understand an abstract without ever having to consult the corresponding article.

This means, for example, that if in the abstract for a paper dealing with chemistry a need arises to refer to a specific compound, the latter must be identified by *name*, not by some number that happens to be assigned to it within the body of the article. Thus, the following abstract (which we actually encountered in a recognized journal— apparently not a very strict one!) provides the uninformed reader with virtually no information whatsoever, and it certainly is of no value for documentation:

> "Syntheses are described for the α-chrysanthemylmethylene-γ- and -δ-lactones **7–9**, **11**, and **12** from lactones **1–3**, **5**, and the product of reaction between aldehyde **6** and compound **10** from 2-oxochromane (**4**). Compounds **7–9** display cytostatic activity."

Serious attention devoted to proper formulation of an abstract can pay handsome dividends. Nonetheless, you should not count on finding that the masterpiece you slaved over is actually utilized by the various official abstracting services (e.g., *Chemical Abstracts*). On the contrary; these organizations generally rely on their own professional *abstractors* to prepare copy consistent with each service's unique quirks and preferences.

By way of reminder, the most important information to incorporate into an abstract would include:

● goals and scope of the investigation – methods employed – key results – major conclusions

In addition, try to imagine, by examining a few precedents, what the editors of your target journal might wish to see. Incidentally, editors of German or French journals, for example, often require authors to submit abstracts in *English* rather than in the language of the article, or it might be that *both* versions would be solicited.

Never write an abstract in such a way as to require illustrations, tables, structural formulas, or other unusual elements that cannot be handled strictly as text. On the other hand, no such prohibition applies to the synopses that certain journals publish as *substitutes* for complete papers (see Sec. 3.2.3).

Finally, we should clarify one important (but certainly not self-evident) distinction. Some journals end each article with a "summary", in which the most significant results and conclusions are collected and restated briefly for the reader's convenience. This is *not* the same as an abstract. For one thing, a summary typically avoids reference to anything connected with methodology, and background is generally ignored. The abstract, situated at the *beginning* of the article, should be viewed as an encapsulated version of the *entire* article, and is in fact likely to be treated as a surrogate for the article by the reader pressed for time (cf. O'CONNOR 1991, p. 73).

3.3.3 The Actual Article

These "preliminaries" are of course followed by the text of the article itself, consisting typically of an introduction, a main body (probably subdivided), and a "conclusions" section. In earlier editions we compared these three elements with the takeoff, flight, and landing stages of a journey by air—an analogy that lends itself to interesting and insightful elaboration.

Building upon thoughts introduced earlier (in Chap. 2), we turn our attention first to the "conclusions" section (although this is not where you would ordinarily begin to write!). Under no circumstances should important aspects of your work appear here for the first time. This is also not the place to remind your reader of goals you originally set, nor of methods employed or individual results. On the other hand, it is perfectly appropriate here to recapitulate consequences you have already touched upon. Notice that a "conclusions" section is quite a different thing from the *summary* described above. The tone should be that suggested by an opening like

> "It was thus possible to show that ..."

Whether or not a summing-up of this sort is appropriate can best be ascertained from guidelines you have presumably received from the editor, or you found printed in an issue of the journal. The same applies with respect to other editorial considerations, such as the kinds of subdivisions appropriate for the main body of the paper: e.g., "Results" ("Findings", "Evaluation of the Findings"), "Discussion", and "Experimental" sections (or in place of the latter perhaps "Methods", "Methods and Materials", "Field Work", "Causality", "Case Descriptions", etc.)—or possibly some different pattern altogether.

You might be tempted to label the various components of your article by number, perhaps using the system described in Sec. 2.2.5 and illustrated once again in Fig. 3–2, but by all means first look for precedents from a recent issue of the particular journal to which you intend to send your manuscript.[43]

In some journals, subheadings within the "Experimental" section (ordinarily *not* numbered) feature functional words like "Preparation" or "Measurement" as a way of clearly indicating the nature of particular experiments. Alternatively, the headings might be something like systematic names for chemical compounds. A special section labeled

---

[43] We emphasize a *recent* issue, since stylistic elements are subject to *change*, and the changes can sometimes be drastic!

1	Introduction
2	Materials and methods
2.1	Mice
2.2	Antigens
2.3	Antibodies and conjugates
2.4	Preparation of Serum IgM and IgG
2.5	Preparation of IgG F(ab')$_2$ fragments
2.6	Immunoadsorbents, antibody and antigen isolation
2.7	Preparation of mouse organ lysates
2.8	Iodination of cell-surface antigens and immunoprecipitation
2.9	PAGE and immunoblotting
2.10	Autoradiology
2.11	ELISA
2.12	Measurement of the IgG affinities
2.13	Computerized data banks
3	Results
3.1	Autoreactivity of serum natural antibodies from various mouse antigens
3.2	Identification of the organ antigens reacting with normal immunoglobulins
3.3	Binding of IgG to MHC antigens
3.4	Polyreactivity of normal IgG
4	Discussion
5	References

Fig. 3–2. Outline constructed as the basis for a paper submitted to the editors of an immunological journal. – Of special interest here is the extent to which the "Materials and methods" section has been subdivided, even though in the published version it amounted to only two pages. It was apparently considered important that an exceptionally high degree of structure be conferred upon this section.

"Statistical Evaluation" (or something similar) would almost always be present in an article from a biomedical journal.

Subdivision of "Results" and "Discussion" sections is more likely to involve labels much more general in character, such as "Reactivity", "Bonding", "Identity", "Behavior", "Optimization", etc.

With a short communication, the only orientation provided for the reader is likely to be the title, though one additional heading might be present, such as "Experimental", or possibly "Methods".

3.4 Preparing the Manuscript

3.4.1 Text

Introduction

Ergonomic factors have a significant effect on the writing process. Drafting a scientific article is surprisingly demanding from an intellectual standpoint, and it requires a clear head, calm nerves, and confidence in your ability to isolate yourself temporarily from the demands of everyday life. Try to escape at least for an hour or two from telephone interruptions and other potential sources of stress; *if* you succeed (a rare triumph, unfortunately) you will find the effort has made a tremendous difference. Moreover, the more completely you manage to circumvent distractions, the easier it will be to come back down to earth when you finish. A private office where you will be out of the reach of "casual visitors" seems to us almost indispensable. Apart from these very basic considerations, it is up to every author to search independently for the precise conditions that permit him or her to work as productively as possible (see also "Writing Techniques" in Sections 2.3.1 and 4.3.1).

Assuming you already have before you a fairly stable, refined outline, together with at least rough sketches for all the illustrations you anticipate including, a general layout idea for every table, and ready access to the appropriate laboratory notebooks and reports, it should take only a couple of hours for you to accomplish most of what needs to be done in the first round.

- Your chances of eventually producing a nicely coherent paper that hangs well together will increase if you can compose the entire first draft in a *single* sitting.

Linguistic and technical polishing can come later. Indeed, that aspect of the job *should* be postponed until you have a chance to distance yourself a bit from your preliminary efforts.

One more piece of general advice:

- Never begin with the most difficult parts—which incidentally are also the parts that are most important.

One good place to start is the "Experimental" section, irrespective of where it fits within the article. From there you should have no problem proceeding to the "Results", and then working your way sequentially through the "Discussion", "Conclusions", "Abstract", and finally the title.

But beware of one important danger:

- Assuming you compose your article by typing it into a computer, there is a serious risk you will conclude far too early that your work is finished, simply because the initial output *looks* so perfect.

Remind yourself of this admonition from time to time. Rewriting and fine tuning should constitute a large part of the overall endeavor.

You should not even *try* revising or polishing your manuscript from the screen, by the way. As we indicated previously (under *Subsequent Drafts* in Sec. 1.4.2 and *Writing Techniques* in Sec. 2.3.1), the screen view of a document affords too little perspective. Instead, scrutinize your work frequently as *printed* output, ignoring the fact that you seem to be wasting a great deal of paper (it is definitely *not* a waste!).

At this point we need to delve into matters that relate specifically to preparation of a *journal article*. (Detailed information along similar lines with regard to scientific text in general will be found in Chapter 5.) We begin with the assumption that the article you are preparing is eventually going to be *typeset*; that is, your work will undergo typographic transformation at the hands of editors and other personnel at a publishing house, rather than simply being reproduced in the form in which you submit it. (The latter situation is discussed briefly in Sec. 3.4.6.)

● Once you finish writing you will be sending your manuscript to the editors of the journal you hope will publish your article (see Sec. 3.5.2).

The word *manuscript* (*lat.* manus, hand; scribere, to write) is of course no longer literally appropriate. It has been a long time indeed since authors furnished editors with *handwritten* documents. For a century or so, *typewritten* copy was the standard, but now virtually all papers are submitted as computer-generated documents. For this reason, the traditional word "manuscript" has to some extent given way to alternative expressions like *typescript* or *compuscript*, and it would probably make sense to declare the term "manuscript" obsolete once and for all. The logical formulation "computer-script" (or "compuscript") has actually never become popular. Suggested substitutes include "manuscript on diskette"[44] and "electronic manuscript".

The most common way today to transport a manuscript from one place to another is to incorporate it into an e-mail attachment. This is frequent practice, for example, of multiple authors working in different places but wishing frequently to exchange thoughts.[45] Perhaps *digital manuscript* is the most satisfactory compact way of embodying the underlying concept in the absence of any universally accepted technical term. The alternative adjective "electronic" puts too much stress on a particular technology, whereas "digital" emphasizes the aspect of this type of "document" that is truly characteristic. Unfortunately, however, "electronic publishing" *has* become a standard part of editorial terminology (see Sec. 5.4), so we must resign ourselves to the fact that the whole matter remains in a somewhat unsettled state.

---

[44] Actually, the "manuscript on diskette" is already a thing of the past. Now it is more common, even for non-professionals, to "burn" information onto a compact disk (CD), in which case the term "electronic" is nearly inapplicable as well because this storage technology instead has an *optical* basis. Stored information here takes the form of microscopic "holes" on an otherwise perfectly plain surface. One disadvantage of a CD is that it cannot easily be edited.

[45] Much of the information interchange between the authors of this book was "electronic" in this sense, and also to some degree "global"!

What we will *not* be discussing here in any detail is the precise nature of a digital document as it actually exists on a particular magnetic (or comparable) data-carrying medium, although from the standpoint of one early standard (DIN 1422-1, 1983) this could be a logical subject for us to explore.

For our purposes, then,

● A *manuscript* will be understood as the version of a document ultimately submitted by an author to a publisher.

Linguistic technicalities aside, paper-based documents—what might loosely be called manuscripts in the "classical" sense—are still important, and they will long need to be a subject of attention. We selected this little detour through a terminological swamp as a point of departure for methodical consideration of the "birth and early childhood" of a scholarly paper intended for a scientific journal; along the way we intend to examine how modern reality has recently reshaped relevant parts of the communication landscape.

Matters of Form

● The editors of most journals now assume that authors will submit their manuscripts in *two* forms simultaneously: on paper but also as digital documents.

The paper version (one or more copies) is generally used for soliciting comments from *referees*, as well as for editing purposes, while the digital version serves as a physical basis for what will ultimately be the published document. An "electronic manuscript" may reach the editor in material form (e.g., on a diskette or CD) as part of a package that also contains printed copies, but an increasingly popular alternative is submission by *e-mail*, one objective being to achieve the earliest possible official submission date.

● Printed manuscripts should normally be prepared on plain, letter-size paper of good quality, double-spaced, and with one exceptionally wide margin (roughly 5 cm).

By "plain paper" we mean paper *without* a letterhead. In the days of the typewriter, "double-spaced" meant four "clicks" of the platen for each new line, resulting in the equivalent of one blank line between every two lines of type. But of course that observation is now little more than a distant reminder from the "dark ages". Today one simply chooses from among several *line-spacing* options displayed in a "menu bar" at the top of a word-processor screen. Alternatively, activation of a menu command like "Format: Paragraph" opens a new window on the screen in which a precise numerical line-spacing value can be entered based on the standard unit "point" (single-spaced text in 12-point type reflects a line spacing of about 15 or 16 points; cf. Sec. 5.5.1). Nevertheless, despite a sophisticated array of options at the modern author's disposal, document "formatting" in general should be approached very gingerly!

● Editors prefer to see as few format specifications as possible in a submitted manuscript: ideally only those explicitly requested.

All visual aspects of a printed page in a published journal (its *format*) are determined by the editorial staff. Efforts on the part of an author to achieve in a manuscript some specific appearance will simply complicate the publication process. (We touch upon this point again below.)

Most editors specify that hardcopy (paper) submissions should be double-spaced. One advantage relative to single-spacing is of course that the text is easier to read, but double-spacing also reduces the likelihood that the reader in a hurry will lose his or her place on skipping from one line to the next. Furthermore,

● Only with generous line spacing is it practical for *editorial markings* to be inserted *between* the lines, where their purpose will be most obvious.

Tight line spacing can also cause serious problems if a manuscript contains *superscripts* and/or *subscripts* (see Fig. 5–3; this illustration is based on typewriter-like output, where the problem becomes especially severe).[46]

Our recommendation that one margin be left exceptionally wide (5 cm, roughly 2 in) is also designed to facilitate editing. A wide margin provides space for lengthy explanations related to major corrections, or extensive hand-written instructions to a typesetter. Editorial guidelines may specify whether the wide margin should be on the left or the right. A manuscript page of the kind we describe provides space for about 30 lines of type with roughly 50 characters per line, for a total of perhaps 1500 characters per page.

Editors normally do not work directly with a "digital manuscript" in the early stages leading up to publication, even if—as is usually the case—a *data-transfer medium* (e.g., CD) supplied originally by the author will later serve as a basis for the published version. Reviewers also usually insist on being provided manuscripts in printed form, because they find it too difficult to maintain the proper perspective with text displayed on a screen. It would obviously be an imposition to ask the reviewers to prepare their own printed copies of digital files, so they receive hardcopy manuscripts from the outset. Similarly, editors have long recognized how poorly screen output lends itself to conscientious, high-quality editing. In point of fact, *three* printed copies of a manuscript are often required by an editorial office, and in most cases it is considered the author's responsibility to provide them. One copy is reserved for editing purposes, while the other two are sent to reviewers. Not many years ago publishers were still forced to make do with faint *carbon copies* prepared on flimsy *onion-skin paper*, but the nearly universal availability today of versatile and rapid photocopiers, coupled with high-quality hardcopy originals from an author's desktop printer, means it is now a simple matter to obtain as many pristine "original" copies of a document as necessary: automatically collated and stapled at that!

---

[46] The standard DIN 1304 Part 1 (1989) makes an interesting terminological distinction: subscripts set to the *right* of a character, and only these, are to be referred to as *indices*. Occasionally there is need to introduce a mark of some sort *directly* above or below a character. This generally requires access to either special symbols or special software (such as the "equation editor" that accompanies Microsoft WORD).

● All pages of a manuscript should be carefully numbered (paginated).

Keeping track of the page count and introducing the requisite *page numbers* are tasks that can be left to one's word processor so long as "page number insertion" has been specified in a *header* (or *footer*) that then becomes part of every page.

● Editors normally prefer that authors *not* apply typographic distinctions (boldface, italics, etc.) to headings—or for that matter any other element in a manuscript.

Official instructions regarding *character formats* will be issued by the editors themselves as part of the *markup* process. Markup entails inserting into a manuscript a great many brief (usually coded) messages, primarily for the benefit of a typesetter. "Character formatting" in this context refers not only to the sort of type to be employed, but also to its style and size (for more information see Chap. 5). As previously noted, restraint is in order with respect to an author's making format decisions about anything: letters, words, symbols—even paragraph spacing and footnotes. To put the matter succinctly: *no* special formatting whatsoever should be conferred upon a manuscript without prior authorization from the editor. The author who disregards this advice will almost certainly make the publisher's job more difficult.

In fact, there should be no extra markup instructions of *any* sort anywhere in an author's manuscript, not even ones consistent with a "standard" editorial system (e.g., SGML; cf. Sec. "'Markup' of an Electronic Document" in Sec. 3.1.3 and also Sec. 5.2.2). This admonition will probably be viewed as redundant, but repetition is in some sense warranted, because the message is such an important one. If you wish to avoid unpleasant confrontations with your editor—and would prefer *not* to be responsible for inflated costs, and possibly for publication delays—pay strict attention to the detailed set of instructions provided every author. The reason we emphasize this point repeatedly is that—unfortunately—too many authors ignore it. Since there are still no industry-wide standards applicable to information transfer, formatting or markup "help" voluntarily dispensed by an author has virtually no potential to be valuable.

3.4.2 Formulas and Equations

The word "formula" in the context of a scientific or technical manuscript may refer to either a mathematical statement or a chemical structure. The former could also be called an *expression* or an *equation*, whereas *structural formula* is the correct technical term for the latter. We begin with a consideration of formulas of the first type (see also Sec. 6.5.2)

● From a typographic point of view, mathematical expressions can be regarded as an exceptionally complicated form of text.

A mathematical expression usually lends itself to dissection into a series of individual *symbols*, many of them unfortunately absent from the standard set of characters in a typical *type font*—including the fonts commonly supplied with word processors. One usually has access to several fonts containing a full range of standard letters, numbers, etc., but relatively few of the *special characters* that distinguish mathematical copy: more specifically, the only ones found are those present in the "ASCII" (American Standard Code for Information Interchange; see also "Type Formats" in Sec. 5.3.3) system. *Greek* letters (both upper- and lower-case) often figure prominently in mathematical expressions as well. These, together with at least some of the most important mathematical symbols, are conveniently available from a widely-distributed special type font *known* as "Symbol".

The problems involved in reproducing mathematical expressions do not all come from a scarcity of symbolic characters, however. Even more challenging is the fact that such expressions rarely have all the required characters on a *single line*. For example, *subscripts* (e.g., indices) and *superscripts* by definition are set slightly below or above adjacent characters, respectively. Expressions that take the form of *fractions* require the definition of even more vertical "levels". High-quality simulation of a relatively primitive mathematical expression can sometimes be accomplished with an ordinary word processor, but it is almost always a time-consuming and awkward proposition. Either the methods at one's disposal prove severely limiting, or the results obtained from a promising "work-around" fall disappointingly short of expectations. There is no obvious solution apart from resorting to special software developed for this express purpose. Dedicated *utility programs*[47] are indeed the best alternative in most cases, which usually entails electronically pasting whatever expressions you create, one by one, into their proper places in the master text document.

The author who works frequently with complicated mathematical expressions may find it worthwhile to master a version of the specialized and highly flexible type-setting (or "formatting") program TEX (originally written "TeX"; see KNUTH 1986, SEROUL and LEVY 1991, or SNOW 1992). TEX allows one to simulate professionally set type remarkably well, and it was designed explicitly to meet the most demanding requirements of mathematicians. It supports impressive implementation of more than 1000 powerful commands, but in its original form it could hardly be described as "user-friendly".[48] We question whether the effort required to become fluent in TEX is actually warranted for the average scientist; physicists and mathematicians are the ones most likely to accept the challenge, in the process learning what amounts to a special programming language.[49] For more information see Sec. 6.6.

A brief general introduction to preparation of the *structural formulas* so central to chemistry (see DIN 32641 E, 1994) is provided in Sec. 7.2.5. Our comments here

[47] Such as the "equation editor" mentioned in the preceding footnote.
[48] Modifications *based* on TEX but sporting a more user-friendly interface have been available for several years. One of the most popular is LaTEX (see, for example, LAMPORT 1994, GOOSENS, MITTELBACH and SAMARIN 1993, or DETIG 1997). A brief introduction to LaTEX is presented in Sec. 6.6.
[49] TEX is especially recommended for use with journals sponsored by the American Mathematical Society. The Society has even encouraged development of a distinct "dialect" called AMS-TEX, which will be put at the disposal of any author interested in submitting a manuscript to one of the Society's journals.

will be limited to a few general observations in the context of dealing with the combination of text and formulas.

- Structural formulas that may be required in an article are usually kept separate from the corresponding text in what is called a *formula manuscript*. One must of course then provide unambiguous instructions explaining which formulas belong where.

Separation has always been required due to the fact that even the most advanced typesetting equipment is incapable of reproducing directly a sophisticated chemical structure (apart from strictly linear representations or simple statements of elemental composition). Thus, structures to be incorporated into a manuscript must be handled just as *illustrations* would be, which means as they wend their way through the publication process they follow a course very different from that taken by text—at least this has traditionally been the case. Restrictions may also apply with respect to the e-mailing of formulas. The extent to which your particular publisher is subject to such limitations is one more area you certainly should explore with the editors.

The personal word-processing system at your disposal is almost certainly capable of combining text and graphics, of almost any type, seamlessly within a single document, so you are probably wondering why that should not also characterize facilities found in the professional realm. Part of the problem stems from a lack of universally accepted standards. The industry is gradually "maturing" in this respect (in large measure due to pressures exerted at the level of the individual scientist), but change— here toward more compatibility—is a ponderous thing. It is quite likely that within a very few years the scenario we are depicting here will seem like ancient history.

Be sure to bear these problematic issues in mind, especially if you are accustomed routinely to working with special software for creating chemical structures (e.g., CHEMDRAW from CambridgeSoft Corp.) and would normally without hesitation incorporate the resulting structures directly into your documents. In many cases all structures associated with manuscripts for *publication*, no matter how they are prepared, will need to be kept separate for eventual capture photographically. Embedding in the master document would then be delayed until after preliminary typesetting.

Returning to preparation of the manuscript in general, all chemical compounds depicted by structural formulas should be assigned sequential numbers in the order of their appearance, taking special care to ensure that no compound inadvertently acquires more than one number.

- In published works such *formula numbers* are usually set in special type, most often boldface.

You will also need to propose a specific location for each structure—invariably a spot *between* two paragraphs, with the structure itself indented (or perhaps centered on the page). The suggested placement is most easily indicated by placing there the corresponding number, enclosed in double parentheses; e.g., ((16)) or ((17–20)). This will automatically be interpreted as an *instruction*, and will therefore not be typeset.

Double parentheses always signify an element that is *not* to appear in the final document: usually a notation directed to an editor, a typesetter, or the author. This is one example of a type of message generated in the course of document processing that many refer to as *metainformation*.

When a formula number is introduced along with a compound name within running text, it is typically enclosed in (ordinary) parentheses and placed immediately after the name; e.g., "2-pentanone (**5**)". Parentheses are omitted when a previously defined number is instead used as a *replacement* for a name; e.g., "…a solution of ketone **5** was …". Similar numbers can be called upon to designate *reaction equations* or *reaction schemes* (*groups* of reaction equations).

3.4.3 Figures

A Figure or a Table?

The equations and structural formulas we have just been discussing are a way of supplying information in highly compressed form, information that would be extremely awkward to express through words alone.[50,51] Scientific communications frequently attempt to relay information in two other compact, non-verbal (or at least non-text) ways: through *figures* and *tables*, which become our agenda in an introductory way for this and the subsequent subsection of the present chapter. Technical details are reserved for Chapters 7 and 8.

Distinctions can of course be drawn between various *types* of figures: from a technical point of view, between *line drawings* and *halftone illustrations*, for example; or from a content perspective between *illustrative drawings* (e.g., of apparatus) and *diagrams* or *graphs*. The latter will largely be the focus of our attention in what follows.[52]

● *Graphs* and *tables* represent two complementary approaches to expressing *relationships*. In a graph the representation is *pictorial* in nature, while a table attempts to communicate through a carefully planned *spatial grouping* of a set of numerical (or occasionally verbal) elements, each referred to as a *cell*.

The "relationship" in question might be a mathematical *function* specifying a strict dependence of one quantity upon another. Such a relationship could be "illustrated" by pairwise juxtaposition of a selection of discrete *values*, resulting in a table. The same overall goal might instead be reached with the aid of a graph based on coordinates

---

[50] The most unambiguous, relatively compact alternative to a structural formula is a formal description of *connectivity* and *topology* generated by a computer program developed for that express purpose.

[51] You may have read that the depiction of a right triangle with squares affixed to each side has been suggested as a "message" that should be understandable by intellectual extraterrestrials in the absence of any explanation.

[52] Graphs can of course be of many types, a few of which are examined in Chap. 7. Here we presuppose in most cases the typical "*x,y*-plot" bounded by Cartesian coordinates.

representing the two quantities. In effect, graphs and tables can be envisioned as equivalent *analog* and *digital* illustrative devices, respectively. (Characterizing them in this way of course glosses over the fact that a graph, too, can be simulated digitally: by treating its content in terms of a—perhaps exceedingly large—*matrix* of discrete values. Incidentally, strictly numerical tables can often be transformed easily into graphs with a few simple keystrokes, even within the confines of a word processor.)

It is often necessary to confront the issues of which of the two alternative approaches would be the more effective in a given case, because in a publication it is almost never permissible to adopt *both* approaches. In a journal article one would *not*, for example, present data on the annual output of the steel industry as a function of calendar year and/or perhaps region in both tabular *and* graphic form.

● The economic realities of publication generally preclude a "both *a* and *b*" approach when dealing with information that could in principle be expressed either in a table or a graph.

When is tabular presentation most effective, and when would a graph be better? No simple answer will be valid in all situations, but a few generalizations may be welcome.

● A *table* is always the simpler to create (although doing so properly is considerably more complicated than composing a piece of text), and table construction does not require the availability of special tools or skills.

A table is probably the best choice if you see merit in accentuating or emphasizing particular *examples* of a relationship (either numerical or verbal).

On the other hand, if the goal is to encourage the reader to *visualize* a relationship, with special attention to qualitative rather than quantitative aspects, a graph is the modality of choice—even if for simplification you resort to *unscaled axes*.

● A well-designed graph or other visual device is frequently able to convey an idea much more effectively than words would.

Before you make up your mind to include figures in your manuscript, however—and this includes graphs!—be sure you realize that first designing and then refining an effective visual device involves considerable effort and imagination, and technical realization of your ideas can also pose problems.

The Processing of Figures

Illustrations are usually kept separate from text (the *text manuscript*) during processing that precedes publication. The set of illustrations accompanying a document is sometimes referred to as an *illustrations* or *graphics manuscript*. Until the majority of publishers have successfully completed the transition to electronic processing for all types of manuscripts, the guidelines that follow can still be considered broadly applicable.

There is no point in trying to embed illustrations directly into a digital manuscript destined to be typeset, because

● Text and graphics follow distinct paths through the production process.[53]

Your text will all be typeset in some routine way, whereas illustrations may demand individual attention on the part of a draftsman or (increasingly) one or more "reproduction specialists". In the event that photomechanical *direct reproduction* of a submitted manuscript is anticipated, it *would* of course be appropriate to combine text and illustrations in a single file. Even this exception assumes a document whose only illustrations are line drawings, however; if *halftones* (cf. Sec. 7.4) are present—and permitted!—there may be a need even here for extra processing steps. The technical measures required vary dramatically from situation to situation, so authors need to be aware that unique circumstances applicable to individual projects may have important implications for processing.

The observations above to some extent still reflect the state of things when we last summarized the publishing landscape: in 1997 as we were preparing the fourth edition of this book's German counterpart. At the same time, we have endeavored to incorporate more recent developments wherever appropriate. In an age in which pervasive change has become the norm, the currency of what we said then—or might say today!—must always be considered suspect.

From personal experience you are probably aware that graphic information on a letter-sized "page" can easily be captured and repackaged for further processing (e.g., scaling). Nothing more complicated is required than an inexpensive desktop *scanner* capable of generating files based on digital graphic data. You also know that various programs you routinely run on your computer, offline as well as online, have the ability to combine graphic material with text. A graphic image containing textual information can be taken to a much higher level of intelligibility by analyzing it with *Optical Character Recognition* (OCR) software. In this way one can detect and extract alphanumeric elements present in a scan file, for example, and save them as a *text data file* for manipulation with a word processor. This represents *one* way in which a page of printed text can be transformed into a digital entity subject to systematic content search. Proceeding in the opposite direction (i.e., from data file to paper), virtually all personal printers now on the market are able to express digital information derived

---

[53] This particular statement is losing some of its validity as printing firms increasingly abandon printing plates and as they become more accustomed to illustrations prepared, stored, and transmitted digitally. The daily newspaper that features photographs shot earlier in the day, possibly halfway around the world, offers vivid proof of the relentless march of technology. Some major publishing houses already work *only* with digital copy, and no longer even have the tools required to "strip in" an illustration. The question of whether published illustrations are "best" prepared from material provided in digital form (as in the case of this book) or from conventional photographic prints now hinges less on technology than on cost. Surprisingly, perhaps, the traditional approach based on photographs tends still to be the more economical of the two! What usually turns out in fact to be decisive is a publisher's preference regarding whether archival should be in the form of individual pages or complete data files.

from both text and graphics files—or combinations thereof—as impressively sharp, durable, and detailed hardcopy for close examination at the user's leisure.

Everything we have described in the preceding paragraph can be accomplished by anyone with access to a well-equipped personal workstation. Why is it, then, that—in general—publishers still urge their authors to treat text and graphics as separate things? Part of the reason, as we have stressed repeatedly, is that the publishing industry still lacks a comprehensive set of standards for information interchange. The wide variety of software employed by different authors is also a complicating factor. The most compelling reason, however, is the need for editors and publishers to maintain strict control over *presentation*. Reconstruction of a carefully prepared set of individual pages containing multiple items (text, tables, figures) and saved in a complex data file sometimes leads to profound and unpredictable layout changes as a professional publication is being assembled, and this cannot be tolerated.

If you anticipate supplementing your article with graphic material in digital form it would be wise to ask early about preferences your publisher might have with respect to file format (e.g., GIF, JPEG, TIFF, etc.; cf. Sec. 7.3.2). Keep in mind the possibility that, in the interest of ensuring optimal results, you will be requested instead to supply high-quality *hardcopy* versions of all graphics for in-house scanning.

Relating Figures to the Accompanying Text

It is extremely important that all graphic elements be *placed* properly in the finished product. "Properly" in this case means every illustration should be as close as possible to the text passage that first mentions it, and each should be situated directly above (or below—or beside!) the corresponding caption (see below).

● A clear, explicit indication of the first text reference to each illustration—in the margin, in the form of a figure citation enclosed in double parentheses; e.g., ((Fig. 12))—will help the production staff achieve optimum placement of graphic elements in the course of preparing *page proofs*.

While markings of this sort—known as "figure callouts"—are greatly appreciated, do not assume that every illustration will necessarily be placed where you anticipate. Should the preceding paragraph end too near the bottom of a page (or column), for example, there might not be sufficient space available for an illustration. Obviously one can't put *half* the illustration there and the other half on the *next* page! This is why it is so important that the *first reference* to each illustration be labeled clearly in a hardcopy version of the text, preferably highlighted by an arrow (perhaps red), in the margin. When the time comes to assemble individual pages as a last step prior to printing, the person responsible for page layout will attempt to identify the best possible (feasible!) arrangement, using an author's (or editor's) arrows for guidance.

There is of course no way for us to foresee everything the future has in store for the publishing industry. Perhaps authors someday will be encouraged to position illustrations—including institutional or corporate logos—in accurate screen images of their copy, which will then be passed along electronically for direct reproduction by a publishing house fully accustomed to dealing routinely with complex files featuring both text and graphics—even full-color graphics.[54] If this should in fact occur, an interesting consequence would be the gradual disappearance of a distinction between text and graphic material. Even now, with a full-featured word-processing program it takes only a "mouse click" on a "format tool bar" to open a dialogue box for recalling from a disk or even creating a graphic image that, with the same software, can be cropped, edited, and finally inserted wherever you wish within a digital document— ready for immediate printing or electronic dispatch to some other location.

Many long-standing "rules of the publishing game" are fast becoming obsolete. For this reason it is essential (at least now, as this edition of our book is about to go to press) that you discuss with your editor exactly how illustrations should be handled. Until recently, most attempts to capitalize on major innovations and "simplifications" like those we have been considering produced only disappointing results. Some publishers unfortunately discovered to their dismay that they were not even in a position to make effective use of *plain text files* (*ASCII files*) received in digital form, with the result that they were forced to continue typesetting everything "from scratch". It turns out to be surprisingly difficult, for example, to translate and implement in a reliable way the various formatting instructions and other codes embedded silently in the digital representation of almost any manuscript prepared with a word processor. Too many different versions of such software are in common use to guarantee absolute compatibility with a publisher's facilities, and transforming (translating) "text data" from one (formatted) digital form to another can be frustratingly difficult and time-consuming— to the point that it is often costlier than conventional typesetting.

We must now return from this rather lengthy digression to straightforward advice and information.[55]

● Illustrations accompanying a manuscript must be *numbered*, and every such illustration must also be *cited*—by number!—at least once somewhere in the text. The figure copy sent to a publisher must of course also be clearly labeled with the appropriate numbers.

In this context, an editor speaks of "anchoring" an illustration in the surrounding text through a citation, analogous to anchoring a buoy in a stream. The analogy is actually

---

[54] This vision ignores one important consideration: *design* as it applies to the pages of a document, which plays a much more important role in effective communication than the average author realizes. Experience suggests that few authors will ever be enthusiastic about taking on as a spare-time hobby the challenge of becoming accomplished designers!

[55] We have tried to remain generally within the boundaries of typical scientific and technical writing and publishing. The illustrator or layout specialist concerned with the art for a children's book is likely to find some of our advice to be counterproductive. Small children have no interest in figure captions, for example; to them text and figures are all part of a single visual entity.

quite good, since illustrations in a sense correspond to graphic objects "fixed" only loosely within an ongoing text stream.

● Figure citation in a text passage simply means making mention of that figure, by number and in parentheses, usually at the end of some sentence or phrase; e.g., "... (Fig. 12)".

Figure citations can also be worked into the body of a sentence—in which case parentheses would be omitted ("... as illustrated in Fig. 12 ..." or "... is represented by the upper curve in Fig. 14").

Again, the general rule is that

● No illustration should be without a number, and every illustration must be referenced, by number, somewhere in the text.

Exceptions to this rule are rarely permitted, and *any* exception would certainly require an editor's explicit sanction. Once in a while, for example—usually in a textbook—a legitimate case can be made for including a small graphic image directly within a line of running text, just as one might an equation, because that image is considered essential to understanding the argument. The alternative of a formal citation would perhaps be too distracting, especially if the illustration were forced to appear on a different page. Small sketches often contribute heavily to understanding mathematical text in particular, and exceptional treatment of the sort indicated can usually be managed without inordinately complicating the process of preparing satisfactory page proofs. But again we stray; mathematics textbooks are after all not our primary area of concern.

Never include in a text passage a "directional" reference to an illustration or other special element, as in "the illustration below", because space considerations might be such that the item in question would actually end up being placed "above".

Miscellaneous Matters

● Every illustration requires an accompanying *caption* (the *figure caption*).

A caption serves as a figure's "title", and like other titles, figure captions should be kept as short as possible. Nevertheless, the combination of figure and caption must provide the reader a clear understanding of the figure and the reason for its presence. The reader should never be forced to dig around in the text in search of an explanation.

A figure caption often serves a second function as well: it adds *details* concerning the figure's content, information too specific to fit comfortably in the body of the text.

● Explanatory material of this type is technically known as a figure's *legend*; its proper placement is directly after the caption itself (i.e., after the "title").

A *block format* is especially well-suited for figure captions. Any legend required is simply included within the block, typically separated from the true caption by a period and a dash (but it need not be deliberately displaced to a separate line).

- Text which is to accompany figures (captions and legends) should be collected in a separate *captions manuscript.*

Traditionally this has been an ironclad rule, but in point of fact some editors now prefer that figure captions instead be included in the text manuscript, near where the corresponding figures belong—but always as separate paragraphs. This is yet another case in which it may be necessary for you to seek explicit guidance from the editorial office.

- The presence of *typographic elements* in figures should be minimized.

Scales and *axis labels* are of course an essential part of certain graphs, and these labels obviously cannot be avoided. Other alphanumeric information, however, should insofar as possible be supplied in abbreviated form (e.g., a letter "V" instead of the word "valve"), or limited to numbers and symbols for distinguishing various types of data points. Abbreviated markings would be carefully explained in the figure's legend. Admittedly this approach is not ideal from the reader's standpoint, since it demands a certain amount of glancing back and forth between figure and legend, but it can produce substantial savings of both time and money in the preparation and editing of complicated figures. The absence of unnecessary words in a figure also simplifies the process of recycling figure copy should an occasion arise, for example, to translate your publication (or a part of it) into a different language.[56]

The preceding argument has actually become somewhat less persuasive now that most figures are prepared with computer software that provides a "text tool" for incorporating—or editing—alphanumeric characters in a graphic object, characters that also can be created in any desired typeface.[57] But quite apart from technical considerations there is a significant aesthetic issue to take into account in this context: a drawing that contains too much lettering tends to look cluttered and "busy", and it loses much of its uniquely graphic character. You should be aware too of a potential problem related to *keyword* searches. It is easy for software to locate a word present in a figure caption or legend, but the same sequence of characters located somewhere within an illustration would be completely overlooked.

Finally, bear in mind that figures—together with the tables discussed below—are important elements in a paper because of their ability to capture a reader's attention. Before reading an article many casual "browsers" try to acquire a general sense of its content by first skimming rapidly through the pages. Apart from headings, it is figures and tables that enlighten them the most. As an author you therefore have a responsibility to devote special attention to these eye-catching features.

We suggest you begin preparing at least some of your figures and tables *before* composing the bulk of the manuscript, since such work can be extremely time-

---

[56] You may have reservations about following our advice here—and understandably so! It is certainly true that a cartoon would lose much of its value if the "balloons" were removed, but we are not suggesting going to anything like that extreme.

[57] Such a tool is often represented in menu bars or tool palettes by a stylized letter "A".

consuming, especially taking into account problems you may encounter securing permission to reproduce material you might want to "borrow" from some other source (see "Legal Aspects—Citation of Figures" in Sec. 7.1.2).

Technical aspects of the preparation of figures are addressed in Chap. 7.

3.4.4 Tables

Much of what we have said regarding figures and their relationship to a text manuscript applies equally to *tables*, including observations regarding recent developments in typesetting. Thus,

● Tables should in most cases be collected in a separate *tables manuscript*, with each table linked to the text in a clear, unambiguous way.

Typesetting devices today are largely computer controlled; consider for a moment the possible implications if a word-processing file representing your manuscript were being typeset and the system suddenly encountered a piece of embedded output from a spreadsheet program. Your word processor no doubt would have handled it easily enough, but the same transition could be much less straightforward for a typesetting device. Even a table created with a word processor (Microsoft WORD, for example) involves a number of (invisibly embedded) special commands that are easily misinterpreted. The resulting ambiguity with respect to any advice we might offer is aggravating—presumably for you as much as for us—but things should be much clearer within at most a few years. Today, what is good advice under one set of circumstances (i.e. in a particular country or setting) might be bad advice elsewhere. Common sense dictates that you should seek definitive advice from the editorial office, and under no circumstances should you assume that guidelines applicable to one journal apply equally well to another.

● Just like figures, tables are always numbered and anchored formally in the text.

It is customary in at least some editorial circles to label the first text reference to a table with a *blue* arrow placed in the margin of a printed copy of the manuscript. (Recall that we suggested use of a *red* arrow in the case of a figure.)

Tables, like text, should be double-spaced (or at least "one-and-one-half" spaced). If constructed too compactly (or reproduced directly from a previously published document), major problems will confront the editor or printer who wishes to introduce instructions for modification, or provide typesetting instructions. We urge you also to pay more than casual attention to the many subtle details responsible for giving the best tables their professional look (cf. Chap. 8).

There is no need to be concerned if one of your tables threatens to extend over multiple pages. It may be that in typeset form it will actually fit quite cozily on a single page—or indeed even in one column—but if not, the editorial staff will deal with the problem. You should also not hesitate to propose that an exceptionally wide table be rotated by 90° (i.e., so it extends parallel to the long side of a sheet of paper). Again, it may prove possible to accommodate its printed version in the normal way, but resolving such issues is in any case an everyday matter for a professional editorial team. Traditionally, table layout has been the province of specially trained personnel— who still find it no easy matter to carry out their assignment on a computer screen, given the expectation that the finished product will measure up to the highest professional standards. Keep in mind, too, that typesetting a table is considerably more expensive than managing running text. For this reason alone, superfluous tables are best avoided.

● Assign to each of your tables an informative, sequentially numbered heading.

In contrast to figure captions, a *table heading* is an integral part of the table itself, and it accompanies the table throughout processing. What makes this possible of course is that both the table and its heading are subject to more or less conventional typesetting.

As noted previously, a good word processor always includes a set of "table mode" commands similar to those in a spreadsheet package. There is thus no need for you to search around for special "table-editing software". A starter set of *columns* and *rows* is specified when a new table is first created. Column widths can be subsequently adjusted as necessary, rows and columns can be introduced or deleted at will, and the size and style of type employed is subject to specification—for each cell (or even character) independently. This makes it a relatively easy matter for the conscientious author to prepare tables closely resembling those created by a professional. Nonetheless, editors or typesetters at the publishing house are likely to make at least some changes in the tables you so painstakingly and lovingly put together.

● Technical considerations related to page layout are likely to be the determining factor with respect to final form—a generalization that applies not only to tables but also other special elements associated with your text manuscript.

If you have questions that seem not to be addressed in the publisher's "Guidelines for Authors", by all means raise them promptly with the editorial office.

One additional point with respect to table *content*. Never abbreviate words in a table in an effort to make some column narrower, for example. Non-standard abbreviations, especially ones not clearly explained somewhere prominent in the article, are an unreasonable imposition upon the reader's patience. If necessary, common *acronyms* or standard *symbols* are permissible, but even these should be explicitly "decoded" somewhere obvious, preferably in a footnote to the table.

More information regarding tables is provided in Chap. 8.

3.4.5 Footnotes and Other "Interjections"

Authors are often tempted in the course of a narrative to interject some informal comment: an elaboration, perhaps, which though directly relevant to the unfolding text, would disturb the flow of an argument, and could in any case safely be ignored by the reader in a hurry. Impulses like these led to invention of the *footnote*, a device just as useful in conjunction with scientific text as anywhere else. Footnotes also offer an author the opportunity to float tantalizing ideas that might not be ready to withstand determined challenge; i.e., casual "personal observations", or bits of bold speculation.[58]

As the term suggests, a footnote is intended to be placed at the "foot" of a page, specifically the page that contains a direct *reference* to the note. Footnote references sometimes take the form of *symbols* entered as superscripts, such as

$$*, **, \dagger, \ddagger, +, \ldots$$

Such a symbol would appear not only at the appropriate point in the body of the article but again (usually smaller) farther down the page, immediately preceding whatever supplementary information the author wishes to provide.[59]

If *symbols* like those suggested above are utilized as indicators for footnotes, and if more than one such symbol happens to be required on a particular page, the *sequence* of symbols typically conforms to a prescribed order (e.g., * first, then $\dagger$, followed by $\ddagger$, etc.). Unfortunately, selection rules vary from culture to culture, and sometimes even from one publisher to the next. A particularly awkward nuisance is the fact that symbols introduced by an author often must be changed when a document is typeset, because footnotes are no longer distributed among the pages precisely as they were in the manuscript. Any such delicate, manual operation automatically means increased cost.

For many reasons we see it as preferable that footnotes be labeled with *numbers*. If one chooses this route there are also advantages in sustaining a continuous numerical sequence throughout the document (or within each chapter in the case of a book[60]).

There remains the problem, however, of actually fitting footnotes on the pages where they belong. Notes are usually set in rather small type and separated from body text by a short (e.g., 4 cm) dividing line extending in from the left margin. If too little space is available to accommodate a complete footnote (e.g., one that is exceptionally long), its text can be continued at the bottom of the following page. The presence of such a *continuation* can be signaled by extending the (normally short) dividing line across the entire width of the page (from margin to margin; we have not made this

---

[58] We obviously succumb frequently to use of this option. At least some readers look forward to footnotes, partly because they offer a rare bit of insight into the mindset of the author, and because a few turn out to be a source of amusing "treats".

[59] The use of multiple signs of the same type (e.g., ***) as footnote references, though common, has the aesthetic disadvantage that it makes it virtually impossible to align the text from several consecutive footnotes in a uniform way. This is one of the reasons we prefer a numerical citation system (typically the default mode for word-processing programs).

[60] In this particular book we have chosen to "restart the clock" with the number "1" with each chapter. Note numbers, just like the symbols described earlier, are introduced as superscripts set in small type. Some authors add a "close parentheses" symbol following each such number in the body of the text— e.g.,[1]—as a way of clearly distinguishing footnote numbers from *other* numerical superscripts, especially exponents (powers; e.g., 2^3). Many journals (e.g., *Angewandte Chemie*) also adopt this or some similar practice with respect to *literature citation* numbers as well, possibly enclosing them in superscripted square brackets to avoid ambiguity, as in [23]; cf. Sec. 9.3.2.

distinction, signaling note continuation instead with an arrow). Sometimes the space available for notes turns out to be too *large*, in which case part of the page simply remains blank.[61]

These are the cumbersome conventions that once sent chills up the spine of the author obliged to prepare a formal document with nothing more "intelligent" than a typewriter. Today's author is in luck: word processors offer the luxury that software manages everything related to notes, with no need whatsoever for outside intervention.

An alternative way of dealing with notes is again to introduce numbered note citations consecutively throughout a document (or chapter), but then collect all the corresponding text at the *end*, in which case footnotes will have become *endnotes*. Endnotes cause fewer layout problems, but they are a nuisance to the reader, who is forced always to *seek* what looks like it might be an interesting note.

● A word processor does such a convenient job of managing numbered footnotes (or endnotes) that it makes little sense today for an author even to consider using arcane symbols as note alerts.

Software can be relied upon to ensure that every note is assigned the proper number, including notes introduced after the fact, which almost always require that existing numbering be revised. If you choose to treat your notes as *foot*notes, appropriate space is made available automatically by the word processor, and necessary "continuation" is also cleanly attended to. This is just one more example of the extent to which word-processing software spares today's author from onerous tasks long assumed to be an unavoidable part of preparing "professional quality" copy.[62]

Should you so desire, with a word processor you can at any time obtain a comprehensive screen overview of all the notes in your document. In WORD, for example, requesting a special "Footnotes View" opens on the screen a separate scrollable notes window below the normal text window. The two windows are coupled in such a way that when one is scrolled, the other keeps pace. Either window can be subjected independently to a "search" for a specific word or sequence of words (or letters).

There is a sharp difference of opinion about whether "notes" (either footnotes or endnotes) should ever be permitted to contain bibliographic citations (i.e., indicators pointing to literature references). Generally speaking, the practice is frowned upon, because it makes locating the citation(s) for a particular reference within an article more difficult (notes become one more place where a citation can hide!). One solution is to *merge* the notes and literature references, using the same type of citation marking for both, in which case the combined pool of information might be entitled "Notes

---

[61] This would occur, for example, if addition of one more line of body text would mandate squeezing onto that page yet *another* footnote.

[62] Even so, if you aspire to create a document that is *truly* "professional looking" it will be necessary to go beyond the capabilities of even the most powerful word-processing software and enlist the help of a "page-layout program", such as PAGEMAKER (Adobe Systems) or QUARKXPRESS (Quark); cf. Sec. 4.4.2.

and References". As the prospective author of a journal article you must simply accept whatever practices have been embraced by the editors of your target journal.

True footnotes are becoming increasingly scarce in publications, largely for economic reasons. It once was standard procedure to place not only explanatory notes but also the separately administered *literature references* at the "feet" of the appropriate pages so the reader would always have immediate access to the relevant information and thus be in a position to decide at once whether or not a lengthy note should be read immediately, or jot down a reference that looks like it might be worth pursuing. This advantage is now provided by only a very few books and journals; literature references in particular are almost always consigned to the end of a document. One can hope that advances in typesetting technology may allow us to anticipate a day in the future when the lowly footnote will be "reborn".

Technical and economic considerations aside, a few strident critics in recent years have been waging a bitter war against the very *idea* of (interjectory) notes, declaring them to be "pretentious academic affectations" on the grounds that anything not suitable for the body of the text must necessarily be superfluous. Similar charges have even been leveled against text set in *parentheses*, since this could be regarded as equivalent to an "in-line note". We view these charges as overzealous and unwarranted nonsense, as you would no doubt surmise from our unapologetic (even enthusiastic) embrace of both "note" devices. In our judgement the average reader is perfectly capable of thinking on two levels simultaneously. After all, a good orchestral or choral conductor is expected to be alert on a dozen or so levels while reading from a musical score!

3.4.6 *Special Considerations Applicable to Direct Reproduction*

Certain journals ("letter journals", see Sec. 3.2.3) publish authors' contributions in precisely the form in which the editors receive them—from the standpoint of both content and appearance. Every submitted (and accepted!) manuscript is simply photographed and reproduced exactly "as is". Manuscripts of this type are described as "camera-ready". We restrict ourselves here to a few brief remarks, since as we indicated earlier we believe publications of this sort have little future; technology has rendered them virtually obsolete.

Should you be planning to publish in such a journal, we remind you that

● No one at the publishing house will attempt to improve or modify your manuscript in any way prior to publication.

This means that pages you submit—after reduction in size, typically by about 30%—will be passed through to the reader in exactly the condition in which they arrive at the editorial office.

Standard procedure for preparation of a "traditional manuscript" calls for double spacing, but this would be inappropriate for camera-ready copy. For both aesthetic and space-utilization reasons the correct spacing is instead equivalent to "one-and-one-half" lines. By the way: size reduction will help compensate for the fact that the characters in *professionally printed* text (typically based on 9- or 10- "point" type; see Sec. 5.5.1) are smaller than those in the usual output from a typewriter or a word processor, which in most cases approximates "12-point" type.

Journals in the "letter" category normally limit illustrations to *line drawings*; that is, continuous-tone photographs are unacceptable. Figure copy obviously *should* be incorporated into one's manuscript in this case—again because no alterations whatsoever will occur at the hands of the publisher. One common practice is to paste photographs of original line drawings into spaces that have been left open for them. With illustrations originally drafted by hand it is usually necessary that such a photograph reflect a *size reduction* (usually about 50%), because the scale of the original is normally too large. Be sure that any *lettering* present in your drawings is relatively large, since it must remain clearly legible after reduction; a capital letter in figure copy for direct insertion into a text manuscript should be about 2 mm high.

3.5 From Manuscript to Publication

3.5.1 Publishers and Editors

Publishing Houses

A journal is the product of a *publishing house*, a commercial enterprise dedicated to preparing and distributing the periodical, but interested in it largely from an economic point of view. Even for a cause as noble as the advancement of science it is improbable one could find a benefactor willing to underwrite and promote a journal without serious attention to the laws of the marketplace. This is not to imply that all journals are the property of independent commercial publishers. Indeed, many belong to scientific societies or similar organizations, but the printing and marketing activities are usually delegated to publishers working under contract. In principle the parent scientific body might assume all aspects of the operation, but if it did so it would *de facto* turn into a publishing house. This would mean dealing not only with the problems involved in acquiring manuscripts—and in most cases advertisements as well—but also printing, circulation, management, financial accounting, etc. All these are foreign to the average scientific organization, although they might prove manageable if the primary goal were simply to provide information to the group's members. A tradition of broad "scholarly publishing" on the part of professional societies is limited primarily to Anglo-American parts of the world.

The identity and official address of a journal's publisher are most easily ascertained by consulting the publication's *masthead* (or "nameplate"), which is printed somewhere in every issue. This is also a good place to look for postal and e-mail addresses for corresponding with the publication's editorial staff (see below), as well as the URL for an Internet source of additional information.

Editors and Editing

No matter how production and related matters might be handled, every journal is dependent upon an *editorial office* to manage the publication's *content*.

● The primary task of a journal's *editor* is maintaining a steady influx of contributions, and then seeing to it that these contributions are transformed smoothly into published papers (often including digital versions).

The editorial staff might consist of one individual or a group of people, working either on a volunteer basis or professionally. Editorial services sometimes are rendered under contract. For a small scientific journal they are often assumed by practicing scientists whose primary professional concerns lie elsewhere: on the teaching staff of a university, for example, or at a medical facility. Entrusting supervision to an active practitioner lends an air of credibility and "seriousness" to a publication, but even the most devoted "part-time" editor is forced to limit his or her involvement, and it is unrealistic to think the affairs of a major journal could be managed successfully "on the side". A more conventional editorial operation consists of a *senior editor* in overall charge, several subordinates assigned to positions as "editors" or "editorial assistants", and diverse support personnel.

● The editorial office of a journal is charged with passing along to a *production department* (or subcontractor) content of sufficiently high *quality* in sufficient *quantity* to permit publication goals to be met in a timely fashion, so all issues of the periodical can be released on schedule.

As unexceptional as this responsibility may sound, the effort entailed in fulfilling it can be enormous—and sometimes targets are in fact not met. Success is heavily dependent upon a senior editor with a solid grasp of the appropriate scientific discipline, considerable publishing experience, and outstanding organizational skills.

● The first responsibility of every editor is to see that manuscripts submitted by authors are made suitable for publication, which usually involves overseeing at least some alterations.

Why must an author's manuscript be subjected to "alteration"? In the first place, editors frequently receive manuscripts which, although describing fascinating work, fall disappointingly short of acceptable standards with respect to *expression*. This might reflect *semantic* problems, for example, or serious weaknesses with respect to *syntax* or other facets of grammar. Even blatantly careless *spelling* is far from unusual despite the almost universal availability to authors of "spell-checking" software. Moreover,

- Manuscripts as submitted almost never meet every requirement from the standpoint of *form*.

Anyone with experience in the editorial office of a journal will confirm how few of the manuscripts received are immune to the foregoing generalization. Standards of "form" apply to many details of far greater concern to an editor than to the average author—fine points regarding the style of literature references, for instance. Editors by their nature are extremely demanding when it comes to *uniformity (consistency)*.[63] This is why they circulate detailed *guidelines* summarizing their expectations. Unfortunately, too few authors pay serious attention to such guidelines as they prepare their manuscripts, and this laxity or perversity is a major source of editorial discomfiture.

Quite apart from stylistic conventions perhaps applicable only to a single journal, which might therefore be dismissed as arbitrary, there also exist a host of official *standards* or *norms* to which publications of all types are expected to adhere, especially national and international conventions defining permissible *nomenclature*, *terminology*, and *units of measure*. In these matters editors are ordinarily much better informed than authors—as would be anticipated, since such knowledge is so central to their work. Authors are too prone to take a more "pragmatic" position, as in: "I frankly have no interest in such trivia; stuff like that has nothing to do with the important news I'm trying to spread". In other words, this provides an especially fertile field for editorial revision—but one could argue that is perfectly reasonable: the burden associated with quality publication is being appropriately shared.

One American dictionary defines "editing" as "**a**. Preparing (written material) for publication or presentation, as by correcting, revising, or adapting. **b**. Preparing an edition of a work for publication. **c**. Modifying or adapting so as to make suitable or acceptable".

Editing a manuscript can also entail *shortening*, in the sense of "condensing", "excising", or simply "jettisoning ballast".

- Several types of editing can be distinguished with respect to a journal article, including *content editing*, *linguistic editing*, and *formal editing*.

An author's manuscript is ordinarily "set right" in a series of steps, and at the hands of several people. In addition to efforts in the areas already mentioned, a certain amount of *language polishing* can be anticipated with a manuscript drafted in English by an author more accustomed to communicating in some other language.

The increasingly intense economic pressures to which scientific publishers have become subject has led editorial teams of late to forego some of the tact they once

---

[63] This despite Ralph Waldo EMERSON's famous observation that "a foolish consistency is the hobgoblin of little minds, adored by little statesmen and philosophers and divines".

prided themselves on. Thus, editorial guidelines—those "rules" we have repeatedly urged you to consult—now sometimes contain a warning that if a submission fails to meet certain minimum standards with respect to form (e.g., organization, numbering conventions, citation rules, etc.) it will not even be shown to a reviewer, but will instead be returned immediately to the author. And this kind of threat should be taken seriously!

- The job of editor involves considerably more than displaying editorial skills.

A senior editor's responsibilities begin with ensuring that an ample supply of suitable manuscripts will always be available, and end with worries about a journal's image and market position. Close inspection reveals that these apparently disparate "outer limits" are in fact intimately related: a highly respected journal will by its very nature act as a magnet for submissions, so in this case the problem of manuscript flow largely takes care of itself. What immediately comes to mind is the famous "chicken and egg" conundrum. Which comes first, a good journal, or an abundance of contributions?

Issues related to the *production* of a journal also command a certain amount of an editor's attention. Technological progress has made it feasible for articles now to be essentially *typeset* before they ever leave the editorial office, circumventing much of the processing that once threatened to cause publishing delays. Merely acquiring the ability to "massage" manuscript copy on a video screen has revolutionized everyday routine in editorial offices. Most serious editing is still carried out on paper, but changes that are required can now be entered immediately by an editor into the master (digital) version of a manuscript. Page makeup, too, is in most cases now attended to by the editorial staff, also as an on-screen process: edited copy is fitted into columns with precisely the required dimensions, tables and illustrations are optimally placed on the page, etc. *Tempora mutantur et nos mutamur cum illis*: times change, and we change with them (we ask the reader's indulgence with respect to this apt bit of Latin one of us just happened to have at his fingertips …).

There is one other important responsibility of an editor still to be addressed:

- The editor of a scientific journal is the one who *evaluates* and in the end *selects* articles to grace the publication's pages.

Every respected journal receives far more manuscripts for potential publication than can possibly be accommodated.

- The "better" the journal, the stricter will be the editor's criteria with respect to the quality of published papers, and the more latitude he or she will enjoy in manuscript selection.

Editors take pride in the role selectivity plays in their work, as might be inferred from the following illustrative synopsis, reported recently by a prestigious chemistry journal:

> 113 manuscripts were received for consideration as full papers, together with 15 short communications. Of these, 38 (30% of the total) were returned to their authors on the grounds that they either did not adequately reflect the goals of

the journal or failed to meet minimum standards for quality, while 6 are still under active review. Of the accepted manuscripts, 38% were published essentially as submitted, 49% underwent some modification, and 13% required extensive revision.

As this excerpt illustrates, editors and reviewers are often faced with situations that defy simple "yes or no" judgements. More subtle courses of action are in fact the rule: manuscript changes or "improvements" are proposed for consideration by an author with the understanding that if the recommendations are accepted, the contribution will be published. Revisions proposed under terms such as these are generally of a more or less formal character, having to do, for example, with *citation* style or *source* attribution, matters related to *nomenclature*, *abbreviations*, or *footnotes*, and similar issues with little impact from a content standpoint. Nothing more is usually at stake than stricter adherence to established policies, and hewing more closely to the characteristic "style" of that particular journal.

On the other hand, sometimes an editor feels the need to be more blunt, leading to a terse directive like "Shorten it!": to perhaps "half the length"; or "8 pages maximum". Or, "This is unacceptable as it stands because you ignored our directions regarding organization!" At this point an author is obliged to rethink the proposed paper in its entirety. In the first case this means reexamining every paragraph—word by word, number by number, and sentence by sentence—in search of passages that are superfluous. If the *logic* or *organization* of a manuscript has been called seriously into question, the only recourse is to accept the editor's evaluation and start over—probably with bitter resignation following a colorful outburst of righteous indignation—or else to try a different journal.

Attempt to view all constructive criticism aimed at your work as what it was almost certainly intended to be: a colleague's sincere attempt to enhance your communication skills and make the significance of your accomplishments more obvious to a reader lacking your familiarity with some particular area of scholarly endeavor. As the person closest to your work you are probably *not* in the best position to judge these matters, so why not give the experience and intuition of a reviewer or an editor the benefit of the doubt?

With some journals, rejection rates can be as high as (or even exceed!) 80%. This means that, in order to be accepted, a paper must be considered so good—in the sense of authoritative, significant, and original (cf. Sec. 3.2.2)—that through its publication the hard-earned reputation of the journal will be, if not enhanced, at least upheld. Editors of the most prestigious journals thus come to acquire a kind of *power*, and it has far-reaching consequences. This editor is after all the single individual in a position to determine whether your wish to have your work published in a particular journal will be fulfilled or not. A good editor exercises the "powers" of the office judiciously, however, and takes seriously the advice of reviewers and consultants when considering whether to accept or reject a manuscript.

Most editors are also under the watchful eye of an *editorial board* anxious to ensure that content selection takes place in a professional and unbiased way: i.e., there *is* a certain amount of oversight provided by the "supervisors of the supervisor". Manuscript review is clearly one of the important arenas for witnessing a phenomenon we alluded to earlier (in Sec. 1.1.2): what has been described as "a powerful system of *values management* in scientific communication".

Ignoring for the moment possible major concerns, an editor will call for minor alterations in virtually every manuscript received, even one that in general is clearly worthy of publication. Until relatively recently, editor/author communication in this respect was quite open and direct: hand-corrected copy, together with a "clean" unmarked manuscript, would first go from the editorial office to a typesetter with the job of preparing an initial printed version of the pending article. Then the same marked copy, in which all requested changes were clearly apparent, was sent back to the author along with the preliminary print (known as a "galley proof"; see Sec. 3.5.4). The author was thus in a position to examine and consider everything that had so far transpired, check the accuracy of the typesetter's work, and lodge any protest that might seem in order.

To the extent that editors now increasingly rework submitted manuscripts themselves (on a computer screen) as a first step in the actual production process, much of the former "transparency" of processing from an author's point of view threatens to be lost. We perceive here a real danger of too much anonymity and "behind the scenes activity" infecting the operation—certainly not something one would have anticipated as the byproduct of an honest effort to *enhance* communication. But authorship so far appears to enjoy relatively little respect in cyberspace.

● In an abstract sense, the most basic responsibility of every editor is to oversee *quality control*.

Early fears that technological advances would spell the end of the publication phenomenon as we have long known it have so far proven groundless. The central role of professional publication in upholding the tradition of quality in scientific communication, and in shielding the scientific public from an uncontrolled deluge of potentially suspect information, remains at this point largely unchallenged. Editorial offices were once described as the "valves" preventing an information overflow. It is exactly this function that few people have ever seriously questioned. Or is there somewhere in the world a scientist who would not be aghast at the notion that his or her work might appear, uncontrollably, in the immediate vicinity of, say, pornography?

3.5.2 Submitting the Manuscript

A manuscript is normally submitted to the journal of one's choice in *duplicate* or *triplicate*. As always, the journal's own guidelines should be consulted for details. One printed copy will almost certainly be sent by the editor to a reviewer for comment (also known as a *referee*; it may be that *two* reviewers will immediately be recruited). Most editors also expect that the printed version of a manuscript will be supplemented with an identical document in digital form (see below).

● Each (printed) copy of the manuscript should be accompanied by a complete set of equations or formulas, tables, illustrations (figures), and other relevant supplementary items.

At least one backup copy of everything should be retained by the author for security purposes.

● Materials destined for the editor should be accumulated in a single package, to which should be attached a formal *cover letter*.

In the cover letter, express your desire that such-and-such a manuscript be published (refer to it by title!), followed by categorical assurance that the paper is not also being offered to some *other* publication. A brief indication of the context of the work would not be out of place (e.g., "This represents a continuation of the studies described previously in YYY…").

● Be sure to pack everything carefully to maximize the chances of the parcel arriving intact.

The appropriate address for shipping is ascertained most easily from the journal's formal masthead, which appears somewhere in every issue (cf. "Publishing Houses" in Sec. 3.5.1). This address may, incidentally, be quite different from the *business address* of the publisher. In some cases authors are requested to send their contributions to national or regional editorial representatives. If instructed to send the manuscript directly to the publishing house, be sure you indicate on the address label the name of the target journal, because many publishers manage the affairs of several journals.

● The editorial staff will soon send you a *confirmation notice* affirming that everything has arrived in good order.

This official confirmation generally indicates as well that the manuscript will be sent away for review prior to any decision regarding publication, and that in the event the work is eventually accepted for publication, all legal rights to its content will pass to the publisher (cf. Fig. 3–1). The letter in effect serves as a *contract* between you and the publisher. If confirmation of receipt has not reached you within two or three weeks of submission you should initiate a tactful inquiry, since there is always a possibility your package might have gone astray, or the confirmation letter could have been lost.

For many authors today, postal communication ("snail mail") is far too leisurely, so they prefer to rely instead on some form of *telecommunication—e-mail*, usually.[64]

---

[64] A manuscript can be transmitted by e-mail halfway across the world in a matter of seconds, and at almost no cost to the sender. Unlike a fax message, an "electronic letter" has the advantage that it can be processed at the receiving point as digital information. Data files of many types—including text— can be "attached" to an e-mail message, as if with a paper clip.

If you fall in this category, be sure to check with the editors in advance to make sure they are prepared to deal with manuscripts submitted in this way (most will say they would be delighted).

Even the economics of dispatching manuscripts electronically can be highly favorable. For an institution accustomed to sending off a great many documents on a regular basis the savings realized can be on the order of several thousand dollars a year! Some institutions prefer a slightly different alternative: making documents accessible to others through an Internet *FTP server*.[65]

It would also be wise to ascertain for sure whether you are expected to submit a backup digital manuscript—on a *diskette*, a *ZIP disk* (Zigzag Inline Package), or a *CD-ROM*, for example. If so, inquire also about the preferred *format* for such a file. (Even today there are publishers that insist on receiving "text-only" or ASCII files, and refuse categorically to have anything to do with digital graphics.) The procedures applicable to a given publication are in most cases directly related to the "intelligence level" of available typesetting equipment.

As an illustration of the way things are changing, consider the following brief policy statement issued a very few years ago by the journal *European Science Editing* (October 1996):

> Please submit all contributions, other than very short ones, on (1) 3.5-inch disks (IBM-PC compatible format) with files saved in ASCII—plain text, no word processor codes or page divisions, etc., or (2) by e-mail. A double-spaced printout should also be provided.

These instructions, in which the e-mail option is designed to act as a subtle barrier to inappropriate codes,[66] at the time represented a sensible approach to avoiding compatibility problems, at least for a journal whose articles rarely exceeded two or three pages. Compare this, however, with the following excerpt from the *current* (2002) guidelines for the same journal:

> Contributions should preferably be sent by e-mail or submitted on disk (see ... below) ... Longer items such as articles may be sent as e-mail attachments, other items as ordinary messages ... *Text* should preferably be produced in Microsoft Word (.doc or .rtf extensions) or in WordPerfect ... If ASCII format (.txt

---

[65] FTP stands for "File Transfer Protocol", a set of standards for facilitating the electronic transport of documents from one networked computer to another. Internet subscribers can in this way request files of all types (text, graphics, or software) from remote archival locations, writing them directly to the hard drives of their own computers.

[66] E.g., formatting information; most e-mail systems reject messages containing code characters—with the possible exception of ones incorporated into attachments. In some cases if you attempt to send a formatted manuscript as an e-mail attachment you may briefly see on the screen a message like "attachment being encoded". This is generally not evidence of a problem, and certainly should not require intervention on the part of an IT (information technology) specialist. On the other hand, difficulties are not without precedent, and there is always the possibility that the recipient of your message will be unable to "open" certain attachments. See Chap. 5 for further information.

extension) is used, however, indicate italics or bold lettering by underlining [the appropriate words] in the printout ... *Tables* set in MS Word may be included in the text. Tables set with other programs must be sent as a file on disk, separately from the text. *Figures* must be sent in separate files, saved in .bmp, .tif, .jpg, or .eps format.

As might be expected, the American Chemical Society (ACS) is especially far along the road toward full utilization of digital tools. While the *Journal of the American Chemical Society* does still accept hardcopy submissions, authors are strongly encouraged to submit via the Web (neither fax delivery nor the use of e-mail attachments is permitted). Web submissions of short communications *must* be prepared with the aid of special templates, available for downloading in versions customized for Microsoft WORD, WORDPERFECT, and FRAMEMAKER for both the IBM and MACINTOSH platforms. Similar templates are available for full papers, though their use is not yet mandatory. Special instructions are provided for TeX users, but submission of ASCII files is explicitly forbidden! Sample articles may be downloaded for inspection in the form of PDF files. In an even more recent development, ACS is in the process of implementing what it calls the "Paragon System", which permits an author not only to submit manuscripts via the Web but also at any time to access comprehensive information about the current status of a pending manuscript. Beyond that, the Paragon system is designed to put reviewers in a position to carry out their evaluative activities online.

For more information pertaining to digital documents see "Technical Prerequisites" in Sec. 5.4.1.

3.5.3 *Manuscript Review*

Most journals send submitted manuscripts upon receipt to one or more (commonly two) *reviewers* (*referees*) for comment. Editors maintain extensive lists of colleagues who have expressed a willingness to serve in this capacity on a volunteer basis, as well as contact information for acknowledged experts available for special consultation as required. Bear in mind that the rigor attending a journal's refereeing practices is a valuable indicator of the *quality* of that journal. If it cannot describe itself as "fully refereed" a journal is assumed by most to be "second-rate" at best. The tenacity of this conservative respect for quality on the part of scientists generally (where quality is defined in terms of peer approval) is one factor that initially impeded the establishment and development of e-journals—justifiably, in our judgement, especially at a time when things were obviously in a state of flux. Interestingly, quality assessments with respect to scientific publication come regularly from two complementary but opposite vantage points: authors use peer review policies to judge the relative worth of a journal, and journals use peer review to judge the worth of an author's efforts. This suggests a fundamentally balanced and healthy state of affairs, even if the judgements themselves may not be 100% accurate.

● Submitted manuscripts are subject to evaluation in a variety of contexts.

Questions likely to be raised with regard to a potential publication include:

– Is the work described both meaningful and significant?
– Would both the objective and the results be classed as original?
– Does the approach taken seem appropriate given the question(s) raised?
– Were the proper experimental methods utilized?
– Do results disclosed unambiguously support each conclusion drawn?
– Does the manuscript include a comprehensive and thoughtful discussion?
– Is the article well put together?
– Does it address issues of interest to this journal's audience?
– Have figures and tables been prepared in a professional way?
– Have rules of nomenclature and terminology been faithfully observed?
– Are there sufficient literature citations, and are references correctly formulated?
– Is the manuscript of an appropriate length?
– Does the article culminate in a concise, pithy summary?
– Does the title accurately reflect the contents?

One other issue is becoming the target of increasingly intense scrutiny:

– Is there sufficient statistical support for the author's conclusions?

The latter criterion has come to assume so much importance in recent years that one can now safely assert that

● Papers are rarely accepted for publication unless quantifiable results they report have been evaluated statistically.

Referees typically assign "grades" to papers they review (e.g., "outstanding", "high-quality", "average", "below average", "poor"). Unfortunately, they often are forced to conclude that, while the work reported is in principle worthy of publication, certain aspects of the presentation are so substandard that changes are absolutely essential (elaboration, for example, or restructuring or abridgement). The editor will promptly pass to the author suggestions and comments received from the reviewers. Only under rare circumstances would someone on the editorial staff attempt to deal independently with a reported problem.

If two referees come to substantially different conclusions, it is the editor who must decide how to proceed. He or she is, after all, ultimately responsible for everything that appears in the journal. The author who in the end must cope with a rejection notice always has recourse to other journals; indeed, submitting elsewhere may not even require a "step down the ladder". It could be, for example, that the content of the article was simply deemed a poor match with the mission of the particular publication approached, which in no sense can be interpreted as demeaning the quality of the work, or indicative of a "snobbish attitude" on the part of an editor.

Refereeing has been the subject of endless debate in circles frequented by both editors and authors, largely because it can have such a profound impact on an individual scientist's career. Even eminent scholars have on occasion expressed serious reservations about the entire "peer review" system, and it will no doubt remain a source of considerable anxiety. Is "tough refereeing" really necessary? Might mandatory peer review be more costly than it is worth in terms of time, effort, and psychological pain? How professional, trustworthy, and independent is the average referee? And how *fair*? Is the system not perilously susceptible to abuse? Wouldn't almost any referee tend to be exceptionally critical of the work of a competitor in his or her own field, or be prone to display favoritism—perhaps unconsciously? Should a referee be allowed to know who submitted a paper under consideration? And should authors be advised of the identities of those who referee their papers? Would more openness make reviewers reluctant to express their true feelings?

Questions like these have been raised again and again in "letters to the editor" columns as well as at scientific meetings, and they have been discussed at excruciating length from a host of perspectives, input coming from people with widely divergent interests and the potential to be affected in very different ways by the outcome.[67] The subject also lends itself to disciplined and objective investigation, however, in terms of meticulously formulated, searching questions that are also *answerable*: e.g., "How often do two referees disagree in their evaluation of a manuscript? To what extent are manuscripts that have been rejected by one journal subsequently accepted for publication in another? If delayed publication in this sense does occur, how likely is it that the second journal will be as prestigious as the one that initially rejected the paper? The most comprehensive study to date along these particular lines is almost certainly one carried out and subsequently described by DANIEL (1993). Based on extensive (unfiltered!) information supplied at his request by the editors of a very highly respected chemistry journal (*Angewandte Chemie International*), DANIEL concluded that the refereeing process, assuming it is properly managed and carefully monitored, on balance performs a valuable service. He found no justification whatsoever for dismissing it as a dangerous source of malevolent side effects.[68]

---

[67] In the mid-1980s, attention was drawn to a gene-technology study submitted for publication in the highly regarded journal *Nature*. The work had to do with the nucleotide sequence in a gene that codes for interleukin I. Prior to eventual publication the manuscript was twice rejected, in part because the work was considered incomplete. The macabre part of the story is that the scientific investigation was sponsored by a genetic engineering company, and one of the reviewers was in the employ of a key competitor! At stake was information potentially worth millions of dollars—and the controversy swiftly passed over into the legal system in a judicial wrangle that lasted several years. A report dated 20 September 1996 in the German newspaper *Die Zeit* presented an account of the case under the provocative title "An Important Lesson: Referees Learn Much—What Can They Legitimately Do with Their Knowledge?"

[68] The distinguished Harvard chemistry professor William von E. DOERING, American born of German parents, once delivered a lecture in the historic Bunsen Lecture Hall of the University of Heidelberg

→

What remains for us to consider in this section is how suggested improvements eventually find their way into a manuscript. In the first place, comments from the referees (as well as the editor's own thoughts) are communicated directly to the author, who is encouraged to respond to major concerns through revision of the manuscript in question. The adapted manuscript is then reviewed once again by the editorial staff to make sure all outstanding issues have been satisfactorily resolved.

Minor corrections are another matter. These are considered to fall under the heading "editorial changes", and would normally be resolved by the editors alone, without active involvement on the part of the author, as noted at the end of Sec. 3.5.1.

It is clear that editorial procedures can be expected to evolve further in the future as the effects of new technologies become increasingly pervasive. One potential scenario we see as attractive would play out as follows: An editor indicates as clearly as possible in a copy of the original manuscript—possibly a digital copy—a set of proposed changes. This copy is then forwarded to the author with a request that he or she incorporate into the master document all changes that appear *acceptable*. The revised manuscript is subsequently returned to the editorial office, where it is reviewed and eventually restored to the production stream.

The nature of procedures *currently* in effect at a target journal should be clarified through direct contact between author and editor.

3.5.4 Editing, Typesetting, and Page Proofs

What happens next? In the paragraphs that follow we describe the traditional course of a typical manuscript after the initial rounds of evaluation and correction. Here and there we make note as well of recent changes brought on by new technologies. Despite substantial progress, at virtually every stage in the process there remain places where the workflow might be streamlined further, and ample opportunity exists for editors and publishers to derive even *more* benefits from the latest developments. It is probable that within a few years virtually nothing will be left of what for decades has been "routine" in the journal world. The most pressing objective for publishers at the moment

---

which began roughly as follows: He said that his intent was to discuss a notion that was currently fascinating him, something he called "valence isomerism". He had in fact submitted an article on the subject to *Angewandte Chemie*, but it had been rejected as not worthy of publication. Since at the time nothing more interesting occurred to him, he wished to make that the subject of his lecture anyway. (This must have occurred in the early 1970s. One of us was actually present to experience it. At that time the editorial offices of the journal in question were also in Heidelberg! Not a soul walked out of the room in protest.) Long before his retirement from Harvard in 1986 DOERING had received accolades from virtually every quarter for his exciting and startling contribution to our understanding of bonding in certain types of compounds. What can we learn from this incident? 1. Even "reviewers" sometimes make mistakes; *errare humanum est*. 2. The editorial staff of *Angewandte Chemie* proved immune to intimidation. DOERING was already a renowned chemist, but what he was proposing was fantasy, pure and simple. 3. All of us are enriched by occasional fantasies originating with special individuals.—A truly new idea? First it is seen as ludicrous, then perhaps as scandalous, and—in rare cases—ultimately declared self-evident!

is to complete the transformation to *digital processing*, but in such a way that quality does not suffer. The challenge has been expressed elsewhere (*Apple Magazine*, issue 5 winter 1997, p. 24) in slightly different terms:

● Copy of every sort—text, graphics, photographs, audio clips—should be digitized at the earliest stage possible.

Ideally, thorough digitalization would transpire before a manuscript leaves an author's desk.

In the conventional approach to publication, a package consisting of text, formulas, illustrations, and whatever else might be connected with a particular manuscript—after all anticipated revision and editing—is first passed along to a *typesetter* responsible for the initial steps leading up to printing of the text. Alternatively, someone in the editorial office might transfer the text into an in-house word-processing system, thereby avoiding separate typesetting. Assuming page makeup is still accomplished manually, what follows next is a series of *photographic* operations resulting in *films* of the text and all subsidiary elements. Finally, *printing plates* are prepared, either on the basis of the films or directly from a computer file that represents a complete set of preformatted pages already containing graphics and any other "special" features. [69] The most modern printing processes eliminate the need for plates altogether.[70] "Wet chemistry" printing is largely a thing of the past, and completely digitized production processes are clearly the wave of the future. It was editors who for the most part set in motion the wheels leading to these changes, and now they proudly lay claim to authority over efficient "electronic editorial offices" (EEO; cf. Sec. 5.4.2). The willingness of

---

[69] The broadest technical term for the tangible entity (regardless of its nature) bearing an image that is to be printed, and also providing protection for that image until printing actually takes place, is *print carrier* (or *print form*). In the case of *offset printing* (more accurately: offset *lithographic* printing) the carrier is a thin metal plate. In the past, images were always transferred to such a plate with the aid of photographically prepared films. In more modern "direct-to-plate" ("computer-to-plate") processing, copy is imposed on a printing plate *directly* without any need for intermediate offset films or manual mounting of negatives. A "preview" of the anticipated results is carefully inspected on a video screen, and—assuming no problems are discovered—the corresponding image is conveyed to a light-sensitive print carrier. In a related procedure, a POSTSCRIPT data file (see "Miscellaneous Peripheral Components" in Sec. 5.2.1) directs the motion of a laser beam which in turn produces an image on the carrier. Much of what only recently was "state of the art" processing even in a modern printing plant has thus been banished to the dustbin of history.

[70] An example is the so-called *risographic* printing process ("digital printing"), in which data from a computer file causes tiny holes to be burned in a *printing film*, through which ink later passes onto sheets of paper waiting to be printed. ("Riso", by the way, is simply the name of the Japanese corporation that perfected the process. Nevertheless, it is a suggestive combination of letters, reminiscent, for example, of "risotto", or simply "rice". Or consider the possible allusion to quality: "resolution" with the Riso methodology is said to be 600 dpi.) The technique is especially effective—and economical—with relatively small print runs (a few thousand copies). The familiar "workhorses" of the offset printing industry over the past quarter-century—photographic films, a skilled "mounting" staff, and offset plates—are fast being reduced to the status of artifacts from an expensive and time-consuming detour off the more rational path of direct reproduction. The *Publisher's Handbook* (*Verlagshandbuch*, PLENZ 1995), from which we drew much of the information in this and the preceding note, was itself produced risographically.

editors to adapt and enthusiastically assume a more intimate role in production has undoubtedly contributed in important ways to the survival of journals as we have always known them.

● By today's standards, the ideal editorial arrangement would allow the forthcoming issue of a journal to be assembled entirely from data files supplied originally by authors.

Under these circumstances, information would never leave the "electronic level" from the time it was first saved on an author's word-processor disk until it emerged in hardcopy form for public consumption. Does this mean we think editors are destined to become little more than "publication disk jockeys", charged only with shuffling electronic data files in the proper way and at the proper time? Certainly not. Many of an editor's current "busywork" tasks may indeed become obsolete, but the most definitive responsibility—maintaining "quality"—is unlikely to change significantly. Automated "electronic assembly" of a journal through the simple merger of a set of profound and technically flawless digital manuscripts painstakingly crafted by conscientious and gifted authors is a utopian dream, not a vision of future reality.

At least one tangential aspect of this unfolding story warrants brief mention:

● There is certainly no assurance that the end product of (electronic) editing of scholarly manuscripts will always be a *printed* entity.

Indeed, the possibility is real that tomorrow's equivalent of many journals will be confined entirely to the digital realm. Any of several media could eventually turn out to be the preferred repository, one option being that information ripe for dissemination will simply be "uploaded" to some secure and widely accessible but largely unstructured database.

But once again we must return from the future to the present—and the past.

Until recently, *figures* accompanying an author's manuscript were almost always reconstructed, in modified form (by hand!), by professional draftsmen on the publishing-house payroll. At the very least, an author's figures were subjected to extensive editing, and labels present were almost without exception replaced by new ones based on a locally standardized alphabet. The next step was filming (or scanning) the redrawn figures as a precursor to incorporating the corresponding graphic images into printing plates. The many steps entailed were among the responsibilities assigned to a publisher's in-house *production staff*. Today it is essentially standard practice for analogous ends to be achieved electronically, and under the direction exclusively of members of the editorial team.

Irrespective of the nature of prior processing steps, *proofs* (or *galley proofs*) of fully edited text and graphics were always prepared next and dispatched promptly to the author for *proofreading*. It was the author's responsibility to examine these proofs as quickly as possible, in minute detail, and flag every error (as described in the section that follows). Inexperienced authors were usually surprised to discover that, at this stage, text, tables, footnotes, and illustrations were all on separate sheets rather than merged as they would be in the final document. As explained earlier, this was because various components made their way through the editorial and production maze along independent paths. It is quite possible that the journal you are working with still does many things in the traditional way, sufficient reason for us to continue our description of the "classical" methods:

- Authors are admonished that *proofreading* has as its sole purpose the detection and repair of processing errors.

Any attempt to exploit this step as a convenient excuse to *modify* a piece of text, or silently *supplement* data originally submitted, will earn a stern reprimand.[71] Changes made at this stage are considerably more expensive than typesetting the same amount of material in the first place, and publishers regard it as legitimate to ask *you* to cover at least part of any unjustified expense—out of your own pocket! Consider too that if you request significant changes this late in the game it could cast doubt on the overall reliability of what you are reporting. Any truly essential last minute changes require explicit authorization from the editor—and even then they may appear in the final publication labeled "notes added in proof". You also need to consult directly with the editor if you wish to take exception to some editorial change apparently made without your knowledge.

Corrected proofs are returned directly to the editor (not to a production department, even if that is where they originated). Authors' corrections ordinarily take the form of annotations in the proofs themselves (see below). Actually, anything you introduce now will be viewed initially as *suggested* corrections. Some of your requests might well be ignored, especially ones that appear to be merely stylistic.

Authors' corrections (assuming they are approved) are sent to the production staff for physical incorporation into whatever will serve as a forerunner of the printed document (e.g., an offset plate). It is only at this point that the various components (text, formulas, tables, illustrations) would traditionally be combined and arranged in orderly *columns* and *pages*, a process known as *page makeup*. This led finally to another set of proofs: the *page proofs*.

- Page proofs of journal articles are normally examined only by the editorial staff.

---

[71] We are all subject to limits when it comes to changes. Consider e-mail: Once a message has been sent on its way it is no longer subject to change, not even in your own files. The structure inherent in an "information traffic system" like *Netscape Communicator* or *Outlook Express* takes care of that: nothing can be altered after the fact (though *America Online* has always featured a unique "Unsend" command in its e-mail facility). IT essentially blocks any temptation you might have to reconsider what you said, let alone attempt some sort of "cover-up". Proofreading is admittedly not subject to such technical obstacles; you are therefore left to practice restraint on your own. If you do insist on proposing a last-minute change, at least indicate it by markings in a distinctive color, because it must remain clearly visible for later scrutiny.

Two main questions must be addressed as page proofs are inspected: (1) Is every piece of every article accounted for, and have components been merged in acceptable ways? (2) Have all authorized corrections been properly dealt with?

Authors (if permitted) tend to develop their own ideas—different from those of an editor, and not always practical!—about what constitutes an "acceptable" page makeup. The final decision always rests with the editor, of course, and the potential for unproductive arguments at this stage is one reason, apart from cost and scheduling factors, why final page proofs are normally *not* shared with authors. We have already emphasized that one of a journal editor's chief obligations is ensuring that each issue is released on schedule,[72] and the risk of delay would be too great if every author were allowed to become involved yet again so late in the process.

Thanks to advances in computer technology generally and computer-based typesetting in particular it has become increasingly feasible to combine what in the past were two stages of proof correction into one (via a procedure known as *direct makeup*), bypassing altogether the familiar galley proofs. An obvious advantage from the author's standpoint is the opportunity this provides to have a preview of the projected page makeup (or *layout*) at proofreading time. Furthermore, the editorial staff is able to direct last-minute attention primarily to the publication in its entirety.

We encourage you to find out exactly what processing steps are anticipated with your particular article. The editor may be able to allay any concerns you have by offering to show you a *preprint* immediately prior to publication, although if the journal is one released at two-week intervals virtually no time would remain for responding to eleventh-hour objections.

3.5.5 Proofreading

The Art of Proofreading

It was noted in the preceding section that authors are usually obligated to *proofread* at least one printed version of their articles. Checking scientific text *effectively* for processing errors demands intense concentration on the part of the reader. More than that, for many it requires rethinking the very process of reading. Professional people in general tend to read rapidly. Most have little choice, given the almost universal mismatch between available time and tasks to be completed. Much professional reading would be more accurately characterized as "skimming". Proofreading, on the other hand, requires minute attention to every detail, which quite literally means reading *letter-by-letter*. It is not enough that one know how a certain word *ought* to be spelled; the question is, has that word been spelled correctly *here*? It should come as no surprise that typesetters think very differently from scientists; indeed they operate in a totally

---

[72] Dates of publication are even the subject of formal agreements between publishers and the Postal Service!

different world. To a typesetter, speed is of the essence, and what the text *says* is completely irrelevant (and probably unintelligible anyway). At the hands of a typesetter, "conformation" can easily become "confirmation", just as "neuron" might be blithely transformed into "neutron", and "valence" could end up as "valance".

Occasionally, typesetting *computers* are responsible for mistakes, especially with respect to end-of-line hyphenation, which is often based on rules that are fallible. The subject is complicated by the fact that British writers and publishers hyphenate certain words very differently from their American counterparts (division by *derivation* as opposed to *pronunciation*[73]). If in doubt regarding the proper placement of a hyphen, look in a good dictionary, and always remember that the "intuitively obvious" place to divide a word very frequently turns out to be the *wrong* place to divide it (at least in the eyes of the "language experts"). As you might imagine, hyphenation errors tend to be especially prevalent when English text is typeset by a publisher more accustomed to working with another language, or when foreign words must be divided.

Tables are where it most often becomes clearly apparent whether a manuscript has been carefully proofread or not, especially with respect to *numerical data* and to *units* specified in headers. Errors in figure captions are also frequently overlooked.

You should always anticipate numerous errors in text that, for whatever reason, has been *scanned* and interpreted with an OCR (optical character recognition; see "Miscellaneous Peripheral Components" in Sec. 5.2.1) system. The editorial office may have discovered, for example, that a defect in one of your text files prevented it from being "read" by the publisher's word-processing program, so an attempt was made to recreate the file from an OCR scan of one of your printed copies.[74] Many of the mistakes introduced along this pathway are, in essence, "visual" mistakes. Like a human, the scanner acquires its raw data through an "eye". Unfortunately, its vision can be a bit fuzzy, and things really become problematic with poor copy (very light print, for example, or pages defaced by smudges or coffee stains). The accuracy problem is of course compounded by the fact that the "brain" connected to the scanner's "eye" has

---

[73] Consider the word helicopter. No one could blame you if you proposed to hyphenate it as he-li-cop-ter, because it is *spoken* that way. Alas, the word is of Greek origin, derived from *helico* (cf. helix), a reminder here of the screwdriver-like motion and function of the propulsion system, and *pterein*, to fly. Hence—helico-pter? An American dictionary actually recommends hel-i-cop-ter: yet a *third* possibility. Some dictionaries, even good ones, omit information regarding hyphenation, and some newspapers appear to be unaware of the problem altogether, terminating lines whenever they become "full", seemingly unconcerned about the nonsensical results or the aesthetic pain inflicted.

[74] The OCR process amounts to asking a computer to "read" text from a graphic image. This entails examining the image closely, point by point, keeping careful track of where it is black and where it is white. From observed patterns, a "guess" is then made about what letters might be where. This recognition stage is extremely complicated, and it relies on sophisticated algorithms developed specifically for the purpose. If one is fortunate and the "guesses" turn out to be right 99% of the time, that *still* means that a typical recreated manuscript page (consisting of some 1500 characters) can be expected to contain about 15 erroneous characters!

relatively little "reasoning power". Examples of letter or symbol combinations frequently transposed by OCR software include:

im/nn, H/II, 'I'/T, ü/ii, (j/G

The "slash" (/; also "virgule" or, if more slanted, "solidus") used to separate symbols in the examples above is itself a source of complication, since it can prove virtually indistinguishable from an upper-case italic letter *I* (as in *Italian*) or a lower case italic letter *l* (*latin*). Furthermore, certain type fonts that are visually very appealing and effective seem almost tailor-made to engender confusion with a scanner (*cf.* rm *vs.* im *or* nn in the case of the "Times" computer font).

You are most likely to spot random mistakes introduced by scanning if you approach the proofreading job with a "trust nothing" attitude and make doubly sure to focus your eyes intently on the printed page. Fortunately, the human brain is equipped with a "visual memory" function capable of sounding a "Something is wrong here!" alarm even before the reader is consciously aware of a problem, let alone its nature. This safeguard is most apt to be operational and primed to issue an alert when you *know* that the text before you is suspect because it was scanned, another justification for trying to be as fully informed about the production process as possible (see also "Strategic Considerations" in Sec. 5.4.2).

A certain amount of effort can be saved if you are sure that at least some of your data files were never altered from the time you prepared them. In that case there is almost no chance that new errors will somehow have crept in. If you are positive your files contained no mistakes in the first place, you can be "99% sure" they are still clean, so there would be *less* reason for proofreading the corresponding material exceptionally carefully. A fairly quick examination should suffice, concentrating primarily on *ends of lines*, special *symbols*, and similar sensitive sites. This is in fact one of the reasons authors relish production methods that never require a manuscript to drop down from the digital level to which it was originally raised. You could probably be persuaded to take greater pains preparing your original "manuscript on disk" if you knew that the extra vigilance could spare you a painful and time-consuming proofreading ordeal later.

The ground rules are unfortunately quite different if you are dealing with an ordinary manuscript that has undergone conventional processing, since errors of virtually every type could have been introduced almost anywhere, and your goal must still be detecting and correcting each and every flaw!

● Special attention should be accorded to the proofreading of *numbers*.

Bear in mind that a single erroneous number in a published experimental procedure could someday be responsible for a devastating explosion, or a gruesome case of poisoning! Assuming you have not had reason to memorize numbers present in your manuscript (and therefore are unlikely to recognize mistakes on sight), there is no alternative to examining carefully every printed digit and comparing it directly with the original manuscript, at the same time thinking about it a bit from the standpoint of plausibility.

● The best proofreading occurs when two people work together as a *team*.

If one member of such a team takes the initiative to read a manuscript slowly and painstakingly aloud while the other compares what is being articulated with a printed copy, the chances are good that (nearly) all errors will be detected.

Checking *page proofs* is largely a matter of making sure nothing is missing. Do all paragraphs begin and end as they should? Are all footnote markings and the footnotes themselves where they belong? Is the placement of tables and figures reasonable? Has every figure been paired with the correct caption? Above all, of course, has every correction specified in the galley proofs actually been made, and made properly?

With page proofs it is especially important to look not only at the precise spots where changes were ordered in the galley proofs, thereby verifying that everything has been done correctly, but also to inspect the general *vicinity* of each such change. Acting upon a galley-proof instruction sometimes causes *new* problems to appear, and this is almost as true with electronic typesetting as it was decades ago when printers worked with movable metal type. Something could well have gone wrong anywhere from the point of change to the end of the corresponding paragraph, because every line of type in that region has probably been reset. For example, these lines are excellent candidates for new end-of-line-hyphenation issues.

Marking the Mistakes

We have dealt at length with the "philosophy" and technique of proofreading, but have so far said nothing about how the editors should be notified of mistakes you find. In previous editions of our book we discussed proof *correction* in considerable detail, but this time we leave it largely up to you to find out what correction scheme the journal working with your paper prefers. Perhaps you will be asked to familiarize yourself with a set of special *proofreading symbols* to be applied directly to the problems and/or nearby in the margin. On the other hand, some editors are content with a much more informal alerting mechanism.

If you submitted a carefully corrected electronic manuscript in the first place, and if this has served as the basis for typesetting, there may well be relatively little for you now to do. Members of the editorial team will almost certainly have made a number of minor changes, bringing your work more closely in line with policies peculiar to their particular journal, but they will have done their job very meticulously, and it is unlikely to have resulted in damage. You may be permitted to use on-screen methods as you go about proof correction, later sending to the editors a fresh copy of the refined contribution, possibly as an e-mail attachment. By the way: be sure you don't overlook e-mail you might *receive* at this stage, because additional instructions or information could be coming to you from the editor.

Carefully file away one complete copy of corrections you have made. All original proofs must of course be returned to the publisher, and you certainly would not want to face repeating the whole job if the package you send the publisher is somehow lost in transit.

There remains one point for us to raise in this context: what happens in a case involving multiple authorship? This is a team-management issue, left more or less up to you. But it impinges on the publisher or editor as well. Who is to receive proofs? A question to be addressed early! If a dozen or more colleagues cooperate with you on a project, who is entitled to coauthorship of the resulting article? Assume for the moment you are the responsible (or "corresponding") author, and the only one to whom proofs are sent. Fortunately, we now live in a digital world! This means you *could*, in principle, send to all your colleagues copies of anything you receive in digital form, asking each for comments. It is of course comforting to know that the recipients of your message can also register *their* comments at the digital level.

Word-processing programs such as WORD include special provisions for just this type of situation in the form of *comment* and *revision* functions. Any amendments, changes, deletions, or supplementary material can be clearly differentiated by color—typically yellow for comments, red for revisions—and tagged with the name of the reviser as well as the date and even the hour of the change. Forerunners of this convenience feature were special "editing programs" (such as DOCUCOMP), as described in some detail, for example, in the 3rd edition of the German counterpart to this book *Schreiben und Publizieren in den Naturwissenschaften*. We felt it unnecessary to provide a lengthy explanation this time, merely drawing your attention instead to appropriate menus in your word processor. Now all you need do is define your coauthors as a subset in your list of (electronic) addressees, after which you can communicate with them by a mouse-click.

Once corrected proofs are on their way to the printer, you should simply sit back and await the pleasure of seeing the perfect article—yours!—in an upcoming issue of your favorite journal.

4 Books

4.1 Preliminary Thoughts

4.1.1 What Is a Book?

There are some who would have us believe that the *book* as a cultural artifact is destined within a few years to disappear somewhere into a corner of *cyberspace*. Indeed, one advocate of this view went so far as to suggest that the Frankfurt Book Fair (the world's largest annual book exposition) should be regarded as "the deck of a sinking ship". Is there in fact no point in our trying to envision new cargo for that ship? It certainly is true that the *scholarly monograph*[1] as a medium of communication encountered rough seas during the 1990s. With few scholars convinced they could spare the time to work their way through a complex book, and many arguing they could no longer afford to *buy* such books anyway, it stands to reason that book production would suffer. Equally striking has been a disturbing scarcity of scientists betraying any interest in *writing* books, or even being active participants in their preparation. Under these circumstances there is certainly every reason to question whether we should have taken the trouble to put together the present chapter.

We think it *was* a worthwhile effort, however, with respect at least to a fair number of our readers. The recently prevalent negative attitude toward books actually shows some signs of moderating, if only as a consequence of the concept "book" gradually becoming subject to broader interpretations. Thus, certain CD-ROMs are now increasingly accepted as in every sense the equivalent of books. Nonetheless, the latter development has in fact had only a minor impact on what we are setting out here to accomplish, since the way a book's content is developed and refined—our chief concern in this chapter—is relatively independent of the *form* the product takes, at least at the creative level.[2]

---

[1] This is the type of publication we generally have in mind when we utilize here the more compact but less definitive word "book". The difference between the two concepts—"monograph" and "book"—is similar to the one we established earlier between "scholarly journal" and "periodical" (cf. Sec. 3.1.1, and also Sec. 4.1.3).

[2] The publisher of this book, Wiley–VCH, is responsible for one of the most impressive electronic books currently available—both in terms of extent and also scientific stature: *Ullmann's Encyclopedia of Industrial Chemistry*. The 5th Edition of this monumental work, consisting of 36 thick volumes and a one-volume index, was completed in 1996. Soon thereafter it became available in its entirety on CD-ROM. Would users of this work in the future concentrate their attention exclusively on the disk-based electronic version—also available in the meantime via the Internet? To the astonishment of the book world, at the 2002 Frankfurt Book Fair Wiley–VCH made the announcement that "Ullmann's 6th Edition is coming in print!" The new edition has been available since early 2003—this time in 40 volumes containing about 30 000 pages—and early sales results have been excellent.

We will begin by supposing that *you*, one of our readers, have been toying with the idea of becoming a *book author*. Bear with us through this little exercise; it may be you will find yourself feeling surprisingly comfortable in the role we suggest, and much sooner than you could have imagined. At the very least you almost certainly will discover that some of what we discuss (e.g., the preparation of an *index*; Sec. 4.5.1) will prove useful to you someday in other contexts.

Interestingly, there is no simple, concise way to capture accurately all the implications of the word "book", even though everyone purports to have a clear personal sense of what the term means. For our purposes, in the limited realm of academic book literature a serviceable characterization might run something like "the essence of a body of knowledge, expressed in symbolic form and sandwiched between a pair of covers". There is, by the way, more to this seemingly vague "description" than might at first be obvious: for one thing, it establishes the fact that a book constitutes an *isolated* block of information, and it conveniently bypasses arbitrary limitations one might be tempted to impose in terms of origin, subject matter, style, or purpose.

Certain assumptions automatically accompany the idea of a "book", or at least a *published* book: that someone (an *author*) must have written it, for example, that numerous copies exist (or *once* existed), and also that it should be possible through purchase or loan to gain access to the work so long as you know the full and correct title. We might also add that even a scholarly treatise is expected to serve as more than simply a repository of information. Every conscientious reader can reasonably expect to benefit from a *transfer of knowledge* through the study of a book, and it is at least desirable that the information be served up with a dollop of entertainment and satisfaction as well.

For a book to come into being, considerable help is required from a *publisher* or a *publishing house*. Later, it is the publisher who ensures that *copies* of the work are offered for sale, and that insofar as possible the most appropriate segment of the public becomes aware of the book's existence: in short, the publisher *publishes* the volume.

- Since, as previously noted, a book can be viewed as an "isolated" piece of information, it falls under the official heading of an *independent publication*.

Another example of an independent publication—albeit one issued in parts, and at regular intervals—is a *periodical*, such as a journal, but the designation "independent publication" is *not* applicable to a single issue of a journal, much less to a journal *article*. The importance of the subtle distinction implicit here will become clear later when we discuss formal ways of citing the literature (in Sec. 9.4).

Books exhibit great diversity with respect to both form and scale, and they can be tailored to suit a wide range of needs. A list of the many different *types* of books would include such (to some extent overlapping) categories as novels, histories, biographies, texts, monographs, technical manuals, dictionaries, atlases, encyclopedias, … pocket editions, hardcovers, paperbacks, spiral-bound or loose-leaf collections, …—and so far we've barely scratched the surface.

Books are classified in many ways: on the basis of genre, for example, or purpose, but also according to physical characteristics. A *published work* (to employ a slightly broader term) sometimes consists of more than one item (e.g., several *volumes*). We have already noted that some CD-ROMs can be regarded as books, and it is easy to make a convincing argument that a book could also manifest itself, for example, as a boxed set of microfiche cards.[3]

Returning to the conventional notion of a book, the corresponding pages can be joined together ("bound") by a gluing technique or through a more lasting—but also more complicated and expensive—thread-based procedure. Alternatively, a set of pages can be combined into a book in a flexible way, between the covers of a ring binder. It is not unusual for a *loose-leaf work* of this sort to be issued in portions, a few pages at a time, where the recipient assumes responsibility for accumulating the individual sets and keeping the whole thing in order.

Many titles today are released in both the classical format (i.e., as print publications) and as electronic data files, with a CD-ROM (compact disk read-only memory) typically serving as carrier for the latter. The boundaries between printed information and digital data files were once very clear, but they are becoming increasingly blurred. Thus, a software manual—a book—is essentially useless except in conjunction with the corresponding (digitized) computer program,[4] while computer files can be an indispensable adjunct to a textbook. Many books now rely heavily upon accompanying supplementary material supplied on CD-ROMs, typically secured to one of the book's covers. A very recent addition to the "book" family, one still in search of its proper niche both in the marketplace and in the average reader's everyday life, is the *electronic book* (*e-book*; we revisit this topic briefly in Sec. 4.1.3).

As a result of such developments, community libraries now find it necessary to issue reports of the number of *media* acquired during a particular time interval rather than accountings based exclusively on books. A quick look at a recent (2002) online version of *Books in Print* under the broad subject heading "Rome" (an example picked at random) produces an impressive list of almost 4400 books, but the OCLC "WorldCat" database adds to this inventory 142 more titles that member libraries have available in the form of computer files, mainly commercial CD-ROMs (49 of them in English, by the way).[5] In the near future we may well be patronizing public institutions known not as libraries, but as *media dispensaries,* or *mediothèques*, perhaps.

---

[3] This observation stems from the frequent use of microfilm methods to simulate or archive visually complex materials of various types—sets of spectra, for example, but also complete books. The technique at one time appeared on the verge of extinction, presumably to be superceded by electronic alternatives. As things have developed, however, microfilming continues to be important, especially in security applications and for archival purposes, as in the preservation of library materials subject to deterioration as they age.

[4] Ironically, software suppliers are lately choosing in many cases to make the relevant *manuals* available only in disk form!

[5] The CD (compact disk) was first introduced by Phillips in 1982 as a medium for storing audio files (which soon earned widespread acclaim from *audiophiles*). As early as 1985 the technology was adapted to accept computer data as well, leading to the "ROM" part of CD-ROM. This acronym has

→

University libraries have witnessed especially profound changes in recent years (cf. our remarks in this regard with respect to journal holdings under "Electronic Publishing" in Sec. 3.1.1). Fifteen years ago less than 5% of a typical science library's budget would have been allocated to "non-print media", whereas by 1998 the corresponding figure in Germany was closer to 15%—and increasing rapidly. The shift has been no less striking in most other parts of the world. Electronic media today unquestionably represent the cutting edge of information transfer. This is not surprising given some of the obvious advantages: CD-ROMs occupy very little shelf space, almost any computer workstation built within the last few years is capable of reading them, and they can easily be made available for direct access beyond the confines of the facility in which they are housed (at any hour of the day or night, irrespective of library hours!) by linking them to an efficient *network file server*. A remote client wishing to delve into the stored material then needs only a PC capable of connecting to the corresponding network, perhaps in conjunction with a valid access code. Even if openness at this level is precluded by the fact that the library lacks a sufficiently flexible "site license" for the source in question (e.g., a copyrighted database or reference work), simultaneous access by several readers occupying multiple workstations *within* the library may still be permissible.

Electronic (better: *digital*) storage is especially advantageous when one wishes to locate a specific piece of information. The appropriate search can be exceptionally rapid and truly exhaustive. Moreover, a database supplier can update a set of digital files quite easily at convenient intervals, so that a comprehensive search can often be accomplished through a *single* digital source, whereas with a comparable book resource there might be a need to examine several volumes, including miscellaneous sets of supplements—which might not even be shelved in one place. Also not to be overlooked is the selectivity that can be achieved with a *multidimensional search* of digital data (that is, a search defined by a logical combination of multiple criteria, which might even be of different types). We revisit in Chap. 9 other special features that make these latest approaches to information storage so attractive.

● A primary goal of book publication within the natural sciences, technology, and medicine[6] is encapsulating, organizing, and disseminating discrete "pieces" of knowledge in the form of manageable packages, where the nature of the package can be optimized to suit specific situations.

This seems an opportune place to restate our firm belief that the "book" as a *concept*—as distinct from a particular *medium*—is most unlikely to become obsolete in the foreseeable future.

---

become almost synonymous with the catchword "multimedia". Commercial CD-ROMs routinely couple text-based information with graphics, sound, video, computer-generated simulations, etc. The fledgling medium was quickly awarded a prominent place on bookstore shelves, in striking contrast to the way the same merchants had earlier distanced themselves from music preserved on vinyl disks.

[6] Note that we have encountered once again the "stm" trinity (cf. Sec. 3.1.1), although we obviously recognize that scholars in other academic areas make similar use of books.

4.1.2 Where Do Books Come From?

Before a book can be published, someone must first take the initiative to write it: that is to say, to become its *author*. Book-writing schemes are sometimes so ambitious that their fulfillment requires the efforts of more than one person, resulting in *co-authorship* and the genesis of what comes to be known as a *multi-author work*. With more than four authors there is usually the further need for a coordinator, or *editor*,[6a] to accept responsibility for assembling and integrating segments composed independently by different members of the authorship team. A book's editor is also expected to serve as spokesperson for the collective of authors.

It is essential that every prospective author, group of authors, or editor establish as early as possible a close working relationship with a *publishing house*, the institution that will ultimately produce copies of the book and coordinate their distribution (cf. Sec. 4.1.4).

- The most immediate goal is achieving a clear set of understandings between the aspiring author(s) and the potential publisher regarding the *feasibility* of producing a particular work based on a shared interpretation of its essential characteristics.

In the beginning, a conversation or a simple memorandum may serve as an "understanding" in this sense, but an informal, nonbinding arrangement becomes less satisfactory as more time, effort, and capital are committed to the project. If in the end you find yourself unable to come to suitable terms with a publisher, any effort already invested in writing would need to be shrugged off as wasted—or at best a noble gesture.

- As a potential book author or editor it is very much in your interest that a *written* agreement with a publisher be negotiated as early as possible.

Publishers are also anxious to conclude formal agreements, since these serve as a necessary precursor to vital in-house planning. If a publisher expresses a willingness in principle to accept a manuscript and then transform it into a published book, those proposing to prepare it can expect soon to be presented with draft *publishing contracts*. Depending on the circumstances, a document of this type could take the form of an *author's contract*, an *editor's contract*, a *translator's contract*, or a *reviser's contract*.

[6a] The word "editor" functions at several levels. In Chap. 3 it was introduced as representing a person responsible for the content of a journal, possibly employed by a publisher. The book division of a publishing house usually includes a number of executives responsible for maintaining the firm's publication program in a planning sense, and they also are referred to as "editors" (sometimes "senior editors" or "acquisition editors"). In the present context the people we are calling "editors" see themselves primarily as scientists—typically at universities or research institutes—but are engaged "on the side" in the preparation of volumes for publication by a publishing house, a task that can be a very time-consuming one. This type of editor assists the publisher in, among other things, not only planning but also the recruitment of authors. With multi-volume works like handbooks or encyclopedias the editorial responsibilities are often shared by several members of a team. To facilitate communication and decision-making, such an editorial team may include representatives of the publishing-house management.

A valid contract must of course be in full accord with the law, especially important aspects of *copyright law* (cf. our introduction to the subject in Sec. 3.1.1).

In earlier editions of this book (EBEL, BLIEFERT, and RUSSEY 1987, 1990) we provided a brief summary of major national and international copyright provisions, but much has changed in recent years as a consequence of technical developments and the advent of the "e-age". Some of the most basic aspects of the copyright concept are the subject of fundamental reappraisal—or have been shelved temporarily pending thorough study. Copyright protection has been modified repeatedly of late, in both scope and substance. Part of the turmoil is a result of the special problems posed by such unprecedented phenomena as "electronic publishing" and "publication on demand". International treaties negotiated in the aftermath of the 1886 Berne Convention for the Protection of Literary and Artistic Works fall far short of current needs.[7] Consider just one example. Suppose that in the course of a search for technical information you were to use the Internet as a way of opening the "pages" of an electronic treatise, the counterpart to a print publication—perhaps the *Ullmann's Encyclopedia of Industrial Chemistry* referred to earlier (in footnote 2 of this chapter). You might well decide it would be helpful to download to your PC from this source an especially interesting article in its entirety, primarily for your own use but also perhaps to share with colleagues, students, or merely a friend. Would you be at risk of breaking the law? If so, at what point? Legal documents prepared in the 19th century certainly cannot provide a definitive response. And what about the case of a "pirate" in some remote corner of the globe who chooses to do the same thing as part of a blatant self-enrichment scheme? Internet piracy has already drawn a great deal of public attention in the music scene, resulting recently in a court decision that the primary activities of the huge "Napster" music-sharing cooperative were illegal. Copy-right law is a fascinating subject, but just as we did in Sec. 3.1.1 we must again resist the temptation to pursue it in depth.

Returning to our original subject—publishing contracts—we note that in general these agreements tend to follow rather standard patterns irrespective of whether they are proffered in printed or "typewritten" (computer-generated) form. In Germany, for example, the nature and substance of an acceptable publishing agreement has been the subject of intense and ongoing negotiation between two national organizations: the *Börsenverein des Deutschen Buchhandels* (German Book Dealers Trade Organization), representing the interests of publishers, and the *Verwertungsgesellschaft Wort* (*VG Wort*; roughly translated, "Corporation for the Exploitation of Verbal Communication") acting on behalf of professional authors. Both organizations gladly make sample documents available upon request. A valuable source of relevant information for authors in the United States is the *Guide to Book Contracts* prepared by the Na-

---

[7] On the international level, copyright issues are now the province of the World Intellectual Property Organization (WIPO, http://www.wipo.org), an agency of the United Nations.

tional Writers Union, an organization that also provides consultancy services for prospective authors in need of assistance.[8]

● A publishing contract spells out in detail the rights and responsibilities of the two complementary forces necessarily engaged in creating a book—author(s) and publisher.

A typical contract is particularly explicit with respect to the nature and scope of *usage rights* that may be subject to transfer from an author to a publisher, as well as terms according to which the author(s) will be compensated. Other clauses provide a broad description of the projected work, including its anticipated magnitude, together with a tentative schedule, very likely specifying a deadline for submission of a completed manuscript.

In contrast to standard practice with journal articles, book authors expect to be rewarded (modestly) for their creative efforts in the form of some sort of *honorarium*. At the present time, a typical honorarium for a monograph would amount to perhaps 7% of the market price on every copy sold, or 12% of the publisher's net revenue after factoring in dealer discounts and taxes. Terms such as these define what might be called a *participatory honorarium*. An author contributing to a collective work is normally compensated through a sales-independent *fixed honorarium*, the level of which reflects the extent of that particular author's individual contribution to the publication as a whole. In recent years authors have been asked to share more and more of the financial risk involved in publication. This is the reason why most agreements now specify a monetary reward dependent upon the number of copies sold, or incorporate a sales-related premium—over and above a modest fixed honorarium—that begins to accrue only after sale (at full price) of a certain minimum number of copies.

A scientific publishing house, like any other publisher, is a commercial enterprise. This is true even if it happens to operate under the auspices of a scientific society. Business activities are expected to bring in at least enough revenue to sustain the operation, perhaps providing in addition a modest income for shareholders. Despite the fact that scholarly publication is arguably an "essential activity" from a cultural standpoint, it is still subject to the harsh laws of the marketplace.

● Only if a publisher can predict with some confidence (based on past experience) that demand for a particular book, when offered at a realistic price, will be sufficient to more than offset production costs—only then is it likely the work will be accepted for publication.

In principle, publication costs associated with a book might be underwritten through a foundation grant or some sort of *printing-cost subsidy*, but this is rare in the natural sciences. Generally speaking, if a scientific or technical book does not appear to be

---

[8] Home page: http://www.nwu.org/nwuhome.htm. Another Internet site with links to a vast array of sources of publishing information applicable especially to the United States can be reached at http://publishing.about.com.

economically viable—if no potential market of sufficient size can be clearly identified—it stands little chance of publication. What we mean here by a book's "market" is the audience to which that particular work is targeted. If the identifiable market realistically exceeds a certain critical size, then it should be possible to prepare a reasonable number of copies of the book and dispose of them at a *market price* the readership will tolerate, in which case—with proper publicity—sales (to libraries and individuals) should more than compensate for production costs.

● The more specialized the topic, the smaller will be the target audience, and thus the more difficult it will be to demonstrate the practicality of publication.

Simply *preparing* a book for publication is an expensive undertaking, and this preliminary work represents a significant fraction of all the identifiable *fixed costs* of a publishing endeavor. Here we are referring to typesetting costs, for example, as well as general editorial expenses, none of which are in any way dependent upon the number of copies printed (the *print-run*). In addition, account must of course also be taken of a number of *variable costs*, expenses that do indeed reflect the print-run: acquisition of paper, the printing process itself, and binding, among others. An appropriate share of the overhead expenses of the publishing house as a whole must also be included in the "fixed-costs" category, derived from such mundane things as rent, taxes, utilities, office supplies, bookkeeping, maintenance, etc. Only if the fixed expenses for a project can be distributed over a substantial number of copies will the overall cost per copy drop to within a reasonable range. Reducing the print-run causes per-copy costs to increase dramatically, leaving the publisher no choice but to establish a correspondingly elevated wholesale price. If this in turn leads to a market price beyond the reach of the target audience, the economic consequences would spell the project's doom.

A publisher is almost never able to break even with a print-run of less than about 600 copies (nearly all of which must then be sold!), and for many technical monographs that figure comes perilously close to the size of the potential market. In other words, stm publishers routinely operate in a precarious environment. It should thus come as no surprise that attempts are made to optimize every available adjustable parameter for each book project. It is of course helpful to find authors willing and able to take advantage of the most sophisticated copy-preparation techniques, submitting their work in a form such that it requires very little in-house processing—ideally as camera-ready copy (as described, for example, in RUSSEY, BLIEFERT, and VILLAIN 1995) or a digital document from which the publisher is able to print more or less directly.[9] (As noted in Sec. 3.5.4, the most modern printing methods completely avoid any need for masters in the form of films, and also bypass print carriers.)

● It is very important that a prospective author and publisher jointly consider carefully the character and extent of a book's projected target audience, noting special opportunities that may exist for effectively *promoting* the proposed work.

---

[9] We ourselves have taken this latter route with all our books in recent years.

Among other key issues for early intense author–publisher discussion are the intellectual demands posed by the text as a function of the anticipated target audience, the desirability or even necessity of incorporating costly special features like extensive color, viable options with respect to binding, and perhaps even alternative production methods.

● In your role as author you should actively encourage the publisher to suggest ways you can facilitate smooth transformation of your manuscript into a printed work, particularly things you might do to minimize the need for expensive editing and refinement of what you submit.

Sometimes it is not a prospective author who supplies the impetus for book publication. Scientists, engineers, and physicians—in sharp contrast to novelists, for example— tend by nature not to view themselves as writers: in most cases they write only because their professional responsibilities occasionally demand it of them. The author–publisher relationship in science is thus substantially different from that in the creative arts. One consequence is that *publishers* are often the ones who propose topics, which then forces them to seek qualified scientists to compose the desired books while carrying on with their normal professional duties. Large multi-volume works in particular almost always originate conceptually in a publishing house.

Most scientists of course see their first priority as the pursuit of research, and they can be slow to appreciate benefits that might accrue from investing a substantial amount of time in the writing of a first-rate book, or joining together with others in the collective realization of a major publishing venture.

● Writing or editing a well-received book can be a source of considerably more peer recognition and prestige than devoting the same amount of effort toward publishing a series of research papers.

This factor alone may be just the incentive required for someone—for you, perhaps!— to contemplate becoming a book author, especially in view of the generally acknowledged obligation of every productive scientist to develop and maintain a strong record of publication.

4.1.3 What Are Books For?

Books serve a multitude of very diverse ends. Some are designed to entertain, some to distract, to motivate, to engender "hero worship" perhaps, to provide insight, to convince, to shock, to indoctrinate … to introduce, to teach, to inform … the list is endless. In the case of this particular book we set ourselves three primary goals, namely the last three cited, ones that clearly are appropriate for a discourse on "writing and publishing in the natural sciences".

In a section entitled "What Are Books For?" we can hardly avoid making some reference to *content*, although generally in this chapter, just as elsewhere, our main interest is in organizational, technical, and formal issues. In other words, we have no aspirations of providing the reader with a detailed guide to the scientific book literature. We will, however, briefly consider some of the major *types* of scientific books. One that immediately comes to mind is

● The "how-to" book: essentially a compilation of instructions, consisting typically of both verbal and graphic components and designed to help the reader accomplish some very specific task(s).

Books in this category are of course prevalent outside the sciences as well: in the context of hobbies, for example, and a great many other aspects of a person's nonprofessional life. Illustrative examples would be cookbooks, tax guides, sports tips, advice for the home handyman, gardening insights, ... *again* an endless list. In English, books like this—including ones relevant to the sciences—often feature the very words "How To" in their titles; e.g., *How to Find Chemical Information* (MAIZELL 1998), also *How to Write Scientific Papers* (DAY 1994), and *How to Write and Publish Papers in the Medical Sciences* (HUTH 1990). "How-to" books occupy one especially important niche in the scientific literature, that devoted to practical training and methodology. Laboratory manuals are an obvious case in point, long referred to facetiously by chemists as "cookbooks". The handbooks that accompany computer programs also consist largely of instructions the reader is encouraged to follow.

A loosely related category—and a fiercely competitive one—is

● The *textbook*: crafted to provide the reader with a systematic general introduction to some limited and clearly defined body of knowledge.

A textbook is in effect a tool to facilitate *study*, where study is one of the chief activities engaged in during a disciplined period of learning. Such study often entails more or less close interaction between a *teacher* and one or more *students* with a desire (or need) to be taught. Most students today find it difficult to imagine "studying" in the *absence* of textbooks, although this was not always the case in the past, and it is certainly not a foregone conclusion that it will be in the future.[10]

A typical textbook is designed for use in conjunction with a classroom experience, although some are deliberately structured to obviate the need for such group encounters. Through the years, students have by and large affirmed that it is significantly easier to master complicated subject matter when a good textbook is available to supplement oral instruction. Especially in an environment distinguished by (or at least tolerant of) overfilled, stuffy classrooms it is hardly surprising that some find it distinctly preferable

---

[10] Education is another area subject to enormous impact from modern technology, and the past can hardly be regarded as a reliable guide to the future. Our uncertainty with respect to the outlook for textbooks is of course related to possible alternative sources of information, including the Internet, but also rapid developments in the area of "distance learning". Personal experience forces us (with considerable dismay; or is it mainly nostalgia?) to take seriously as well the fact that student willingness and even *ability* to learn from the printed page has declined markedly in recent years. As a teacher, one must decide whether to do battle with this situation, or simply accept the consequences and move on.

to direct most of their attention to a good textbook—in comfortable surroundings of their choice.

What are a few of the factors that distinguish a "good" textbook? Actually, the informal definition with which we began provides a partial answer to the question. In the first place, a good textbook must offer a *systematic* treatment of the subject matter at hand. It should also deal with the material as *comprehensively* as the circumstances warrant—in sufficient depth, for example, to permit one who studies the book conscientiously to perform well on an impending examination. The treatment might or might not be designed to go beyond the essentials, or to encourage and assist the interested student in finding additional information elsewhere.

Students today feel under intense societal pressure to succeed, and there seems to be a growing tendency for them to interpret education as vocational preparation rather than an end in itself. As a result, many increasingly equate learning with memorization, even in the disciplines most dependent upon the application of logic and independent creativity. Textbook authors vary greatly in the extent to which they attempt to combat this trend by stressing the classical goal of acquiring *understanding*, and also in the strategies they pursue to foster independent thought.

It goes almost without saying that a good textbook must reflect the current state of the discipline in question. Equally important, however, it must present the subject matter in as transparent a way as possible—transparent from the *student's* perspective! Students also react positively to a book that is *appealing*. Thanks to advances in printing technology, little stands in the way of the author who wishes to exploit typographically complex layouts, or text liberally interspersed with graphic elements, in an attempt to meet the latter challenge. Nearly every textbook now on the market features extensive use of color as a way of highlighting important aspects of the presentation and enlivening illustrative components.

Illustrations have long been considered an indispensable feature of textbooks for the more descriptive sciences, like biology and geology, but visually oriented content recently has made important inroads in educational materials for essentially all fields. This has certainly come as a response at least in part to the impact of computers in education at every level. Unlike books, desktop and laptop computers make it possible through astonishing feats of animation to simulate phenomena that cannot possibly be exhibited directly, at the same time introducing into the educational process a whole new dimension of *interactivity*. Furthermore, computers can also enrich text-based instruction through the media of animation and *sound*, at the same time exploiting the far-reaching potential of *hypertext* linkages (cf. "Further Ramifications of the Digital Revolution" in Sec. 3.1.2).

One of the greatest challenges confronting today's textbook author is identifying the most productive ways of combining a printed document with computer-related enrichment derived either from a CD-ROM or the Internet. The first pairings of auxiliary CD-ROMs and textbooks to provide a hint of the wonders that might be accomplished through such a partnership made a tentative appearance in the early 1990s, and now the combination is to be found everywhere.[11]

Textbooks over the years have assumed many shapes, sizes, and costumes. Indeed, every prospective author should give serious thought to the question of the *form* that would best fit the intended application, and not simply rely on a publisher's independent judgement. Is a paperback or a hardcover edition preferable, for example, given the way this particular book will be used? What about a loose-leaf collection, with some provision for customization at the hands of the instructor? Or a spiral-bound volume capable of lying flat on a laboratory bench top when opened? Some authors lean toward weighty tomes explicitly designed to withstand long-term use as reference texts, whereas others favor slim, efficient presentations better adapted to one-time perusal by the student necessarily preoccupied with surviving on a tight budget.

The design, development, and production of a first-rate textbook is always an expensive undertaking, requiring close attention to financial implications. It is not unusual for a publisher to suggest test-marketing a low-cost "trial version" prior to authorizing full-scale production of a "world-class" masterpiece. One opportunity for economizing that sometimes presents itself in the case of a publisher with international ambitions—producing savings through a certain amount of "recycling", especially of graphic content—is a decision at the outset to release a textbook in two or more languages, where the "flagship version" would normally be in English (the language for which demand should in principle be greatest).

Students today have available to them in most disciplines an incredibly broad array of texts. Those who are highly motivated would thus theoretically be able to supplement the textbook recommended by an instructor with other materials more suited to their own tastes, interests, and/or abilities—but, alas, we have now begun to indulge in an educator's wishful thinking. In point of fact, the textbook market is almost entirely instructor-driven. As a result, publishers everywhere are obliged to expend a great deal of effort (and money!) trying to seduce individual instructors into "adopting" their particular offerings.

It was once commonplace, especially outside the English-speaking tradition, for specific textbooks to acquire a reputation as indispensable "classics". These were typically books associated with two or more authors, and they continued to be regarded as benchmarks over a span of many years—even decades. Updates would occasionally appear in the form of "new editions", but little fundamental change would be discernable. Though widely respected and certainly familiar to everyone trained in the corresponding discipline, stodgy workhorses of this sort were in fact seldom exemplary from the standpoint of currency of outlook or instructional merit.

---

[11] Regrettably, however, many highly touted CD-ROMs developed as text supplements—and announced with great fanfare and hyperbole—have proven in the event to be of disconcertingly limited practical value, testifying all too often to a minimal show of creativity on the part of the author (or some hapless associate). One is sometimes tempted to conclude that a "marketing specialist" at the publishing house insisted the author conjure up some sort of CD-ROM primarily as evidence of "up-to-dateness".

A very few pedagogically gifted and enthusiastic university professors have managed through the years to make names for themselves as truly outstanding textbook authors, revered by generations of students from many linguistic backgrounds—but we doubt very much that any textbook author has ever profited financially from his or her work in a way that begins to rival the fortunes that accrue routinely to creators of popular comic-strip characters, for example. We next turn to the type of book of greatest interest to us, namely

- The *monograph*: a highly structured, timely, and utterly comprehensive source of authoritative information with respect to some narrow specialty.

The monograph derives its name from being devoted essentially to a *single* topic (cf. the Greek μονοσ, single)—not from the absence, say, of multiple authors. Works of this type are to be found at the pinnacle of the "knowledge pyramid" represented by the world of books. With any pyramid, a platform at the apex—in this case the one reserved for monographs—must be narrow indeed. Among other things this means that if a good monograph already exists on some topic, writing another makes sense only if the facts have changed, or if a different point of view needs to be reflected.

A monograph is almost always directed toward an extremely select audience (and thus produced as a very limited edition[12]), so the books themselves are inevitably expensive. This is regrettable not only from the perspective of the potential customer, but also the scientist contemplating writing such a book. No one in his right mind would undertake the task in the hope of reaping a financial windfall (or of rescuing a publisher from an impending bankruptcy). The fact that monographs on almost every conceivable topic continue to be written and published suggests there must be other powerful motivating forces at work. Among them is almost certainly a desire on the part of at least a few scientists to alert the world (more accurately: fellow specialists) to their personal expertise in or passion for some corner of knowledge. Others find that writing a book on an abstruse topic is a good way of forcing themselves to *learn* as much as possible about it, perhaps as preparation for launching a new research initiative. Also not to be overlooked is the seductive power of the old adage that "a book carries more weight than ten journal articles", meaning that authorship of a well-received book can be a valuable asset in the quest to "publish and survive", or simply gain acceptance into the "Who's Who" elite.[13]

On the other hand, no one has yet been rewarded with a Nobel Prize for the writing of a book (with the exception of the prize for literature, of course!), and authorship will certainly not guarantee a place in the Happy Hunting Ground of eternal wisdom. Fortunately there is no shortage of reasons (including altruistic ones, such as wishing

---

[12] The word "edition" is used in two different senses: both in reference to a particular *version* of a book (e.g., a *hardcover edition*, or "the 3rd revised edition") and in the context of a *print-run* (such as an edition of 5000). It is obviously the latter usage which applies here.

[13] We can honestly profess not to have fallen victim to any of these rationales. Our sole intention has been to touch the reader's mind: nothing more, and nothing less.

to close a perceived gap in the existing literature) for taking on the challenge of composing a monograph.

● Monographs differ from textbooks in several important respects.

Textbooks, unlike monographs, typically include sets of exercises and other enticements to engage the reader in a program of "self-testing" over the material presented (solutions are often provided for at least some of the challenges, or hints about where insight should be sought in the literature), but they seldom pretend to be useful guides to original sources. At most one normally finds in a textbook only a sparse, generalized list of "additional readings". Rarely are the "facts" that constitute the substance of a text supported by any documentation. By contrast, the author of a monograph feels obligated to document virtually every assertion, as a result of which the book becomes an important gateway to the primary literature (see Chap. 3).[14] Such entryways into the repository of scientific evidence are extremely useful, and the serious professional would be most reluctant to see them walled off through disappearance of the monograph as a medium of communication.

Continuing with our cursory survey of the informational scientific book literature we would be remiss if we failed to pay tribute to the category of monumental multi-volume reference works. These generally appear on a library's shelves one volume at a time over the course of several years. Decades ago, many of the important works of this sort were in German, but now almost all are in English, just as is the case with most of the important monographs (as well as journals, regardless of where they originate).

● These *encyclopedic works* (sometimes illogically dubbed "handbooks") attempt to bring together in one place all the reliable information available at the time of publication pertaining to some relatively broad subdiscipline (e.g., enzymes, organometallic chemistry, nuclear physics, etc.).

Encyclopedias can render invaluable service, especially with respect to information acquired long ago and so far not digitized for efficient retrieval. They result from the efforts of countless authors—specialists—first recruited and then encouraged and supervised by small editorial teams.

It would be virtually impossible to publish simultaneously all the volumes constituting such a compendium, so some parts of a scientific encyclopedia are typically more up-to-date than others. A reference work of this sort might conceivably be kept more nearly current by issuing it in loose-leaf form, so that individual pages could occasionally be replaced, or the text selectively supplemented. But that represents a tiresome prospect, most especially from the consumer's standpoint—who in effect is transformed into a "subscriber". Such an organizational form is perhaps best suited to

---

[14] The distinction between a text and a monograph admittedly becomes rather fuzzy near the interface defined by "advanced texts" and "introductions" to narrow fields of specialization.

tax-related compendia, since taxes are a disagreeable topic anyway, and subject to frequent (seemingly random) change. Much more common with encyclopedias has been a practice of releasing, perhaps every decade, sets of "supplemental volumes".

In the "e-world" of today publishers have found it indispensable to focus their attention on constructing *digital* reference works. One significant rationale is that even a large encyclopedia can be compressed into a single CD-ROM (or two at the most), a format far more convenient and less wasteful of resources than a conventional printed set. More to the point, a reference work in this form is easily updated as the need arises, yielding a new, completely current "edition" in the guise of an inexpensive replacement disk, ideal for "shipment" through the ordinary mail. Digital storage also simplifies the task of later issuing a revised print version, as illustrated recently by the *Ullmann's Encyclopedia*.[15]

● Convenient *online access* may well become commonplace with published works in the future.

We find ourselves navigating here in mists separating the world of essence from that of appearance: between the real and the virtual. Only two years ago it seemed safe to say with some conviction that the elegant bound volumes which once proudly identified the latest edition (or a major supplement) of an encyclopedic work were predestined to give way, even if reluctantly, to unobtrusive compact disks, distinguished visually only by cold, anonymous labels proclaiming something like "Version 2.1". Now we must ask: Will this really happen? How typical will the unanticipated *Ullmann's* odyssey prove to be? For the moment we can only say "Que sera, sera …".

● Publishers everywhere are now busily reformatting—and preparing to market—all their flagship "information banks" in the form of digital ("electronic") collections.

The development is one that promises the reader (perhaps we should instead say "user") vastly improved access, greater flexibility, and a currency never realizable with print media. Many digital reference works are already available for perusal 24 hours a day from almost any spot on the globe via the Internet (albeit in most cases only after

---

[15] *Ullmann's Encyclopedia of Industrial Chemistry* is a major reference work with a long tradition. The 5th Edition saw its transformation into an English-language work from the original German, a CD-ROM version of which was proudly unveiled by our own publisher at the 1997 Frankfurt Book Fair. A large international audience was thus presented with the rare opportunity to assess first-hand a striking example of what is possible with electronic database technology. Visitors expressed delight at discovering how easy it could be to carry out a rapid, comprehensive, targeted search within an encyclopedic database of this magnitude—certainly a far cry from the always frustrating challenge of locating something in a large printed collection, helped only by a massive, complex, awkward (and at the same time surprisingly impotent!) set of conventional indexes. This ambitious step into the future was at the time unparalleled, and most regarded it as an irreversible, one-way venture, but that turned out not to be the case. As mentioned in footnote 2, a new 6th Edition of *Ullmann's* has very recently become available—once again in print! An electronic-database version will continue to be distributed, of course, but the surprising part even for publishing insiders is that a print version not only survived, but experienced a heralded comeback.

payment of a usage fee). Hopefully it is a realistic assumption that they will be conscientiously *maintained* by their sponsors as the years pass.

We are reluctant to conclude this section without at least some mention of the concept of the "e-book" in general. For some years there have been wide-spread predictions that the "book" as a printed work was a doomed species, soon to be replaced almost entirely by digital compilations one would download as desired (or required)—for an appropriate fee—to highly sophisticated, eminently convenient "e-book readers", probably resembling the so-called tablet computing devices. The reading public would thus be outfitted with attractive, easy-to-read, high-resolution viewing screens, possibly with sufficient space for displaying two or more pages simultaneously, straightforward and powerful controls for "paging through" an "electronic volume", handy provisions for highlighting and perhaps note-taking—all in the context of a thin, attractive, light-weight shell to be powered by long-life batteries, a presentation almost as well suited to perusal anywhere and under any circumstances as a paperback book. A parallel assumption has of course been that every conceivable title would be available for rapid and convenient downloading to such a device. Prognoses like these still come forth regularly, but the corresponding *products* have proven elusive. Several "e-book readers" have indeed been the subject of marketing campaigns, but none has yet "caught on", presumably because the optimal combination of features has not yet been achieved. Online digital libraries have been launched as well, but in the absence of the "right" reading device they have largely stagnated. Will the (admittedly attractive) vision ever become reality? Once again, only time will tell.

4.1.4 Collaborating With a Publishing House

Many scientific publishers have taken on the responsibility of producing both books and journals. This actually makes considerable sense, in that experience gained with one of the media can pay important dividends in the other. This has especially important consequences with respect to a publisher's single most important and valuable resource: authors. While oversight in the two areas of publication at a given publishing house usually rests in separate hands, a single pool of scientists is the source of manuscripts for both. As a result, a journal editor who has been deeply impressed by a particular scientist's research contributions is likely to suggest to the book division that that person be approached about possibly writing a book.

● The book division at a scientific publishing house typically consists of several distinct planning groups, each specializing in a unique subject area and each under the supervision of its own *acquisition editor* (or *senior editor*).

An important part of an acquisition editor's job is acting as a source of ideas for new publication projects, ultimately setting such projects in motion and overseeing them at least through the initial stages of planning. Successful launch of any book endeavor depends upon considerable close interaction between an editor and the prospective author(s) or volume editor(s), and an acquisition editor is normally the author's most direct point of contact with a publisher.

Acquisition editors also shoulder much of the responsibility for the economic well-being of publications, which in turn requires that they be intimately familiar with all the steps involved in both production and promotion, and enjoy good working relationships with key personnel in the two areas. Of necessity they also participate in the budgeting process. Finally, an acquisition editor serves as the publisher's representative in negotiations to establish the specific terms of a publishing contract (cf. Sec. 4.1.2).

At many publishing houses, acquisition (or senior) editors hold terminal (or at least advanced) degrees in the disciplines with which they have the most contact. This puts them in a strong position to engage in meaningful dialogue with potential authors (usually at the level of the generalist conferring with the specialist) and at the same time maintain currency with respect to important developments and trends in the field—able to recognize in a timely fashion that a need exists for a particular type of book. Acquisition editors turn to copy editors and other personnel for assistance in the day-to-day processing of manuscripts and ensuring that publication proceeds as efficiently and smoothly as possible.

● The aspiring book author should expect to have contact not only with the publisher's editorial department, but also with people in both *production* and *marketing*.

Most outsiders picture a publishing house, especially its production department, as an organization engaged mainly in mechanical operations such as typesetting and printing, where finished products (books and journals) steadily emerge as if from an industrial production line. There is certainly some kernel of truth in the characterization, but in general it falls wide of the mark. *Physical* production of books is not in fact the concern chiefly of a publishing house, but rather of more technically oriented establishments (*printing companies*, or *typesetting shops*). The corresponding activities *might* reflect one aspect of a publisher's operation, but even the largest publishing houses have for several years been moving increasingly in the direction of contracting such work out to others—in what might be regarded as a classic example of industry choosing to become more focused by "slimming down", and "outsourcing" inessential activities.[16]

● "Production" in the most general sense encompasses everything that transpires from the time a manuscript leaves the editorial office until copies of the finished product (whether a book or a journal article) are in the hands of distributors.

---

[16] "Inessential" is a harsh word. No one would be enthusiastic about investing much attention in an endeavor that was dismissed as "inessential". We are speaking here of operations that class as "inessential" from a *management* point of view: management in this case of a publishing house. Anything that could in principle be done elsewhere, outside the company, would probably be labeled as "inessential" by at least some of those whose major preoccupation is with the budget. Such people sometimes need to be reminded that absolutely nothing worthwhile will be published—or successfully marketed!—if these particular "inessentials" are not dealt with properly.

In other words, production includes (1) completing the last steps in technical refinement of a text manuscript and all its accompanying "special" components, to the point that everything is ready for typesetting and subsequent printing, (2) preparing specifications and soliciting bids for manufacture of the product, (3) negotiating all necessary contracts and overseeing their satisfactory fulfillment, (4) monitoring deadlines and keeping the project on schedule (at the proofreading stage, for example), (5) purchasing and warehousing essential supplies (e.g., paper and cover stock), and much more besides.

A publisher's production department usually bears primary responsibility for developing dependable, detailed cost and revenue projections based on information received from an acquisition editor. Production department personnel are also assigned one or more seats at the table when matters of book *design* are discussed, particularly in the case of a work expected to be unusually demanding from a manufacturing standpoint (requiring extensive attention to color, for instance, especially if color accuracy is crucial, or involving a large number of uncommonly sensitive or unusual graphic elements).

The qualifications required of the typical production specialist at a publishing house have changed dramatically in recent years. What was once a workshop manned by a group of skilled technicians has evolved into a negotiating center staffed with trained information specialists.

● The *marketing department* also makes a critical contribution toward a publisher's success.

Producing good books is one thing; selling them effectively is quite another. Publishers rely upon authors to help by supplying as much information as possible concerning the target readership for their books. A first attempt is made to "pick an author's brains" with the aid of a carefully designed *author's questionnaire*. There is obviously no one in a better position than the author to describe those people most likely to benefit from purchasing or at least consulting a pending book. Other useful insights an author might contribute during the development of a marketing campaign include lists of scientific organizations or industries where interested readers tend to be concentrated, of scientific meetings or trade fairs that should be considered priority venues for displaying the book, and of the periodicals most likely to publish useful reviews—or where advertisements should be placed.

We have so far had little occasion to comment on the *management* aspects of a publishing house. Direction of the enterprise as a whole might be in the hands of one or several people, identified officially as "the *publisher(s)*", who might in turn be included among the firm's owners, partners, or major shareholders.

● When important publishing decisions are pending, especially if major works are at stake, authors occasionally find themselves drawn into discussions at the highest echelons of management.

For readers who have already had extensive contact with the publishing world, much of what we have been describing probably seems elementary and even self-evident, but for others it will undoubtedly have come as a surprise to learn, for example, how many different people contribute to transforming a book-publishing scheme into reality—and a *success*.

● Scientific publishing houses today are in most cases large, complex enterprises of which the scientific public in general is almost completely oblivious, despite the central role publishers play in the furtherance of science.

We conclude this section by redirecting our attention to the author, without whom there will *be* no book. Authorship can be a remarkably rewarding activity, but committing oneself to writing or editing a book also has far-reaching consequences. Under no circumstances should you make such a commitment until you are sure you understand just what will be entailed, especially the enormity of the undertaking—and until you have also become reasonably well acquainted with people at the publishing house with whom you will need to work most closely.

● The success of a book-publication venture depends largely upon the level of trust and respect established between an author and an editor, but also between the author and key support personnel in other departments.

Spend some quiet time considering very carefully how comfortable you feel dealing with the specific individuals who will be charged with turning your manuscript into a book of which you can justifiably be proud, and think also about impressions the firm as a whole has made upon you. Some of our observations in Sec. 3.2.4 regarding selection of a target journal for a research article may help you in choosing a book publisher. In particular, be sensitive to how your book appears to fit in with the company's overall publication program. A logical place to begin your inquiry would be recent catalogues, but Web sites are valuable substitutes or supplements. Other useful input may be derived from colleagues, especially ones who have themselves written books. By no means downplay impressions you gain informally while visiting book displays at scientific meetings or congresses, or from the book announcements that undoubtedly come your way through the mail. Most important of all is the human factor, however: the potential for your forging warm personal relationships with those who will necessarily become your partners.

4.2 Planning and Preparation

4.2.1 First Drafts of the Title, Outline, and Preface

Just like the publisher, you as author- or editor-to-be will feel much more comfortable once you have put together, in writing, a detailed description of the book you have in mind.

● The first step should be establishing a *working title*.

This first approximation to a title may bear no resemblance whatsoever to the book's eventual designation, but it should provide a reasonable capsule summary of the book as you envision it (cf. the discussion of dissertation titles in Sec. 2.2.2). This is a good place for us to stress once again the importance of titles generally. A good book can effectively be destroyed by a poor title, since it would prevent the work from receiving the recognition and attention it deserves. Titles that are too long are just as bad as ones so brief they give no information. Similarly, a title can suffer as much from being too all-encompassing as from being overly restrictive.[17]

At some point your title will need to be the subject of serious discussion involving various people at the publishing house—but not yet!

● Next, you should prepare an *outline*.

The outline is to be a sketch of what you want to cover in your book, clarifying the *order* in which you propose to make your points. You will later be using this outline as a rough template for putting the book together.

Your outline should take the form of a sequential, hierarchical list of the sections, chapters, and important subdivisions you anticipate the book will contain, which of course mirrors the work's projected structure. At a more detailed level of organization you might already want to add here and there a few *keywords* to indicate your current thinking about how certain arguments ought to be framed. Eventually you will almost certainly implement a *numerical system* (Sec. 2.2.5) in conjunction with the book's headings, and since your outline amounts to a tentative formulation of those headings you might as well begin employing numbers now.

Obviously, nothing you prepare at this point is set in stone. Changes can always be made later—not only in wording but also in structure, through the introduction or elimination of subsections.

● As a last step, take the time to annotate your outline with an estimate of the number of manuscript pages you expect each chapter title and subhead to represent.

---

[17] Even a subtitle like "for Engineers" can prove disastrous if in fact there are many non-engineers who should become aware of the work. Such a title might be dismissed merely as an "understatement", but that is probably the most charitable thing that could be said about it.

With this additional information what has until now been a *qualitative* description of the project begins to give way to a more *quantitative* one. As you proceed, try to make sure that individual chapters and sections are appropriately "weighted", with those parts you intend to emphasize—theoretical considerations, perhaps, or methods and applications—being allotted the most pages.

The combination of working title and tentative outline should be shared as soon as possible with your contacts at the publishing house. Among other things, this will make it easier for them to be sure the proposed book has been assigned to the right *disciplinary categories*. From the publisher's standpoint it is important for every work to be properly positioned within the firm's overall publishing program, partly in the interest of devising an effective *publicity campaign*, but also to ensure that the title will appear in all the appropriate *catalogues* and *advertising fliers*. Marketing personnel may also want to begin planning how they will alert as much as possible of your anticipated *target audience* to the fact that such a book is on the horizon.

The tentative outline is of special interest to the acquisition editor since it spells out quite precisely what it is you really have in mind for the project. An outline also makes it easier for the editor to follow how work is progressing. Moreover, an outline helps pinpoint the most appropriate *reviewers*, or *consultants* whose reactions could be useful. Don't be surprised if you receive in response to this submission a few suggestions of things you should reconsider: changes or adjustments the editor thinks could enhance the book's quality, timeliness, or market appeal. Editors are by profession people able to think in broad categories. You may regard yourself as *the* expert on this particular topic, but you must still remain receptive to considering seriously the input of others.

● Perhaps surprisingly, another useful element to prepare at this stage is a preliminary *preface*.

One purpose a preface serves is to provide a broad characterization of the proposed book along with your reasons for writing it. Unlike the rather conventional preface associated with a dissertation (cf. Sec. 2.2.3), a well-conceived book preface affords insight into an author's thinking, and after reading it one should have a better idea, for example, of why certain choices were made with respect to content. The "official" purpose of a preface is of course to prepare the *reader* of a book for what will follow, but at least some of the content of a good preface can help sell the book as well, through its inclusion in promotional material. Selling the book is probably not foremost in your mind at the moment, but there are worse goals—and not only from the publisher's standpoint! No matter how good a book may be it is useless if it simply gathers dust in a warehouse, its content and very existence unknown to the public.

● Ideally, a book's preface combines a thumbnail self-portrait of the author with observations regarding where this particular document fits into the author's professional and/or personal agenda.

Properly handled it can actually have a significant impact on how widely the book is read. If you stop and think for a moment you will surely see the logic of including in prepublication descriptive material and advertising copy words drawn directly from a preface prepared by the book's author, especially since

- A preface is where the author attempts to explain to others, in his or her own words, something about a book's *background*, *purpose*, *content*, *organization*, and *limitations*.

These are all important subjects, and you should address each of them as meaningfully as you can. Again, there will be ample opportunity later for you to modify this first draft of the preface—perhaps even change it drastically—as well as to fill in the required personal touches, such as special words of gratitude.

4.2.2 Sample Chapter

It is not at all uncommon for an author to be asked—sometimes even before a formal contract is drafted—to put together a *sample chapter* from the proposed book. In making such a request an acquisition editor is attempting to meet several objectives. In the first place,

- A sample chapter can be quite revealing about an author's linguistic idiosyncrasies, style, and expressive prowess, insight into which can contribute to a better assessment of how well a proposed book project is likely to mature.

Furthermore,

- The author who is forced to sit down and actually compose a substantial block of text comes away from the experience with a much clearer sense of the challenge ahead.

Be sure to select for your sample a chapter that would truly class as representative.

- The sample chapter should include examples of all the "special features" the book will contain—formulas (equations), tables, illustrations, etc.—so you can begin to work out and demonstrate how you will actually deal with them.

The acquisition editor will take personal responsibility for reading and editing your sample chapter, even if subsequent work on the manuscript will be delegated to others. For an editor this is a valuable opportunity to examine and comment on the author's attitude and approach to writing, as well as to address issues related to *nomenclature* and *terminology*, correct usage with respect to *quantities* and *units*, and various other matters we address in Part II (Chapters 5–9).

- While the preparation of a book manuscript is first and foremost an exercise in individual creativity, in most cases it also involves at least some interaction with others.[18]

Most authors tend to be reluctant to subject themselves to a "writing test" so near the outset of a book project. If you are really troubled by the prospect it may be that most of the desired ends can be achieved by your providing the editor instead with one or two reprints of papers you have published on a related topic.

The more formidable and complex the projected work, the more important it becomes for author and publisher to agree at an early stage on a wide range of issues. Mutual understandings arrived at now will greatly reduce the likelihood of unpleasant confrontations later.

In the case of a journal, the standards against which an author's work is measured have been developed internally by the staff, and authors who wish to see their papers published have no choice but to accept existing guidelines and prepare their manuscripts accordingly. With a book things are somewhat different: each work is unique, so there is the potential for considerable flexibility, although it is still important that your book fit in well with the company's overall publication program, both conceptually and in terms of style.

- One of the main objectives in assuring consensus as you begin to write is increasing the probability that you will direct a sufficient amount of attention toward *uniformity* (*consistency*) in your output.

The "uniformity" to which we are referring is something one actually takes for granted—almost unconsciously—in a first-rate book, extending to such seeming minutiae as conventions regarding the hyphenation of compound adjectives (is a certain plan "half-baked", or "half baked"?) and spellings in those rare cases where a good dictionary recognizes legitimate alternatives (e.g., "judgment" or "judgement"?). Setting up unambiguous guidelines now, and then carefully adhering to them, can eliminate the need for investing large amounts of time later in "search and correct" forays—no small consideration with a work the size of a book.

The preparation of a sample chapter is especially important with a manuscript coming from someone whose native language is not English. Only when a significant piece of the author's writing has been examined will it be possible for the editor to project the extent to which *language polishing* may be necessary (obviously another cost factor).

A sample chapter also helps a publisher from a technical standpoint, because it may provide a first clear indication of potential production problems. Is this manuscript

---

[18] We note in passing that in recent years entire novels have been written on a collective basis within the "confines" of Internet chat rooms, but we think it unlikely this modern approach to composition will influence how you write your monograph. In all probability there *is* no one else in a position to deal with your topic as authoritatively as you. Still, the occasional piece of proffered advice is not something to be spurned out of hand.

going to be awkward because of the large number of special symbols or complex equations it contains? How much of a role will illustrations play, and in what condition will they arrive? Answers to questions such as these, taken in conjunction with what is already known about the probable overall extent of the manuscript, make it possible for the first time to develop a meaningful estimate of overall costs. No publisher can be enthusiastic about concluding a binding publication agreement until at least a preliminary economic analysis has been conducted.

Finally, the sample chapter presents an opportunity to address one of the most important areas of technical concern:

● To what extent will it be possible to use diskettes, CD-ROMs, or the like, submitted by the author, as a basis for typesetting?

The question is an important one, and the answer is unfortunately far from predictable. Most scientists today have had considerable experience with word processing in the course of everyday activities—but they also have acquired their own unique sets of ingrained habits, including reliance on specific auxiliary software tools. In addition, many take advantage of every opportunity to showcase their expertise as "document designers". They relish exploiting the full capabilities of a personal computer not only because of the benefits in terms of manuscript preparation (especially with respect to such "add-ons" as equations), but also because in their experienced hands the resulting output can be so visually appealing. The publishing industry, on the other hand, has experienced a history of serious problems with the interfaces used to link authors' word-processing files with computer-based typesetting equipment, mainly due to the lack of standardization. As a result, publishers find it very difficult to profit as much as one would hope when favored with an "elegant" digital manuscript.

Fortunately, this situation is becoming less problematic with time. From the standpoint of both author and publisher it would clearly be ideal if a manuscript, once committed to disk, never again had to leave the "electronic environment", and typesetting would thus never require that text be "keyed in" a second time. The goal, in other words, is an essentially continuous digital production process (cf. Sec. 3.5.4). Any potential "interface problems" attributable to peculiar features of an author's word-processing system and work habits on one hand or shortcomings in the publisher's typesetting facilities on the other need to be identified as soon as possible and somehow circumvented.

● A *sample diskette* (or data file) based on an author's first exploratory chapter is an ideal starting point for testing the compatibility between the author's word-processing output and the publisher's typesetting equipment, in the interest of working toward something approximating "diskette/disk typesetting".

We return to this subject later (in Sec. 5.4) for closer examination.

Publishers frequently go one step farther at this stage and prepare a *test print* from the sample chapter. This helps them determine such things as optimal type fonts and sizes and an appropriate page layout, in turn opening the way to establishing the book's probable *page count*. Experience has also shown that offering a stressed and nervous author a glimpse of what a finished book will look like can have a distinctly calming effect, at the same time serving as a stimulus to proceeding as expeditiously as possible with the monumental chore that lies ahead.

● The sample chapter should be accompanied by a comprehensive *list of symbols*.

This list should include a precise definition of every symbol you can reasonably anticipate requiring in the text—those representing *physical quantities*, for example, but also miscellaneous symbols of other types (e.g., ones likely to play a role in mathematical formulations). From a technical standpoint, your symbol catalog may influence a publisher's choice of type font—or even the typesetting process. Later, the same list, updated and expanded to include *abbreviations* and *acronyms*, will probably be incorporated into the book itself, near the front, where it will help the reader interpret the text.

A publisher sometimes suggests using this occasion to work out a system of *codes* for representing in the text important symbols not readily available as a part of the author's word-processing system. For example, it might be agreed that the frequently required mathematical symbols Δ and $\times$ will routinely be simulated in your manuscript by the rarely encountered \$ and §, respectively—a perfectly sensible scheme so long as neither the dollar sign nor this particular section symbol will actually play a role in the text. Similarly, there may be reason for devising a way to signal the need for a special font, perhaps, or subscripts and superscripts—especially in the event that the publisher expresses a strong preference for plain, unformatted text.

4.3 Developing the Manuscript

4.3.1 Organizational Considerations

Composing a journal article entails assembling material for a maximum of about 10 printed pages, but the challenge we are now dealing with is larger by at least one and perhaps even two orders of magnitude. A typical *monograph* runs to something like 200 or 300 pages, and a comprehensive *textbook* could easily reach—or exceed!—1000 pages. You are asking for disaster if you embark upon such a monumental challenge without extensive thoughtful and focused *preparation*.

● The first thing to consider is *how* you propose to capture your thoughts as text.

Would there be advantages in starting with a *handwritten* manuscript for later transformation into "typed" copy? If so, who will do the "typing" (presumably with a computer), and with what word processor? The basic issues are essentially the same ones we raised in the context of dissertations (see *Writing Techniques* in Sec. 2.3.1), with one important distinction: the larger objective of creating a full-fledged book requires even more attention to efficiency. Despite all the recent advances in writing technology, some authors still prefer to begin with handwritten drafts, and the occasional writer with *stenographic skills* (and thus conversant with *shorthand*) may in the beginning wish to continue exploiting this dying art.[19]

If you do decide to prepare a handwritten manuscript, the medium should be a stack of sequentially numbered sheets of reasonably high-quality, letter-format paper, for inscription on one side only. As pages are filled they should be transferred to a loose-leaf binder as the easiest way of maintaining a reasonable semblance of order, for which you will become increasingly grateful as you transition into editing and revision.

- Dedicate a separate binder to each chapter, and place in it copies of *every* version you create, from the earliest sketch to the finely tuned document you eventually send the publisher.

Each segment of your manuscript will inevitably evolve through several stages over the course of time, and a *dated* copy of each should be saved, because now and then you may wish to make comparisons. Separate the various versions within a given binder by plainly visible divider sheets.

Orderly management should also characterize evolution of a *digital* manuscript. Thus, file names should be chosen to reflect not only content but version numbers as well, and you will want to be able easily to ascertain when a given version was prepared (the computer automatically maintains a record of when a file was last *saved*). Each time you decide to make changes, first save a fresh copy of the existing file—in its current state, but under a new name—and confine your editing activities to the *new* copy. In this way the original will still be there if you wish later to reexamine it. There is no good reason for *not* saving multiple versions of your manuscript, by the way; the process takes almost no time, and with today's personal computers it is most unlikely you will ever find yourself short of hard-disk storage space.[20] Each chapter (better: every major *subdivision* of each chapter) should be treated as a separate entity. Just like documents in a filing cabinet, computer files that are too large become a major nuisance, and they certainly cannot class as "user friendly".[21]

---

[19] Some of our own earlier books in fact took shape through this approach: a stenographic sketch, lightly polished, was read aloud by the author into a tape recorder, and a secretary then transcribed the "dictation" as a classical typescript. In the meantime, however, we have become enthusiastic converts to relying from the outset on word processing. Surprisingly, we find this to be not only a more technologically advanced approach, but in some senses also a more personal one.

[20] Recall we have also urged that you retain *backup* copies of all your files—stored in a separate location!

[21] Admittedly, small data files have the disadvantage that they make it difficult to retain a clear overview of a project. The word-processor's "Outline" function also loses much of its value, and searching for places where you may have employed certain terms becomes impractical as well unless you know in precisely which file to look. You may therefore find it worthwhile at some point to create for special purposes one or more truly inclusive files, despite their inconvenience. We return to this issue later.

A first draft of any long document, whether handwritten or preserved in the form of computer files, is certain to be rather crude no matter how experienced the author. At least some reorganization and several stages of revision can be anticipated, so there is little point in laboriously transforming a handwritten effort into a typescript immediately. On the other hand, once your document shows signs of becoming mature you will surely want to redirect your attention to an electronic version. Given the magnitude and complexity of a book project it would be folly today to proceed very far on the basis of handwritten drafts or classical typescripts. In the unlikely event that—through lack of experience—you do not yet feel comfortable personally entering text into a word processor, your only recourse is to recruit someone to help.

- No outside "typist" should be asked to cope with hard-to-read handwriting, let alone shorthand; at least provide your assistant with a supplementary *oral transcript*, prepared with some sort of recording device.

This indirect approach to an electronic manuscript is clearly workable, but it unfortunately condemns you to accepting the fact that, for example, a reference you may make in your text to "genes" is likely to show up in the first computer-based version as involving "jeans". We have alluded to this problem earlier, albeit in a rather general way. One thing is clear: the strategy author–secretarial service, familiar from the world of business, should be ruled out at once, because office help unable to understand reasonably well the technical jargon so prevalent in scientific text will be incapable of producing even an approximation to error-free copy.

- The tremendous advantages inherent in your engaging on your own in word processing activities become especially apparent during the *development stage* of composition.

It is hard to overemphasize the value of digital technology when you wish to make changes in an evolving manuscript, especially a large one, a subject we expand upon in Chap. 5 (specifically Sec. 5.3). Indeed, the decision to write a book can be a powerful rationale (if one is even needed!) for your finally coming to terms with a good word-processing system and learning to work with it as effectively as possible. So long as you don't set your initial standards too high, or try to take immediate advantage of too many of a program's "bells and whistles", the learning curve is surprisingly gentle, something most of our readers presumably discovered long ago.

- A "writer's computer" serves primarily as a highly sophisticated electronic typewriter, but it displays incredible tolerance with respect to mistakes.

What most fundamentally distinguishes the writing process today from its counterpart as little as two decades ago is the fact that errors or other shortcomings in a document, as soon as they are recognized (on a display screen, perhaps), can be eliminated swiftly without leaving any trace. Immaculate printed copy becomes available at the touch of a button after each round of editing, and there is almost never a need to reenter significant amounts of text manually. The author so fortunate as to possess reasonable typing skills of course enjoys the special advantage of being able to capture text very rapidly—indeed, significantly more rapidly than the average person can write by hand. Moreover, during text entry almost no conscious attention need be squandered by a seasoned "touch-typist" on what the fingers are doing. Nevertheless, even a complete novice with respect to keyboard techniques can enjoy most of the benefits of a word processor, and after very little training.[22]

Especially with a large document, other more subtle strengths of the computer-based approach to writing soon make themselves felt. For example, much can be gained from maintaining, alongside files representing individual chapters (or subsections) of an evolving book, a fully *integrated* copy of the current state of one's work as a whole. Assuming you have at your disposal sufficient *working memory* (Random Access Memory, RAM)—almost a given with today's computers—an entire book manuscript can in this way be available in its own file for active processing: 400 pages of text, say, which is the equivalent of perhaps 1.5 MB of stored information. With a single search operation one can locate in such a file literally *all* the places in the book where a particular symbol or word appears. For one thing, this allows you, as you write, quickly to determine precisely what you may already have said about a topic you propose to address once again, and also to examine the various contexts in which those comments appear. (The suggestion is one we have already made briefly in passing).

By taking advantage of a word processor's "Replace" function with an integrated file you can also *change* the way something has been expressed—again, throughout an entire work, and in a single pass. Furthermore, by switching the screen display of your file to *outline view* (cf. *Developing a Concept* in Sec. 2.3.1) you have at your disposal an overview of the complete manuscript, which you can then exploit to gain rapid access to a specific topic, subsection, or chapter, a much more appealing prospect than fumbling randomly through multiple files or several hundred sheets of paper. When dealing with a single file the size of a book the computer will probably take a few seconds longer than usual to do your bidding, but the price is a trivial one compared to the benefits.

The matter of revision requires a bit of special consideration when several authors are collaborating on a work, particularly if files are being passed back and forth. Who proposed what changes when? Which file is currently the "authentic" version? Fortunately, today's word processors are equipped to lend an organizational hand here as well, in the form of "Insert Comment" and "Track Revisions" commands, which automatically make note of who is responsible for each appended comment or proposed revision, and when it was recorded (date and even time). If colleague Y spends an afternoon working over what colleague X wrote in the morning, this latest effort becomes fully documented with each keystroke Y enters. To achieve the maximum

---

[22] One of us had the occasion to become deeply interested in Thomas MANN and his writings, and has been musing about how *he* would have reacted to the availability of a word processor. And what about GOETHE? We find it almost impossible to imagine *his* literary life in the absence of his influential secretary ECKERMANN, but such "human adjuncts" to writing must probably be accepted as a thing of the past.

benefit from these facilities, be sure to assign an identifying code to each member of the team in advance to rule out unintended anonymity.

It has been interesting to watch from the sidelines as scientists in particular have in recent years come to discover how it can literally be "fun" to become deeply immersed in a word-processing project. Publishers report that this fascination with a technique has translated into a new level of willingness on the part of researchers to at least consider engaging for the first time in a major piece of writing. Given their dedication to the promotion and oversight of book production, publishers are of course delighted to capitalize on this unanticipated but highly welcome spin-off from computer technology.

But now we turn to another major issue for the aspiring author to resolve early on:

- *When* will the envisioned writing activity occur?

Do you anticipate devoting evenings to it, after work? Or perhaps weekends? Or vacations? Possibly *all* the above![23]

Virtually every productive scholar complains about a lack of time to get things done, and the typical scientist's professional day extends well into the evening hours, jealously reserved by others for leisurely pursuits. Life for the avid researcher is stressful almost by definition. This is why the timing question we are now raising demands urgent attention as you make your plans for becoming a book author. Before you make a firm commitment remember too that your family must also be willing to accept some of the ramifications. Like it or not, agreeing to write a book means you will be obligating yourself to put other "priorities" on hold for the next two or three years!

- Be prepared to subject yourself to a rigorous *work schedule* almost from the moment you sign a formal contract with a publisher.

We strongly urge you to formulate at once a rigid, detailed *plan* committing you to meeting specific writing goals by specific dates. Start also to keep an accurate and comprehensive record of your progress—a sort of "work log". If at any point you find yourself falling behind on the schedule you set, take remedial action immediately, and in the process seek effective ways to avoid similar problems in the future.

- Begin serious, determined work as soon as you agree to a concrete *deadline* for completion of a manuscript.

Publishing contracts typically provide for writing periods of two or three years, or occasionally even longer, although for a minor contribution to a collective work this would probably shrink to a matter of months. It is in any case a serious mistake to

---

[23] The English physical chemist Peter W. ATKINS, who has written half a dozen textbooks, claims that his preferred approach to writing is to isolate himself in his studio at 5:00 A.M. in the company of a pot of strong coffee. The late American chemist Louis F. FIESER wrote his world-famous organic chemistry textbook from beginning to end with the constant help of his wife Mary. We are not privy to the organizational approach they used, or their daily schedule, but whatever their method it would be a gross understatement to describe it as successful!

regard a proposed set of terms as generous. More likely, the period envisioned will barely suffice even if you begin concentrated work today—as you should! Indeed, to do otherwise would be irresponsible.

Postponing a daunting chore never makes it easier to complete. In the present case, the reverse is true, because every postponement forces you to launch yet another meticulous search for new literature that may have appeared: reports you will need to study, evaluate, and then acknowledge in some way in your manuscript. Nothing is worse than finding yourself condemned to what seems an endless series of rewrites simply to keep your cherished document from being obsolete before it ever reaches the public.

There is, by the way, another peril to guard against, one to which too many writers fall prey:

● Allowing long pauses to punctuate your writing puts you at risk of losing the thread of whatever arguments you are attempting to formulate.

A related pitfall is setting aside blocks of time for your creative efforts that are inherently too small. As a result you will find yourself wasting a major part of each work episode trying once again to immerse yourself fully in your thoughts.

Along the same lines,

● Take a few minutes at the *end* of each writing session to organize the materials you know you will need during the *next* session, as the most effective way of minimizing unproductive start-up delays.

The latter recommendation is analogous to a suggestion we once heard that if your (standard transmission) car develops the trait of frequently refusing to start, you should try always to park it in a *downhill* manner to facilitate ignition by coasting! Personal experience has taught us that taking to heart this particular organizational suggestion can greatly increase the efficiency of one's work, but note that it requires your consciously *refraining* from that which comes so very naturally: setting the pen aside or bidding farewell to the keyboard—with a huge sigh of relief—as soon as the day's goal has been achieved: formulation of the crowning sentence marking the end of a particular subsection, perhaps.

The chances you will actually succeed in putting the finishing touches on your masterpiece within the allotted time period depend heavily upon your taking seriously from the very beginning the awesome nature of the workload you face—and promptly *undertaking* some of that work. A good first step is pulling together copies of review articles you may already have written on the subject at hand, or dusting off and arranging relevant sets of lecture notes. Next, locate and gather up all the books and journal articles you expect to be consulting regularly, and refresh yourself on the content of each. It is important that you develop a good grasp of all the key literature sources before you officially begin writing. When you know you are fully primed to begin pouring out your message in the form of a steady stream of pithy sentences, and are confident of your overall mastery of the material, then by all means charge ahead in the critical race against time.

If in a few weeks, despite your best efforts to the contrary, you find you are not keeping up with the writing schedule you set, contact your publisher at once and be absolutely candid in describing the situation. Any other course of action marks you as an unreliable partner. Perhaps more sobering, remember also that

- Publishing contracts reserve to the publisher the legal right to withdraw from all prior commitments to an author who fails to produce a completed manuscript within a certain specified period of time!

In other words, the lackadaisical author runs a serious risk of literally wasting precious *years* of a career by taking a writing assignment too lightly, because in the end there may be absolutely nothing tangible to show for efforts expended. Unpleasant consequences also follow when one member of a *team* of authors falls significantly behind in an assigned piece of writing:

- Deadlines not met by one *coauthor* of a multi-author publication, or an individual contributor to a collective work, exact publication delays that work to the detriment of everyone involved.

Honesty and contrition, painful as they may be, become a negligent author's only recourse, for what is at issue is flagrant disrespect of the interests of others. Prompt admission of the fact that you will be unable to fulfill the terms of an agreement to which you are a party at least opens the possibility that a substitute author can be recruited to fill your place.

It seems fitting to conclude the present subsection by urging you to

- Reserve for your writing activities a special dedicated *work room*—or at the very least a reasonably isolated desk together with a commodious set of cabinets and drawers.

Failure to devote sufficient attention to logistic matters of this type invites chaos from which you may find it very difficult to recover.

4.3.2 Assembling the Background Literature

Let us assume you have already gathered together the stock of reference material that will need to be available as you write. Where should this treasure trove be kept? Some of your writing will undoubtedly take place at home, but at least part of the "essential" document collection probably needs to stay at your place of work—for use not only by you but also by your colleagues. It is obviously impractical to create a duplicate library simply to meet your needs as an author, so we suggest you consider doing the next best thing: compiling for reference purposes a *synopsis* of your working library.

● It is a relatively easy matter for the author with access to a personal computer to construct a highly portable, personalized *literature compendium* for use at home or wherever else it may come in handy.

A general-purpose database program (cf. Sec. 8.5.2) will serve the purpose nicely. The first step is to define, format, and conveniently organize an appropriate set of *data fields* (see below). In effect, each record you create for your new database will be the equivalent of an electronic "file card", subject to manipulation in all sorts of useful ways. You can readily manage hundreds of individual *data records* of this type, encapsulating in the process an almost unlimited amount of information. Such a database can be an invaluable asset for the author preparing a monograph. A "laptop" computer is especially convenient for housing the corresponding file, which can then easily be transported from place to place. Within seconds, wherever you happen to be, you can locate and examine any piece of information in the database. A targeted search within such a file can be set up in a multitude of ways: on the basis of *author*, for example, or by *institution*, *source journal*, *catalog number* (which might in turn point to where the original document is stored), *bibliographic characteristics*, *date of origin*, *keywords*, and perhaps some aspects of *document content*—indeed, on the basis of any data field you choose to provide. *Groups* of records can also be *designated*, *isolated*, and *sorted*—temporarily or permanently—in any way you find helpful: numerically, for example, or alphabetically, or chronologically.

There are advantages in establishing several different data fields reflecting document content. This might include fields for *abstracts*, *summary comments*, *excerpts*, key *facts*, *interpretations*, or even interesting *quotes*. Text passages you have incorporated into such a file can prove surprisingly useful as you try to formulate what it is *you* want to say in your own treatment of a subject. You might even decide on occasion to *copy* material from your literature file directly into the emerging manuscript to serve as a starting point for context-related adaptation.[24]

Some authors incorporate into their electronic literature files entire *documents*, capturing them in digital form through scanning and subsequent interpretation with a computer-based *OCR program* (cf. *The Processing of Figures* in Sec. 3.4.3). In this way they obtain convenient access—from anywhere!—to the full text of key references, avoiding any need for sorting through shelves or filing cabinets filled with *photocopies* or *reprints*, or undertaking inconvenient pilgrimages at odd hours to the nearest library.

A related but more sophisticated approach to literature management takes advantage of commercial software developed explicitly for the purpose: programs like ENDNOTE or PROCITE (cf. Sec. 9.2.2). One argument in favor of pursuing this more structured route is that it adds another useful tool to assist in manuscript preparation: the ability to formulate instantaneously any formal literature citation you may require, knowing that it will conform in every detail to virtually any imaginable set of format

---

[24] Obviously we are *not* encouraging plagiarism! The point of this suggestion is quickly building into your developing manuscript something to *consult*, *reflect upon*, and *elaborate*.

specifications you establish. Professional literature-management software also facilitates downloading into your files (e.g., from the Internet) information derived from various external databases. We reexamine this subject at length in Sec. 9.2.

4.3.3 The Structure of the Book

Now that you are ready to commence serious work, you should study once again the tentative outline you prepared earlier for the publisher and make sure you still regard it as an appropriate guide to your proposed masterpiece. One of the several organizational techniques we discussed in our treatment of dissertations—the cluster method, manipulation of idea cards, or mind mapping (see "Developing a Concept" in Sec. 2.3.1)—would now be worth putting to the test. By this point at the latest you should also begin developing a stock of *keywords* that will suggest content for the various headings in your outline.

● Be conscious of the risk that certain subsections may become too long or complex, and feel free to create additional subdivisions any place you think they would help.

The rule proposed earlier that a section (subsection) should never be allowed to exceed a few pages (see Sec. 2.2.5) is important to remember, especially with respect to its influence on *cross-references*. Cross-referencing is meant to help the reader find tangentially relevant content located elsewhere in a document. From the reader's standpoint, such references would ideally point to specific *pages*, because it is obviously rather easy to locate material of interest within the narrow confines of a single page. Unfortunately, creation of a set of true *page references* imposes an enormous burden on the author, because durable page *numbers* don't even exist until after page proofs have been prepared, which means giving references their final form would need to be delayed until shortly before publication. You can perhaps imagine the enormous problems page references pose when a new edition of a work is to be prepared! In principle one could perhaps work initially with temporary numbers, referring to *manuscript pages*, but even that becomes terribly cumbersome because of the number of discrete drafts that are sure to be involved, all paginated differently. The effort entailed in creating and monitoring true page references for an entire book is so great that it can rarely be justified; therefore:

● See to it that your manuscript is organized around a multitude of *numbered headings*, and construct your cross references so they point to *section numbers*.

The reader wishing to take advantage of this type of cross-reference is assisted by the fact that numbered titles identifying the current chapter and the current section will appear in the form of *column titles* at the tops of all the book's left-hand and right-hand pages, respectively.[25]

---

[25] We note in passing that information professionals include column titles, page numbers, and cross-

→

Our parenthetical discussion of cross-references has caused us to jump somewhat ahead in our story and allude briefly to page proofs and some of the limitations imposed by publication, but we felt it important that you give early thought to this particular service you will want to offer your reader. It is in any case not premature to keep cross-referencing in mind as you begin to associate keywords with your various subheads.

● Cross-references should always direct the reader to the *lowest subdivision level* possible, which would ideally encompass no more than about five or six pages.

Numerical subdivision more than four levels "deep" becomes unwieldy, especially with respect to the labeling of cross-references. We have already addressed ways to circumvent overly complex numbering in the context of dissertations (Sec. 2.2.5).

● Finally, this is a good time to reexamine yet again your preliminary outline and subdivision plan, specifically in terms of deadlines you have set for yourself.

The internal logic and structure that characterize your book need not be a determining factor as you plot the best course to follow in *writing* it. You will almost certainly recognize that some chapters will be "easier" or "harder" to compose than others. A few you are probably prepared to begin drafting at once, while with others you may need to do considerable preliminary reading and pondering. This insight is what should shape your decisions about what parts to begin drafting first. As author you are clearly the one in the best position to devise an optimum action plan, with which now you should fine-tune your work schedule.

4.3.4 Developing the Content

First Draft

Once you have in hand a mature, carefully adjusted subdivision plan, together with a reasonable schedule for completing the various sections—and not before!—it makes sense to begin writing a *first draft*. The immediate goal is to express for the first time in *sentences* the content implied by your subdivision headings and whatever keywords you have so far associated with them.

● Write quickly and in broad sweeps!

More important than anything else is that you rigorously maintain your train of thought. Linguistic niceties and minor details should all be ignored—completely!—for the time being.

Much that you write initially will ultimately be discarded, as we know only too well from our own experiences. This stage in the writing process is especially

---

references, among other things, in the special category of "metainformation" associated with a published work.

challenging for the perfectionist—the person who strenuously resists doing a "halfway job" of anything—but aspirations to literary perfection simply must be deferred. Consider the following analogy, drawn from a totally different art form: On a first pass, the output of even the most gifted sculptor is always unspeakably crude. Chiseling out the details and refining the rough edges begins only after considerable time has been devoted to achieving a harmonious reconciliation of overall shape and form. All thought of developing the desired surface finish is postponed until the very end.

It would be a complete waste of time for you now to fret over the formulation of a specific text passage only to discover on a first serious reading that the sentence in question is superfluous. It will be quite a while yet before you really know what it is you in fact want to say and how those messages would best be expressed. In all probability you will even conclude that some aspects of the outline itself are still a disaster. Every author finds the fleshing out of a book's content to be a profound learning experience; indeed, it has been asserted that this exhilarating phase of unanticipated enlightenment should be acknowledged as one of the primary rewards for undertaking to write a book in the first place.[26] Committing oneself too early to a particular way of phrasing some thought can in fact seriously interfere with the overall creative process and impair your ability to gain new insights.

The "Special Features"

Almost every scientific "text" consists of much more than simply text. Verbal assertions gain important support from such subsidiary—or perhaps we should say auxiliary—features as *equations*, *illustrations*, and *tables*.

● You should devote a significant amount of attention to your book's "extras" now, even as you are assembling the very first draft of the text.

Once you have established the nature of the requisite special features they too can be "drafted", and appropriate identifying numbers can be assigned [e.g., Table 3, Equation (14)]. Parallel to composition of the text manuscript, work can thus commence, for example, on the associated "illustrations manuscript". Once again, nothing "polished" is required; more appropriate for the time being is a set of crude sketches, or simply indications of promising sources of inspiration. The same principle applies to every part of the manuscript.

● Begin now also to construct accurate *lists* of figures, tables, and other subsidiary material you intend to include.

---

[26] We were impressed some years ago upon first encountering Al GORE's widely discussed book *Earth in the Balance: Ecology and the Human Spirit* (1992), in part as a consequence of the message it contained, but also because of the author's refreshing candor about the writing experience. The later vice-president of the United States and narrowly defeated presidential candidate declared that one of the reasons he embarked upon that particular writing project was to gain a clearer understanding of and appreciation for views and positions he already held, while at the same time testing his beliefs against the facts as best he could ascertain them.

The lists will help you keep some perspective on the project and at the same time contribute to consistent assignment of correct numbers (not only for illustrations, but also for tables and equations), making it less likely that you will inadvertently overlook something. The lists will later aid you in putting together collections of copy for figures and tables, as well as in constructing a *caption manuscript*.

● Even in your first draft you should attempt to employ correct nomenclature and terminology, and to associate generally accepted symbols and units with all physical quantities.

This is a good time to direct renewed attention to the list of symbols you prepared earlier (see Sec. 4.2.2). You should also begin making notes of ways you wish to formulate particular recurring expressions (with respect to wording, for example, or the use of hyphens) so you will be consistent throughout the book. Assuming you are working with a word processor it would be helpful now to develop at least a few *AutoText* entries for introducing with a few keystrokes certain complex expressions that are sure to arise repeatedly, but which are a nuisance to type.

● Only when you have in hand a representative, consistent, and complete manuscript for the entire book (!), one that encompasses all the associated special features, should you begin serious polishing and correction, finally paying close attention to linguistic considerations and enhancing the clarity of your writing.

Revision

Refining a crude manuscript until it is in every way ready for publication is a time-consuming, intellectually demanding process.[27] The conscientious writer is sure to become somewhat frustrated as one revised version after another proves still to have major shortcomings: seemingly just as many as the previous version, albeit different ones. At first you will probably find yourself focusing primarily on phraseology within individual sentences. Later you will recognize that your perspective has imperceptibly shifted and become broader, and that as a result you have begun raising new questions. What might you do to make certain small groups of sentences work better together? Where have you inadvertently introduced phrases that are repetitive, or overused various words? Finally you will recognize you have begun evaluating large blocks of text in their entirety—and probably discovering that certain paragraphs are in fact misplaced from the standpoint of a clear line of argumentation.

At a distance of several days or even weeks from what you suppose has been the definitive reworking of your text you can expect once again to recognize—regrettably—

---

[27] We are well aware that there are poets, composers, song writers, etc., to whom this does not apply. Some make it a firm rule never to revisit what they have once set on paper. On the other hand, others are so critical of themselves that they become almost suicidal. The first proof of BEETHOVEN's 9th Symphony—to be found in Berlin—contains over 1000 corrections from BEETHOVEN's own hand. Every one of them is cherished, even if some are unintelligible. There are many gates leading to eternity, but all provide limited access.

that parts of your writing are not as good as you had imagined. Here and there you will discover a word or phrase that is clearly superfluous. Places will stand out where you have skipped over or taken inappropriately for granted some important intermediate step in the formulation of an argument, or you will suddenly be struck by a more effective or dramatic way of expressing yourself. Some trusted colleague may also point out to you weaknesses that you as the author were too close to the work to identify.

Nevertheless, take comfort from the fact that you have nearly reached the finish line. You will be pleasantly surprised to discover that the number of changes to be made this time around is relatively small, and that the text as a whole has become something of which you can (and do) begin to feel proud.

● At this point you should be working with printed text that is *double spaced*, with a wide right margin (ca. 5 cm).

During final steps of editing and preparation of the manuscript for typesetting, the extra space provided by these formatting suggestions will prove very welcome.

● Corrections or comments related to further processing are best entered between appropriate pairs of lines, or nearby in the margin.

Assuming you are working with a word processor, format adjustments of this sort can of course be made at any time, and almost effortlessly.

4.3.5 Final Copy

Text

Step by step, your manuscript is gradually approaching the state known as final copy, suitable for sharing with the publisher. In preparing the final copy it is especially important that you pay close attention to all the various guidelines you may have received.

● It is your responsibility to ensure that your final copy conforms in every detail (from a *form* standpoint) with the publisher's wishes.

The publisher will most probably expect you to supply at least one clean *hardcopy* of your text, even if there is every intention that typeset copy will ultimately be generated directly from your data files (disk files). Take some pains to provide copy that is reasonably attractive. Even editors are affected by aesthetics. Fortunately, primitive computer-based manuscripts of the past, derived from dot-matrix printers with their infamous "computer script" and printed on fan-fold, perforated-edge paper, are now only an unpleasant memory; indeed, no publisher today would be likely to tolerate something so crude.

● Before printing the final copy, be sure you clarify with the publisher the extent to which formatting is desirable (including invisible commands embedded in your word-processor file).

"Formatting" can be taken here to include indications that symbols for certain physical quantities should be typeset in italics, or that italics (or boldface characters, or "small caps") are required at certain places in the literature references (cf. Sec. 9.4.1). It is important to be clear on whether instructions such as these are to come from you or will be introduced later by staff at the publishing house.

The occasional author who for whatever reason chooses not to take advantage of word processing should submit the cleanest manually prepared manuscript possible (although despite one's best efforts this will probably still contain a few last-minute hand corrections). No matter how clean the copy, however, the publisher will almost certainly establish as a first priority having someone prepare an electronic version of the document as expeditiously as possible.

● Handwritten corrections in a typed document destined for typesetting should always appear either between the lines or directly at the site of the problem.

This type of correction is used only for calling attention to places in *text to be typeset* where changes are required. Additional line-by-line correction of this sort may be added by members of the editorial staff in the course of in-house refinement of your document. This is *not* a suitable occasion for employing traditional proofreading marks, usually set in the margins (cf. Sec. 3.5.5); these are reserved literally for work with *proofs* (galley proofs or page proofs). The rationale for marginal correction in this case is that once copy has been typeset it becomes very difficult to indicate changes directly in the body of the text due to space limitations, and there would in any case be a considerable risk that in-text markings would be overlooked.

When requesting alterations in a *manuscript* that has yet to be typeset it has always been customary to apply the annotations in such a way that they facilitate the arduous duties of the typesetter. A typesetter forced to glance back and forth repeatedly between text and margin is necessarily working inefficiently, and a relatively large number of mistakes is certain to creep in. In other words,

● From a typesetter's standpoint, the only sensible place for signaling manuscript corrections is directly at the source of the problem: *within the text.*

Manual typesetting is of course exceedingly rare today.

Interestingly, the optimal way to approach correction has in some ways become more ambiguous as manuscript development has migrated increasingly to personal computers. Word processing tends to blur the distinctions between a manuscript and typeset copy. Thus, printed output from a computer still corresponds to a manuscript from the standpoint of maturity, but its appearance is at least strongly reminiscent of "typeset" copy. What is to be made of this paradox? That could well depend on what will happen next.

The mechanics of correction serves as a convenient backdrop for us to consider briefly the overall process of copy editing and its consequences at a typical publishing house. Most copy editing is still conducted with *paper printouts*, for reasons described earlier (Sec. 3.4.1). One of two routes might be followed by copy once it has been marked by the editors. Sometimes the edited manuscript is returned to you, the author, with the expectation that you will implement all necessary changes with your computer, subsequently returning to the publisher both a revised printout and a corrected data file, either on a diskette or disk or as an e-mail attachment. It thus becomes your responsibility to examine carefully all the places where changes have been proposed. You would obviously want these to be as clearly apparent on the printout as possible, so marginal correction using formal symbols could make some sense.

Alternatively, the marked manuscript and an original diskette might instead be passed along to a printing firm for preparation of the revised data file. The same considerations still apply, but the responsibilities will have been distributed differently.

Actually, there is no real need for a set of formal procedural rules if a brightly colored pen or pencil is used for correction purposes (e.g., a red ballpoint). The resulting markings will be easily recognized no matter where they are placed.

Modern word processing includes valuable tools for *highlighting* changes that may have been introduced into a text file. With recent versions of WORD, for example, it is easy to "track" even a complex sequence of changes, allowing direct comparison at any time between text as originally prepared and its edited or revised counterpart. Upon request, changed places can be made to stand out very clearly, not only on the screen, but also on specially designated hardcopy printout. For example, WORD—if so directed—typically *strikes through* text that has been deleted, and *underscores* anything newly added. The program also permits various annotations and comments to be introduced, and in a distinctive *color*. While these can be dramatically apparent on the screen, they remain utterly invisible in ordinary hardcopy printouts.[28]

● Editing functions like the ones described offer the great advantage of adding transparency to the editing process and enhancing constructive interaction between authors and editors.

Another point supportive of "tracked" electronic manuscript correction is the fact that changes are thereby *documented*, so it becomes clear when and at whose request each alteration was made. All of this would appear to represent a significant step along the path toward true *electronic editing*, and it can certainly help alleviate an author's concerns that his or her manuscript might be subject to "tampering" of a sort that would be difficult to monitor or even recognize (see the comments near the end of Sec. 3.5.1).

Other Elements

In "the old days", many a clever author skilled at manipulating a typewriter platen, took pride in managing (with difficulty) to embellish a manuscript with reasonable typed approximations of simple mathematical equations, even ones featuring super-scripts and subscripts. Whether today's author, equipped instead with a sophisticated

---

[28] This set of capabilities was mentioned briefly in Sec. 4.3.1 in the context of multiple authorship.

word-processing system, would be equally successful depends not only on the flexibility of the word processor, but even more upon the user's patience, experience, and determination.

It can be exceedingly awkward—and frustrating!—to prepare all but the most elementary equations within the confines of a word processor alone because of the complex typography required. Even more serious problems can arise, however, when a publisher attempts to transform such makeshift equations into their typeset equivalent, because tricks to which authors cleverly resort in the preparation of *special symbols*, *subscripts*, and *superscripts* (with or without the help of supplementary software) can produce gibberish with professional typesetting equipment.

● Even today there is much merit in preparing equations for a manuscript in the traditional way, by hand, and keeping them separate from the actual text to ensure that the body of a document will lend itself to automated (computer-based) processing at the publishing house.

Be sure to seek the publisher's advice in this important area as early as possible, preferably with the aid of samples submitted for preliminary evaluation (cf. Sec. 4.2.2).

● Final copy for a book manuscript is likely to include a variety of components; e.g.:

  – the *text manuscript*
  – an *equation (formula) manuscript*
  – a set of *tables*
  – a *captions manuscript*, and
  – *illustration copy*.

Responsibility for merging everything into a single typeset entity is normally assumed by the publisher, often with the assistance of outside professionals.

● The text manuscript itself should commence with a *table of contents*.

A meaningful and reliable table of contents can only be prepared after the text manuscript is otherwise complete in every way, because prior to that point various chapter and section headings are subject to change, and page numbers remain very much fluid. Upon request, a full-featured word processor is typically able to generate a table of contents automatically (e.g., on issuance of an appropriate "Insert" command) *provided* the program can recognize headings for what they are on the basis of their *style* assignments.[29] At the publishing house a typesetter would of course immediately remove an author's page numbers from the table of contents, since numbers can only be specified properly for the published edition after page proofs have been prepared.[30]

---

[29] Automatic table-of-contents generation has the special advantage of ensuring that all headings listed in the resulting table will be absolutely identical to those in the text . Editors were long able to claim they had never seen an author-prepared table of contents that was completely accurate in this respect!

[30] The situation is obviously quite different with a book manuscript presented as camera-ready or "print-ready" copy; in that case the author will produce a definitive table of contents. This book, for example,
→

- An author is obligated (usually through the terms of the publishing contract) to retain at least one complete copy of a submitted book manuscript, along with backup copies of all associated computer files.

A text manuscript is normally supplied on sheets of ordinary letter-sized paper, numbered sequentially and printed on one side only, with all copy double-spaced. Double-spacing, together with a wide right margin like that recommended previously, results in pages that average ca. 30 lines of roughly 50 characters each (including blank spaces): i.e., approximately 1500 characters per page (see Sec. 3.4.1). Two or two-and-one-half such pages lead to about one printed page in a typical book. A reference work arranged in two columns and set in rather small type might correspond to as many as three manuscript pages per printed page, whereas with a pocket-sized paperback the conversion factor is likely to be two or even a bit less. Book publishers once took pride in their ability to develop complex formulas for estimating accurately the final page count of a book from the extent of the submitted manuscript. Today, however, they generally content themselves with a report of the total number of characters present, a piece of data typically compiled by the word processor with which the manuscript was prepared (in the case of WORD, by invoking the "Word Count" command in the "Tools" menu).

A few publishers still prefer that authors provide final copy on special *manuscript paper*. Guideline markings on such paper specify those areas in which text is to be typed or printed. The markings themselves are in a faint blue ink invisible to a photocopier. The formal *page template* upon which paper of this sort is based normally incorporates the wide right margin we recommended earlier, in part as a way of ensuring that text will not be obscured by subsequent correction requests (as introduced by a right-handed editor!). The *left* margin should also be at least 25 mm (1 in) wide, incidentally, to permit a small number of editorial marks there as well (e.g., type-size instructions). The practice of supplying custom manuscript paper is rapidly dying out now that virtually all manuscripts are prepared with word processors rather than typewriters, which makes it all the more important for authors to observe on their own any formatting instructions they receive.

- A manuscript intended for *direct reproduction* (by photomechanical means)—a conference report, perhaps—must be the subject of even greater care with respect to rigid compliance with a publisher's instructions.

This matter has been touched upon previously in the context of "letter journals" (Sec. 3.2.3).

- Once it is truly complete, your book manuscript (typically a single copy) should be carefully packaged and dispatched to the publisher by the most reliable means at your disposal.

---

is almost entirely a product of its authors; the same was true as well for the first edition, published in 1987/1990, and all editions of the German counterpart.

If there is an understanding that you will also supply a *diskette* or similar data carrier containing an "electronic manuscript", this should be enclosed in the package with the hardcopy version. You might be surprised to learn that one double-density, double-sided 3.5-inch diskette, with a capacity of about 1400 KB (*kilobytes*) of information, would normally suffice for the text of a 300-page book! Diagrams or other illustration copy destined for direct reproduction should be provided in the form either of *originals* or high-quality photographic copies.

4.4 Typesetting and Printing

4.4.1 Processing the Manuscript

The fact that your manuscript has been received at the publishing house will be confirmed in writing by the editorial office. Initial stages in further processing will be under the supervision of an editor assigned specifically to your project, who will begin by becoming intimately acquainted with your work and any challenges it threatens to pose. The manuscript will probably then be passed on at least fleetingly to in-house subordinates (or perhaps a freelance assistant) for preliminary editing. Some of the technicalities associated with manuscript correction and copy editing in general have already been touched upon in Sec. 4.3.5, especially in the subsection "Text".

Language polishing may be required in the case of an English-language manuscript prepared by an author whose native language is not English. If so, it would generally be entrusted to a professional with special experience in this area. Ideally this would be someone at least broadly familiar with the subject matter, although technical terminology is normally not the chief area of concern.

Should serious problems of any sort be discovered, the author will of course be contacted promptly in the interest of expediting agreement regarding possible corrective measures. Actually, no such discussion should be necessary if there has in fact been a thorough exchange of ideas and review of a sample chapter well *prior* to manuscript submission (cf. Sec. 4.2.2).

It is the editorial staff's assignment to ensure that changes deemed absolutely necessary be implemented as smoothly and efficiently as possible.

● For economic (and public-relations) reasons, editors and their staffs strive to meddle as little as possible with submitted manuscripts, but they will not hesitate to raise questions whenever they feel it absolutely necessary.

Editing may at various stages be directed toward concerns in any of several broad categories, including content, language, and formalities associated with style (cf. the subhead "Publishing Houses" in Sec. 3.5.1).

Content problems typically involve such things as illogical or awkward subdivision of a manuscript, imprecise or misleading definition of some important concept, incorrect presentation of formulas or equations, inappropriate technical terminology (a sensitive point especially in the case of translated works), illustrations that fail to support the text, and the like.

Linguistic problems range from unprofessional modes of expression, through serious shortcomings in syntax or grammar, to improper spelling and/or punctuation. Some authors are extremely grateful for assistance in these areas, but others are quick to go on the defensive as soon as they suspect an editor wants to "tamper" with their unique "style" of writing. Most editors thus tend to exercise considerable restraint. The "service function" an editorial team is called upon to perform has been likened to the mission of a laundry. Just as a customer legitimately expects to receive back in good order garments entrusted to a launderer's care—free of damage in the form of shrinkage, discoloration, or other physical abuse, but at the same time immaculately clean and expertly ironed—so an author is justified in harboring comparable expectations with respect to a submitted manuscript.[31]

Formal problems in a manuscript most often relate to *inconsistencies,* or insufficient attention to details—in references to the titles of subsections, for example. Other common pitfalls in this category include erroneous figure captions, overly complex tables, or carelessly constructed literature citations.

There is no reason for an author to be ashamed about weaknesses an editorial staff uncovers with respect to any of these matters, or about taking advantage of an editor's help. After all, editors are by definition "pros" when it comes to writing. They also enjoy the benefit of considerable experience, as well as the fruits of membership (and interaction with colleagues) in national and international professional organizations like the Council of Science Editors (CSE, formerly the Council of Biology Editors, CBE) and the European Association of Science Editors (EASE).

Since problems of so many different types may need to be addressed in freshly submitted copy, it is not unusual for several individuals to participate in the review of a single manuscript, with the workload divided so as to take maximum advantage of each person's strengths. We refrain from discussing editing here in greater detail given the availability of several outstanding books devoted exclusively to that subject (e.g., O'CONNOR 1978; O'CONNOR 1986). Let us simply assume that within a reasonable period of time your edited manuscript will be passed along to those responsible for *production*: either in-house personnel or representatives of a subcontractor.

In the production department the edited manuscript first undergoes another broad review—this time from a more mechanical and structural perspective—in anticipation of what lies ahead. Prior to the "electronic revolution" a text manuscript would next

---

[31] The analogy is admittedly a bit forced in at least one important respect: in the former case the items in question—though rumpled and soiled—were at least clearly serviceable when they left their owner's hands!

have been assigned to a typesetter, figure copy would have been dealt with in the domain of professional draftsmen and reproduction specialists, and chemical formulas and the like would have been attended to by other experienced technicians.[32]

4.4.2 Page-Proof and Galley-Proof Correction

The most basic considerations in proof correction have already been introduced: in Sec. 3.5.5 in conjunction with our discussion of journal articles. There the emphasis was on the first stage in conventional proofreading—that devoted to *galley proofs*[32a]— because authors of journal articles rarely participate in the second phase, correcting *page proofs*. The situation is somewhat different with books, however. A book is regarded more nearly as the personal property of its author, so the author has a part to play throughout the proofreading process. For this reason there are a few more points we need to cover.

● Our earlier discussion revolved around comparison of an edited manuscript with a set of printed *fragments*, the galley proofs, but now the task includes examining pages in essentially their finished form as they relate to *corrected* galleys.

The latter part of this assignment in fact takes on a rather different character from that described earlier. Careful attention must of course still be directed toward the possibility of "typesetting" errors (which today usually means errors introduced during the editorial process into what were originally an author's—largely flawless?—computer files), but in proofreading the typeset pages of a book one must be especially concerned with how the work as a whole has been *assembled*.

At the galley-proof stage of traditional book production, text, tables, captions, and illustrations are still isolated components, just as in the manuscript itself. The whole point of an author's submitting the pieces separately is to permit independent processing of each segment by specially trained personnel at the publishing house. In the days

---

[32] Much has changed in the assignment of responsibilities and the available methods that characterized "the old days" (i.e., prior to perhaps fifteen years ago). But a look into the past and at traditional structures still has merit, even if professional draftsmen working with pen and ink at drawing boards are today almost nowhere to be found. The processing of graphs and diagrams has become a matter of manipulating images on a computer screen, for example. Indeed, all the information that constitutes the average book— text as well as graphics of every type—now finds its way onto printing plates through computer-guided laser beams. Wet-chemistry processing can be regarded as virtually obsolete, and "dry" methods are everywhere the norm.

[32a] No one has yet succeeded in explaining to us the origin of the word "galley" in this context. We offer here the results of our own speculation. A "galley" was a type of ship made famous by the slaves that propelled it. The word later took on the meaning of a kitchen situated on a ship, for the preparation of meals, and in this sense it is still commonly employed today with respect especially to aircraft— "ships of the air". We might thus picture the following scenario: preparation of victuals for subsequent delivery to a table, or at least to the folding tray in front of an economy-class airplane seat, has been interpreted as a kind of slave labor. Galley slaves were consigned to the vessel's lowest deck, with no one paying much attention to their treatment. Perhaps adoption of the expression in the printing trade amounts to a droll play on words.

prior to electronic typesetting, these various pieces eventually came back one by one in printed form to the author for official approval. The text would arrive as one seemingly endless column (cut arbitrarily into pages of varying length), accompanied or supplemented later by individual tables, initial prints of redrafted figures, etc. As explained earlier, "proofreading" these galley proofs entailed looking for possible errors in material being printed for the first time, mostly errors of the typographic sort introduced during manual typesetting.

At the subsequent page-proof stage the situation was quite different: galley corrections had presumably all been attended to, and the various parts had finally been combined into a single entity. The author was thus presented with a first opportunity to see pages essentially like the ones that would grace the forthcoming book. Illustrations were now incorporated in the proper places, all (hopefully!) with the appropriate captions, and equations (both mathematical and chemical), chemical formulas, and footnotes would also presumably be where they belonged. Headings, too, were there, typically in type of a distinctive size and style, all precisely as they should be. *Page numbers* were also making a first appearance, and assuming "user-friendliness" had been taken into account, correct *column titles* would have been added.

- Several of the newly included features necessarily became the object of special attention on the part of the person delegated to examine page proofs.

The first challenge in checking conventional page proofs is of course ensuring that every error identified in the forerunner galley proofs has been correctly dealt with. If the very first proofs you as an author happen to see are page proofs—typically the case with modern processing—it is especially important that you pay close attention to manuscript changes that may have been ordered (and implemented) by someone other than yourself. In many cases, an editor will only have *flagged* corresponding problems, with changes designed to eliminate the problems being executed independently by a technician with little or no understanding of the document's content.

Improper correction is not necessarily a sign of careless work on the part of a repair artist. The person who *proposed* the correction may actually be the one at fault if the correction marking was illegible, for example, or positioned incorrectly (or ambiguously). In any event, problems still present must now become the subject of *another* set of correction orders. In a complicated case (such as one involving an equation, formula, table, or illustration) it would be advisable to express your wishes very explicitly, perhaps with extra elaboration in the form of a restatement of that which is supposed to appear. All such notes should be placed as close as possible to the site of the flaw, and enclosed in *double parentheses*; e.g.:

((*Line like that on the right !*))

- Assuming page proofs have been prepared on the basis of corrected galleys it is important always to inspect carefully the general *environment* of newly set type.

As already mentioned in Sec. 3.5.5 under the heading "The Art of Proofreading", sometimes new problems are introduced even with use of the most modern computerguided phototypesetting equipment. Not all the extra "glitches" are attributable to human error. Additional hyphenation problems are a case in point, resulting from a need to divide for the first time some word that falls at the end of a line. New errors of this sort may even develop *ahead* of a correction site if the computer suddenly decides it would be advantageous to reconsider the entire pattern of existing line breaks. In other words, one should look meticulously with a set of page proofs at any printed line that ends differently than the comparable line in earlier proofs.

● A major responsibility assumed by the party examining page proofs is ensuring that all elements of the manuscript have been *combined* in the proper way.

Are all newly introduced components precisely where they belong, and free of errors? Has every subsidiary feature been placed conveniently? Are you quite certain nothing has been omitted?[33]

Various observations relevant to "correct" placement of special elements as well as the organization of pages in general, and difficulties associated with creating optimal page breaks, were included in our treatment of journal articles (Sec. 3.5.4). With any complex work—in which text is accompanied by tables, figures, equations, and formulas—distributing copy over specific pages is almost always problematic, especially since both formal *and* aesthetic factors must be considered. Compromises are inevitable. For example, it is often impossible to arrange for every illustration to appear on the very page where it is first cited, in which case some less satisfactory alternative must be accepted. Authors are of course anxious to see the best possible layout achieved for their work, but they are also notorious for overrating their own powers of *improving* upon a layout the publisher has proposed. Should you find yourself in a particularly creative frame of mind as you examine your proofs, and thus inclined to urge significant layout changes, at least check the feasibility of each of your ideas with a ruler! Lines of text must be counted carefully to make certain that a seductive alternative arrangement will in fact work. Be sure to bear in mind too that a subhead is never acceptable as the last element on a page, and no page may commence with a *one-line* residue from a paragraph that began on a preceding page. Particularly with a complicated work, page division quite literally becomes an art.[34]

---

[33] In the event that you yourself are the one responsible for the book's layout (because material you submitted was reproduced unchanged, as in camera-ready copy), such questions will of course have long since been addressed.

[34] The job title traditionally associated with the person responsible for ascertaining where page breaks should occur is *makeup specialist*. "Makeup" is in fact the English term that comes closest to describing the task of arranging pages in a forthcoming publication. The somewhat broader notion of "layout" includes overall page *design* as well. (Typical dictionary definition of "layout": "**a.** The art or process of arranging printed or graphic matter on a page. **b.** The overall design of a page, spread, or book, including elements such as page and type-size, typeface, and the arrangement of titles and page numbers. **c.** A page or set of pages marked to indicate this design".)

No matter how worthy you believe your ideas to be for restructuring a set of page proofs, you must not underestimate the associated technical and financial consequences.

● Dividing a document into pages is a process that demands considerable human involvement if the best possible arrangement is to emerge in terms of a given set of design parameters; *revising* a layout to accommodate even a minor change can be extremely difficult despite the benefits of computer technology with respect to on-screen makeup.

The author bears a heavy responsibility for corrective measures needed at the page-proof stage, and for the detection of previously identified problems that may remain. It is very much in your own interest to submit at the outset a manuscript that is as orderly as you can possibly make it, and to search diligently through galley proofs and other early printouts for errors. Careful analysis must *not* be postponed until the page-proof stage! By all means feel free to alert your publisher to problems you do find now, however, and to express your opinions regarding layout issues, but at the same time firmly resist any temptation to insist upon frivolous changes at the last minute.

● The later in the production process alterations are made, the higher are the associated costs.

With traditional production methods, the available corrective measures vary widely depending on whether galley proofs or page proofs are involved. Modern typesetting (or its equivalent) is accomplished almost entirely with computer-controlled, laser-based optical equipment. So long as the content of a publication remains within the computer environment it can be regarded as relatively flexible. In principle, therefore, changes of every sort might be effected fairly easily. At least in the case of books, however, final transformation of raw copy into an integral set of pages often still entails some use of photographic methods. As soon as information leaves the digital realm it acquires the "static" qualities inextricably associated with light-sensitive media. There is little doubt that as computer capacity and speed continue to increase even as relative costs decline, many of today's technical limitations will disappear. Evolution in this general direction is clearly evident, but progress in comprehensive computer-based book publication continues to lag somewhat behind that in seemingly related areas due to the enormous volume of information to be processed, especially with extensive use of dot-matrix versions of high-resolution graphics. At present most book publishers still find themselves somewhat dependent on photographic processing methods, particularly for work of the very highest quality.

Even in the future, when technology is no longer a severely limiting factor, change orders issued late in the book production process will continue to be frowned upon. After all, considerable effort is expended in the process of devising a pleasing, balanced arrangement of text, headings, footnotes, and all the other elements constituting a book. Sometimes a publisher will even request that minor changes be made in the *text* of a passage simply in the interest of facilitating the makeup of an especially troublesome page. Last-minute insertion of a new paragraph—or even a single line—can spell the utter demise of a carefully crafted page layout, a truism anyone can confirm who has attempted to create professional-quality layouts as part of a desktop publishing project.

This is a good place to mention once again the possibility that you might want to consider extending your efforts as author to the point of camera-ready or print-ready copy. This can actually be quite feasible even if nothing more sophisticated is available to you than a full-featured word-processing program and a high-quality laser printer.[35] If your manuscript includes illustrations or other graphic elements, however, the desired ends will be met much more easily and professionally with the aid of *page layout software*, such as the programs PAGEMAKER,[36] FRAMEMAKER or INDESIGN (Adobe), or XPRESS (Quark). Software in this category relies heavily on specialized *page description languages*, which in turn provide the key to achieving at least semi-professional results.[37]

The author who already has some experience in desktop publishing will readily appreciate the problems last-minute correction can pose. At the professional level every late modification dramatically increases the cost of book production—and thus the market price for the work in question, thereby reducing the likelihood it will reach as wide an audience as one might wish. Indeed, authors themselves are sometimes required personally to cover at least a portion of the expense associated with excessive last-minute revision (e.g., anything above 10% of the financial burden otherwise attributable

---

[35] Significant numbers of authors do now present their publishers with books in "camera-ready" form (cf. Sec. 3.4.6). In some cases this means submitting a printed document carefully prepared on clean, letter-size pages for capture photographically or electronically and subsequent reduction in size to make it more nearly consistent with a standard book format. Alternatively, the author might instead supply appropriately formatted electronic data files to serve as the input for a computer-driven printing operation. Any such aspirations must of course be discussed at length—and very early!—with the prospective publisher. Some publishers now offer a special honorarium to the author so inclined, as compensation for extra services rendered. We hasten to point out, however, that a policy involving compensatory payments of this sort could be perceived as an ominous harbinger of a disconcerting prospect: the possibility that in the future authors *not* interested in playing "the layout game" might find themselves penalized. Shades of a *Brave New World* to come?

[36] PAGEMAKER is generally regarded as the "grandfather of all DTP software". Perhaps no other single program has had such a decisive influence on the desktop publishing movement. Paul BRAINERD, who was responsible for the development of PAGEMAKER (specifically for the Apple MACINTOSH platform), is in fact credited with invention of the term "desktop publishing". He began his long professional career as a typesetter in the classical tradition, and one might interpret his later achievements as an impressive tribute to this honorable profession—although they also contributed significantly to its demise. BRAINERD's work garnered international recognition in 1994 when he was awarded the coveted GUTENBERG Prize by the city of Mainz (Germany), once home to the legendary printer for whom the prize is named.

[37] The boundaries between page-layout programs and word processors have in recent years become increasingly blurred: comprehensive text-editing packages like WORD now offer a remarkably wide range of layout features, supporting, for example, multicolumn page design and graphic material embedded directly in text passages. Meanwhile, layout programs have been enhanced to permit not only extensive processing of a document's text, including both search-and-replace and spell-check functions, but also at least limited editing of illustrations.

to typesetting): one more reason to devote extra attention to submitting a mature, polished manuscript in the first place.

● At the page-proof stage, significant changes in the text (including, for example, revised terminology) will normally be permitted only under the most compelling circumstances.

This is another matter requiring extensive consultation with the acquisition editor supervising the project. In the absence of a persuasive argument to the contrary there is every likelihood that an author's late requests for changes will simply be rejected, and any instructions added for that purpose will be deleted before final proofs are forwarded to the printer. The same of course applies to changes affecting illustrations, which tend to be especially costly.

Typeset copy often exhibits a few instances of what are known as *blockades*. These typically take the form of black rectangles, ■, or rude interruptions like "**XXX**". Such a signal indicates that some piece of information is known to be missing, incorrect, or otherwise deficient. Most blockades are "lifted" at the galley-proof stage, but at least one kind can persist into the page proofs: the warning that a *page number* is required to complete a cross-reference. The numbers obviously cannot be provided until page-makeup is complete, one of the major reasons for our injunction earlier (in Sec. 4.3.3) that page references be *avoided* whenever possible.

● Once the page proofs have been thoroughly inspected and no unresolved problems remain, the author is expected to confer his or her *imprimatur* (official approval) upon the document and promptly return all page proofs to the publisher.

The author's right to grant a definitive release in the form of a personal imprimatur is an important privilege, one explicitly acknowledged in publishing contracts. In the case of a multi-author publication it is the editor who issues the imprimatur. (With a journal, the head of the publisher's editorial office is usually the one with an exclusive right to authorize publication.) The first page of a set of page proofs typically displays a distinctive stamp with a special place for the author or editor to enter a dated signature declaring that the corresponding material is ready for publication. Even after an imprimatur has been provided, however, actual printing will not take place until the editor in charge has also conferred his (or her) formal blessing upon the work.

In practice, an imprimatur conveys a somewhat more qualified message than the previous paragraph implies: "printing may begin as soon as final *corrections* I have requested have been satisfactorily attended to". In an ideal case there would of course *be* no further corrections required, so that the Latin imperative could in fact be taken literally: *imprimatur* = it shall be printed!

4.5 Final Steps

4.5.1 Index Preparation

Every academic book requires an *index*—more specifically, a *subject index*. Better stated: this type of work requires at least *one* index. In contrast to a novel, a scholarly book is designed for more than simply reading; it is also likely to serve a *reference* function, and for this reason at least some potential users are entitled to special assistance. Through a book's index, you as author ensure that others will succeed in finding specific information they want, in that you assemble a systematic, alphabetized list of all the included topics of significant interest together with the pages on which these are treated. Consulting a book's index is typically the first step for someone with a question regarding that work's content, and the index must provide definitive help in locating answers.

● A good index serves as a convenient guide to precise information applicable to a broad spectrum of questions.

The common observation that a scholarly book is "only as good as its index" is hardly an exaggeration. You should therefore become familiar with the criteria an index must fulfill if it is to meet such high expectations. In particular, what should an index include?

● The most important type of index for a book, the general subject index, is an orderly list of the concepts dealt with in the text, each entry being accompanied by a select set of relevant page numbers.[38]

Every regular *index entry* incorporates at least one page reference. On the other hand, not every concept alluded to in a book necessarily acquires its own entry, and not every page with a potential connection to an included entry earns a citation. This is where the art of index development comes into play—an art based on the indexer's ability to make a host of important decisions wisely.

The author planning to prepare an index for an academic book needs to be sensitive to the basic distinction between a *concept* and a *designation*. A concept might best be described as the *idea* underlying something—the essence of the thing—whereas a designation (*appellation*, *denomination*) is a specific word or group of words, often language-dependent, used to evoke the corresponding concept. Within a particular discipline, a word (or set of words) of this sort may rise to the level of *terminus technicus*[39]—that is, it might achieve the status of an official *technical expression*, or come to be regarded as a piece of *technical terminology*. The role of such terminology

---

[38] Other special indexes found in some books include, for example, chemical compound indexes, reagent indexes, and molecular formula indexes in the field of chemistry, or species indexes in biology. *Author indexes* are a common feature of collective works, comprehensive monographs, etc., and also within the universe of scholarly journals.

[39] The plural form of the Latin expression is "termini technici".

is to minimize the tendency for professionals to invent their own unique ways of characterizing established concepts and, insofar as possible, to encourage development of an internationally sanctioned standard vocabulary. As an author in the sciences you have long been accustomed to the regular use of standard terminology for concepts central to your own discipline. "All" that remains for you to do as you construct the index for your book is to point as many readers as possible, as exhaustively as you can, toward information you have included regarding the concepts you treat, primarily on the basis of appropriate technical terminology.

● We assume you have been careful while *writing* the book always to express yourself in ways that are absolutely correct; now it is time for you to incorporate into your index the technical terms you have chosen to utilize (together with much else, of course; see below).

The complexity of index preparation will become fully apparent at the very latest when you begin seriously to take *synonyms* and *homonyms* into account. For example, the frustrated reader who fails to find in the index to a computer book any page references for "PC" needs to be directed toward what the author knows will be the more fruitful alternative "Personal computer"—or vice versa. One of your tasks as indexer is to select the specific places in your index where the various sets of page references will actually be collected. Notice the implication here that particular sets of numbers should appear only *once*; at the same time remember that the reader will unfortunately not be in a position to know intuitively where that will end up being. Take the case of someone interested in ascertaining what a particular book has to say about "desktop publishing". If no such entry appears in the corresponding index—and there is also no *cross-reference* associated with the term—the person seeking information might very well give up, despite being in possession of what could in fact be a valuable resource, and simply return the book to the shelf, dismissing it as useless simply because the author had failed to suggest turning to the index entry "DTP". A *good* index would have supplied the absent helping hand.

● Be sure to incorporate a substantial number of cross references into your index.

These references, sometimes in the form "→ XXX" rather than "*see* XXX", help the reader becomes a *user*; that is, find the explanations sought even when they fall under some term other than the one that first came to mind. Obviously such a reference can never be allowed to lead to a dead end; there must actually be something to *find* under "XXX". Certain sophisticated indexing software (see below) in fact helps ensure the absence of such "empty" references.

Large publishing houses often rely on the help of trained *indexers* to increase the likelihood that works they publish will prove as accessible as possible for readers of many types, with ample attention paid to both technical and linguistic considerations.[40]

---

[40] Database managers confront an indexing challenge very similar to that faced by publishers.

The editors of professional journals also regard compilation of a serviceable annual index as an activity of the highest priority. *Lexicography* (the description and inter-pretation of a vocabulary, as in preparation of a dictionary) may not be a particularly well-known discipline, but you should still try to live up to professional lexicographic standards as you go about indexing your masterpiece. The task clearly demands much more than simply cobbling together an alphabetized list of words you happen to have used in your text.

- A good index reflects every *important* concept discussed in the corresponding book.

At the same time, however, page references should be limited to sources of *meaningful* information about the concept in question. Ideally, the indexer would also make a serious attempt, at least sometimes, to identify *contexts* within which referenced concepts are treated.

Successful interpretation here of the words "important" and "meaningful" of course depends upon goals that have been established for your particular book, something about which there needs to be very close agreement between you—the author—and the sponsoring editor.

Important concepts in this sense are ones that contribute in some way to the thematic structure of a book. A *meaningful* reference is in turn one that points the reader toward a place where an important concept has been defined or otherwise treated in some detail. Certainly any concept reflected in one of the book's *chapter titles* or *section headings* warrants a place in the index. If this were the only source of entries, however, the index would amount to little more than a reconfiguration of the table of contents, which clearly is not appropriate. Other potential sources of "important concepts"— too often overlooked!—are the captions accompanying tables and figures. Nevertheless, most of your index entries will probably come from the text itself.

How *extensive* should your index be? There is no hard and fast rule we can quote. An index occupying one-tenth as many pages as the body of the book would probably be approaching an upper limit, but in some cases as little as ten percent of that might suffice. On the other hand, a 5-page index for a 500-page book would in most cases class as rather meager, and certainly suspect. Any book the author expects or hopes will become a reference source should be outfitted with an especially comprehensive index. The index is actually one of the first features a scholar examines when trying to assess the reference potential of a work. One of our goals as we prepared *our* book was to offer the reader an exceptionally dependable and convenient index.

- In selecting index terms, let yourself be guided by questions you could envision being logically raised by a reader or scholar upon encountering your book, and by how such questions might be framed.

This suggestion may at first seem trite, but it can actually be a source of considerable insight. An example or two may illustrate the point. If you were unfamiliar with the content of your book, and randomly saw its title in the course of a library visit, what sorts of things might you expect to find in it, and what are some of the words you would naturally associate with its probable themes? What about your colleagues? Might they be inclined to interpret it somewhat differently, or be drawn to it for different reasons, or turn naturally to different terminology? Could someone reasonably reach for this book as a possible source of information regarding subjects *peripheral* to your primary reason for writing it?

Consider a concrete case. Imagine someone interested in finding the melting point of a particular chemical compound, one that your book—given its title—almost certainly will have mentioned. A search for the desired information would probably begin with the index and an attempt to find there the *name* of that compound as an index entry—ideally with the subentry "melting point". It is much less likely the visitor would start instead from the premise that the compound's name would appear as a subentry under "melting point". Were the latter to be in the index at all—or become the object of a search—it would almost certainly be in the context of a *definition* of "melting point", or perhaps ways of measuring melting points—possibly even in conjunction with something as abstruse as the relationship between melting point and structure. Now suppose further that your book could legitimately serve as a source of the melting point in question. It might be through a table reporting data of other types as well (boiling points, densities, etc.), very possibly for a number of substances. It would nonetheless be unreasonable to expect you to include "melting point" as a sub-entry under the compound name in your subject index. The unit of information in question is too small and—from your point of view—unimportant. More justifiable might be a broader reference, such as one to "physical constants". Surely someone consulting your index for the reason cited could be expected to recognize that melting points fall in the "physical constants" category, and this less restrictive entry has the advantage of being equally useful to a reader interested in solubilities.

One principle the foregoing example illustrates is the importance of *grouping* related concepts whenever possible in the interest of preventing an index from becoming "bloated". By the same token it would be silly to take a topic addressed only *once* in the entire book and subdivide it into a series of subentries—all pointing to precisely the same page! About the only thing that would accomplish is attracting attention to the plethora of information you had managed to work into a single page of text. Unfortunately, absurdities like these are altogether too common in indexes. Subdivision of an entry makes sense only when the resulting subcategories point to *different* locations.

● Each full-fledged index entry must be accompanied by one or more *page numbers*— or *page ranges*, such as "315–320".

Appending "f" or "ff" to a page reference (e.g., 17f, 290ff) is a shorthand way some authors use to avoid specifying (or perhaps ascertaining!) an exact page range, where "f" carries the meaning "(and) the following page" and "ff" stands for "(and) subsequent pages".[41] Note, incidentally, that a passage beginning on p. 36 and ending on p. 37 is referenced with "36–37", whereas noncontiguous sites of information on the same two pages would be reported as "36, 37".

Your most important concern for the moment should be establishing the *nature* of entries that need to appear in your book's index. In choosing entries you must constantly remind yourself that readers may try to find information on the basis of terminology different from what first comes into *your* mind (cf. the rather simplistic examples presented earlier). This problem is actually one to which you should begin directing attention in the course of *writing* your book. Be alert as you go along for potential *synonyms* for whatever you happen to be discussing: alternative terminology you should jot down somewhere, because it deserves a place in your index, at least in the form of a "see …" entry pointing to the alternative term you prefer to stress.

As mentioned earlier, it is inappropriate to provide identical page-number information multiple times in conjunction with more than one term, since this is a redundancy that unjustifiably wastes space. You may, of course, encounter situations in which two distinct concepts are so closely related that the corresponding index entries will necessarily *share* certain references. This is perfectly legitimate, but under these circumstances *each* of the entries should also be provided with a "see also" reference to the other. Calling attention to the fact that a relationship exists between two terms is not only permissible, but desirable; employing multiple terms interchangeably, on the other hand, is bad form.

You will undoubtedly discover certain concepts that come up repeatedly over the course of the book, probably in a variety of contexts. This in turn is likely to generate a great many page references, all of which would seem logically to fall under a single index entry.

● No index entry should be allowed to accumulate an excessive number of page references, mainly because it is an unfair imposition to expect a reader to examine all of them, one by one, in search of some particular scrap of information.

A reasonable upper limit for the number of undifferentiated references accompanying a given entry might be five. If, as you proceed to construct your index, you find this limit occasionally being breached, take the time to divide the offending lists into several *subentries*, or to supplement at least some of the references with *contextual* information. Either approach will be welcomed by future readers, because you will be making your index more "user friendly".

---

[41] The automatic indexing routines supplied with word-processing and page-layout programs (see below), normally make no provision for this option, but an automatically generated index could be suitably modified manually.

● Often a concept lends itself to subdivision according to multiple schemes, corresponding to different ways of approaching it, each in turn giving rise to its own unique form of index entry.

For example, appended *adjectives* can be useful for indicating something about the context in which a particular concept is treated:

> Chloride determination
> –, gravimetric   *x, y*
> –, potentiometric   *z*

Alternatively, reference to a concept might be brought into sharper focus by adding a *noun* denoting a second concept, thereby creating a more complex expression:

> File
> –, Card-   *x*
> –, Data-   *y*
> –, Personnel-   *z*

In yet another approach, supplementary nouns (concepts) are linked to a primary index entry through suitable prepositions:

> Data
> –, collection of   *x*
> –, overwhelming an audience with   *y*

Sometimes the role played by the preposition is better reversed; e.g.:

> Substitution
> – with bromide   *x*
> – with chloride   *y*

Note that in the latter case linguistic logic dictates there should *not* be a comma following the dash. Also, a dash like those shown above could be replaced by the rather vague *tilde* (~) without in any way altering an entry's information content or its interpretation. Actually, punctuation is largely superfluous in most indexes.

Many times nothing is lost even if one dispenses with prepositions as well. What results is a set of index entries and associated subentries resembling the *concept pairs* that play a prominent role in the "keyword in context" (KWIC) indexes provided with certain review journals.

The very fact that subentries are indented serves to distinguish them clearly from associated principal (or "higher-order") entries. Indentation is an important device for making the content of an index easier to interpret;[42] e.g.:

---

[42] This helpful feature is normally incorporated even in indexes generated semi-automatically by word-processing or page-layout software.

   vehicles
    land conveyance
     passenger *u*
     freight *v*
    water conveyance
     motorized vessels *x*
     wind-driven ships *y*
  (etc.)

An index of the type illustrated is sometimes described as *interlocked*, *nested*, or *hierarchical*.[43]

 You will discover that the index accompanying this book has been constructed along the general lines described above. In most cases we have elected, for example, to omit punctuation, prepositions, and optional linking words. Our approach to index preparation has in the past been repeatedly singled out for praise by reviewers and others.

 One other convenience feature is worth considering as a way of further enhancing the user-friendliness of your index:

● Deliberate *inversion* of at least some of the concept pairs you create can significantly increase the index's accessibility, a worthwhile addition despite the resulting redundancy.

Thus, the concept pair

   table
    heading *x*

could advantageously be supplemented with its inverted counterpart

   heading
    table *x*

This would facilitate a search irrespective of which of the two concepts is closest to the reader's primary interest. In the one case, page *x* would be brought to the attention of someone researching various aspects of the concept "table", whereas the other entry would be encountered first by the person interested in "headings" and their treatment in general.[44]

 Sometimes the first word that comes to mind in conjunction with a particular text passage leads to an index entry that would in fact offer very little insight into the associated content, at least for many potential users of the index, in which case one should elaborate a bit even if subdivision is inappropriate (rejecting, for example, the

---

[43] Systematic indentation of subentries is a standard feature of a hierarchical index.

[44] The foregoing example of course assumes that other subentries will also be present under both "table" and "heading". Otherwise formulations like "Table, heading *x*" and "Heading, table *x*" would have been preferable.

unembellished entry "Iron" in favor of something more pointed, like "Iron, biocatalysis"). At the same time, never let an index entry expand into the equivalent of a heading, or into a long descriptive phrase (e.g., "Iron as the active center for oxidoreductases"— let alone an even more bizarre "clarification" that included "Participation in electron-transport processes in energy metabolism"). Scrupulously avoid using your index as a place for telling a story. The sole justification for an index is creating ready access to the concepts and concept clusters that shape a book.

Computer software commonly employed by authors (e.g., WORD, PAGEMAKER) allows one to mark certain words within the text for automatic inclusion as entries (or subentries) in an associated index. This involves embedding in the document "hidden" (i.e., non-printing) codes to indicate both the beginning and the end of whatever it is you wish to single out. You must also specify how the corresponding excerpt should be treated (e.g., whether it is to represent a primary or a subsidiary entry). Relevant vocabulary not actually present in the document itself can be covered as well (e.g., grammatical variants of important words that *are* in the text), and there is provision for subentries at multiple levels (see below). Judicious use of such tools can greatly accelerate index preparation, especially since the index evolves along with the text. Eventually, the program is directed to gather up all marked index components, format them, and arrange them in the proper order. In a straightforward case there may be almost no additional work required on the part of the author.

If you choose to pursue this route to an index you will of course want to verify that the marking process is functioning properly. This is easily accomplished by rendering the markings temporarily visible on your monitor. Be sure to hide them again prior to printing, however.

● If a given term ends up being tagged more than once, the program understands that the various references are to be collected and incorporated as a single entry.

An index generator of this sort also allows you to introduce into the text (in hidden form) *alternative* words, separated by commas as needed, that cause additional index entries to be created, representing one way of dealing with synonyms. Separation by a colon instead of a comma typically causes whatever follows the colon to be treated as a *subentry* under what precedes it. The developers of WORD showed such zeal in engineering this particular option that they made provision for concept hierarchies up to *seven* levels deep: a clear example of overkill, since one would almost never encounter a compelling need for more than three levels, even with a complex reference treatise or a handbook. An example of the way a text passage looks before and after introducing index markings—based on MS-WORD, and with "hidden text" displayed— is provided in Fig. 4–1.

The foregoing rather detailed account was intended mainly to illustrate another of the many ways a personal computer can contribute to reducing an author's burden. With any "semi-automatically generated" index it is of course likely you will at the very least find yourself wanting to fine-tune the formatting (though WORD in particular

a

Die *Auflösung (Punktdichte)* solcher Geräte, also die Zahl der „Elementarschwärzungen", aus denen grafische Elemente zusammengesetzt sind, wird normalerweise in der Anzahl von *Punkten* (engl.: dots; nicht zu verwechseln mit den im Abschn. 2.4 beschriebenen typografischen Maßen Punkt oder Point) oder Linien pro Länge (seltener pro Fläche) angegeben. Im deutschen Druckereigewerbe spricht man bei der Auflösung meistens von Linien/cm.

b

Die .i.*Auflösung* ; (.i.*Punktdichte* ;) solcher Geräte, also die Zahl der „.i.Elementarschwärzung ; en", aus denen grafische Elemente zusammengesetzt sind, wird normalerweise in der Anzahl von .i.*Punkt* ; en (engl.: .i.dot ; s; nicht zu verwechseln mit den im Abschn. 2.4 beschriebenen .i.Punkt : typografischer; n Maßen Punkt oder .i.Point ;) oder .i.Linien : pro Länge ; (seltener .i. Linien : pro Fläche ;) angegeben. Im deutschen Druckereigewerbe spricht man bei der Auflösung meistens von .i.Bildpunkt ; .i.Punkt : Bildpunkt ; .i.Linien/cm ; .

c

Auflösung 1	Linien/cm 1
Bildpunkt 1	Point 1
dot 1	Punkt 1
Elementarschwärzung 1	Bildpunkt 1
Linien	typografischer 1
pro Fläche 1	Punktdichte 1
pro Länge 1	

Fig. 4–1. Creating an index. – **a** Text fragment; **b** The same text fragment containing index codes (code symbol at the beginning of a term .i. , at the end ; , and for indicating a subentry :); **c** Excerpt from the corresponding index entry. Index preparation with Microsoft WORD. Embedded code symbols and text that do not appear in printed text are shown against a dark gray background.

offers a remarkable number of format options), especially if you are preparing camera-ready copy. This final touch-up is best accomplished with page-layout software rather than a word processor, because assignment of definitive page numbers to the references becomes possible only after all layout adjustments have been made. Indexing functions in PAGEMAKER, by the way, are quite similar to those described above (based on WORD), and PAGEMAKER can interpret WORD-generated text coding and process it to the point of extracting a sophisticated index.[45]

● As already noted, an index can never be put in finished form until the corresponding document receives its final pagination.

If arranging the layout for your book will be the responsibility of a typesetting or printing establishment, you should attempt to ascertain as early as possible whether equipment available there is capable of reliably processing embedded index codes. If

---

[45] PAGEMAKER even allows one to prepare a comprehensive index for a document distributed across multiple files—ones devoted to individual chapters, for example.

not, it is probably best to forego their use to avoid creating problems. You will of course then face the prospect of building the index *manually.*

Any early work you undertake to this latter end will at best give you practice, and possibly a small head start, since no meaningful page numbers exist yet. Indeed, most authors find it preferable to postpone manual index work until they have in hand a final set of page proofs.

Manual indexing typically begins with colored marking pens used to annotate a set of proofs reserved exclusively for that purpose. Thus, terms in the text you wish to be indexed are first highlighted and then collected in some type of data file where they can be subjected to the necessary processing. For example, you might highlight proposed first-level index terms with red, associated subentries with blue, and sub-subentries with green, missing components being written in by hand as needed. Interrelationships among the various labeled items can then be clarified with arrows. This "classical approach" to index preparation was for a long time the method of choice.[46] The advent of modern technology actually facilitated its implementation. Thus a *computer* file is a very convenient repository for collecting and manipulating a set of manually designated index entries—and an enormous improvement over the earlier use of file cards!

Unfortunately there would be little point in trying to program a computer to perform the indexing function on its own. No computer could be expected to distinguish reliably between important and trivial references to a particular term, nor would it be capable of analyzing effectively the context in which an index term arises, let alone identifying relevant text passages in which the term in question *fails* to appear. What a computer *can* accomplish, however, both flawlessly and rapidly, is the *alphabetization* of a set of entries and their *arrangement* to reflect ascending page numbers.

A very different way to approach indexing is to begin by creating for yourself a skeletal *database* of terms you already *know* your index should contain. In other words, utilize a database program as a tool for gathering together in a suitably constructed file a comprehensive collection of index-worthy terms (together with qualifiers that might be appropriate for subentries) as they come to your attention while you are examining your book's page proofs. The relevant page numbers should of course be noted as well. If you suspect that another page number needs to be added to what you believe is an existing entry, simply search the file for that entry on the basis of a suitable

---

[46] Should you in fact wish to try preparing an index in this way the following supplementary suggestions may help. If an entry is to consist of a combination of several words from the text, these can be joined by a colored underline that bypasses intervening (irrelevant) words. Relationships between primary entries, subentries, and sub-subentries are conveniently symbolized with straight or curved arrows pointing from the higher-order term to its lower-order counterpart (e.g., a red arrow directed from a word designated in red toward an associated word in blue). Required minor grammatical adjustments (e.g., ending modifications) can be indicated between lines of text. If you discover that a given index term occurs more than once on a single page, in different contexts, consider the possibility of creating concept pairs accessible from two directions. The margins are a handy place for introducing supplementary instructions or comments, such as the need for a "see …" or "see also …" entry.

word or word fragment. The same database program can of course also be requested at any time to *sort* the entries so far accumulated. One of the advantages of this method is that it helps you maintain some perspective on the project as you work. All database programs provide the necessary tools, and your file itself will almost certainly prove compatible with whatever word-processing software you use, so you should have no problem later treating the material as text. In fact, many database programs (e.g., ACCESS from Microsoft) actually originated as components of "office packages" built around word processors (e.g., the Microsoft OFFICE suite,[47] associated with WORD). Good results can also be achieved with somewhat less ambitious office packages, like APPLEWORKS or Microsoft WORKS.[48]

4.5.2 Title Pages

Indexes are generally grouped together at the *end* of a book. We turn our attention now to the very *first* pages in the volume. These, too, will come to the author in proof form for correction immediately prior to final printing, although their preparation is the publisher's responsibility.

● The first four pages of a book (the *title pages*) are referred to by special names derived from the roles they play.

The inside front cover (which is often attached to a heavy, folded *endpaper*) is followed immediately by what is known as a *half-title* or *bastard title* page. The origin of the name is not entirely clear, but it is presumably meant to suggest that this first identifying page is somehow less worthy of respect than the "true" title page, perhaps because its placement makes it more subject to abuse. In any event, it *may* present the name (typically only the last name) of the author—your name!—but it *always* provides the primary title of the work (subtitles are usually omitted here). If there are multiple authors, the corresponding names will appear in an order previously agreed upon, on sequential lines or separated by commas. In an edited volume, the editor's name— identified as such—may be provided in place of an author. An editor's name often *follows* the title, whereas authors' names usually precede it. It is not unusual for this page also to display the *logo* (or *colophon*) of the publishing house.

The first left-hand (*verso*) page, which comes next, serves as a *series title page*. Thus, for a book that is part of a series, this is where information to that effect would be presented (series title, series editor, previously published volumes, volumes still in the planning stage, etc.). Otherwise, the page would normally be left blank.

---

[47] This suite also includes the popular spreadsheet program EXCEL as well as POWERPOINT presentation software.

[48] WORKS from Microsoft is currently offered as version 7.0 for the WINDOWS operating system. APPLEWORKS, from Apple Computer (the latest version is designated 6.2), is an analogous package for the MACINTOSH family of computers. Both are multipurpose collections featuring several applications of interest to authors.

What immediately follows (a right or *recto* page) is the actual *title page*.

● A title page provides most of the *bibliographic information* required for *citing* the volume.

As the name implies, the principal role of this page is to communicate (generally in large type) the official, unabbreviated *title* of the work, including an *edition number* (if applicable), and any applicable *volume number* information. Above or below this, one would expect to find the full name(s) of the author(s) or editor(s): e.g., "Edited by …", in the latter case. Others who played an important part in preparing the manuscript (such as a translator) might be identified here as well. Finally, the page reveals the *publisher* (perhaps again in conjunction with a logo) and the location(s) of the publisher's principal office(s).

You may want to direct some thought yourself to the *placement* of the title on the main title page, and perhaps even to the size of the type in which it and the authors' names will be set in preparation for compromises in case your ideas differ greatly from those of the publisher.

● The subsequent left-hand page is known as the *copyright page* or the *impressum*.

This serves as a place for a considerable amount of "fine print", including of course a *copyright notice*, which consists of the copyright *symbol* ©, a year of publication, and the identity of the copyright owner. Other information on this page typically includes an official characterization of the volume as it appears in the records of the appropriate national library (e.g., the "Library of Congress Cataloging in Publication Data" in the case of a book published in the United States), as well as an *ISBN* (see below), a list of trademarks that may have been utilized, and the full name and mailing address of the publisher. (You will notice that the impressum of this particular book provides even more information).

The ISBN (*International Standard Book Number*) plays a role similar to that of an ISSN in the case of a periodical (cf. Sec. 3.1.1): in other words, it serves as an unambiguous identifier of a particular book publication. Each ISBN consists of exactly ten digits, divided into four groups linked by hyphens. The first group (often only a single digit) specifies a *linguistic domain* (e.g., "0" for books published in the United States or Great Britain, "3" for books from German-language countries, etc.), the second identifies the publisher, and the third the particular title in question. A single *control digit* constitutes the fourth "group", a number computed on the basis of an arcane formula from the numbers preceding it.[49] The appropriate numbers for the *title group*, as well as the control number, are assigned by the publisher.

---

[49] The control number [more formally: check digit; cf. standard ISO 2108-1992 *Information and documentation: International standard book numbering* (ISBN)] is computed as follows. The first digit of the ISBN is multiplied by 10, the second by 9, the third by 8, etc., which means that the last digit is multiplied by 2. The sum of the resulting products, S, is then incremented until the next largest multiple of 11 is reached. The augmentation could thus be by any of the numbers 1 … 10, and *this* is the value assigned as a check digit. If the value 10 is required, it is represented by the corresponding roman

$\rightarrow$

● A different book, or even a different *edition* of the same book, always carries a different ISBN.

The official ISBN, preceded by the acronym "ISBN", must always be prominently displayed in the book itself. Once an ISBN has been registered with the appropriate national library (e.g., the Library of Congress), it also becomes a part of the corresponding entry in the national catalogue, which in turn markedly facilitates acquisition (ordering, shipment) of that volume. It is in fact difficult to imagine how today's system of book distribution could function in the absence of ISBN identifiers, and much effort has been devoted to encouraging the inclusion of ISBNs in all formal references to books. The ISBN has incidentally outgrown its original nature as a "book number". In virtually every bookstore it is now possible to order a host of diverse "media" by reference to appropriate ISBNs (e.g., CD-ROMs).[50]

The copyright page may be followed by additional pages designated for a *dedication* and perhaps also an *epigraph* (a pertinent quotation—literary or otherwise—carefully selected by the author). Next would come the *preface* (sometimes also a *foreword*) and the book's *table of contents* (or simply *contents*), although not always in that order. Taken together, this entire collection of pages constitutes a book's *front matter*. The front matter might be extended to include such things as a *directory of authors* (in the event there are a great many contributing authors), and possibly a *list of symbols*. The latter is a summary of special devices employed in the book to represent quantities and units, for example, as well as important (or unfamiliar) acronyms and abbreviations, in each case accompanied by a brief explanation (see also Fig. 6–1 in Sec. 6.1.3).

● Pages comprising the front matter are usually paginated with *roman numerals*.

The rationale for introducing unique roman pagination is not only that this helps distinguish front matter from the main body of the work, but also that it permits this section to be separately and independently numbered. The book itself can therefore be paginated *before* preparation of the table of contents and thus before one knows how many pages must be assigned to the table of contents itself. Separate numbering also facilitates last minute changes in the front matter, such as unexpected insertion of a foreword.

● Authors and editors are urged to proofread front matter especially carefully.

---

numeral, X. Expressed in mathematical terms, the check digit is 0 if S is divisible by 11, otherwise it is $11 - S$ mod 11. For the first (United States) hardcover edition of *The Art of Scientific Writing*, 0-89573-495-8, issued by VCH Publishers (New York), the check digit 8 was computed as follows:

$$S = 0 \cdot 10 + 8 \cdot 9 + 9 \cdot 8 + 5 \cdot 7 + 7 \cdot 6 + 3 \cdot 5 + 4 \cdot 4 + 9 \cdot 3 + 5 \cdot 2 = 289$$

The next multiple of 11 greater than 289 is 297 (27 · 11), so the number to be added—and thus the check digit—becomes 8.

[50] Within the last few years it has become common in the book trade to make extensive use also of *barcodes* (for electronic scanning) in addition to ISBNs. Printed on the cover, dust jacket, or shrink-wrap packaging, these can further facilitate the handling process in warehouses, bookstores, and libraries.

Errors present in the first few pages of a book are especially critical, since they are more likely than errors elsewhere to be noticed by more people—who might in turn quickly form a negative impression of the book! The most troublesome component is usually the table of contents, where errors are overlooked with surprising frequency. For safety's sake, *all* headings and their associated identifying numbers should be examined individually in the actual pages of the book, and compared letter for letter with corresponding listings in the table of contents, with each reported page number carefully verified.

As a last gesture, the publisher may send you a set of *specimen sheets*. At this point the book will have already been printed and *signatures* (groups of sheets, *advance sheets*) will have been assembled, but binding will not yet have taken place. A fatal mistake discovered now might mean several sheets need to be reprinted, but in principle correction is at least still possible. Specimen sheets must be inspected very rapidly, since now no major interruption is likely to be tolerated in the production process.

4.5.3 Binding

To complete the project the book must of course be outfitted with some sort of *binding* (cover), and perhaps provided with a *dust cover* (*dust jacket*). At some point the decision will have been made whether the work is to be released as a *softcover* (*paperback*) or a hardcover edition. In either case, modern technology makes it possible to incorporate into the cover elaborate, artistic design characteristics, potentially including extensive use of color. This recently acquired flexibility is in fact one reason why it has become common practice to refrain from outfitting scholarly books with decorative dust jackets.

- An author's blessing with respect to a proposed cover design is even more important than consensus about the title page.

Irrespective of what creative ideas you may harbor, the cover design ultimately selected must be consistent with standards and traditions set by the publisher. Considerable time may elapse before all the interested parties come to complete agreement in this area, and several very different design proposals may be considered. For this reason alone, discussions concerning a book's cover should begin as soon after manuscript submission as possible (or even before!).

Assuming the cover will consist of more than simply *typographic* elements, consensus will be required regarding *illustrative material*. This represents a valuable opportunity to showcase an illustration drawn from the book, for example, or to develop a new artistic interpretation of one of the book's central themes. Choice of the most appropriate illustration, assigning of proper weight to aesthetic factors but also characteristics valuable from a marketing standpoint, determination of the role color should play—these are all matters with the potential to unleash serious differences of opinion. A scientist/author inevitably sees things from a perspective quite different from that of a production manager or marketing director. The acquisition editor responsible for the project may find it a challenge to function effectively as a mediator in the ensuing discussion, let alone successfully defend his or her own notions. We believe, incidentally, that the cover design adopted for this particular book is a good example of a very successful collaborative effort.

A book's *back cover* is usually reserved for what is essentially advertising copy, promoting either the book itself or possibly a related title from the same publisher. Text employed may well play a part also in publicity material (cf. Sec. 4.2.1).[51] In recent years it has become quite common for text from this source to be posted on the publisher's (or a book dealer's!) Web site, often accompanied by a miniature reproduction of the front cover.

● To be effective in helping sell a book, cover copy should speak to the volume's content, goals, and strengths in the hope that targeted readers will in fact perceive that the work is of substantial interest.

Most authors are quite willing to accept guidance in this area from professionals at the publishing house. It may even be, for example, that the first draft of the promotional text will be the work of the acquisition editor, submitted for review to other members of the staff charged with the design and release of product information in general. The author will nevertheless be allowed to have an important voice in final wording.

Once agreement has been reached on all aspects of this auxiliary text material, essentially nothing more stands in the way of publication.

In this chapter we have repeatedly stressed the magnitude of the challenge faced by a potential book author. It thus seems appropriate to conclude with a message of a rather different sort:

● The moment you as author of a newly published book hold in your hands a first printed copy will almost certainly class as unforgettable.

The event might even prove to be among the happiest moments in your career. We hope you will seriously consider our invitation to seek for yourself this kind of reward in the foreseeable future.

---

[51] If the book is to be provided with a dust jacket, the advertising material is likely instead to be relegated to one of the jacket flaps.

II
Materials, Tools, and Methods in Scientific Writing

5 Writing Techniques

5.1 Introduction

In previous chapters we have repeatedly stressed preparatory issues and the "organization" of one's work, especially in Sections 2.3 (dissertations) and 4.3 (book manuscripts). For most of the remainder of the book we will be concerned more with specific tools and techniques useful in the preparation of a manuscript, possibly for publication, in the natural sciences or engineering, essentially elaborating on topics alluded to first in Sec. 3.4.[1]

We will be utilizing the term "copy" in referring to written sketches of all types, irrespective of whether or not they are intended for publication.

- In "computerese", what we mean by a piece of copy could equally well be called a (digital) *document* or a *data file*.

Until relatively recently—roughly 1980, perhaps—there would have been little need for us to comment on "techniques" of writing per se. Almost everyone was familiar with the use and general capabilities of a typewriter, leaving room only for suggestions regarding how a first-rate manuscript should *look*, at which point the work itself could commence. Things began to change in this regard as manual typewriters were gradually displaced by their electric counterparts, which offered a few extra features and thus introduced new capabilities.

A true revolution transpired, however, as *computerized* "typing" became dominant, in the form of integrated word-processing (text-editing) systems based on personal computers. The effects of this revolution have transformed virtually every phase of manuscript preparation. Indeed, change can be said in the meantime to have become a constant, ushering in ever more dramatic advances in the potential for the individual scientist to prepare scientific documents that meet the most rigorous standards for technical quality, and there is certainly no end yet in sight. The key concept in this context is *word processing*, but recent developments have moved the front lines far beyond a mere "processing of words" to the elegant fashioning of entire documents replete with tables, formulas, equations, and diverse graphic elements (including both *line drawings* and *photographic illustrations*—possibly even movies or other forms of animation). In other words, every aspect of manuscript preparation is today handled

---

[1] Many readers are certain to find some of our remarks unnecessary—roughly as welcome as cold coffee. If you often have the sense that what we are discussing is hopelessly elementary, then by all means skip ahead. The pointed comments designated with bullets (●) should help you maneuver more easily. Simply from the standpoint of logical development we have not found it easy to eliminate large amounts of material from previous editions of this book and its German counterpart. A "democratic" mindset has also played a role: we suspect that some readers may not be as advanced in their knowledge as you, and it is always wise when making a presentation to try to ensure that the entire audience is "on board".

quite differently from what was standard practice in the very recent past. Text and graphics—but also typography and layout—occupy center stage, culminating in such novel manifestations as *electronic publishing* (Sec. 5.4), and all sorts of matters now deserve a healthy measure of our attention.[2]

● We have attempted in our book to do justice also to the historical dimension. How is it that what is today "state of the art" evolved from "the traditional"?

Providing at least occasionally a historical overview seems to us a more important responsibility today than even a few years ago: computer software becomes ever more inclusive, and its range of applicability more extensive, but at the same time the corresponding users' manuals increasingly threaten to shrink to the point of non-existence. "Introductory" CDs and online help systems cannot always be regarded as satisfactory substitutes. In most cases these are designed to offer advice directed to specific problems rather than presenting ideas in context—which means that one must already know what *sort* of advice is required. We do not therefore see it as inappropriate to assemble here, in one place—a book dedicated to writing and publishing in the natural sciences—*supplementary* information and background, especially at a time when many readers and users of the book rely upon the professional journal as their primary source of current (often short-lived!) knowledge. We will of course try to address the *latest* developments and products as well, to whatever extent it seems warranted based on our own collective experience.

In this chapter we restrict our comments largely to the subject of computer-supported text processing; other technological aspects of preparing a scientific report (including issues related to graphics, spreadsheets, and database management, for example) are touched upon only briefly, in many cases simply in the form of references to other relevant chapters.

In the previous edition of this book we still felt it appropriate to provide a brief treatment of manuscript preparation with a typewriter, but this topic has now been omitted entirely, and for one simple reason: scientists and students of science throughout the world write almost exclusively today with word processors, and virtually everyone has a way of accessing a multifunctional personal computer. Universities have in the past few years come to regard familiarity with computers and their everyday applications—e-mail, Internet "surfing", word-processing, etc.—as a given for *all* students, and curricula have been redesigned to take full advantage of this background. It seems inconceivable that the trend will ever be reversed.

● Nearly every aspect of modern science is heavily dependent on computers in some way.

---

[2] Such subjects were dealt with at length in a book—unfortunately now out of print—by RUSSEY, BLIEFERT, and VILLAIN (1995) that emphasized specifically the preparation of scientific and technical documents.

A basic understanding of *information technology* (IT) and the principles of *data processing* are essential to the practice of science today. Students are expected to acquire in addition a certain amount of experience in *computer programming*, as well as intimate acquaintance with a wide range of computer hardware and software. Just as computers are nearly indispensable in every laboratory, so a scientist's writing station must now be deemed incomplete in the absence of a "PC". To some extent the "office" computer is used in support of experimental efforts, through data interpretation and evaluation, but it is mainly the thing with which one prepares polished, professional-looking descriptions and presentations of the *outcome* of research—and for the first time virtually at the site of the action. No longer is the scientist or technician obliged to reproduce equations for formal purposes with India ink, vellum, and specialized templates: equations of every sort are instead constructed directly "on-screen", with invaluable help from sophisticated, commercially available computer applications. Much of the important software for organizing and presenting scientific documentation has been developed through cooperative interaction between information specialists, programmers, and the scientists who hope to utilize the results. In some cases scientists working alone have devised innovative computer applications in response to unique pressing needs, with essentially no thought directed toward potential "marketability" or financial gain. Some abstruse applications are so specialized that significant profits would in any case be out of the question.

● The modern scientific report is almost always a digital entity, combining verbal, numeric, and graphic information assembled in a form that lends itself well (as required) to sophisticated enhancement, drastic modification, or creative reuse.

The final product turned out by today's writer is perhaps best described as an *electronic manuscript*.[3] But this digital entity is of course not an end in itself, nor should it be regarded merely as a way-station for information that will later be transferred to paper. Far more is involved:

● The electronic manuscript is the most fundamental manifestation of a new dependence on electronic data processing at every stage in the formulation, archival, retrieval, and reproduction of scientific and technical information.

The information revolution has attracted at least as much attention and encouragement from within purely scientific circles (universities, research institutes, government agencies) as in the offices of publishing houses specializing in scientific works. Some have set themselves very narrow goals, such as promoting the concept of the *electronic book* (cf. Sec. 4.1.3); more to the point, however, there is an ongoing quest on the part of many to identify the most effective media conceivable for encapsulating, presenting, and archiving the latest results of scientific research and development.

---

[3] Recently, the more appropriate term *digital manuscript* has begun to gain popularity (cf. Sec. 5.4.1, however). The alternative expression *compuscript* appears destined to be forgotten.

- Whereas the principal objective of research itself is acquisition of information, the primary reason for *publication* is shaping that information so it becomes widely accessible in both the intellectual and the physical sense.

Publishers have a special interest in encouraging writers to develop text in digital form, because this gives their authors the power to submit more polished manuscripts for publication, which can in turn be processed with minimal effort. Information specialists have long regarded effective word processing as a key ingredient in the transformation of manuscripts into printed form—nearly "error-free" and in the most cost-effective way possible. Developments in this direction have taken a variety of shapes, the most extreme perhaps being *desktop publishing* (cf. Sec. 1.1.1), a subject to which we return in Sec. 5.4.2, restricting our attention at the moment to the word-processing phenomenon itself. While some acknowledged it sooner than others, nearly all publishers now concur that electronic data files derived from word processing constitute the most desirable "bearers of the message", and are of far greater overall significance than any particular form of publication to which they may lead.

5.2 Word Processing and Page Layout

5.2.1 Hardware and Operating Systems

The Personal Computer

From a formal standpoint, scientific text must in general be described as quite complex. Taken together with increasingly exacting demands and elevated expectations with regard to quality in the transfer of information, this complexity means that the person proposing to prepare a scientific document today must of necessity devote a certain amount of attention to computer technology and the various tools available for computer input and output.

In what follows, some readers may criticize us for being inordinately preoccupied with the past, but we believe it inappropriate and indeed counterproductive to lose sight entirely of historical development. Nevertheless, if you are someone who grew up with computers and have long enjoyed their dismantling and (successful!) reassembly, our feelings certainly won't be hurt if you skip over the next few paragraphs.

Especially in the early days, attempting to use a *PC* (Personal Computer) in a writing project meant mastering a great many skills foreign to the average researcher, who at the same time of course felt quite at home with a typewriter and a drawing board. It was necessary above all to come to terms with the fundamental concept of *text-editing*— which meant exploiting a data-processing facility for the input, manipulation, and output of text: what eventually came to be known as *word processing*. The tools available at the outset would strike us now as primitive. They were after all conceived primarily to facilitate such essential computer-related tasks as editing the "syntax" and sequence of programming commands. Things have in the meantime changed dramatically, and powerful multi-talent PCs have become familiar faces in every office. In our discourse we will occasionally employ various terms in referring to an office computer, such as "workstation", "network node", or "server", depending on what it is we wish to emphasize, but in most cases the simple abbreviation "PC" will suffice.

The acronym PC originated, incidentally, with IBM ("Big Blue"), the world's largest computer manufacturer. Initially it was used to denote a specific computer model, introduced in 1981 and designed explicitly for personal use, whether in the home or in the office. This relatively compact device (at least by the standards of the day!) can be regarded as the progenitor and prototype for all of today's "computers for the individual user". Over the course of time the implications attached to "PC" expanded greatly. One now applies the abbreviation quite freely to any high-performance, readily transportable computer located somewhere along the spectrum between *workstations* at the high end and a variety of "minidevices" (handhelds, etc.) at the bottom. Most are actually treated as relatively stationary devices, but the category is flexible enough to encompass so-called laptop and notebook computers as well. In the most general sense, the term PC covers a wide range of devices, functioning mainly under either the WINDOWS or the Apple MACINTOSH operating system, although a clear bias has developed as well toward regarding WINDOWS-based devices (once referred to as "IBM-compatibles") as the "true PCs", with MACINTOSH models constituting a subspecies of their own. A rather recent addition to the lineup is the workstation operating under the UNIX system (or a variant thereof, such as Linux).

The typical PC is a modular array, whose principal components are collected inside a single compact housing. These include a *processor* (or *central processor*), a generous amount of *working memory* (or *RAM*, from Random Access Memory), and some sort of *bus system* to support signal transfer. All the latter reside on a so-called *motherboard*, which can be regarded as the true heart of the PC. In order to provide video display of data subject to processing, especially graphic data, there is also need for a *graphics card* embracing its own processor (the *graphics processor*) together with additional dedicated memory (*graphics RAM*). Long-term data storage is the province of a *hard disk*, typically (in 2003) with a capacity on the order of 60 GB (gigabytes). Generally located within the master computer housing as well, but accessible from outside, is some combination of diskette and/or CD-ROM drives, although it is becoming increasingly common for the role of the former to be assumed by a CD burner. Thus, the most important portable storage medium at the moment is clearly the CD, but this will probably give way soon to the *DVD* (*Digital Versatile Disk*), offering considerably greater capacity (4.7 to 17 GB vs. 650 MB).

Indispensable components related to data input and output include a keyboard, a "mouse", and a video display. Since these are usually not incorporated directly into the main computer chassis they are commonly referred to as *peripherals*. Another extremely important peripheral device is the printer. The latter is not normally treated as an integral component of the PC itself because, like a scanner, for example, its presence is not strictly required for operation of the system. At the same time, an author in particular would be unlikely to forego a printer entirely, because close examination and careful proofreading of word-processing output is practical only with copy in printed form. In this sense a printer, while not *essential*, must be considered an extremely critical peripheral device for the user with a serious interest in PC-generated documents.[4]

Beyond this, virtually every modern computer can be characterized as an "expandable" device, in that it provides "slots" for the addition of accessory "cards" of various sorts. As noted above, one such slot is typically occupied by a graphics card (though equivalent facilities are sometimes deployed instead on the motherboard). Another could house a sound card, for example, or a network card to facilitate use of the Internet.

The complete range of computers employed in laboratory and office applications extends from the essentially stationary mainframe computer (which of course would probably be charged with various more complex assignments as well), via *workstations*, to *microcomputers*, desktop personal computers, and finally *laptop* and *handheld* devices. Unquestionably the most popular choice is the PC,[5] a configuration that could never have been achieved without the microprocessors and memory chips invented in the 1970s. Perhaps the biggest advantage the early PC enjoyed over its predecessors was the independence with which it operated, at the same time supporting an astonishing array of applications. Laptops have added to this the advantage of true portability (in one's briefcase, for example), encouraging their utilization not only in the office, but also at home and even in transit. It is noteworthy that today's miniscule laptop systems are able to accomplish things in terms of both capacity and speed that only a few short years ago would have required a cumbersome mainframe system.

● The actual "brains" of a computer reside in its *central processing unit* (*CPU*) or simply *processor*.

The CPU is responsible for all computational and organizational activities. Among other things it is here that arithmetic and logical operations are performed and commands are decoded. The CPU is in turn controlled by a *clock-pulse generator*, usually situated along with the RAM on the motherboard platform. Mention should

---

[4] An important supporting role in this context is played by so-called printer *drivers*, small auxiliary programs we will have occasion to consider later.
[5] At one time the computer giant IBM (International Business Machines) attempted to establish "PC" as a company trademark, albeit unsuccessfully (at least in terms of everyday speech). Nevertheless, other companies have been somewhat careful about using the abbreviation, with IBM retaining, for example, exclusive rights to employ and license the term "PowerPC". As already noted,"PC" is widely understood to imply at least an IBM-compatible system.

also be made of another component located here: a read-only memory (ROM) chip[6] into which will have been programmed—in at least semi-permanent form—various instructions essential to managing and controlling a host of basic hardware functions, including the so-called *BIOS* (Basic Input/Output System).

The first widely adopted operating system designed for microcomputers, and thus for personal computers, was developed in 1979 by Seattle Computer Products and distributed under the name *Disk Operating System* (DOS). All rights to DOS soon migrated (in 1981) to the Microsoft Corporation in nearby Redmond, Washington. On the basis of this one acquisition, the founder of the latter company, William ("Bill") GATES, advanced to the ranks of the richest men in the world—a reminder that desktop-computer-related activities, including above all word processing (much of it carried out under some version of the program WORD, also a Microsoft product), are of truly global significance, exerting an impact in many ways comparable to that associated with the introduction of the automobile or the airplane. The DOS operating system was quickly renamed MS-DOS, where "MS" of course stands for Microsoft.

● A computer's "operating system" is usually regarded as including not only system software, file-management capabilities, and command interpretation, but also fundamental support for an input/output system.

Developers and promoters of operating systems obviously have a vested interest in overseeing and managing the extent of *compatibility* between their systems and ones developed by others. This unfortunately translates into the fact that one cannot assume specific programs or data files will be universally applicable, a circumstance that has been the source of much discomfiture among users, especially with respect to data sharing, software purchase and adoption, establishment and maintenance of effective networks, and incorporation of hardware enhancements.

The year 1994–95 witnessed Microsoft's introduction of the dramatically simplified *user interface* known as WINDOWS, followed by WINDOWS NT (NT: New Technology), which signaled a decisive and long overdue break from most of the awkward, restrictive communication formalisms imposed by MS-DOS. These new interfaces quickly became the most important relatively fixed stars in an otherwise rapidly changing computer firmament.

Computers in the MS-DOS/WINDOWS family, all based on processors developed by the Intel Corporation (including the legendary "PENTIUM"), have encountered their most persistent competition in the home and office market from products of another American computer firm, Apple,[7] configured around processors from Motorola. By

---

[6] In most cases today this would actually be in the form of EEPROM (Electrically Erasable and Programmable Read-Only Memory).

[7] Apple was founded by Stephen JOBS and Stephen WOZNIAK, two young pioneers in the "chip technology" revolution, in a garage in Cupertino, California, near the heart of "Silicon Valley" on the West Coast of the United States. Long before IBM and the other major players followed, these visionaries devoted their attention to the idea of a computer tailored specifically to the needs and mindset of the ordinary "man in the street". More than anything else, the system they envisioned was to be distinguished

→

emphasizing "user-friendliness", Apple managed in the late 1980s to capture and retain a not insignificant (and strikingly loyal!) share of customers, especially in the education and desktop-publishing environments and other fields marked by extensive involvement with graphics (e.g., marketing and design).

Virtually all personal computers today depend upon an internal "hard disk" (HD) as their primary venue for data and software storage, a device at whose heart is a rapidly rotating disk capable of reliably "remembering"—over very long periods of time—a wealth of information encoded magnetically into a vast array of tiny, discrete zones on the disk's surface. Stored material pertinent to a particular work session is accessed by first transferring the corresponding file from the hard disk to the computer's active memory (a process known as "opening" a file), and—if they should happen to be modified—those files are later returned to the same disk for long-term retention, either joining or replacing their predecessors. The first hard disks introduced were capable of accommodating perhaps 10 or 20 MB (megabytes) of data, but today even a laptop computer is expected to offer a hard disk with a capacity of perhaps 40 GB (gigabytes) or more. Indeed, storage space of such awesome dimensions is essential for accommodating the complex software users are now accustomed to calling upon. Where it was once self-evident that data and programs of all types (even entire operating systems!) could routinely be moved from one computer to another via "floppy disks" (whose capacities at the time were on the order of 800 KB), it became necessary with increasing frequency to resort to larger carriers, such as "ZIP" disks (designed to accept 100 MB of data; see below). Unfortunately, within a very few years even this highly reliable, relatively inexpensive alternative began to show clear signs of obsolescence.

All the storage media so far described operate on the basis of *magnetically* preserved information, but now it has become commonplace to utilize *optical* means of storage as well. With the aid of a so-called *CD burner* one can thus impose up to 650 MB of data onto a compact disk (CD), where it is expected to prove much more secure over prolonged periods than data entrusted to any magnetic medium. Once data files have been "burned" onto a CD in this way the information becomes impervious to damage by electromagnetic fields, for example. Only over a matter of decades—or even centuries—might there be a risk of data loss due to material breakdown, assuming the disk is not physically abused or subjected to intense heat. An even more formidable "storage monster" is the DVD, a disk nearly indistinguishable visually from a CD, but

---

by a graphic, highly visual command structure that essentially did away with the rigid, strictly encoded and formalized control regimen so fundamental to DOS. Apple in many ways laid the groundwork for what became a "popular computer culture". Utterly revolutionary at the time of its introduction, Apple's "Macintosh" line in particular was defined by the unique characteristics of its operating system, the latest incarnation of which (as of 2003) carries the designation "System OS X" (where "X" refers to "10"). From the outset, transparency and ease of mastery were the hallmarks of the "Mac" system, and there is no question but what the obvious advantages it offered (in the eyes of many) powerfully influenced the genesis of the "Windows" interface for the PC, now the industry standard. There were in fact several complex court cases pitting Apple against IBM that dragged on for years revolving around the extent to which the Windows environment was essentially "borrowed" (without permission, of course) from Apple.

which with today's technology can house up to 17 GB (!) of data. The biggest problem still to some extent plaguing the latter medium has been the high cost of DVD burners.

Both the CPU and the RAM of a PC are under the strict supervision of a *controller*, an electronic switching system for coordinating the activities of such external devices as the *keyboard* together with various *drives* (e.g., hard-disk units). The corresponding duties are so complex that they demand the availability of a special *microprocessor* dedicated to this one function alone.

● Anyone proposing to utilize a computer routinely for manuscript preparation should have at least a general understanding of the way such a device goes about its work.

The preceding observation reflects the fact that the person who is totally uninitiated in this respect is likely to find even the operating manuals supplied with a computer to be only marginally intelligible. This can lead to serious problems when a non-routine situation is encountered, especially one demanding a certain amount of educated "trouble shooting".

As previously noted, input devices, output devices, and data-storage systems are commonly classified as *peripherals*. More or less "standard" peripheral components include the *keyboard*, a *monitor* ("video screen"),[8] *printers*, an *external hard drive*, often some sort of *floppy* or *ZIP drive*, and a *CD-ROM drive*.[9]

● Modest amounts of data as well as command sequences tend to be introduced into a computer directly, as required, via a keyboard (i.e., rather than through a storage medium), usually with support from the convenient portable switching device known as a "mouse".

Other widely employed input systems include electronic "tablets" (drawing boards), scanners, and digital cameras. It is also becoming increasingly common to tap databases accessed via the Internet as sources of input.

We next consider keyboards in some detail, although the subject will come up again under the heading "The Basics" in Sec. 5.3.1, together with comments related to optimal use of the mouse (in a subsection entitled "Mouse Techniques").

Keyboards

Keyboards have been constructed to reflect a wide variety of design concepts, including some tailored specifically to word processing. In most cases, however, the basic key layout follows the "QWERTY" system,[10] introduced in the mid-19th century and

---

[8] In the case of a laptop, the screen is typically based on a *liquid crystal display* (LCD), whereas most desktop-computer monitors still rely on cathode-ray tubes analogous to those familiar from television sets (although LCD displays are becoming increasingly popular here as well).

[9] Availability of a CD-ROM drive has in recent years become nearly essential, since this is the medium almost always employed for distributing commercial software.

[10] The name "QWERTY" simply reflects the arrangement of the first six letters of the topmost row of letter keys on a standard keyboard. That this keyboard arrangement has endured is ironic, since it

→

eventually accepted as the standard for mechanical typewriters. Although typewriters themselves are essentially a thing of the past, the acquisition of classical "touch-typing" skills based upon them is almost more valuable now than ever due to the ubiquitous nature of the PC and the extent to which typing fluency can accelerate data input. The discussion that follows is limited almost exclusively to keys important in word-processing applications, despite the fact that the complete inventory of keys (varying in number, but often approaching 100) is assumed to be available for all computer applications.

Generally speaking, keys are used to perform two distinct but complementary functions in word processing:

- The keyboard is the intermediary through which *characters* are introduced into developing text, but it also represents one means of generating *instructions* directed toward the system.

The first assignment is of course attended to primarily with keys clearly associated with specific *symbols*. The same keys may also be involved occasionally in issuing commands, but activity of this type typically entails concurrent use of certain less self-explanatory keys as well. The symbol keys can in turn be subdivided into three sets: keys for *letters*, *numbers*,[11] and *punctuation*.

- Whenever one of the symbol keys (perhaps together with *shift*) is depressed in a word-processing environment, the corresponding symbol immediately appears on the screen and is entered simultaneously into the computer's active memory—the RAM.

In the case of a MACINTOSH computer, each symbol key is potentially a source of as many as *four* symbols (only two of which can be inferred from the *keycap*) depending upon whether it is pressed alone or in combination with one or both of the keys labeled *shift* and *option*. For example, the four symbols y, Y, ¥, and Á are all available from the MACINTOSH "y" key. Under the WINDOWS operating system only two or three alternatives can normally be specified from a given key (e.g., the pair y and Y, or the set +, *, and ~).

The *spacebar* (which generally lacks a label) serves a purpose just like that of a symbol key, albeit for introducing a *blank space* rather than a visible symbol.

---

originated in an attempt (by Christopher Latham SHOLES) to devise the most *inefficient* arrangement of the alphabet possible—because early typewriters (including the popular one SHOLES himself invented) tended to jam if operated too rapidly. Many more favorable layouts have been devised over the years, the most familiar being the Dvorak System, patented in 1936. This system, which was approved by the American National Standards Institute in 1982 as an alternate standard to the QWERTY arrangement, has been shown unambiguously to increase user productivity and decrease fatigue relative to the latter. Dvorak keyboards are available from several sources, and a standard keyboard can be modified to reflect the Dvorak system through use of appropriate software.

[11]*Number keys* are commonly represented twice: once in a row above the letters, and again—usually off to the right—in the "3×3+1 pattern" familiar from calculators and telephones.

Several keys (referred to as *special keys*) fail to produce any symbol whatsoever on the screen, since their role relates exclusively to the issuance of *instructions* of some sort: e.g., establishing which of several possible symbols is in fact to be created next, defining the style to be assigned to one or more ensuing symbols, or perhaps specifying a modification applicable to a symbol already entered.

Issuing instructions or commands to the system from the keyboard typically entails pressing an appropriate *instruction key*, either alone or in combination with a symbol key. Names assigned to a few of the instruction keys (dependant to some extent upon the make or model of the computer) include *tab, caps lock, option, string, alt*, and *Fx* (where *x* is a number).[12] Most of the latter have no counterpart whatsoever on a classical typewriter, and for this reason they are deliberately situated outside the bounds of the "QWERTY" symbol-key field. Pressing an instruction key in most cases produces no immediate visible effect: the computer instead recognizes that some instruction is *about to be issued*, or that its behavior should change in some way.

Providing an instruction via the keyboard occasionally requires that the operator depress two or more keys simultaneously (e.g., *Alt-B* or *Option-Shift-Clear*). Any letter associated by default with such an instruction will usually have been chosen for its mnemonic relationship with the desired effect, making it easier to remember (e.g., *Command-S* for *save, Command-P* for *print*, or *Command-F* for *find*).

● An instruction from the keyboard is executed efficiently as soon as the command is issued, but the true utility of this form of communication is a function of the effort the operator is willing to devote to mastering it!

The actual effect of a particular keystroke combination almost always depends upon both the immediate circumstances and the nature of the program currently in use. Every operating system, and every word processor, spreadsheet, or database program, is associated with its own set of standard (default) keystroke commands, although these often are subject to customization by the user.

The alternative way of issuing instructions (commands) involves designating an appropriate option from among many listed in one of several *menus* of commands, which are available for temporary screen display upon request. The mouse is usually used for communicating a particular choice to the computer (and summoning the desired menu). This route avoids the need for memorizing keystroke combinations, with menus in general carefully designed to be quite self-explanatory. On the other hand, the experienced user invariably finds use of menus to be a slower and more cumbersome signaling approach than taking advantage of the keyboard.

The WINDOWS operating system provides the initiated with yet another technique for issuing commands: "three-key combinations", a descendent of shortcuts of a similar nature dating from the now nearly obsolete DOS operating system but still popular

---

[12] Another such key, found only on a MACINTOSH, is a command key labeled with a character resembling a *four-leaf clover* (as well as a small apple). In conjunction with appropriate symbol keys its use again results in issuance of instructions.

with computer "junkies". Thus, depressing the "Alt" key all by itself causes the *menu bar* to become active. Striking a second key leads to full display of one specific menu (e.g., the "Format" menu in the case of "t"), and pressing yet *another* key serves to designate the particular item of interest from that menu.

The set of "special keys" also includes the *return* (or *input*) key—a direct descendant of the electric typewriter's *carriage return* key—together with the *backspace* (or *delete*) and *tab* (*tabulator*) keys. The most common function of the return key is to confirm the validity of some preceding keystroke entry. On those occasions when a program is presenting the user with a *dialogue box* to which a response is required, pressing the return key generally serves as a way of saying "okay!" (or confirming some default option). In the absence of such a dialogue box, this key's principle task is to generate a *new line* in the context of text or data entry.

- With word-processing software, introducing a "new line" in this way actually signals a desire for a new *paragraph*.

The *backspace* or *delete* key, labeled on many computer keyboards with an arrow pointing to the left, has as its primary function elimination of whatever symbol (e.g., letter) appears immediately to the left of the cursor.[13] Some computers also provide a key with an arrow pointing to the *right* for eliminating the character *following* the cursor. The effect in both cases is rather like selective application of an eraser to penciled text.

- The backspace key can also be used to eliminate a larger fragment of text provided the latter has first been *marked*—by "passing over it" with a mouse while a particular mouse button is depressed, for example [Sec. 5.3.1, "Marking (Highlighting)"].

The *tab* key provides a convenient way of moving text to the right of the cursor *forward*: to a spot previously designated on a *ruler* displayed at the top of the screen, or to the next default stopping-point recognized by the program itself (typically spaced at half-inch intervals). Tab settings are especially useful in the construction of simple, orderly tables, since they facilitate establishing the straight left edges required for defining columns.

Six other special keys are available for selectively moving the cursor—representing an alternative to cursor placement exclusively with the mouse (cf. "Mouse Techniques" in Sec. 5.3.1). Four are labeled with arrows suggesting the consequences to be anticipated: displacement of the cursor one "space" up, down, to the right, or to the left. The other two, the "home" and "end" keys, move the cursor to the beginning and the end, respectively, of the current line.

Another special key found on most keyboards is labeled *esc* (for *escape*). The corresponding keycap is sometimes sarcastically represented by a stylized casket.

---

[13] The "cursor" or "insertion point" is a visible—blinking—device on the screen that indicates where newly typed text will appear. In most cases it takes the form of a vertical line. The cursor can be repositioned by using the mouse to "click" on the preferred site (see "Mouse Techniques" in Sec. 5.3.1).

Applications involving this key affect operation of the computer as a whole (as opposed to some currently active document). Other valuable keys are ones that cause a body of text to be *scrolled* vertically with respect to the screen (i.e., *page up* and *page down*).

Additional information regarding the computer keyboard and its function is best taken from the user's manuals supplied with the computer and specific pieces of software.

Miscellaneous Peripheral Components

Some of the items typically classed as computer "peripherals"—e.g., the keyboard or a disk drive—occasionally appear as integral parts of the computer itself, even though they are more often freestanding and joined to the computer through one or more cables. At the opposite extreme, a single computer can in principle be associated with *multiple* keyboards, drives, or output devices (screens, printers), perhaps through a "networking" arrangement (such as an "intranet"—cf. Sec. 1.1). It is also not unusual today to find data input supplied to a computer directly from a laboratory instrument, such that the latter in some sense represents a high-level "peripheral device". Information can of course also be acquired by a computer from an *external database*, which also thereby functions as a sort of "peripheral".

Storage units associated with a computer are arguably among the most important peripheral devices. Most are based on *disks*, which may be of two types: *fixed* (or *hard*) disks and systems that accept *interchangeable* disks. Virtually every computer today includes at least one of the former, as noted previously, but this might be supplemented by an independent *external* hard drive.

For many years the standard interchangeable disk—referred to as a "diskette" or "floppy"—was one with a diameter of 3.5 inches, encased permanently in a square plastic housing. A window in the housing, protected by a movable cover, granted the corresponding drive system access to the active surface of the disk itself. Such a diskette of the "HD" (high-density) type is designed to hold up to 1.4 MB (megabytes) of data— equivalent roughly to the text content of a 700-page book.[14] *Data-compression techniques* make it possible in most cases to increase this storage capacity by a factor of two or more depending on the nature of the stored information. Nevertheless, software has gradually become so complex that 1.4 MB cannot begin to house today's typical program, even in compressed form, so larger interchangeable storage media have for years been in high demand. One of the first alternatives to be widely accepted was the SYQUEST disk cartridge, soon to be displaced by the so-called ZIP disk capable of holding 100 MB of data (see below). A true "quantum leap" was achieved in the 1990s, however, with development of the CD-ROM. One consequence of this evolution is that certain computers no longer offer any provisions whatsoever for 3.5-inch "floppies". A CD-ROM drive has in contrast become almost obligatory, if for no other

---

[14] A single text symbol requires 1 byte of storage space, and a typical printed "page" corresponds to roughly 2000 such symbols.

reason than to permit access to commercial software distributed in this form only. Newer computers tend to include facilities for *storing* user-generated data on CD-ROM disks as well, through a process known as *burning*. Data storage involving this medium is based on *optical* techniques rather than the traditional *magnetic* read/write operations otherwise associated with disks of all types.[15] As noted previously, a single CD-ROM carrier can accommodate as much as 650 MB of information.

- As with files on other removable media, software or data obtained in CD-ROM form is generally copied to a computer's hard disk for everyday use, since hard disks not only offer enormous storage capacity but are also associated with access times significantly more favorable than what has so far been achieved with alternative media.

Progress in the realm of removable storage is ongoing. The newest "star" is the DVD, an optical storage medium with an outward appearance identical to that of a CD, and based on similar principles, but providing much greater storage capacity: between ca. 5 and 17 gigabytes per disk! DVD drives—which are also able to read conventional CDs—are already commonplace PC accessories, but the same unfortunately cannot yet be said of DVD burners due to persistent high cost.

Just like hard disks, CD-ROMs, floppy disks, and other removable media are valuable for storage of both data files and programs—indeed, large numbers of each simultaneously.

Data files of all types, and even entire programs, are now commonly distributed with the aid of yet another "storage medium", the Internet![16] One important cautionary note must be raised in this context, however: files downloaded from foreign sources, even familiar ones, sometimes prove to be responsible for the spread of "computer viruses".[17] It is vital that you make every effort to prevent your computer from

---

[15] Rather than relying on selectively magnetized surface sectors, CD-ROM storage entails use of a laser beam to selectively create ("burn") tiny depressions within defined tracks on an otherwise perfectly smooth disk surface, pits that later can be "read" (interpreted) with the aid of a second laser beam. Apart from vastly increased storage capacity, CD-ROMs offer the additional advantage of greater *lifetime* relative to magnetic storage systems. Whereas magnetically stored files are considered to have a reliable life of at *most* 10 years, a reasonably well-maintained CD should show no signs of data degradation for at least 50 years. Virtually all computers manufactured since about 1996—laptops included—accept CD-ROMs, and this is the medium on which most software is now distributed. Optical storage has arguably opened the way to an interesting new chapter in "desktop publishing", by the way, in that soon all computer users will be in a position to both "write" and "read" massive literary works as a matter of course, permitting whole libraries to be circulated in an exceptionally convenient format.

[16] Many programs downloadable from this source are either "freeware" or "shareware" packages that have been placed at everyone's disposal by generous, public-spirited programmers. "Shareware" differs from its cost-free counterpart in that the interested user is requested voluntarily to remit to the developer a small fee. Ordinary commercial software can often be downloaded from the Internet as well, generally by supplying (via a "secure" network connection) the number of a credit card to which the purchase can be charged.

[17] Viruses are relatively small, malevolent programs designed by their authors to take up permanent residence on any host computer they encounter. Once established, such a program then proceeds to carry out, at a predetermined time, one or more preprogrammed operations. These might lead to the →

becoming "infected" in this way. If one is fortunate, a "disease" so acquired will turn out to be little more than a nuisance, but it can also lead to total loss of enormous amounts of stored information—even loss of one's ability to control the computer itself—and an "infected" computer represents a menace to *other* computers with which it comes in contact, through either a network or files transferred by means of removable media.

Commercial programs have been developed over the years to meet seemingly every conceivable user need. The number of such programs able to coexist simultaneously and harmoniously within any particular computer is largely a function of storage capacity. Unfortunately, the size of the average program has grown exponentially in recent years in part as a response to relentless increases in PC hard disk (and RAM) capacity together with the provision of faster processors, developments that have in turn discouraged treating "efficiency in programming" as a high-priority commodity. At the same time there has also been a steady increase in the number of "features" associated with the typical program, often bearing little relationship to the true extent of the corresponding advantages. Thus, installation of the latest version of the Microsoft OFFICE package, for example, requires the availability of roughly 200 MB of hard-disk space (ten times the total capacity of the first hard disks offered!), and with all the optional "extras" this balloons to over 600 MB.

A few additional words are in order here regarding the hybrid storage medium sometimes referred to as a "removable hard disk". Systems in this category combine the flexibility of diskettes (albeit at substantially higher cost) with access times nearly as favorable as what one would expect from a hard disk. The Iomega Corp, probably the most important long-term source of such drives and disks, developed the popular ZIP and JAZZ drives, which utilize special carriers with a capacity of 100 or 250 MB in the former case and 1 or 2 GB in the latter. A recently developed type of "removable disk", completely devoid of moving parts (and not even round!), takes advantage of the "flash memory" principle and is available in the form of both flat cards and "sticks". These are based on the principle of the *electrically erasable and programmable read only memory* (EEPROM), and have been widely adopted for use in such devices as digital cameras and PDAs ("personal digital assistants", mentioned near the end of Sec. 1.3.3). Since the latter are quite capable of being used for composing brief text messages (and in some cases offer the bonus of outstanding handwriting-interpretation capability), it would be inappropriate for us to ignore such memory devices completely. Flash memory systems are currently marketed with a capacity of up to 512 MB. They

---

display of unexpected, humorous messages, or—much more ominously—the erasing or corrupting of major parts of the host hard disk! In addition, viruses frequently propagate by infecting *other* computers— ones accessible through Internet addresses stored on a host computer's mailing lists, perhaps. Special software exists for detecting (known) viruses the moment they arrive, and blocking or eliminating them before they can cause harm. Every computer user is strongly advised to purchase such protection (and *update* it regularly—which is usually possible free of charge—as a way of ensuring current "state of the art" vigilance with respect to these nefarious pests.

can be induced to interact with a computer either through insertion into a specially designed interface or over a *universal serial bus* (USB) connection.[18]

Hard drives, drives based on interchangeable media, and computer-compatible memory cards can, like keyboards, be regarded as members of the diverse family of *input devices*. Other examples include the *scanner* useful for digitizing hardcopy, which we consider next, and the *tablet* (for freehand drawing, among other things).

The principle underlying a scanner is passage of a fairly intense beam of light (usually from a fluorescent tube) slowly over a surface (a photograph, a drawing, or a page of text), where a set of light-activated sensors is used to examine systematically the spatial distribution of light reflected from that surface. Non-reflecting areas (such as the site of a printed letter) are carefully noted and recorded, leading to the equivalent of a detailed digital image (a "bit map") of the scanned copy. This image is made up entirely of dots, whose size and spatial proximity are a function of the scanner's *resolution*. Each dot represents information stored in one particular "bit" of memory.

A distinction can be made between two types of scanning activity. In the simpler of the two the scanner merely ascertains where a piece of copy is black and where it is white, providing a sufficient amount of information to permit barcode scanning, for example. The information content can be enhanced significantly by adding sensitivity to gray-scale gradations, or even color, thereby permitting faithful mapping of *continuous tone* images (cf. Sec. 7.4) and/or multicolored material. For many purposes at this level, the *digital camera*, captured images from which can easily be downloaded into a computer, has evolved into an attractive alternative to a scanner.

The second, more "intelligent" implementation of scanners revolves around the concept of interpreting areas of reflection vs. nonreflection in terms of *patterns*. Sophisticated software is marshaled to seek parts of an image that display characteristics typical of specific letters or other symbols, permitting the image to be "translated" wherever a match is detected. *Optical character-recognition* (OCR) software of this sort has matured to the point of being remarkably effective, making possible the rapid interpretation of entire pages of printed text with a very low error rate.

● When utilized with appropriate software, a scanner can therefore be a valuable tool for examining printed matter and capturing its verbal content in a digital text file.

Having surveyed the most important tools supporting computer input, we now turn our attention to the category of *output* devices.

● One key element of every PC is obviously some sort of *display screen* (*monitor*), and we will take this as our starting point in the following brief survey.

The screen has the task of providing a user with a visual record of information currently subject to processing. The information in question might have been recently introduced

---

[18] USB protocols have recently been acknowledged as the preferred approach to connecting peripheral devices to a PC generally.

through the keyboard, but it could also have been retrieved from a file stored long ago. The screen serves as the computer's principal medium for "communicating" with the user, who would otherwise be "working blind".

Depending upon screen size and quality, only a few lines (perhaps at most a "page") of text is on display at any particular moment. Efficient work with a computer, especially over a long period of time, demands the availability of a relatively large, high-resolution screen, one that is as free as possible from "flicker" (see below). Screen size is expressed in terms of a *diagonal* measurement, calibrated traditionally in inches. Prolonged use of a screen smaller than 15″ is at best inconvenient, and should be avoided if at all possible.[19] The typical PC monitor today provides a 17″ display.

The apparent "stability" of an image on the screen of a cathode-ray tube is a function of the frequency with which images are refreshed. With a refresh rate lower than about 50 Hz the replacement phenomenon is perceptible with the human eye, so the screen appears to "flicker". The higher the frequency, the less the distraction potential. Acceptable results require refresh rates above ca. 75 Hz. As mentioned earlier, portable computers are generally equipped with flat displays of the *liquid-crystal* type (LCDs, Liquid-Crystal Displays), which are immune to the "flicker" problem.

Printers

The term "output device" in the narrowest sense is applied primarily to *printers* and *plotters*, though plotters are in fact rarely encountered today. Devices of both types operate much of the time fairly independently of the computer itself, but they of course rely upon the latter for information, so "connectability" (compatibility) is an important issue.

Hardcopy output of text and graphics primarily intended for screen display (whether newly introduced or recalled from memory) actually presents a complex technical challenge. Modern printers are outfitted with their own memory banks, typically in the megabyte range, and are so designed that they manage the output process largely on their own. This entails transforming every required letter or symbol, and every graphic element, into an appropriate "dot pattern",[20] which can in turn drive a source of either an ink stream or light rays.

Because of the complexity of the process and the amount of time it consumes,[21] it is impractical to use a desktop printer for preparing more than a few copies of a

---

[19] The actual *usable* extent of a screen is somewhat less than the nominal size quoted in monitor descriptions. Screens built into laptop computers are usually smaller than the recommended 15″.

[20] The technique of representing letters and other symbols by compact arrays of dots has long been practiced in reprographic printing methods, and in "halftone" reproduction of photographs, which relies on a *screening* process. The most common measure here of output quality is the *point density* (*resolution*) achieved.

[21] Printing in this way a page of text may require only 5 to 10 seconds, but reproduction of a full-color bit-map image the size of a page can easily consume several minutes due to limitations imposed by the computer and restricted memory capacity of the printer.

document. Indeed, *one* copy should in most cases suffice; duplicates can be made much more rapidly (and economically) using a dedicated copier.[22]

- A printer constitutes the most important interface between a computer and the external world at large, and it is usually the first such interface with which a user becomes intimately familiar.

For anyone with aspirations of generating high-quality printed computer output, acquaintance with the fundamental principles of data interchange between computer and printer is essential. The better one understands the phenomena involved, the more effectively and efficiently the desired results are likely to be obtained.

Fundamental to word processing is the general concept of a "symbol", since all the text files one sends to a computer-based printer are processed as long strings of individual symbols (and/or the corresponding dot matrices)—not as composite text entities, although a file obviously could serve as a source of *graphic* images that the reader will *interpret* as text.

The novice generally assumes that printer output from a computer will be identical with what appears on the screen, consisting of symbols displaying all the same visual characteristics, line- and page-breaks in the same places, etc. WYSIWYG ("What You See Is What You Get") is supposed to be the byword, but it is a promise fulfilled only to a first approximation. Actually, in many respects that is a good thing, because printed output usually looks much *better* than a screen representation! This is true because most printers in use today offer significantly higher *resolution* than what can be achieved by a monitor, and for this reason alone images from the two sources will not be identical. For example, typical screen resolution is 72 dpi (dots per inch), whereas a laser or inkjet printer can easily provide copy with a resolution of 1200 dpi or more.

Both on the screen and in print, each image must be constructed as an array of dots: a *dot matrix*. In the case of a screen image these dots are referred to as "pixels", whereas with print they are "matrix points". Numerous techniques have been developed for generating the matrix points constituting printed images. Early PC printers relied on ink transferred in response to selective impacts from within clusters of small pins or needles, but these devices have long since been superceded by systems based on *inkjets* or lasers. Print output of the very highest quality requires access to an *image-setter*, a device that produces high-resolution images on film or some other light-sensitive medium rather than paper. Such films are used for preparing printing plates.

The limiting factor with respect to output quality on paper (or a light-sensitive surface in the case of laser imagesetters or cathode-ray tubes, CRTs) is the *point density* or *resolution* of the corresponding dot matrix. A symbol represented by a small number of relatively large dots inevitably looks crude. Lines running at an oblique angle become especially prone to reveal the "stepping-stone" quality so characteristic of computer

---

[22] It should be noted that multifunctional office copiers are increasingly coming to serve as shared *printers* for networked computers, a development that greatly facilitates preparation of multiple copies of newly composed documents.

printout years ago. The finer the matrix of dots, the more nearly a printed symbol will mimic the "engraved" look from classical printing techniques. The output of laser and inkjet desktop printers on the market today is superior to what the best electric typewriters ever achieved, approaching "near typeset" or even "typeset" quality. There is perhaps no better illustration of the progress in writing technology than the fact that high-quality printers ideal for office or home use now cost only a few hundred dollars.[23]

In the case of an *inkjet printer*, a stream of extremely fine, electrically charged ink droplets is directed with the aid of electric fields to produce the desired symbol or pattern on a piece of paper. With a *laser printer* it is a laser beam that is pointed with high precision toward a target, with subsequent processing similar to that employed in a xerographic copying device. Most desktop printers now in common use are based on one of these two basic principles. Printers in both categories are noteworthy for their (comparatively) rapid throughput and extremely low noise level.

With inkjet printers, two techniques predominate for directing the ink droplets to a sheet of paper. In the "drop-on-demand" approach, ink droplets are generated only as they are required. The two chief manifestations of this technique are *piezoelectric* printers and *bubble-jet* printers. The alternative, the "continuous-drop" process, represents the basis of a "true" ink*jet* device. In this case a constant flow of ink droplets emerges at high velocity from a jet: at a rate of up to a million drops per second, each with a diameter of only a few micrometers. This stream is passed first through the electric field of a "charging" electrode. Most of the drops are subsequently deflected by a second, rapidly pulsed electrode into a *collector* of some kind. Only in the absence of such deflection can ink reach the paper to produce an image. Printers of both types have been engineered to achieve very high resolutions (up to or exceeding 2500 dpi).

Any of a wide variety of type forms—or *fonts*—can be conferred upon a piece of electronic (digital) text, and the corresponding design specifications are preserved when the document is stored. Here we are of course dealing not with the lead-based, *metal* "type fonts" so important for centuries in the printing industry, but rather with sets of electronic patterns, which are however in many cases intended to simulate highly regarded metal fonts from the past. The simplest such patterns are the ones used to define *bitmap fonts*.

● In a *bitmap font*, every symbol (letter, number, etc.) that might be called for in a document must be represented by a specific stored matrix of *dots*, in which the dot *density* is appropriate to the desired resolution.

Whenever a symbol is to be reproduced, the required matrix is simply transferred from a master file to the relevant device driver. This was long the standard approach to

---

[23] A color laser printer can be obtained today (2003) for as little as $1000, and a quality ink-jet printer for $100 or less. Modern laser printers easily produce resolutions on the order of 1200 dpi, equivalent to roughly 0.02 mm per dot. A resolution higher than this would surpass the resolving power of the human eye and thus lead to no visible improvement in output quality. It is quite remarkable that the resolution even of ink-jet printers now extends into the 5760 × 1440 dpi region—high enough to mandate use of special paper. Important suppliers of printers include Hewlett-Packard, Epson, and Lexmark.

creating dot-matrix images, especially for screen display. Unfortunately, versatility can require the availability of an enormous number of bitmaps, including discrete patterns for symbols of every size one might conceivably need.

A more efficient approach to symbol representation was devised in the early 1980s: *outline* (or *contour*) fonts, where the matrix for a given symbol is "created" only when called for—to the desired scale, of course. The matrix in this case consists of dots aligned according to a prescribed set of mathematically defined curves (e.g., Bézier curves).

● Outline fonts are based on symbol *boundaries*, expressed mathematically, that are defined and "filled" upon command with dots at the proper resolution.

Examples of outline fonts are the PostScript fonts, originally introduced by Adobe, and TrueType fonts, first announced jointly by Apple and Microsoft in 1991. "Times New Roman" and "Arial" are familiar examples of widely employed TrueType fonts. The corresponding size-independent, "scalable" models need only be defined—mathematically—once (or once in each of several styles: italic, bold, etc.) for a given font. Actual symbols are created only as needed in the form of high-quality bitmaps at the proper resolution and to the desired scale. The designer of such a font need not take directly into account possible limitations due to the resolution of a specific output device: the appropriate printer driver or processor takes responsibility for generating custom bitmaps consistent with the designer's wishes at the highest resolution available with the output device in question.

Although a very similar principle underlies PostScript fonts, this system requires the availability of special software for creating display-screen representations (e.g., Adobe TypeManager).

Not surprisingly, achieving output of the highest possible quality (with a laser imagesetter, for example), based perhaps on an author's DTP diskette, entails an enormous amount of computational effort, not only in the case of graphic images but also text, especially since standard resolution for professional printing is 2540 dpi. This "intellectual effort" is provided by a device known as a *raster image processor* (RIP), whose task is transforming computer data derived from word-processing, graphics, or page-layout software into a form suitable (and at the desired resolution) for the particular output device in which it resides. "Without the invention of the RIP it would not be possible to prepare output comparable to phototypeset copy on the basis of an author's DTP files" (*Fachwörter-Lexikon*, Plenz 1995 f). A processor of this type is designed to deal routinely with enormous PostScript files as well as the data required for reproducing full-color illustrations, which of course means it must be capable of performing its tasks extremely rapidly.

Many printers (e.g., the Hewlett-Packard DeskWriter) are supplied with special *font disks* that incorporate font files of both fundamental types. These must then be installed in the associated computer: bitmap fonts for screen-display purposes and outline fonts to support printing. The DeskWriter, for example, includes eleven different versions of type, ranging from "Avant Garde" through "Bookman", "Courier", "Symbol", etc., to a set of special characters long commonplace in professional print shops and now collected in the font "Zapf Dingbats". The font-manipulation demands imposed on a printer can be reduced considerably by the use of TRUETYPE fonts. Assuming the print menu in the active program is properly managed, type information already in matrix form is sent to the printer as needed during a print operation in the form of so-called *soft fonts* prepared by the computer. It is important to bear this in mind when preparing POSTSCRIPT files containing TRUETYPE fonts if you propose to submit these later to a professional printing establishment for high-quality output.

"POSTSCRIPT-compatible" printers, available from all the major printer suppliers (Hewlett-Packard, Epson, OKI, etc), are ones specially equipped for processing POSTSCRIPT files; that is to say, they have the inherent ability to generate output of all types under the direction of the POSTSCRIPT page-description language (see below). In this case, text emerging from a word-processing program, together with all graphic elements and formatting information that may be present, arrives at the printer as a series of POSTSCRIPT commands, and these must be interpreted before the document can be printed.[24] Every POSTSCRIPT-compatible printer is equipped with a powerful processor for this specific purpose, which in turn explains why such printers tend to command a premium price.

● Almost every commercial printing service today is prepared to reproduce in hardcopy form (e.g., as a publication) documents submitted as POSTSCRIPT files.

The POSTSCRIPT language was devised in 1985 by Adobe Systems, and then licensed for use by other computer firms.[25]

Because it describes printer output on the basis of vectors instead of pixels, a POSTSCRIPT … file is able to provide instructions for reproduction at any desired scale with no sacrifice in terms of quality. Essentially all DTP and professional graphics programs are covered by Adobe licenses, allowing them to generate POSTSCRIPT files. This means that a publisher dealing with authors engaged in DTP activities has the flexibility of being able to work with almost any supplier of imagesetting services. (*Fachwörter-Lexikon*, PLENZ 1995 f)

One precautionary note, however: it is always advisable before committing an entire manuscript to production to request a few test pages for close inspection.

One additional piece of terminology requires mention at this point: *Encapsulated POSTSCRIPT* (EPS). Encapsulated POSTSCRIPT is a condensed version of format infor-

---

[24] Instead of printing directly from within your word-processing or page-layout program you can also cause POSTSCRIPT instructions to be written to a data file. This file will include a separate description for each page of the final document, corresponding precisely to what would have appeared had you printed the document in the usual way. The unique thing about POSTSCRIPT is that it is a *device independent* format for reproducing text and graphics with printing and typesetting devices of the matrix type (e.g., laser printers and imagesetters).

[25] For more information the interested reader should consult the specialized literature; e.g., ADOBE 1986 and 1990; MCGILTON and CAMPIONE 1992; RUSSEY, BLIEFERT, and VILLAIN 1995.

mation as it might be expressed in PostScript. If a DTP data file containing integrated EPS files is to be sent elsewhere for phototypesetting, it is essential that it be accompanied by the *original* EPS data as well. Otherwise the output that results may prove to be no better than a low-resolution screen image. Moreover, all fonts called for must be present and active in the host phototypesetting system in order for the file to be correctly processed.

● In contrast to the dot-matrix methods that are the basis of most desktop-printer output, *plotters* (*drawing devices*; *x,y-recorders*; *coordinate-* or *curve-plotters*) function instead on the principle of *vectorial representation*.

While plotters are not particularly well suited to extensive text output, they can be very useful indeed for reproducing *line art*: drawings, diagrams, and structural formulas, for example. In vectorial output, a line is described as a true line, and geometrical figures (e.g., circles) by mathematical descriptions of the appropriate type. The quality of a particular presentation depends exclusively on the responsiveness of the drawing head and the widths of whatever lines are generated (or the breadth of analogous light paths across a photosensitive medium). As has already been noted, it is perfectly possible also to describe letters and other symbols in terms of vectors, but the instruction sets tend to be rather complex, leading to agonizingly slow output from a plotter compared to a printer. For this reason, plotted text elements are generally limited to such things as the labels required to accompany a graphic image.

● Both printers and plotters lend themselves to categorization on the basis of their paper-feed mechanisms and demands related to paper form or quality.

Automatic sheet-feed is now the rule with almost all printers. Introducing paper for a long document one sheet at a time would obviously be impractical, and acceptable only in unusual situations, such as when material is to be printed on especially heavy stock (e.g., an insert, or a cover). One occasionally still encounters computer-driven printers designed for perforated, continuous ("fan-fold") forms that after printing must be separated into pages. While definitely inappropriate for preparing a report or a booklet, for example, under some circumstances this type of device could make sense for printing things like lists or financial reports. Plotters sometimes incorporate print heads capable of traveling in two dimensions for printing on standard (e.g., letter-size) sheets, but often they are constrained to motion in a single dimension for use exclusively with paper in roll form, which can be drawn smoothly both forward and backward under the printing element.

Finally, two other peripheral-device categories warrant mention in passing: *modems* (both standard 56K devices and so-called DSL modems) and *ISDN cards*. These serve the purpose of facilitating *long-distance data transfer* via the telephone network, for both text and graphics (cf. Sec. 3.5.2).

With this summary overview of the technical components comprising a typical personal computer system, we next turn our attention briefly to the various software-based activities such a system is called upon to perform.

5.2.2 Word-Processing and Page-Layout Software

Details regarding use of specific programs one might wish to exploit in preparing scientific documents are best derived from manuals distributed with the programs, and would be out of place in a book like this. Instead, we begin by approaching the matter in a more general way:

● A computer program can be viewed as a complex set of instructions for the execution and integration of a series of discrete logical steps that in the aggregate permit the user to accomplish some defined task.

Application programs of this sort, together with a computer's *operating system* (its "internal intelligence") and sundry data files constitute the *software* aspect of one's computing facility.

It needs to be emphasized at the outset that—unfortunately—not every program can be induced to run successfully on every computer. *Compatibility*—or lack thereof—can in fact be a major source of headaches for the ambitious user (especially the novice). Problems in this regard often become apparent when the occasion arises to pass a newly created data file along to a colleague accustomed to working with a *different* computer or software system. This potential complication represents one of the most important reasons for conferring with experienced associates when contemplating an investment in new software you expect to be central to your work.

The same advice of course applies to hardware as well, seemingly the most costly part of an author's digital working environment (at least initially). In some respects, however, it makes sense to approach hardware decisions after the various alternatives have been examined from the standpoint of software. The first step is to make sure you know precisely what it is you want to accomplish, and what routes may exist for achieving that end. Only then will you be in a position to establish the optimum configuration in terms of both software and hardware, including peripheral devices.

Generally speaking, the faster the computer and the greater its RAM capacity, the more likely it is you will avoid unpleasant surprises like frequent program crashes and the resulting loss of data. This applies to work with programs of all types: word processors, spreadsheet packages, graphics applications, etc. Furthermore, you can safely assume that software memory requirements will continue to increase with the release of each new version of a program, offering more "features" amidst mounting complexity.

In the field of word-processing software, recent years have witnessed a brutal war of attrition that has left a single clear winner: WORD from Microsoft. WORD's only remaining serious competitors are WORDPERFECT (Corel), WORDPRO (Lotus), and STARWRITER (Sun).

● WORD has come to be regarded as the "standard" for word processing, with the consequence that programs in this category are now essentially indistinguishable.

Despite familial similarities, however, if one wishes to rule out as completely as possible the chance that a text file will prove to be a source of problems when others try to work with it, the safest course is to stay in the mainstream and rely on Microsoft WORD.

Word-processing programs are so transparent in their basic operation that it is possible to begin working with them immediately, without extensive preparatory study of the user manual ("learning by doing"). This of course assumes a user with a certain amount of experience with computer software in general. The programs also offer wide-ranging help options through which targeted advice can be called directly to the screen as needed ("online help").

Nonetheless, no matter how user-friendly a program may be—or *seem* to be—we regard it as a serious mistake to ignore the accompanying documentation (which may assume the form of files on the CD from which the program is installed rather than a printed document). It is also well worthwhile to purchase from a well-stocked bookstore at least one detailed program guide prepared by an independent author for each of your "workhorse" programs, since these are likely to be more lucid and practice-oriented than "official" materials you received. Such a combination of resources is certain to reward you with considerable insight that is *not* self-evident, and you have, after all, paid good money to acquire all the capabilities each program can provide. In the case of a true beginner we strongly urge enrolling somewhere in a formal intro-ductory word-processing course, perhaps at a local community college or career center.

● Apart from word-processing software, the most important type of program for the ambitious author to acquire and master is one that facilitates preparation of profes-sional-quality *page layouts*.

With programs in the latter category you can establish the maximum degree of control over design aspects of the pages constituting your documents. This is especially helpful if you aspire to work with multiple-column text, for example, or treasure precise positioning of illustrative material accompanying your text. With a page-layout pro-gram one can "flow" the text from a word-processing file into predetermined spaces on custom-designed pages that feature absolutely consistent treatment of recurring elements (column titles, etc.). This often involves *linked* spaces extending over multi-ple pages (even non-adjacent ones), where the program ensures that if changes occur in one text block, blocks that follow will automatically have their content adjusted accordingly. Provisions also exist for tight control over placement of *graphic elements*, and for managing text that surrounds an illustration—even one with a complex shape. Formatting instructions can be supplied in conjunction with defined "styles", just as with a word processor, and these can be coordinated with such style assignments as already exist in the original text file (see "Styles" in Sec. 5.3.3). The positioning of all elements, spacing standards applicable to lines and even characters, and a host of other parameters can be adjusted with incredible precision. The modern author comfortably in command of a powerful layout program effectively treats the computer screen as a canvas for establishing all aspects of the appearance of a document, in the process creating (and "tweaking" as necessary) a sophisticated set of "page proofs". It is also a simple matter selectively to introduce colored or patterned *backgrounds*, for example, and accommodate figurative elements of every conceivable sort.[26]

A trend has long been apparent for both word-processing and page-layout programs to become progressively broader in scope, with word-processors acquiring ever more layout capabilities and layout programs incorporating classic word-processing tools— such as automatic table-of-contents generation, for example, or table-editing features. In essence, the two categories appear gradually to be merging. Many professionals have expressed considerable dismay with respect to this development, decrying the appearance of "monster programs" altogether too ponderous for convenient everyday use. Microsoft WORD in particular has been criticized for becoming unjustifiably versatile (and correspondingly expensive). With nearly 300 distinct commands and keystroke combinations, coupled with a huge stock of dialogue windows, the potential for intimidation is indisputable. Even some graphics editing is now supported, resulting in an ungainly hybrid with word-processing, page-layout, and pictorial characteristics.

With regard especially to layout programs we should note in passing that

● The most popular systems are not necessarily the ones that are most powerful.

The program TEX (cf. Sec. 6.6), developed by KNUTH in 1985 with special emphasis on the layout of mathematical copy, is unquestionably one of the most remarkable tools on the market from the standpoint of flexibility, quite apart from its formula-generation and formula-manipulation power, and the DTP program FRAMEMAKER is recognized as singularly adept at coping with challenges posed by academic documents characterized by large amounts of text but relatively few illustrations. TEX is widely supported by professional typesetters (sometimes invisibly, within the context of other typesetting programs distributed under names of their own, such as "3B2"), but like FRAMEMAKER its exploitation remains something of a rarity in the DTP world relative to the market leader QUARKXPRESS and other familiar packages like PAGEMAKER and INDESIGN.

Market success cannot be completely ignored. You must obviously take into account whether files you prepare will be accessible to others with whom you work. Another significant issue is the extent to which the means exist to integrate your program with others present in a *network* setting you share—in particular with respect to both the WINDOWS and MACINTOSH operating systems. The rapid pace of technical development coupled with constantly shifting consumer attitudes makes it extremely difficult to select the optimum software suite, especially given the far-reaching consequences. Once you commit yourself to a particular set of software packages you will be unlikely to want to change, because if you do you may find that existing files will become useless or salvageable only with considerable effort.

---

[26] The latter must obviously be in digital form, so one would also be well-advised to aecquire high-quality software for creating and/or editing graphics of various types (see Sec. 7.5).

After you are reasonably comfortable with a particular piece of word-processing or page-layout software, before deciding to switch abruptly to something different from another company you should ascertain whether a more recent version of what you already use might suffice, in which case you would be able to maintain many of your characteristic, ingrained work habits. It is beyond our planned scope to offer concrete advice at this point, or to delve more deeply into the advantages and disadvantages associated with specific options, but we do feel it important that you be conscious of the issues.

The closer an author's work comes to publication, the more apparent it becomes that the road from the first data files to a final product is a long and wearisome one. One important step near the end is developing a "master plan" for how a typical page of the final document should look, including adoption of standard formats for both text and the "special" elements, which in turn facilitates detailed design of each individual page.

- *Page layout* is the art of distributing text and graphic elements harmoniously over the surface of a page.

The original connotation of the word "layout" was more nearly that of a "sketch" or a "concept": i.e., a "general design" of something (a city, or a garden, perhaps), particularly with respect to spatial distribution. It is thus little wonder that its application in the publishing world is closely related to the notion of format in general.

- For our purposes, a layout can be defined as the sum of *all* the format decisions that affect dividing a given document into *pages* pleasing to the eye and meeting the criteria of uniformity and internal consistency.

"Layout" as a term has also become a favorite with programmers, for example in the context of one particular mode of presentation within a database application. Specifically, it is used to refer to the structure of the data sheet around which some database has been organized, analogous in certain respects to the "style sheets" in a word-processing environment.

Structuring a document on the basis of a markup language like SGML (Standard Generalized Markup Language; cf. also Sec. 3.1.3 and the headings "Anticipating the Need for an Index" in Sec. 5.3.3 and "Structured Markup Systems" in Sec. 5.4.1) can also be viewed as a "layout-related activity", one that takes particular account of *content*. Thus, before one can begin associating specific text passages with recurring markup categories (in terms of *structural characteristics*, *elements*, *tags*), the required "standard categories" must first be defined. Taken together, these in turn constitute the basis for a layout.

Technical personnel at publishing houses (typesetters, printers) of course have their own interpretation of what "layout" means with respect to a printed work: the host of dimensional definitions that characterize a standard page, size and arrangement guidelines for columns and illustrations, placement rules for titles, captions, and legends, etc. Included would be such typographic features as the margins around a page, heading style, indentation conventions, and the treatment of footnotes—in other words, essentially all the parameters responsible for a particular page design. When the focus is on a layout scheme for one specific page (as opposed to an entire work), the more appropriate (i.e., more limited) expression is "page makeup". Very general standards like type size and heading styles at various levels would be covered within a detailed set of *print specifications.*

In summary,

● The most central consideration in publishing a work—whether of the professional or desktop variety—is not the nature of the text, but rather the design of the printed page.

5.3 Writing and Formatting with a Computer

5.3.1 Becoming Accustomed to Your System

The Basics

Readers already steeped in computer games and other computer-based activities can probably skip Sec. 5.3.1 altogether. We urge the rest of you, however, to resist plunging into your first word-processing project completely unprepared. Opportunities abound for enrolling in classes related to computer-based writing, for instance, often with separate sessions for beginners and those with some experience. There is a surprising amount to learn—and very good reason why manuals describing word-processing programs often run to eight hundred or a thousand pages.[27]

● Attempting to master everything entirely on your own can require an enormous commitment of both effort and time.

The first step is of course becoming intimately acquainted with your *computer*, which means spending a few hours with the corresponding set of manuals, getting a feel for how such a device goes about its tasks and how one guides it. This is the place to learn, for example, how to

- *configure* the PC, and connect it to various peripheral components (mouse, printer, etc.);
- install and update *system software;*
- install *programs*;
- *communicate* with the system;
- operate your particular *mouse*;
- work with the *windows* that are so ubiquitous in computer programs;

---

[27] We are almost certainly not the only ones disturbed, incidentally, by the knowledge that even experienced secretaries ("executive assistants") rarely take advantage of more than about 5% of the tools available with the current generation of sophisticated word processors.

- create and save *documents* and *folders*;
- open, close, and save *files*;
- *manipulate* files in various ways, including *copying* and *moving* them;
- work with *diskettes* and *CDs*;
- *print* a document;
- exploit the latest techniques for exchanging and acquiring information (where the keywords are *e-mail* and *Internet*)

along with much else, of course, including the most sensible and productive steps to take when something goes wrong (e.g., you experience a "system crash"), and various ways your computer can assist you by assuming the roles of office clock, notepad, calculator, address book, calendar, and alarm.

Word processing begins with introducing, generally through a keyboard, all the discrete *symbols* that in the aggregate constitute a document. The nature and number of potential symbols at your fingertips will in part be a function of your computer's operating system: MACINTOSH computers differ somewhat in this respect from the more prevalent WINDOWS devices. You can significantly expand your arsenal of available symbols—for use not only in word processing but other applications as well—by securing access to additional *fonts*, since character sets are to some extent font-specific (cf. "Character Sets" in Sec. 5.3.3). You should by all means familiarize yourself as soon as possible with the basic symbol set common to most fonts, as well as with various keystrokes and/or keystroke combinations that can be used to summon the various symbols, not all of which are obvious from the keyboard itself. Further information in this regard is of course to be found in the manuals accompanying your computer and its software.

The precise appearance of a given symbol, whether on the screen or in print, is a function (among other things) of the *style* applied to it. Any symbol can thus be expressed in *italics* (or *oblique characters*), for example, or *boldface*, as adjuncts to the usual (*roman*) treatment (see also the subsection below entitled "Formatting").

When introducing text it is especially important that you correctly utilize the *return* key (see "Keyboards" in Sec. 5.2.1). Pressing "return" should be regarded as the equivalent of *closing a paragraph* through issuance of a "new paragraph" command. This in turn causes a non-printing *paragraph mark* (¶) to be added below the current text block (possibly indented—see below), with a text cursor to its left. Unless you give instructions to the contrary, this new paragraph will inherit the format specifications of its predecessor.

Terminating *lines* with the (carriage) return key would be a major mistake—but not an unusual one in the case of beginners, especially those who have had extensive experience with a typewriter, which *does* require that its "carriage" be "returned" to the left margin after each text line.[28] Overuse of the return key in a word-processing

---

[28] The very name "return key" is obviously anachronistic in the context of computers; "*paragraph key*" would be more apt.

context not only prevents you from taking full advantage of some of the important benefits electronic documents offer, but it can also be the prelude to a rude awakening later if you decide to rearrange the text, or, worse yet, request that your "electronic manuscript" be typeset. Adapting text to conform to columns with revised dimensions, or switching to type of a different size or style, is sure to mandate a new line-break pattern, but your document will continue to reflect the ("hard") line breaks you manually imposed upon it earlier, even though these no longer make any sense and will in fact prove a source of considerable aggravation (and probably much irritating busywork for someone!).

- Except under very special circumstances, all line-termination decisions should be left to your word processor.

Once you feel comfortable with your computer and its keyboard, the next step as you contemplate preparation of your first word-processing document should be mastery of the most essential *commands* at your disposal.[29]

- Computer commands are generally issued via the *mouse* in conjunction with a list of items displayed in an on-screen *menu*, although in many cases the same end can be achieved from the keyboard—often more conveniently!

Most programs are designed to interpret specific keystroke combinations as equivalent to mouse-selected menu commands. It is quite possible you will eventually decide the default set of keyboard substitutes is deficient in some way, probably because a particular command you have frequent reason to employ has been overlooked. In that case you should consider creating a *custom* keystroke combination addressing your specific need (cf. "Customization" below). There would be little point in amassing a large set of such personal shortcuts, however, because by the time you managed to recall or look up the relevant "trick" you could already have achieved the desired result with the mouse.

Mouse Techniques

Your most important communication tool with respect to information displayed on a computer screen is the *mouse cursor* (or simply *cursor*; from the Latin *currere*, to run), a special symbol on the screen whose location is a function of information supplied by the computer's "mouse". As described below, you will usually employ this cursor to indicate where a future event is to occur in a developing manuscript (by singling out some character or group of characters you wish to change or delete, for example, or signaling where the next typed character should go).

Depending on the computer, the program, and the situation at hand, a mouse cursor can assume any of a number of forms. Within a *text field*, for example, it normally looks rather like an oversized letter "I". Under other circumstances it might resemble

---

[29] The latest MACINTOSH version of WORD, for example, supports roughly 350 different word-processing commands.

an arrow, a question mark, a miniature clock, a cross, a hand, a magnifying glass, etc. Each of these manifestations is related to the particular role currently assigned to the cursor.

● As already suggested, the mouse cursor can be moved freely about on the screen with the aid of the clever mobile switching device known universally as a *mouse*.

This accessory was given the name "mouse" because of its appearance. It is usually a rather small, ovoid device, relatively flat on the underside and with a smooth, convex top. Its link to the computer is a long, flexible cord, suggesting a mouse's tail. Until recently the typical mouse had a hole in its base through which protruded part of a rough-surfaced sphere (the "mouse ball") subject to free rotation in any direction. When a mouse of this type is slid about on a flat, horizontal surface, its ball of course *does* rotate. The ball's rotary motion is monitored by two bearings mounted perpendicular to each other and in direct physical contact with the ball. As these bearings turn, electrical signals are generated, which the computer is able to interpret as quantitative reflections of horizontal and vertical motion of the mouse, motion that is then simulated on the screen by the mouse cursor.[30]

● The mouse is thus used to guide its screen counterpart, the mouse cursor, to a location of interest: somewhere in a text block, for example, where a signal might be sent to the computer requesting that an *insertion mark* ("insertion point"), also known as a *text cursor*, be created there.

To be more precise, after a text window has been so adjusted (with the aid of "scroll bars"; see below) that the relevant text is clearly visible, the mouse cursor is moved into position over the target, and the mouse is "clicked" by pressing a specific button on its back.[31] This has the effect of "fixing" an insertion point at the site in question. In the context of a *table*, "clicking" on a particular *cell* is a common way of activating that cell for alteration, a principle also applicable to the various *data fields* in a database. "Double-clicking" and "clicking and dragging" are closely related operations employed for selecting ("marking") *groups* of characters (words, for example) for modification in some way.[32]

---

[30] A newer type of mouse depends instead on an *optical* motion-detecting system centered around a light-emitting diode and appropriate sensors. This has the great advantage of avoiding the frustrating accumulation of dust that regularly interferes with proper functioning of the crucial bearings in a mechanical mouse. An even more recent development is the "cordless mouse", which communicates with the computer by optical means rather than through a cord.

[31] Several such buttons may be present, each designed to produce a different effect, although MACINTOSH computers have always relied on a "one-button mouse". There may also be a thumb wheel to provide for more convenient scrolling.

[32] The terms "clicking", "double-clicking", and "dragging" have become standard components of PC jargon, eloquent testimony to the importance attached to these signaling techniques. The word "click" mimics the sound that issues from a typical mouse when its switch is activated (i.e., one of its buttons is pushed). "Double-clicking" means sending two "click" commands in rapid succession. "Dragging" entails first clicking somewhere and then *moving* the mouse (and thus the mouse cursor) across text characters or graphic elements while the mouse button is still depressed.

Notebook or laptop computers generally incorporate mouse buttons as an integral part of the keyboard, set in close proximity to either a *track pad* or a *track ball*[33] for moving the mouse cursor.

Clicking anywhere within a text block causes the insertion point to jump from its previous site to this newly specified location. An existing insertion point can also be displaced with keystroke commands: up, down, to the right, or to the left, always one "notch" at a time, depending upon which of four "arrow keys" (also known as *cursor keys*) is pressed. Special keystroke combinations are typically available for advancing a text cursor directly to the beginning of the next word, for example, or to the end of the line, or to the next paragraph.

An insertion point is represented on the screen by a vertical line[34] situated between two text characters. It differs from other screen symbols in that its existence is strictly transient, and it never appears in a printout.

● The location of the insertion point is made more apparent by the fact that it "blinks".

The principal purpose of the insertion point is to indicate precisely where one should expect to find the next symbol entered through the keyboard (or "pasted" from the "clipboard"; see below). As various symbols (letters) are introduced one after another the insertion point marches relentlessly "forward" through the developing text.[35] Put somewhat differently, anything "inserted" into a document always appears immediately to the left of the insertion point.

● As soon as more characters have been introduced than can be accommodated in the form of *complete words* on the current line, the insertion point automatically jumps to the *next* line.

"Surplus characters" (i.e., a *fragment* of a word that in its entirety would apparently extend past the physical limits of the current line) always accompany the insertion point as it moves to the next line. This explains why there is ordinarily no reason for one explicitly to request a line break: the word processor effectively issues the appropriate signal automatically. (Recall the admonition earlier that the "return" key should only be pressed at the end of a *paragraph*.)

Initially, entered text is consigned to a portion of the computer's memory best characterized as *volatile*: the working memory (Random Access Memory, RAM). In other words, if the power supply were suddenly to be interrupted, all recently typed

---

[33] A "track pad" is a sensor field capable of transforming the path one's finger traverses across its surface into instructions for analogous displacements of the mouse cursor. It thus constitutes a sort of integrated substitute for a mouse. An alternative and equivalent type of motion sensor, the "track ball", takes the form of a sphere, set in a stationary holder, that can be "rolled" with the fingertips.

[34] The line may appear slanted in the case of *italicized* text.

[35] "Forward" here implies to the *right* on a particular line of text: *toward* a hypothetical viewer positioned along the right-hand edge of the page.

text would vanish instantly without a trace![36] Secure storage requires that the document under development be explicitly *saved*: i.e., copied onto either the computer's internal hard disk or some removable storage medium. Alternatively, text can be dispatched by means of a network to some *other* computer for preservation.

Obtaining a *hardcopy* version of one's work of course entails the separate step of sending the document to a *printer*. This strict separation of *input* and *output* operations might at first seem odd, or even a nuisance, but it is actually quite advantageous, because it permits the writer to exercise full control over the information content of a document and its format until the very last moment.

● One way of reducing the amount of waste paper generated as seriously flawed printout is to edit newly composed text in at least a crude way immediately, directly from the screen.

Thus, if you should happen to recognize at once that a certain sentence should be eliminated, by all means select it (by "dragging" across the appropriate words with the mouse) and then press the "delete" key. All characters you have "marked" in this way will be instantly "erased". The space they formerly occupied will be filled automatically by whatever text follows—and no ink will have been wasted on the superfluous sentence! Many obvious faults can and should be dealt with in this way (see the subhead "The Benefits Conferred by Word Processing" below).

Document preparation as we have so far described it offers advantages both economic and ecological. From an *economic* standpoint, never in the past was it possible to write and edit so efficiently and rapidly as today. With respect to *ecology*, none of the efforts described to this point have involved any direct consumption whatsoever of materials; everything has transpired in a "virtual" realm. The fact that many potential documents are never transformed unnecessarily into material entities has far-reaching consequences. Word processing could in fact be seen as an important milestones along the path toward the "paperless office", where filing cabinets are obsolete.[37]

● The mouse enjoys a special relationship with one particular portion of the computer's RAM known as the *clipboard*, which functions as a sort of "short-term memory".

---

[36] WORD offers some protection against such a disaster in the form of an "AutoRecover" system, through which a copy of the current document is saved automatically upon the elapse of a specific time interval established by the user (e.g., every three minutes). This at least limits the amount of work that can be lost through a power failure.

[37] The responsibilities traditionally assumed by a filing cabinet can in large measure be met by defined segments of one or more computer hard drives. In a "utopian office" based on this model, new documents of the traditional sort arriving from the outside world would be "translated" at once into electronic form (using a scanner and OCR software) and then incorporated into the organization's digital records, where in many ways they would prove to be far more accessible than hardcopy counterparts (cf. also the subsection on "Strategic Considerations" in Sec. 5.4.2). The true paperless office is still far from a reality, however, and paper consumption in fact continues to climb worldwide. It is thus difficult to foresee the day when all official communication will be electronic in nature.

Among the most frequently used commands in word processing are "cut", "copy", and "paste". When either "cut" or "copy" is ordered with reference to a marked piece of text, a duplicate of that passage is written to the computer's "clipboard". If an insertion point is now created somewhere else in the document and the "paste" command is issued, the corresponding material is instantly introduced there as if by magic. The text in question could even be written to several different locations, one after another, each time reproduced as faithfully as if with a custom-made rubber stamp. The computer's clipboard is obviously a far more potent accessory than its physical counterpart in an office.

● Text that has been "cut" or "copied" remains available on the clipboard indefinitely, vanishing only when the latter is again called upon to provide temporary storage.

"Cut" and "copy" perform identical functions with respect to loading content onto the clipboard, but with a "cut" command the passage is *deleted* from its original location, whereas with "copy" it is retained (hence the *expression* "copy", implying duplication).

Windows and Toolbars

At this point we should pause to consider certain bits of terminology we will frequently find useful. Our didactic approach will be noticeably different from that of the typical computer manual (which in any case relatively few users take the time to read carefully!), and we make no pretense of rigor.

● A *window* will be understood as a discrete rectangular portion of a computer screen defined by the distinctive *frame* surrounding it.

Screen displays often incorporate several windows concurrently, which can be of several different types. It is commonplace for one such window to behave as if situated "behind" another, with some of its content obscured by the companion that apparently resides "up front".

● Although several windows may be *visible* simultaneously, only one will at any moment be *active*, and thus functional.

The *active window* is the one nearest the "front" (i.e., "closest" to the user), and the only one in which an embedded insertion point will be seen to blink.

In a word-processing environment the user almost always has direct access to one or more *text windows* for composition or editing purposes. Regardless of the presence or absence of text windows, however, at least one *menu bar* will ordinarily be visible, typically stretching across the top of the screen.[38]

The user's manual is of course the place to find out in detail how your screen should look as you set out to compose text, including what menus (together with related *palettes* and *toolbars*) will definitely be present and what additional features are

---

[38] An exception to this generalization is the "Full Screen" view available in the latest versions of WORD.

accessible from titles on the menu bar(s). The manual will also explain the precise consequences you should anticipate upon issuing any of the available commands.

- A default (empty) text window is automatically created when a word processor is first "opened" (activated) to begin a writing session, or when the "New Document" command is issued to establish a new text file. An existing (stored) document called to the screen always appears in its own separate window.

Documents prepared in the past will reside on the computer's hard disk in the form of independent *files*. Hopefully you will have given thought to the best way of organizing these, assembling them in specially created—and informatively labeled!—(electronic) "file folders". Folders can also be grouped: in the equivalent of virtual "file drawers", which might even be associated with several different "filing cabinets"—all in the interest of ensuring that in the future a *particular* document will be easy to locate based on a general knowledge of its content.

Assuming more than one text window is visible, the window of immediate interest can be brought to the fore and thereby made "active" simply by "clicking" somewhere within its boundaries. Alternatively, the main menu bar in most programs includes a title called "Window". Clicking on this produces a list of all windows currently "open" in that program, from which, using the mouse, a particular one can be singled out for activation.

A typical window is subject not only to "opening" or "closing", but also to displacement on the screen and/or size adjustment, where length and width can be altered independently. The entire file associated with an active window is subject to processing, not simply that portion which happens to be visible. If changes are made, however, these will only become permanent when the file is again *saved*. The particular portion of a text file that at any time is the focus of a window can be established by "scrolling", as described below.

- Commands currently at your disposal are collected together in *menus*, which are in turn accessed from a menu bar.

A menu is a special type of window that appears as a result of clicking on the appropriate title in a menu bar. Most menus are of the "dropdown" type, which means they descend like window shades from the menu bar itself. Here you are presented with a list of all the command possibilities provided in the context of that particular menu, from which one can be selected by highlighting it with the mouse. Certain commands in the list may be accompanied by the symbol "..." (ellipsis), indicating that that selection will lead to at least one additional menu (or dialogue box) containing options applicable to the corresponding command.

- The arrangement of commands within a particular menu, and even the set of choices offered, is generally subject to selective modification, opening the way to a certain amount of "custom design" of one's software.

Through this feature you have the opportunity to personalize to some extent the program's "desktop" and general work environment. For example, you might wish to make more accessible certain features you tend to use frequently—perhaps also assigning to them custom keystroke shortcuts.

● Another type of window, the *dialogue window* (or *box*), represents the most important conduit for user input to a program.

Windows in the latter category often take advantage of special communication facilities beyond those so far mentioned, including *check boxes, input fields, "radio buttons"*, and *popup menus*.

Text-Window Properties

The *frame* that surrounds a text window does more than simply isolate and distinguish the window from its surroundings. The very fact that one can activate a particular window from among several stacked on the screen simply by clicking somewhere within its boundaries is illustrative of the point.[39]

A typical text window is surrounded by four "bars" that together constitute its frame. The one across the top is called the *title bar*, because it includes the official designation of the document currently on display. The title bar is also used for moving the window around on the screen: by simply *clicking* somewhere within the bar and *dragging* while the mouse button is still depressed. Upon release of the button the position of the window again becomes fixed. Another feature of the title bar is a small, square area called a *close box*. Clicking here causes that window to disappear, along with any auxiliary features associated with it (and the corresponding file to be "closed"—but *not* automatically saved!). Other small box-like symbols could be present as well, for modifying the size of the window, for example (e.g., toggling between its current dimensions and something approximating a maximum, or between the current state and a window consisting *only* of the title bar).

Frame segments at the right and on the bottom incorporate *scroll bars* in those situations where the associated file content exceeds the capacity of the display area. Thus, clicking on an integral "down" arrow below the right-hand scroll bar, or dragging downward with the distinctive small box that apears within the scroll bar itself, serves to "move" the document within its frame such that text lines disappear one at a time from the top as new lines appear at the bottom. An analogous scroll bar at the bottom of the window "moves" the document in a horizontal fashion "behind" the window frame.

---

[39] Much of what is described here applies to other windows as well, such as those for editing *graphics* or even the windows associated with a *Web browser*. On the other hand, certain of the features described may vary depending upon the nature of the computer's operating system. Thus, the "close box" cited below has been replaced in MACINTOSH "System X" by red and yellow "buttons", where the former *closes* a file while the latter only makes the window *seem* to disappear, consigning it in fact to an unobtrusive "parking place" in what is called the "dock", from which it can be quickly recalled.

Depending on the program involved, the frame surrounding a window may also offer tools for access to certain functions, or information ordinarily reached through menus, a handy feature too often overlooked (see your users' manual for details). In the case of some versions of WORD, for example, the frame segment defining the bottom of the window is a source of useful information about the open document (page number of the text currently visible, character count, cursor location, etc.), as well as the means for switching document "views".

An especially valuable feature in WORD is a provision for *splitting* a window; i.e., transforming one text window into *two* smaller ones, each capable of displaying independently its own selected portion of a document. The two parts of a split window can be *scrolled* independently as well. This is extremely useful if one wishes to compare a text passage with one located elsewhere in the same file.

● Split windows can be particularly helpful when working with long documents.

For example, the upper part of a split window might be used for display of a full-text copy of the document currently under development, with the lower part presenting the same document in *outline* form (cf. "Subdivisions" in Sec. 5.3.3). With such an arrangement you can conveniently verify, as you compose, that what you are writing is actually consistent with your original intention. As noted in an earlier chapter, the outline view of a document is particularly well suited for rapid navigation through a lengthy work. An outline view can also be expanded to the point that it includes, in addition to all headings, the first few words of each *paragraph*, which makes it easier to detect redundancies and perhaps discover an opportunity to rearrange an argument so as to increase its strength or impact. Split windows also facilitate a "cut-and-paste" campaign, with the "cut" effected in one segment of the window and subsequent pasting in the other.

● Before closing a document's last open window be sure you *save* that document in its current form, since otherwise recent changes will be discarded and thus not reflected in the stored file.

Marking (Highlighting)

We have already in this section mentioned in passing the notion of "marking" or "highlighting" a portion of text, but its many potential applications warrant additional comment.

● "Marking" refers to tagging a particular symbol, word, text fragment, graphic element, etc., for future reference, in the process altering its screen appearance such that it clearly stands out from the surroundings.

A "symbol" or "text fragment" in this context could even be something as abstruse as a (non-printing!) paragraph mark (¶), or the space separating two words.

Anything that has been marked can subsequently be "copied", which is to say replicated in the clipboard for utilization elsewhere. Application of this technique to a paragraph mark obviously could be a precursor to creating a new (empty) paragraph but, more important, the new paragraph will have associated with it all the formatting characteristics of the particular paragraph from which the symbol was taken.

● A word processor ordinarily provides no explicit menu command "Mark", but the effect itself can be achieved in several ways.

The most straightforward way to mark a text passage entails placing the mouse cursor immediately ahead of (or behind!) the characters of interest and "dragging" over whatever you wish to be marked or "selected". The immediate consequence is a dramatic change in the looks of the marked text, representing a color reversal (replacement by a "negative" image, with white characters against a black background).

Other marking techniques can be divided into two categories: those involving the mouse and those accomplished strictly from the keyboard. Rather than elaborating here we suggest you instead consult the appropriate sections of your word processor's documentation, or launch a targeted inquiry through the program's built-in "help" system.

● Once a passage has been marked, it becomes subject to selective alteration in any of several ways depending upon how one proceeds, opening the way to many interesting possibilities.

One common reason for marking a passage is to impose upon it a different *type face*, expressing the selected text in italics, for example, or boldface. Alternatively, you might wish to introduce selectively a different font altogether, or perhaps adjust the *size* of the marked type. The marking of a passage could also precede its *removal*, easily accomplished simply by pressing the "delete" key. Or you might have reason to move a text fragment or a graphic element from one location to another with a "cut" and "paste" operation. These few examples are merely illustrative of the wealth of options.

To *replace* a marked passage with an alternative piece of text one simply begins typing the new material! Text entered under these circumstances will appear at precisely the site of any previously established "mark". This incidentally points up the need for an urgent word of caution:

● If a symbol key is *inadvertently* pressed while some passage in a document is marked, the marked material instantly disappears, to be replaced by the specified symbol!

An error of this sort need not be fatal, however. Recent actions of almost any kind can be reversed with the "Undo Typing" command from the "Edit" menu—which may even be powerful enough to allow you to negate *several* recent steps, one at a time. Again, the user's manual should be consulted for details. An "undo" request should be issued as soon as possible, however, to minimize the chances of a mistake becoming permanent. The important thing to remember, however, is that whenever text is in a "marked" state—especially if a lengthy passage is involved—it should be regarded as vulnerable, and markings should not be left in place any longer than necessary.[40]

● As previously indicated, the marking technique is also applicable to certain types of "metainformation".

With WORD you have the option of making the presence of much "metainformation" visible on the screen, like explicit indication of paragraph breaks (¶), for example, or clear evidence of a blank space (a small dot situated at half the height of a capital letter). Displayed indicators of this sort appear as ghostly, gray images.[41] One of the important reasons for revealing them is to facilitate their selective inclusion or exclusion with respect to marking.

We pointed out earlier that a "¶" symbol signals not only the end of a paragraph, but also the fact that the "return" key was pressed at this point. It is the latter action that actually results in incorporation of the important metainformation in question: an embedded summary of all the formatting characteristics responsible for the unique appearance of that particular paragraph, a summary that conveniently remains with the paragraph mark if the latter is copied onto the clipboard. The corresponding parameters are normally applied automatically to a *subsequent* paragraph as well, immediately upon its creation (assuming no instructions are issued to the contrary).

Several other non-printing ("virtual") symbols are also covered by the display option described. In a sense, requesting screen display of these ordinarily invisible features could be regarded as implementing a new "view" option, one whose main purpose is to facilitate editing.

Formatting

The most common reason for marking a piece of text is to set the stage for a *format* change. The Latin root *formare* can be interpreted variously as "to shape", "to fashion", "to arrange", or "to form". When something is "formatted", its appearance is altered. The term is employed in several very different contexts in conjunction with computer applications.

● With respect to the word-processing environment, formatting involves choices regarding the way text is *presented*.

One example of such a format decision would be the assignment to a piece of marked ("selected") text of special *attributes*, such as the boldface or italic *style*.

Format attributes, perhaps of a more general nature, can also be applied to large *blocks* of text—even an entire document. An example is the treatment of the right margin: whether text is subjected to "full justification", producing a clean, straight

---

[40] Extra care is important when working with older software that may lack the godsend of a "multiple undo" function.

[41] Even if visible on the screen they will still be ignored in terms of *printed* output.

line along the right-hand edge, or the "ragged right" alternative, in which different lines are allowed to have different lengths depending on how the corresponding words and (uniform) spaces happen to fit in the line length available. Another illustration of formatting at this level is deciding how many *columns* to use for a piece of text. At the very basic level of individual symbols or letters the chief format options relate to *font*, *type-style*, and *character size*. *Paragraph formatting* (e.g., standards for bulleted passages, indentation of block quotations, etc.) can be regarded as operating at an intermediate level.

The menu actually *entitled* "Format" is one of the most important of all for the writer. This is typically subdivided into categories like "Paragraph ..." "Document ...", "Bullets and Numbering ...", etc., where "..." implies as usual a need for considering various options. It is here that stylistic preferences are declared both on a *global* basis (applicable to an entire document, including margins, line spacing, etc.) and in terms of individual words or even letters.

Many formatting instructions actually can be issued in a variety of ways: not only through menus and submenus (and related dropdown lists), but also by "clicking" on specific "active" elements on the screen (such as a special symbol incorporated into a "tool bar" or a "ruler"), or with an appropriate keystroke signal.

Customization

The novice learning to master a word-processing program for the first time is likely to marvel at its inherent versatility and the ingenious ways provided for accomplishing various tasks; indeed, such a user would not be inclined at first even to consider trying to improve the program. But imagine how convenient it would be if you were able to restructure your personal writing system in such a way that it was custom-tailored to your specific needs, with each menu, for example, presenting just the right options arranged in the ideal sequence. In fact, as we have already suggested, this wish need not be regarded as an idle fantasy: your program almost certainly has already granted you the necessary powers.

● Adapting a default set of program "preferences" to match one's unique needs can greatly facilitate the creative process by minimizing distractions.

Suppose, for example, you regularly have reason to introduce text elements set in a distinctive font other than the one assigned as the "default" ("Verdana", say, in place of "Times"), and with red lettering rather than black. It would be well worth your while to create custom keystroke commands for establishing these parameters (and later restoring the default conditions), because that would mean you could work continuously from the keyboard without interruption instead of repeatedly fumbling around with your mouse. All the information necessary for accomplishing such a customization is present—as usual—in the documentation accompanying your program.

5.3.2 The Utilization of Word-Processing Software

A Writer's Dream Come True

We have indicated that the "Format" menu of your word processor is a source of important tools for establishing in advance a number of characteristics you wish to be applied to specific paragraphs or sections, or even to a newly created document as a whole, as you engage in composition. One such characteristic is the preferred spacing between lines within a paragraph. This can now be adjusted much more flexibly than under the obsolete (typewriter-derived) limitations of "single-line, line-and-a-half, double-line", and your wishes can be expressed with great precision in terms of "points" (alternatively: millimeters); cf. the heading "Fonts and Units of Measure in Typography" in Sec. 5.5.1. The same applies to spacing *between* paragraphs.

Another paragraph characteristic open to specification is the extent of *indentation*. This can refer of course to a paragraph's first line, but also to the paragraph as a whole; indeed, uniform indentation throughout can be a useful and subtle way of calling attention to a particular paragraph. Both parameters are freely and independently adjustable—in centimeters, for example, to a precision two places beyond the decimal point.

- The most striking aspect of text preparation with a word processor is the flexibility it provides with respect to the way characters are expressed and where they are placed—and the fact that formatting established visually on the screen is faithfully preserved when a document is printed.

One formatting characteristic with an especially dramatic effect on the apparent "polish" of a document is how the text is *aligned*. Four alignment alternatives are generally provided: *left*, *centered*, *right*, and *justified*. Expressed more definitively, the latter option produces text that is "fully justified" in contrast to "right-" or "left-justification". In order to achieve full justification a computer is required to determine the total horizontal space requirement of all words constituting a particular line and then adjust *word spacings* such that that set of words consumes precisely the distance available between the left margin and the right margin, leading to the impression that "sharp dividing lines" are present on the two sides of the text block. With *left justification* text is carefully aligned only along the left margin, leaving what is called a "ragged" right margin. *Right justification* of course produces the opposite effect (used only rarely), whereas with *centered* text, lines are distributed evenly about the block's central axis.

Fully-justified text usually conveys the most "professional" look, because in the days before word processors this effect could *only* be achieved through professional typesetting. Nevertheless, full justification is not always the best choice, especially with text set in narrow columns (where long words can result in grotesquely distorted word spacing). It should also be avoided with text still subject to "on-screen" editing, because variability in word spacing makes it difficult to recognize places in which *extra* spaces have inadvertently been introduced between pairs of adjacent words. Some authors deliberately avoid full justification with *drafts* of their documents to eliminate the risk of these being mistaken for finished products. In any event, truly refined copy should be as professional-looking as possible, and the combination of a word processor and a reasonably high-quality printer makes such an effect easy to achieve in a convincing way in the DTP context (see "Desktop Publishing" in Sec. 5.4.2).

We have noted that full justification almost never improves the appearance of text set in narrow *columns* due to the exaggerated spacing required in lines characterized by long words. The spacing problem can be alleviated somewhat through extensive use of *end-of-line hyphenation*, making available *fragments* of long words, but overuse of this supposed solution can also produce unsightly results, as well as text that is difficult to read.

Judicious, *limited* hyphenation often is advantageous, but it can be a source of new difficulties as well. Thus, while a word-processing program may blithely promote its "Hyphenate" command as a panacea for justification problems, "automatic" hyphenation is generally carried to an unacceptable extreme—and may well prove erroneous in cases admitting of ambiguity.[42] Choosing the program's option of "manual" hyphenation will force you into making an enormous number of individual decisions, some of which you would think the program *should* be clever enough to consider obvious. Computer-proposed hyphenation of foreign words is almost always faulty, because it is based on imperfect algorithms rather than authoritative dictionaries.

The overall benefit to be gained from justification of printed text ultimately depends upon *aesthetic* judgements, an observation that also applies to choice of a particular *type font*. While some fonts are inherently more legible than others (as a result of the presence or absence of *serifs*, for example, which are discussed briefly below), most variations are purely stylistic. One interesting and often overlooked advantage of word-processing systems over the typewriter in this context, by the way, is the availability with the former of *proportional fonts* (cf. Figure 5–1 in Sec. 5.5.1). These are fonts in which each letter occupies an amount of horizontal space consistent with its true nature. Thus, a letter "i" should logically be less space-consuming than an "m". In pre-computer days, the absence of this characteristic made it easy to distinguish a document prepared in the typical office from professionally printed text, since mechanical considerations more or less dictated uniform letter spacing with a typewriter.[43] Most would agree

---

[42] Consider the example of "present": is this group of letters intended to signify the verb "pre-sent" (to offer something formally), or the adjective "pres-ent" (taking place now)?

[43] Strictly speaking, this generalization applies only to the "classical" typewriter in its long familiar form, now rarely encountered except in old movies. Later, more advanced "electric" typewriters were in fact capable of producing text containing letters with different widths, but still subject to the limitation of a small set of "standard" widths (usually four). An evolutionary stage between the electric typewriter and the word processor—the *electronic typewriter*—offered greater sophistication in this regard, but its impact was minimal. A rapid decline in computer prices—coupled with increasing "user-friendliness" of computer operating systems—doomed this "hybrid" to a relatively trivial role in the history of writing.

that text set in a proportional font is more pleasing to the eye than copy based on a "typewriter-like" monospaced font (as with the computer font *Courier*).

But we should address at least briefly and more directly the heading assigned to the present subsection ("A Writer's Dream Come True"):

● In a sense, many of a writer's fondest desires have indeed been realized through word processing, with the important result that thoughts can now be captured with striking efficiency, in part because no attention whatsoever need be diverted initially to *presentation*, an issue that can be dealt with independently and extremely flexibly at a more convenient time.

What scribe in the past would not have been thrilled by the prospect of expunging unwanted letters or words with the touch of a button, or effortlessly and instantaneously replacing an offending passage with one that was more suitable—let alone watching as an entire paragraph was transported cleanly from one location to another? Feats like these are child's play for the writer equipped with a word processor, and might even be dismissed as trivial relative to other more exotic possibilities.

● The modern approach to text generation is especially noteworthy for its extreme *versatility* and *adaptability.*

The potential that word processing has unlocked is exhilarating, but at the same time there is a darker side. Consider for a moment the marvelous hand-illuminated manuscripts we have inherited from the Middle Ages, in which paragraphs are commonly introduced by stunningly diverse, sagaciously conceived, and painstakingly rendered pictorial capitals that confer upon the associated passages a profound individuality. We on the other hand—especially as desktop publishers—by and large operate from the premise that printed text is assembled in a strictly utilitarian way from letters drawn from unexceptional collections of standardized bitmap patterns, selected almost at random.[44] But the presentation possibilities actually inherent in word processing are infinitely varied, at least in principle, and there is ample scope for a desktop designer functioning in the electronic age to display creativity, imagination, and passion in addressing the reader's too often ignored aesthetic sensibilities.

The Benefits Conferred by Word Processing

As we have seen, electronic data-processing methodology can be supportive of the writer in a great many ways. Among the most important are without doubt the following:

– Facilitating the *insertion*, *deletion*, or *substitution* of text elements—letters, words, sentences, paragraphs, even entire chapters—as well as the *transfer* of text from one location to another, whether in the same document or between documents;

---

[44] Professional publication designers usually exhibit more refined tastes in this respect, giving close attention to the unique effects achievable with different typefaces, some of which have been cherished—and have evolved—over the course of centuries.

- Selective *grouping* and *organization* (sorting) of text elements under some hierarchical or structural scheme (with reference to headings, for example, but also footnotes or citations);
- *Quantitative* manipulation of selected text elements (for numbering or counting purposes);
- Systematic *retrieval* of specific letter combinations—in the interest of correction, certainly, but also indexing (utilizing "Find" or "Find and Replace" routines);
- *Merging* of text with other elements, such as graphics, potentially derived from a host of sources (data import);
- *Output* of documents in countless ways to conform to almost any spatial distribution and format constraints.

The ability to achieve control over such diverse factors—with remarkable ease!—puts the modern writer in a truly unique position with respect to his or her work. An evolving document need no longer be perceived as a relatively static entity predestined to pass through a finite number of discrete "stages" on the path toward perfection. One's work instead takes on a *plastic* character, susceptible to as much or as little minor adaptation or major overhaul as the author at any time feels moved to undertake. Gone are the days, too, when a haunting suspicion that some subject may have received faulty treatment meant embarking on a meticulous, often frustrating manual search through a stack of paper, or when a complex reorganization required chopping a document apart with scissors and then reassembling the pieces with transparent tape. Changes initiated from a computer screen take effect instantaneously, leaving no tell-tale evidence whatsoever, and—at any stage of document development—a printed copy of all or part of a project in progress can be obtained within a matter of seconds in a form so pristine that it appears ready for immediate dissemination. Those of you so fortunate as to take the new technology for granted cannot possibly imagine how tedious and difficult the refinement of a document was for your forebears.

To pursue but a single example, the true value inherent in an automatic system of footnoting will be clear only to someone who has experienced the agony of introducing into a document a host of literature citations in the old-fashioned, manual way. Almost every manuscript revision required that numbers already assigned to at least some of the citations be changed, and that the whole document be retyped, which of course served as an open invitation to mistakes. Now all aspects of footnote management are attended to by the trusty word processor. Perhaps equally remarkable, it has become trivially easy to revise *how* citations are numbered: replacing a sequential numbering scheme with a "resetting of the clock" at the beginning of each section, for example.

● We have seen earlier how a word processor can lend invaluable assistance in the *structuring* of a document in terms of subdivisions, and in assembling an accurate *table of contents*.

These two tasks might at first appear unrelated—even though both fall under the general heading of "organization and sorting"—but the relationship is in fact a very intimate one. After all, a table of contents amounts to little more than a sequential list of the titles assigned to the various subsections constituting a document (cf. "Subdivisions" in Sec. 5.3.3), elaborated to include the associated page numbers.

- Another arduous task in this same category rendered substantially easier by a word processor is the *indexing* of a complex work (a subject already addressed in some detail in Sec. 4.5.1).

The flexible layout options available with a word processor also make it relatively trivial to construct at least a first approximation of a professional-looking *table* (cf. Chap. 8). Thus, use of the mouse to adjust the "paragraph ruler" at the top of the screen transforms *column definition* into a quick and straightforward operation, where fine-tuning with respect to width and precise horizontal disposition can be postponed until the various columns have been filled with data.

- One other beneficial aspect of word processing surely worth mentioning is the fact that digital-document files are remarkable for their exceptionally high *information density*.

No longer need vast amounts of space be devoted to the preservation of multiple printed copies of important material, and the simplified "transportation" of electronic documents relative to their printed counterparts can result in sizable savings of both time and money. Thus, instead of forwarding to a granting agency a cumbersome parcel containing perhaps ten copies of a complex proposal it may well suffice to dispatch instead a digital data file via the Internet—easily accomplished with a standard e-mail program.

As revolutionary as the word-processing approach to writing clearly is, however, traditional methodology has by no means become entirely obsolete. For example, just because all editing could in principle be accomplished directly from a display screen it certainly does not follow that this is necessarily the *best* way to proceed. Proposed changes will of course ultimately need to be implemented with keyboard and mouse, but nearly all writers agree that the intellectual challenge of determining *how* a manuscript should be altered should still be tackled with a pencil and a clean, *printed* version of a document—in all probability far removed from the computer, quite possibly in one's favorite easy chair. Transferring penciled corrections and revisions to the corresponding computer file constitutes a separate, essentially mechanical step, which might even be entrusted to someone else.[45] A fresh new copy is then printed, after which the process is repeated: as many times as necessary until the author becomes convinced the latest document is a masterpiece.

---

[45] This of course presupposes that the author's proposed corrections are *legible* in someone else's hands—whereas in fact they will probably have been scribbled so hastily that the *author* will sometimes find them baffling (an observation drawn from considerable personal experience!).

Computer technology has brought many advantages to the creative act of writing, but one subtle danger also needs to be acknowledged: On a computer screen (and especially in printed form) a "rough draft" prepared with a word processor can *appear* so polished and professional that the temptation is almost irresistible to short-circuit the editing process and settle for a second-rate document. SCHNEIDER (1989) is particularly forthright in his castigation of those who fall into this trap, and in decrying its malignant influence on "electronic editors" (whom he disparagingly dismisses as "editronicers").

On the other hand, clerical personnel schooled originally in traditional secretarial practices are among the most ardent converts to word processing, despite initial qualms, in part because of the dramatic increases they experience in their productivity. Confident mastery of the wide range of powerful tools comprising the latest word processor can also be for such a person a source of considerable pride—and justifiably so!

Finally, a word-processing station quickly becomes much more than that, assuming the character of a multifaceted assistant eager to help you structure your thinking, manage your appointment calendar and address lists, perfect and maintain your bibliographic records, sort and merge the broad spectrum of documents confronting you, and generally function more effectively as a professional.

We defer to Sections 6.7 and 7.5 our consideration of special supportive software that can simplify the preparation of *equations* and *figures* (*graphics*) for your digital documents.

5.3.3 Advanced Aspects of Text Editing

Formats, Patterns, Styles, and Templates

We have dealt at some length with details regarding the ways word processing can contribute to the writing process, but now we need to step back a bit and shift our attention to certain preliminaries that *precede* composition, looking in particular at how these are (or should be) affected by the decision to adopt a computer-based approach to writing. In particular, what role do thoughts about *format* or *presentation* play as writing commences?

When one first applies a pencil to paper in the "classical" mode of composition, the medium of expression is quite literally a "blank sheet", and the writer exercises and enjoys complete freedom in every sense. Where will the first marks be made? Should one *print* the letters (for increased legibility), or would cursive writing be better (e.g., faster, for example)? The words that come forth will probably occupy lines, but should these lines be close together, or widely spaced—"airy"—to encourage later adjustment of the text? Interesting questions, perhaps, but not ones that necessarily require conscious decisions. Let the manuscript develop however it will!

The situation is rather different for the person working with a word processor. Whether one applies much thought to it or not, a relatively rigid "pattern"—a "template" or a "format"—dictating the appearance of the anticipated document is firmly in place (if only by default) before a single word is composed. The most important piece of technical terminology in this context is *style*. We will consider a little later the general subject of "styles" and how they are useful. Suffice it to say for the moment that in a word-processing environment "style" is a technical term carrying a great deal of significance. Perhaps the most obvious feature of a style in this context—what most influences how a given document will look—is the nature of the lettering. Actually, more is involved than simply "lettering" per se; at the very least there is need for taking into account numbers and punctuation marks as well. Thus, the most fundamental decision to be made has to do with the *character set* to be employed.

Character Sets

The prototype of a *character set* is an array consisting of a finite number of *symbols*, generally understood to encompass all the basic alphanumeric elements, the usual punctuation, and at least a few useful miscellaneous signs (e.g., $, &, @, etc.)—all conceived in a strictly *abstract* sense, however, and thus subject to an infinite number of specific stylistic treatments.

- A particular *font* (or *face*) can be regarded as one discrete manifestation of a character set, to which has been assigned a name meant to stand for the unique design characteristics responsible for that font's distinctiveness.

A word-processing system normally offers the user access to a wide variety of fonts, which are in turn stored as sets of codes that make possible electronic representation of all the individual characters: for display on a screen, of course, but also for output by printers of various types. The codes themselves reside in fixed, predetermined locations in the host computer's long-term memory. Different computers and different operating systems vary to some extent with respect to the character sets and fonts they support, but several fonts (e.g., "Times", "Arial", "Symbol", etc.) have come to be regarded as "standards" assumed to be available on virtually every system.

- The need for a fundamental level of character-set compatibility from one computer system to the next led to establishment at an early date (1968) of what is known as the *American Standard Code for Information Interchange (ASCII)*, acknowledged in the meantime as something of an international standard.

Almost every computer takes advantage of the ASCII coding system, which equates the 128 numbers 0 to 127 (in binary form) with a specific set of symbols.

- Through ASCII, the "essence" of certain symbols has been captured in a universal way in terms of equivalent code numbers: one number, one symbol.

The ASCII character set thus corresponds to 128 distinct, standardized symbols. The first 32 of these actually represent *instructions* rather than displayable characters (e.g., code 13: CR or Carriage Return). The next 16 (codes 32 to 47) represent punctuation and a few special signs ($, %, &, etc.). Code numbers 48 through 57 are assigned to the *numerals* 0 to 9. Codes 58 to 64, 91 to 96, and 123 to 127 encompass more punctuation and signs. Finally, codes 65 through 90 are equated with the (Latin) *capital letters*, while 97 to 122 specify the equivalent *lower-case letters*. To take but a single example, the *code* for "zero", irrespective of how that zero might be depicted, is the *number* 48; that is, the precise *shape* of a zero is free to vary from font to font, but the code equivalency is valid regardless of the font, and 48 can never represent anything *other* than zero.

In fact, a total of 256 ASCII codes have been assigned, but only the first 128 (0 to 127) are accorded universal recognition as *standards*. Beyond 127 the coding may be interpreted in various ways, resulting in variable character sets, which inevitably makes *conversion* problematic in some cases. In order to avoid misunderstandings, codes 128 to 255 are thus described as belonging to an *extended ASCII code*, so that the expression "ASCII code" can be interpreted in the strictest sense as applicable only to codes 0 to 127. It is in the extended ASCII symbol set that one finds such special characters as *accented* letters and letters bearing *diacritical marks*, including the *tilde* (˜) and the *cedilla* (¸).

With reference to a document, *ASCII* (or *text-only*) *format* means that the content is expressed entirely in the particular limited set of characters that all computer systems and all software can be expected to understand in the same way. By definition, then, such a document may contain *only* letters of the Latin alphabet, ordinary punctuation, numerals, and a small number of miscellaneous symbols—together with a *very* few output instructions (including "carriage return" and "new line", used for signaling, in the former case, the start of a new paragraph, and in the latter perhaps the end of a heading or a fragment of free-standing text). No instruction codes exist for indicating unambiguously most aspects of format, as for example the use of italic or boldface type, or characters of increased size, and all such document features would disappear upon saving a word-processing file in "text-only format".

"Styles"

Styles as the term is used in word processing are nothing more than complicated collections of formatting instructions. They can play an extremely important role, however, when word processing is practiced at a sophisticated level. In particular, styles greatly simplify construction of a complex document and ensure that formatting features will be applied in a consistent way. If one wishes to guarantee faithful adherence throughout a document—or even a group of documents—to the descriptive characteristics envisioned, for example, for a typical "paragraph" of body text (or any other text element, such as a line in a table), including type-style, font, margins, line spacing, etc., the sensible course of action is to confer upon that set of characteristics a unique *style name*.

- Styles should be thought of as comprehensive instruction sets, to some extent applicable to individual characters, but also governing the appearance and behavior of larger text units, like paragraphs.

Definition of a style is a very straightforward process. You begin by creating a sample text element to serve as a "model" (a paragraph of running text, a heading at a particular level in your hierarchical scheme of headings, or a "model footnote", for example). Next you *select* (mark) that text element and issue the requisite command for calling up a special "style definition" window (in the case of WORD, the "Style ..." command under the "Format" menu). Here you will see already summarized all the adjustable characteristics currently applicable to the selected text, and you are given the opportunity to add additional specifications if you so desire, or perhaps change specifications you would prefer *not* be honored as they stand. Once you have established a "characteristics set" satisfactory in every way, you assign to it a unique name, which initiates storage of the information in the form of a "style" for future use. Thus, if you later associate some *other* paragraph with this newly defined style name (or the name of any stored style), the result is that all the characteristics implicit in that particular style name are instantly conferred upon the text in question.

> For example, the style specifications applicable to the paragraph you are now reading are reported to be: "Times 12 point, justified, line spacing exactly 12 points, 6 points of space following the paragraph, left indent 1.98 cm".

(Note that this characterization does *not* take into account a size reduction—to 80% of the original—implemented during the publication process.)

Shorter text elements—even special individual symbols (e.g., the footnote marker)—can be "standardized" in a similar way to avoid the nuisance of repeatedly modifying them after they are typed. It is important to realize, however, that even though an element may have had a style conferred upon it, that element is still subject to "customization", which would override the "standard treatment". As a general rule it is fair to say that—ideally—every paragraph in a manuscript should have associated with it the constraints corresponding to some predefined style.

- The use of styles greatly simplifies the process of formatting a document, and it is a great help in ensuring consistency.

Style definition is especially important with an electronic manuscript that might later serve as a basis for typesetting (see Sec. 5.4.1). As the author of such a document you can contribute in a significant way to maintaining consistency[46] in the finished product

---

[46] *Consistency* is revered as a sort of First Commandment by almost all serious editors. Although some aspect of a document's format might differ from what the reader would prefer, a formatting scheme once adopted will be strictly applied throughout a given work. Readers in fact tend to become uneasy when they sense *inconsistency*—new terminology suddenly applied to something already discussed, for example, or a peculiar design of a section heading. It is a bit like the uncomfortable sensation that →

despite the fact that the latter will probably differ significantly in appearance from the manuscript you submit, and that you are unlikely even to be involved in the various design decisions (choice of font, type size, spacing, indentations, etc.). Many characteristics will be established independently by the editor(s) and typesetter, and the final treatment will almost certainly deviate from formatting you would have envisioned, but the fact that you have associated distinctive style names with all the various text elements means that uniformity will not be sacrificed during further processing: all "Second-order headings" will continue to be handled identically, for example. What is more, you will have taken the first steps toward conferring SGML treatment on the document (see also "Structured Markup Systems" in Sec. 5.4.1).

AutoText Entries

What we here refer to as an "AutoText entry" (the terminology is from Microsoft WORD) is vaguely related to the concept of "styles" in that in both cases the user has taken advantage of an opportunity to devise a "personal shortcut" in which a definitive name is assigned to a specific set of instructions, reminiscent of what is sometimes called "construction of a macro". In the case of AutoText entries, however, the "instructions" in question are sequences of symbols (typically alphanumeric characters) that will come to be regarded as equivalent to other more convenient (typically shorter) sequences.

● Any string of characters you anticipate needing repeatedly in the course of composition is a potential candidate for transformation into an *AutoText entry*, which you will then be able to insert instantly wherever you wish simply by entering that string's unique user-created "abbreviation".

The text element in question might be a group of words (possibly including punctuation; e.g., "Sincerely yours,"), or it could be a set of digits (e.g., a Social Security or telephone number), or one or more specially formatted symbols (e.g., an oversized bullet surrounded by a pair of two-point spaces). In a document requiring frequent reference to some difficult name, like "Wojtowicz", or a complicated piece of technical terminology—e.g., *exo*-bicyclo[2.2.1]heptan-2-ol—or merely an awkward but distinctive and often-encountered phrase, such as "integrated digital bipolar switch", much time can be saved by typing the material in question *once*, and immediately giving it a "name" so that it becomes an AutoText entry. This "name" will then be incorporated into the word-processing program itself and thus available for use any time you wish—in any document. You will probably find that a few common phrases have already been subjected to AutoText treatment by the software manufacturer's programmers, and have therefore been at your disposal all along (things like "Best wishes", "CERTIFIED MAIL", etc.).

---

can accompany a visit to a disorganized house. And if an author is careless with respect to presentation, might there not also be problems with some of the arguments?

Anywhere in a document, whenever you type either deliberately or purely by chance one of the keystroke combinations associated in this way with a defined set of symbols, the latter will suddenly appear on the screen in a special little box that seems to float over the text. Pressing "Return" causes immediate introduction of this text at the site of the cursor, where it replaces that which you actually typed.[47] Alternatively, one can issue an "Insert AutoText" command with the mouse and then select the particular AutoText entry of interest (previously defined) from a menu that lists in alphabetical order all the available possibilities.

● Use of the AutoText feature obviously simplifies and accelerates the writing process, but perhaps equally important it also reduces the risk of errors.

A related consideration is the fact that if you take regular advantage of AutoText insertion it is less likely that during a subsequent search operation some occurrence of a pertinent word or phrase will be overlooked as a result of its having been mistyped.

Subdivisions

Today's word-processors also offer valuable tools to assist in the *organization* of a manuscript, most importantly an *outlining function*. This is a feature you should be sure to take advantage of whenever you are faced with preparing a complex document. Developing an outline is much more than a pedagogic nicety insisted upon by high-school teachers: It is a regimen that can contribute significantly to one's basic thought processes and thus influence the logical development of a set of arguments.

Typically, a word processor makes available to the author at least three different (interchangeable) ways of viewing a document:

– The "normal" way, which presents the work as one long continuous block of text, through which one can scroll rapidly from beginning to end. This may include indications of where page breaks would occur if the document were printed, but if so they are generally in the form of unobtrusive dotted lines that do not interfere with the text flow. Examination of footnotes in this viewing mode is possible only in a window dedicated to that purpose.
– A "page layout" view (equivalent to what is also known as a "print preview"), where an attempt is made to simulate on the screen, in every detail, how a document would look if printed: a clear-cut representation of individual pages, complete with all headers, footers, footnotes, and special formatting.
– An "outline" view.

● The *outline view* is designed to emphasize characteristics reflecting a document's *hierarchical organization*.

In the earliest stages of document development, an outline can serve as a convenient place for making notes about the concepts you wish to cover, and then organizing

---

[47] Text inappropriately and *inadvertently* introduced in this way is easily eliminated by immediately selecting "Undo AutoCorrect" from the "Edit" menu.

them sequentially. In WORD, for example, activity of this sort is facilitated by the fact that in outline view a new toolbar appears containing, among other things, bold arrows pointing up, down, left, and right. Clicking on the *right* arrow causes the outline entry in which the cursor rests to be "demoted" one level in the document hierarchy, whereas the *left* arrow similarly leads to a "promotion".[48] *Up* and *down* arrows interchange the active entry (the one containing the insertion mark) with its neighbor above or below, with all the subsidiary text that entry implies shifting as well (including subordinate headings, of course). Changes of both sorts can be applied as often as one wishes. This flexibility continues to be available in "outline view" even after a document has been fully fleshed out, making it exceedingly easy to accomplish a thorough reorganization at any stage of writing.

In effect, what we have just described could be interpreted as an example of employing the "mind mapping" strategy introduced in "Developing a Concept" in Sec. 2.3.1.

● Outline view represents an exceptionally powerful tool for categorizing material from a hierarchical standpoint and then arranging the elements in a systematic way.

You will probably find it most convenient while creating the various blocks of text comprising your document to be working in either the "normal" or the "page layout" view, but for major organizational changes it is best to switch to "outline" view, especially with a long document.

When "outline view" has been selected you have the option, incidentally, of specifying the "depth" within the hierarchical structure (i.e., the lowest *level*) to which you wish text to be displayed. You can thus elect at any time, for example, to examine only material in the first or uppermost level—normally limited to major headings—or everything through the third (or some other) level. Keep in mind, too, the option of displaying a "condensed document", in which you see the *first line* of every paragraph of body text.

So long as a document is based on a properly prepared outline, creating a *table of contents* is trivially easy. You simply place the cursor at the point in the document where you wish the table to appear (typically near the beginning) and then issue the appropriate menu command (in the case of WORD, from the "Insert" menu through its submenu "Index and Tables"). Be sure to examine the various options the program provides with respect to the *format* of a table of contents. The table-of-contents feature alone makes it worthwhile to pay very careful attention to one's outline, especially for a complex document like a book manuscript.

Anticipating the Need for an Index

Another valuable feature offered by today's word processors is help with preparation of a *subject index*, a chore that otherwise is much more burdensome and demanding

---

[48] As is the case with many tables of contents, the farther to the right an entry begins in "outline view", the lower is its standing in the hierarchical scheme.

than you might imagine. In the past an author was required to construct an index on the basis of a formidable stack of file cards, all painstakingly prepared by hand (and highly subject to error!). The rudiments of the corresponding procedure are described in Sec. 4.5.1.

Somewhere among your word-processor menus you will find a menu option (perhaps with a keystroke equivalent) that allows you to designate a selected piece of text as an "index entry". There should also be provision for locating and similarly flagging, if you so choose, *every* occurrence of that particular word or word combination, wherever it appears in the document. The flagging process serves as a convenient way to begin establishing the set of words and phrases you wish your index to contain. Later, an "Insert Index" command will be used to compile the many flagged passages, together with their page numbers, and incorporate an alphabetized version of the result into the document wherever you wish (typically at the end). The "flags" indicating what should be regarded as index entries are embedded in the text file as *hidden text* (see also Sec. 4.5.1, especially Figure 4–1, as well as "Special Considerations ... " in Sec. 5.4.1). Incidentally, these flags constitute another example of metainformation, this time conveying the message "what appears here is understood to be part of an index entry".

It is of course also possible to designate for indexing purposes relevant text passages that do *not* include literal use of an official index term. One need only supply appropriate linking information with the aid of a special dialogue box, whereupon the requisite term is *generated* at that point and embedded again as "hidden text".

Distinctions can be made among otherwise identical index entries such that they will be distributed under a series of *subentries* as a way of resolving the problem of a list that has become too extensive. Once again, a dialogue box facilitates making the necessary distinctions.

● Just as with a table of contents, WORD offers several *design* options with respect to the index.

It is important prior to actual creation of your index to ensure that all "hidden text" is indeed currently hidden (again, through issuance of the appropriate command), since otherwise the page numbers listed would not conform to pagination in your final printout, because the text is in a "swollen" condition due to the presence of superfluous material.

A good index of course contains entries other than those pointing directly to places where key subjects are discussed. *Cross-referencing* in the form of "*see ...*" entries is equally important (e.g., "Subject index *see* Index"), and a word processor offers provisions for these as well. On the other hand, any "*see also ...*" references you may wish to include will probably need to be attended to manually.

Table-of-contents and index preparation can similarly be accomplished with page-layout software. PAGEMAKER, for example, offers the interesting advantage that a single command can be used to create an index reflecting the combined content of *several* document files (corresponding to the set of *chapters* constituting a book, for example).

Alphabetization of index entries is conducted on the basis of rules built into the software, rules consistent with prevalent bibliographic standards. Thus, blank spaces and such punctuation as a hyphen, period, comma, or slash will *precede* letters of the alphabet, consistent with their lower ASCII numbers.

You should not expect a word processor to be equipped for dealing explicitly with an index entry whose subject extends over multiple pages (typically because the referenced passage is so long that it bridges one or more page breaks). Thus, transforming the index entry "Fuel   9, 10" into "Fuel   9f" would need to be accomplished manually (cf. Sec. 4.5.1). Be prepared in any case for a certain amount of "custom massaging" of any extensive index, because a preliminary version created automatically is bound to prove unsatisfactory in at least minor ways.

More complex and reliable automatic indexing systems have been developed for use with professional software, but their management is correspondingly demanding. For example, a large handbook would typically be indexed with the aid of a *database* subject to manipulation independently of the associated text, where the database itself is created with special software exclusively for that purpose. As you would suspect, the initiative for such developments has come from large commercial publishers.

Spell Checking

Once you complete a first draft of a document it is time to begin editing. A useful initial step is allowing the computer to check for spelling errors, which proceeds on the basis of a rather comprehensive dictionary integrated into the word processor. This occurs when you invoke a "Check spelling" command, which will produce either a message saying that the program has found no problems, or a dialogue box highlighting some apparently misspelled word (typically with a bit of the surrounding text to indicate context). You may also be presented with one or more suggestions of what *might* have been your intent. It then becomes your job to tell the program what to do: replace the suspect word[49] with one of the suggestions, replace it with an alternative *you* prefer, or leave everything alone. Once this has been done the search can resume.

- Seriously consider taking advantage of the opportunity to create your own supplementary *user's dictionary*.

Without this option, spell-check systems are of limited value, especially for scientific text, because the verification process is interrupted altogether too frequently by technical terms and personal names that obviously would not appear in a standard dictionary.

---

[49] It should be noted that a "word" in this context is any contiguous string of alphanumeric characters separated from its neighbors by blank spaces. A hyphen is treated as equivalent to a space, and in some cases a period will act like a letter (as in "e.g."). This of course means that a spelling check is an excellent way to locate places where the mandatory space between two words has inadvertently been omitted, one of the many "tricks" the alert user is sure to uncover in the course of working with a word processor.

With some programs, spelling can be monitored automatically as you type. Thus, whenever the computer is presented with an unfamiliar word, that word might immediately be highlighted (e.g., underlined in red), perhaps in conjunction with an acoustic warning, permitting elimination of most errors as soon as they are made. Moreover, you may find that certain common errors magically disappear without any intervention whatsoever on your part thanks to the vigilance of an *AutoCorrect* system.[50]

Search Operations

Somewhat related to spell-checking are a word processor's *search* and *search-and-replace* routines (commonly offered in page-layout programs as well). The active writer will soon recognize that these tools are among the most valuable ever devised.[51]

● The command "Find" initiates a search throughout the current document for the presence of a specified sequence of characters or "target character string".[52]

Issuing this command produces a special dialogue box in which one types the desired target string and then clicks on a "Find next" button. The search process is interrupted upon detection of the first occurrence of the target (working forward from the cursor location), and the screen display scrolls automatically so that this "find" (now highlighted) can be examined in context. Like all highlighted text, the latter is subject to immediate change if you so desire, or a search for the *next* case of the target can instead be initiated.

The "Find" (i.e., search) function can be made to serve a number of useful purposes. For example, you might have reason to ascertain whether or not (or how often, or in what contexts) a particular word appears in your document. Or you might be interested in identifying places where reference has been made to the work of some particular scholar, possibly to ensure that you have remembered to include all the appropriate *citations* (cf. Sec. 9.3). A search based on the word "Figure" or "Table" could help you detect incorrect or misplaced supporting elements with respect to an assigned numerical order, or references that point to the wrong things—a result perhaps of a major reorganization in which a necessary sequence change was overlooked. As a final example, you would be able to pursue a fairly comprehensive search for explicit mention of sulfur-containing compounds by targeting the word fragment "sulf". Here it would obviously be important that you *not* demand a "case match", by the way, because at least some legitimate hits might have the first letter capitalized. Similarly, you would not want your search to reflect the option requiring that the target ("sulf") constitute a "whole word", since you *expect* it to appear as a word *fragment*. Search

---

[50] Any *wrong* implementation of this correction feature, if noticed soon enough, can be cancelled with the "Undo AutoCorrect" command.

[51] This holds true in terms of other types of software too. It has frequently been noted, for example, that a database program is only as good as the search tools it includes.

[52] Be aware that when such a search is initiated from within a *footnote* the process may be confined exclusively to the document's footnotes, whereas a search commencing within the body of a document will be all-encompassing, including scrutiny of headers and footers.

routines can actually be customized in a variety of ways, even to the point of limiting the process exclusively to occurrences of the designated string that display specific format characteristics (font, type size, etc.).

The "Replace" command works very much like "Find", except that each time a "hit" is registered you are given the option of removing the target string and introducing instead some alternative you have specified in advance.

- The potential effectiveness of a "replace" strategy will of course depend to some extent upon how consistent you have been in your choice of words during composition.

On the other hand, "Find and Replace" can also be turned into a device for *increasing* consistency. If, for example, you suspect that you may have randomly utilized two different spellings for the same word, it is a simple matter to substitute the preferred orthography in all the incorrect cases. The same applies with respect to a concept you realize you have mistakenly assigned two different "names" (an instance of inappropriate use of synonyms in a technical setting).

"Replace" is very handy for dealing with more subtle problems as well. For example, it is not uncommon to introduce "extra" spaces inadvertently in the course of typing (i.e., *two* spaces between a pair of words rather than one). This cosmetic flaw is easily attended to by an operation that replaces every instance of "double space" with "single space".[53] Or you might decide after the fact to introduce a special *narrow space* every time a number is followed by a "percent sign", again easily accomplished with a properly managed "search and replace" operation. As suggested earlier, the "search" machinery can even be used to pinpoint distinctive *formats*: words that are *underlined*, for example, or all the *superscripts* in a manuscript. This could be helpful if you for some reason wished to replace every use of the font "Helvetica" with "Verdana".

It is usually better, incidentally, *not* to risk taking advantage of the option labeled "Replace all" with which you may be presented once a first example of the target has been located. Taking the time to look at each proposed substitution is far safer, because there may be alleged "hits" that really *shouldn't* fall in this category (a place where the target happens to be present as a *fragment* in some totally unrelated word, for example).

Most search (or search/replace) systems also permit the use of "wild cards". Thus, a "?" in a target string is typically interpreted as standing for *any* symbol, whereas the more selective "#" could represent "any integer". The latter would be useful, for example, in locating all entries in a reference list related to publications from the 1990s, which should be detectable in a search with "199#" as the target.[54]

---

[53] Actually, at least *two* passes should be made through your document in this case to take care of possible instances of *three* adjacent spaces—which on the first pass would have been transformed into double spaces.

[54] Note that in all likelihood the opening "1" here is actually superfluous.

As a last example, consider the potential advantage of deliberately introducing into your manuscript during composition specific "flags", sometimes called "blockades", at points where you know subsequent editing will be required. For example, if "xxx" were consistently employed to represent missing information—facts that need to be looked up in the library, perhaps—then one could locate all the corresponding "holes" quickly and reliably when the time came to construct a list of things to be sought in the literature. Indeed, you might even invent multiple "flags" to represent different *types* of missing information or data.

Remember, too, that

● Digital documents prepared with a word processor can be subjected to content searches long *after* their preparation.

Admittedly this has little to do with the subject of writing *per se*, but it serves as a reminder of how much easier it is with digital archives to retrieve something from your personal "information cupboard". There is considerable merit in preserving electronic documents of all types—reports, lectures, publications, etc.—in a secure place over prolonged periods of time. You can never tell when something you wrote a long time ago might suddenly become relevant again, and an electronic search is always far more reliable (and much less frustrating!) than shuffling through piles of paper manually.

Occasionally you may find reason to copy a text passage, or a reference list, for example, from an earlier document into one currently in preparation, in which case one is of course dealing with a central aspect of the principal theme of this book: "writing". If you can reasonably anticipate frequently finding yourself in such a situation, then you should seriously consider organizing your text files as subunits in a *database*. Indeed, the recent literature includes quite a few references to an even more elegant application of this concept: *database publishing*, sometimes referred to as *digiset publishing*.[55]

Editing Functions

We have already addressed ways in which a modern word processor can assist in documenting *modifications* experienced by a manuscript, under the subheading "Text" in Sec. 4.3.5. Thus, when you yourself have introduced changes, or if others have tried their hand at "polishing" your work, it can be useful later to examine a complete record of who has done what, and when. The account can be consulted either on the screen (where selective use of color helps provide clarity) or in the form of hardcopy printout.

The latest versions of programs like WORD provide other editorial features as well that can be of great value when multiple people are involved in work on the same

---

[55] This implies the administration and periodic updating of structured file content based on resources collected in a comprehensive database. Relevant data are identified and extracted using carefully established criteria, then compiled, and ultimately formatted, all at least semi-automatically, as a way of preparing newly updated publications (e.g., catalogues, transaction summaries, etc.); cf. Sec. 8.5.2.

document. For example, a document can be "password protected"—indeed, at two separate levels, such that one password opens the way only to *reading* a document whereas a different one authorizes the reader to make *changes*. Moreover, a reader can be empowered to introduce *comments* into a document, notations that are linked directly to specific sites and visible in a separate footnote-like window, again with identification of the source in each case. The author of course retains full responsibility for determining which suggested changes (if any) are actually adopted.

These particular features are especially valuable with respect to a document under development by a *team* of authors. If you are ever involved in the writing of such a work you should make it a point early in the process to ascertain what special editorial features your word processor supports and then devise an effective strategy for their effective utilization, ensuring as well that the proposed course of action is understood and backed by everyone concerned.

5.4 Digital Data and Electronic Publishing

5.4.1 The Digital or Electronic Manuscript

Basic Considerations

So far in this chapter we have addressed computer applications from the standpoint of manuscript preparation, but devoted little attention to how the resulting files are manipulated or utilized. The time has come to correct this deficiency, building upon groundwork laid in Sec. 5.1.

If the plan is to *publish* the content of an author's digital files, two general approaches are available. One is the "classical" route: the file in question is simply directed to a desktop printer, resulting in a *hardcopy* version of the document that a publishing house can subsequently transform into a professional product in the traditional ways. In this case the word processor will have served only as a tool for helping the author assemble a clean, attractive manuscript.

Alternatively, the publisher might choose to work directly with the digital material, in which case the only role a hardcopy version prepared at this stage would play is that of a *control* for ensuring that nothing goes amiss. This second scenario clearly takes far better advantage of the potential inherent in computer-assisted writing.

In what follows, whenever we refer to a *digital* or *electronic manuscript* we mean data files representing a document, prepared with the intent that the document be published, where the files themselves contain all the required text and subsidiary materials.

● The process of converting an electronic manuscript into a published document is an important aspect of what is known as *electronic publishing*.

The expression "electronic publishing" (or "EP") has become something of a talisman in the publishing industry despite its being subject to a host of diverse interpretations. For example, some understand EP to encompass all forms of information distribution by electronic means, and would thus include under this heading many of the activities associated with hosting a modern *database*. It is nevertheless not customary to link database purveyors and publishers in a common category despite the fact that the two now overlap to some extent.

On the other hand, the label "electronic publishing" is sometimes reserved exclusively for forms of publication in which information is not only processed and stored digitally, but also passed along to the user in a strictly digital form (via CD-ROM or the Internet, for example). Information distribution centered around the Internet is also closely associated with more specialized terms, like "Web publishing" or "online publishing".

Against this background, incorporating DTP tools into the process of preparing traditional printing forms would qualify as "electronic publishing" only in the broadest sense. Nevertheless, we will pursue our consideration of electronic publishing largely from the vantage point of hardcopy production. Despite the "Internet boom" (from which much of the early euphoria has, incidentally, evaporated), from the scientist's perspective most "publication" continues to revolve around print applied to paper. The book or traditional journal as a medium of exchange is far from dead, and it appears unlikely to succumb in the years or decades immediately ahead.

Digital data (whether presented on a CD or similar portable carrier or transported over the Internet) can be upgraded to a set of printing forms—avoiding the traditional typesetting process—in any of three ways, involving three types of manuscript text files. We are able to offer here only a cursory sketch of the alternatives, but urge the interested reader to consult more specialized literature for additional information on the subject.

● We cannot emphasize too strongly the importance of establishing and maintaining close author/publisher contact regarding exactly what will transpire between manuscript submission and public distribution, and of initiating the corresponding dialogue *before* manuscript preparation begins.

Many publishers provide explicit background materials describing the implications of each of the three fundamental production processes, including potential variants, and instructing authors on how to proceed.

1. The uncoded electronic manuscript: An electronic manuscript classes as "uncoded" if it consists exclusively of text, and is completely devoid of embedded control codes and formatting instructions (apart from the "line feed" or "new line" signal).

Such a manuscript is by definition based entirely on the letters, numerals, punctuation, and other symbols that comprise the ASCII character set (see "Printers" in Sec. 5.2.1). The alternative expression "text-only manuscript" is also frequently applied in this context. A manuscript prepared with word-processing software can be reduced to this primitive status from within a writing program by selecting one of the options "text only" or "text only with line breaks" under the "Save as ..." command. A considerable amount of information is of course discarded when a word-processing file is converted to "text-only" format, so one should resort to this expedient only if there is absolutely no possibility of utilizing the available data in a more productive way.[56]

● The ASCII-file approach is normally avoided with complicated scientific manuscripts, because too much effort would need to be expended in upgrading the data for publication.

For example, a text-only file provides no indication of where subscripts or superscripts might be required, or distinctive type-styles (italic, underlined, etc.), quite apart from the absence of format instructions related to such things as paragraph indentation or table layout. The only offsetting advantage is that a file of this type can serve as input for virtually any DTP or professional typesetting system, so it will almost certainly not be a direct source of processing problems. One might say that starting with an ASCII file represents a "lowest common denominator" application of an electronic manuscript, one upon which all programmers and software suppliers are content to compromise.

Formatting instructions must in this case eventually be introduced by hand into a printed copy of the crude document (manual markup). Such instructions will not be executed until after the text has been "read into" the publisher's computerized typesetting system. Initial refinement of this sort would normally be kept at a fairly basic level to minimize the demands on whoever is responsible for the processing (which is to say, for example, that no attempt would be made at this stage to impose a particular *layout*).

● With a text-only manuscript it is important that the author refrain from applying any end-of-line hyphenation (word breaks between syllables), and the "new-line" instruction should be issued only for signaling *paragraph* termination.

2. *The manually encoded electronic manuscript*: Sometimes type-style markings and other formatting instructions are actually embedded directly into what still classes as an ASCII electronic manuscript. This is accomplished by the insertion of standardized *codes*, which can later be interpreted and implemented automatically by a publisher's typesetting software.

Anything beyond what would be permitted in a true text-only manuscript (e.g., a specific request for italics, indication of a subscript, a margin adjustment, etc.) must be strictly consistent with an *agreed-upon* coding system (as described in detailed guidelines provided by the publisher's editorial office), and a given instruction must

---

[56] Before implementing this type of "save" be sure you have *also* saved the latest version of the *formatted* manuscript—under a different file name!—in keeping with the standards of your word processor, so that none of the carefully applied stylistic features are irretrievably lost.

be repeated each time it is to be implemented. Codes of this type are constructed entirely on the basis of standard ASCII symbols. Not surprisingly, the printout of a technically complex manuscript containing a host of cryptic encoded instructions would strike the uninitiated reader as gibberish.

To take but a single example, specifying that a particular character is to be set as a subscript requires that one give both "start" and "stop" commands for the subscripting process.[57] You can surely imagine the tangle of instructions that would accompany a complicated mathematical expression! A fully encoded manuscript prior to execution of the instructions can become extremely difficult for even a professional to proofread or edit. Nevertheless, certain computer-driven typesetting equipment thrives on such copy, transforming it flawlessly into elegant, polished documents. Just as with a text-only manuscript it is important here as well that the author scrupulously avoid end-of-line hyphenation, and coding must be restricted to explicitly approved instructions.

Adopting this second approach to preparing and refining a technically demanding electronic manuscript presents a far greater challenge than the average author is likely to anticipate. Since generating extra work is hardly the goal of "high-tech" innovation, a publishing scheme that presupposes availability of a manually encoded document is justified only in the case of a project that is structurally quite simple so the copy will require little adaptation.

3. *An electronic manuscript based on conventional word-processor files*: Most authors with access to a high-end word processor might expect—at least in principle—to be spared the nuisance of adding coded format instructions to a manuscript to explain type-style requirements, margin adjustments, etc.

After all, a simple keystroke command or a click of the mouse normally suffices to produce impressive results on a display screen and—presumably—it should be possible to carry these adjustments over automatically into the typesetting stage. Software has in this case turned wishes into reality through the medium of *hidden* (encoded) instructions. Unfortunately, flawless interpretation of such instructions by a publisher's typesetting equipment can seldom be guaranteed, and serious problems sometimes arise. The decisive factor is the degree of *compatibility* between the relevant typesetting and word-processing systems.

One acquisition editor at a publishing house addressed the issue in the following terms:

> Many of today's authors are equipped with first-rate computers and software, and they supply us with manuscripts that appear ready in every sense for publication. But no matter how promising a piece of copy may look on the surface, underneath could be lurking a tangled web of problems and a concealed requirement for extensive revision before publication is feasible. The potential debacle might reveal itself in-house, but sometimes it first becomes apparent on

---

[57] Thus, to achieve the formatting for "a_3" one might type something like "<I>a</I>₃".

the premises of a contract printer. Each new situation must be subjected to careful, individual attention before the decision is made to undertake direct processing of an author's (formatted) digital files. Electing to proceed means it will still be necessary to work out the most efficient and economical approach to introducing essential editorial changes, for example, and to establish the extent to which the author's extra efforts should be rewarded ... (ADAMSKI 1995)

This analysis continues:

A truly professional layout can only be achieved through the use of genuine layout software. Word-processor files invariably fail to give results that meet professional typographic standards. For example, it is commonplace with inferior data files to find special features like meticulously constructed tables simply disappearing. The slightest change in type size causes line breaks to be shifted to new places, with unpredictable consequences, and there is frequently a need to remove—by hand—scores of superfluous carriage returns, hyphens, tab commands, and blank spaces (the latter introduced in an author's crude attempt to simulate indentation) as a precursor to associating "styles" with the individual paragraphs.[58]

From personal experience we can report that, as recently as 2000, major scientific publishers were still issuing multi-volume treatises produced (*deliberately!*) on the basis of authors' manuscripts submitted exclusively in hardcopy form—for precisely the reasons suggested above.

Many authors are astonished to learn that a professional publisher cannot successfully accomplish something they themselves are able to do so readily (most of the time) at their own desks—that is, faithfully reproduce on paper an image as it appears on a screen. They overlook the fact that, unlike a publisher's equipment, their personal printers were designed from the outset to be driven by word-processing software. As a matter of fact, it is authors who are to blame for many of the problems that arise, in that they tend to employ multiple techniques (e.g., tab stops vs. several blank spaces) in an effort to achieve (*approximately!*) a single effect—indentation for example. Under these circumstances data-conversion problems with professional typesetting systems are almost inevitable.

On the other hand, state-of-the-art is now such that a POSTSCRIPT file properly prepared for a POSTSCRIPT-compatible laser printer can in fact be successfully processed with a laser imagesetter as well, almost as a matter of routine and at a resolution as high as 2500 dpi, comparable to output from a CRT imagesetter. Complete compatibility between files from Apple MACINTOSH computers and laser imagesetters from Linotype had in fact already been achieved by the late 1980s in the context of the pioneering

---

[58] Some publishing houses refuse even to consider a manuscript as a candidate for direct processing unless it is submitted in the form of PAGEMAKER or XPRESS files, with all illustrations—especially those prepared as vector-graphics—already integrated into the text, thereby eliminating any need for mounting ("stripping in") films subsequent to imagesetter treatment.

page-layout program PAGEMAKER (Aldus). The same applies in the meantime with respect to a wide variety of other hardware/software combinations.

In recent years a considerable amount of effort has been invested by printing establishments in what is known as "PDF workflow". This involves transforming submitted POSTSCRIPT files into the much more compact *PDF files* (Portable Document Format, conceived like POSTSCRIPT by Adobe), a conversion that can be accomplished with absolutely no loss of information. Relatively small size is of course one major advantage of what might otherwise seem a pointless detour, but PDF files offer another important benefit too: they are subject to refinement and error elimination up to the very last minute. An "error" in this sense need not be a manifestation of a content problem: more common is recognition that an inappropriate type font has been used, for example, or the wrong color system applied [e.g., RGB (red–green–blue) instead of CMYK (cyan–magenta–yellow–black); cf. Sec. 7.4], or that a line has been made so narrow that it breaks up when printed. The advent of PDF workflow has in many ways revolutionized the printing industry. Periodical and book publishers have enthusiastically embraced innovative approaches like this to product development, and they are cautiously receptive to a variety of constructive initiatives on the part of their authors. Indeed, this book itself is an interesting example of what a publisher can today accomplish—at least sometimes—with an author's digital files (see also footnote 65 below).

● Despite a great deal of progress, however, compatibility problems can still complicate the path linking an author's PC with a high-performance typesetting system.

The principal stumbling blocks continue to be insufficient standardization in the areas of data technology and telecommunication, lack of consensus between hardware and software vendors, and fundamental differences between authors' computers and those that drive the equipment in professional printing establishments.

Taking full advantage of the potential inherent in DTP is unfortunately not a simple matter. Publishers have on occasion expressed serious misgivings about the rapid development of word-processing tools, because progress along this front seems always to bring with it new compatibility problems. We urge anyone seriously considering experimenting with DTP manuscript preparation to devote special attention to the "20 Commandments of Electronic Manuscripts" presented in Appendix C, treating the suggestions there with the same respect due a set of instructions accompanying a dieting regimen.

File Formats

Programs for the manipulation of text and graphics generally support a number of *format options* for data storage and exchange. As an author you should definitely familiarize yourself with the particular file-exchange option known as RTF ("Rich Text Format"), which incorporates into an ASCI-type data file a complete set of ASCII based output-format instructions. The system has been accorded widespread recognition, and is readily interpreted by software of many types, with the RTF instructions being transformed into formatting commands of the type ordinarily employed by the program in question. RTF is extremely useful for converting files created by one piece of software into ones more suitable for another, as well as for exchanging files across dissimilar operating systems (e.g., to a WINDOWS environment from the MACINTOSH world, and vice versa). RTF files are also the perfect vehicle for manuscript transmission via the Internet (as e-mail attachments, for example), since they contain none of the special (non-ASCII) characters otherwise prevalent in native word-processor files, characters that very often lead to hopelessly garbled messages.

Authors rarely have occasion to utilize other alternative file-storage formats, or to work with special software designed exclusively for interconverting files among formats native to various programs. Indeed, few even have *access* to the latter. For this reason it is especially important that publishing houses equip themselves with at least one work station offering a sophisticated set of data interconversion facilities supervised by someone both knowledgeable and capable of dealing with unexpected situations and difficult challenges.

● *Conversion* (or interconversion) in this context is best described as the translation of digital data from one program format into another.

Conversion is accomplished most often with the aid of special program modules known as *converters* or *filters*. To understand how they function it is necessary to recognize that every symbol resident in a data file or a computer memory is present in encoded form as an eight-digit *binary number*. Thus, the capital "B" has been assigned the ASCII code number 66, so it is stored and processed as the equivalent binary expression: 01000010 ($2^6 + 2^1$). As previously noted, ASCII values above 127 are subject to various meanings, so sometimes an "interpreter" may be required at the level of symbols themselves. But this is only the beginning. A true "language" consists not only of letter symbols that combine to form words, but also rules addressing grammar and syntax, and "computer languages" are no exception. In other words, a word processor must employ systematic, grammatically sound provisions for the unambiguous internal coding of such things as type-styles (e.g., italic, boldface, superscript, etc.), font preferences, and tab settings, to cite but a few examples, all expressed in terms of discrete *number combinations*. As you might anticipate, different word processors and different file formats utilize their own distinct languages—and even dialects—and it is this that creates the need for interconversion: i.e., *translation*.[59]

Assuming the availability of an appropriate *import* or *export filter*, all the encoded information present in a "foreign" file can be adjusted on the basis of conversion tables

---

[59] Since in essence the translation process consists only of manipulating numbers, computers are able to execute it very efficiently, typically translating text at a rate of 50 pages or so per minute. The only absolute requirement is that the computer be fully conversant with the "vocabularies" and "grammars" characterizing the respective "languages". The overall challenge can in principle be met by developing a specific converter to deal with each of the possible combinations [program A ↔ program B] or [file format X ↔ file format Y].

until it is consistent with a particular program's own storage format. Export filters have been described as

> … interpreters typically [supplied as] integral modular components of specific PC text or layout programs. In the course of saving data to a diskette or other storage medium in the format of another program a user explicitly activates the particular interpreting module of interest. Since virtually every program—indeed every *version* of a program—is associated with its own unique data-storage format, a great many filters are required to ensure that every potential recipient of a data file will be in a position to process it accurately. (*Fachwörter-Lexikon*, PLENZ 1995 f)

One important prerequisite for successful communication between two parties at this level, however, is a high level of computer competence at both ends of the exchange.

An example of a *dedicated* conversion program is CONVERSIONS PLUS from Data Viz, Inc., which is capable of dealing not only with various *text formats* (e.g., MACINTOSH WORD → WINDOWS WORD) but also a wide range of *graphic formats* (e.g. TIFF → GIF). A software package like this enables a PC user to interpret MACINTOSH diskettes, compress and decompress graphics files, and translate a broad assortment of text files, including ones containing "embedded graphics"—either with or without extra attention to the graphic elements themselves.

Assuming the appropriate filters have been installed, every state-of-the-art word processor is today able to work relatively freely with data files prepared using other mainstream software, and also to save text in forms compatible with alternative programs. Thus, recent versions of WORD can preserve an author's efforts in a text-only state (with or without line breaks), in rich-text format (RTF), or in files compatible with EXCEL, WORDPERFECT, or MICROSOFT WORKS. Even POSTSCRIPT files can be subjected to transformation—into RTF data, for example—perhaps in the context of PDF workflow.[60] Often the only thing required to accomplish efficient file conversion is a certain amount of technical "know-how". The essential machinery is frequently inherent in a computer's operating system, or in some conveniently available application program. For example, a WORD file created on a MACINTOSH computer with WORD 98 or WORD X opens with no problem whatsoever on a WINDOWS PC running WORD 97 or later, and all formatting is preserved intact. One potential stumbling block could be the absence of an unusual type font the author employed, but this of course has noth-

---

[60] POSTSCRIPT is not actually an appropriate topic for consideration here since it is a language whose purpose extends beyond mere formatting (cf. "Printers" in Sec. 5.2.1). In fact, the expression of POST-SCRIPT data in RTF form is far from trivial, and represents from a programming point of view a challenge significantly greater than, for example, converting WORD files into RTF. If such a transformation is to be undertaken at all, it normally proceeds via PDF; indeed, a filter for this purpose is included in Version 5 of ACROBAT, although with Version 4 it was still necessary to utilize an expensive "plug-in". There would be little reason to attempt the reverse transformation (from RTF to POSTSCRIPT) because RTF is purely a vehicle for *format exchange*. A "transformation into POSTSCRIPT" would necessarily require that the file first be tailored to suit a particular word processor or page-layout program.

ing to do with conversion *per se*. It simply means the missing font must be acquired and installed.

Transfer of a word-processor file into a page-layout environment also proceeds quite smoothly in most cases, although certain "tricks" may facilitate the procedure. Even formulas, tables, and graphics embedded in a WORD file are correctly incorporated into a FRAMEMAKER project, for example. A similar level of success is attainable with the combination QUARKXPRESS/WORD, the leading programs in their respective fields, but in this case certain aspects are less than intuitive, and close study of online help files is required, preferably in conjunction with a good reference guide (see also Appendix D).

Structured Markup Systems

Efforts were initiated many years ago to develop a truly universal code for adding markup information to a text file. The most important outcome so far has been SGML, short for Standard Generalized Markup Language (for details see the standard ISO 8879-1986; cf. also Sec. 5.2.2),[61] which has been used to facilitate a wide range of publishing activities, including successful presentation of entire encyclopedias via CD-ROM.

There have more recently been at least two other major developments along similar lines. One is intimately connected with the Internet: HTML, Hypertext Markup Language, a universal document-description language compatible with all computer operating systems (WINDOWS, UNIX, MACINTOSH, etc.). HTML was developed in the late 1980s at CERN in Geneva as a giant step toward what eventually evolved into World Wide Web (WWW) pages. HTML is at least vaguely familiar to anyone who has prepared a personal homepage for the Web. The sole purpose of the language is to facilitate Web-page design and display. An HTML document might be described as an ASCII text file augmented with (often a great many!) embedded format-related specifications (tags) derived from a limited set of narrowly defined categories. All are expressed in coded form and set in angle brackets (< >).

HTML files are sometimes prepared with the aid of a Web browser (e.g., NAVIGATOR from Netscape or INTERNET EXPLORER from Microsoft), but a word processor can serve much the same purpose. In fact, anyone can easily create a very basic Web page without bothering to learn anything at all about HTML. A standard document is simply prepared in the usual way with WORD, for example, and then saved to a special file by selecting the format option "HTML" or "Web page" from the "Save as …" menu . What results is automatic incorporation of all the HTML format codes that would be

---

[61] SGML, introduced in Sec. 3.1.3, "Markup" of an Electronic Document, provides a system of coded instructions for conveying sophisticated *metainformation* (cf. Sec. 3.4.2) regarding a text document. The focus is almost exclusively on *organizational* matters as opposed to typesetting details. Thus, SGML would make a clear distinction between figure captions and table headings even if one's immediate intent were to format both in precisely the same way. (The definitive reference is GOLDFARB 1990. Also recommended are JELLIFFE 1998, BRADLEY 1996, and MALER and ANDALOUSSI 1996.)

required for posting that particular document on the Web: <p>, for example, to signal a new paragraph (¶), or floor to indicate a point in the document where the word **floor** is to appear set in boldface type. Admittedly this minimalist approach to Web-page generation is extremely limited in the extent to which it permits one to achieve "elegance" or special effects, but it does eliminate the need to master a host of specialized and arcane HTML commands.

In addition to text, Web pages can of course also incorporate other elements, such as lines, special background effects (a distinctive color, for example, or what is referred to as a "watermark"), and the wide variety of hyperlinks for which the Internet is famous (see "Further Ramifications ... " in Sec. 3.1.2). Other possibilities include tables, graphics of all sorts, elements that introduce "motion", and data of a "dynamic" nature (e.g., research results drawn from the most current version of one's personal database).

You are unlikely ever to refer to HTML guidelines in the context of preparing an "electronic manuscript" for a publisher—a good thing, because a *publisher's* guidelines are quite enough to deal with! On the other hand, should you wish to try your hand at developing a flashy personal homepage, one that combines an array of colorful features with a professional presentation of more formal material—a description of your research interests, for example, perhaps accompanied by a sample of recent findings—then you would be well advised to invest some time directly in the official "Web language" (the most recent version!) and in learning about techniques others have employed for establishing an effective "Web presence". Compact sources of relevant information include NIEDERST 2001 and WILLARD 2001.

HTML is often described as a subset of the medium-independent SGML document-description language. This is in a sense accurate, but also somewhat misleading. HTML certainly was developed on the *basis* of SGML, but unlike SGML its purpose is not fundamentally that of description in general; rather, it provides a particular *kind* of description, one that a Web browser can translate into formatting information for screen display of a document. In other words, it is much more directly linked to the *appearance* of a document than is SGML, although the displays produced by different Web browsers may vary somewhat. SGML, on the other hand, is completely divorced from how a document looks, and also from the medium of expression, because it deals with formatting only in a *logical* sense. At the same time, SGML is sufficiently flexible to permit the rigorous encoding of essentially any document, although this may entail a fair amount of programming effort in a complex case. If one wished to publish an SGML document on the World Wide Web there would first need to be extensive modification, with the removal of many SGML tags or their replacement by ones more in keeping with HTML. While HTML specifications sometimes end up being combined in rather complex ways, when considered as a language, SGML is far more flexible. A programmer would describe HTML as a "frozen" language, with a vocabulary suited only to screen-display applications, primarily in conjunction with the World Wide Web.

Some years ago it was recognized that there would be merit in trying to couple the rigor of SGML with the ease of use of HTML, a visionary challenge that led to *XML*,

Extensible Markup Language. XML is in fact *the* latest standard for device-neutral file formatting in general.[61a] Virtually every purveyor of word-processing or page-layout software is actively engaged in developing and promoting products with XML-savvy, and much has been accomplished in this respect. The current versions of QUARKXPRESS and FRAMEMAKER, for example, offer the alternative of treating new documents in either of two ways: as "normal" or as "structured" files, where the latter implies that XML conventions are respected.

The engaged reader will recognize that already in this section we have touched on a number of technical issues with far-reaching consequences for the prospective author, so it is perhaps appropriate for us to stress once again the fact that *early* meaningful contact between author and publisher is more critical today than ever before. Sample data files play an extremely important role in forecasting and (hopefully) avoiding problems, especially ones with the potential to involve third parties (outside firms supplying imagesetter services, for example, or contract printers in the case of direct-to-plate production).

The latest technology is being adopted in the publishing industry at an increasingly rapid rate despite the fact that it has proven very difficult to implement in the case of multi-author works, handbooks, and journals. In large measure this is because the requisite close consultation among all those involved is virtually impossible to establish. The serious publisher's only recourse has been to develop and circulate exceptionally comprehensive guidelines, written so transparently that they should be readily inter-preted and flawlessly implemented—*assuming* every author makes a conscientious effort to study the material carefully. Multiple sets of guidelines may be required by an editorial office dealing with various service bureaus.

Special Considerations Applicable to Electronic Manuscripts that Require Typesetting

This section was originally entitled "Diskette Files Destined for Typesetting", but the heading has been modified in recognition of the fact that in most cases data files are now transported from place to place not by diskette but as e-mail attachments.

We have already noted that if an electronic manuscript is to serve as the basis for a typeset publication, end-of-line hyphenation[62] is a nicety one must absolutely forego, because at least some line breaks are bound to change during processing, potentially leading to words improperly printed in hyphenated form. There is one exception to the broad generalization. It *is* permissible to introduce for this purpose "soft" or "conditional" hyphens that appear in print only when a word actually must be divided as the result of a line break.

[61a] A good introductory guide is XML by Example (MARCHAL 1999). Much more thorough are MEGGINSON 1998 and the encyclopedic GOLDFARB and PRESCOD 2002.

[62] "End-of-line hyphenation": The breaking of terminal long words into syllables, practiced in the interest of a more efficient or a more aesthetically pleasing layout.

A normal hyphen (-) is of course created with the key bearing the corresponding label, ordinarily the second one from the right in the top row of character keys on the keyboard. A "soft hyphen" can be produced (in WORD, for example) by holding down an additional control key when the hyphen key is depressed, as explained under the "Special Characters" tab on the "Insert Symbol …" dialogue box. For editing purposes the WORD user has the option of specifying either that soft hyphens *always* be displayed on the screen (in a faint, ghost-like way), even though they would perhaps not be printed, or that they should be visible only when an end-of-line location makes appearance of a hyphen there mandatory.

There exists yet a third type of hyphen used occasionally in an electronic manuscript: the "nonbreaking hyphen", employed for the hyphenated linkage of two elements that should always remain together. An example is an expression like "25-fold", which is understood to represent a single entity, so it should not be subjected to the fragmentation that an ordinary hyphenated expression might incur at the end of a rather tight line. This type of hyphen can of course *never* itself terminate a line. A nonbreaking hyphen is created in much the same way as a "soft hyphen", but with a different special keystroke combination.

We have noted that routine screen display of soft hyphens is optional, but if one does choose to see them, then additional otherwise invisible symbols will be revealed as well. These include the *paragraph mark* (¶) signaling that a new paragraph is about to begin[63] [cf. "Marking (Highlighting)" in Sec. 5.3.1], as well as small, elevated dots that represent blank spaces.

● The conscientious author should be as acutely aware of the *metainformation* present in a document as of the text itself.

Unlike so-called "hidden text", which despite being concealed is still "text" in the usual sense, the symbols we are discussing here are better described as *control characters* related to text *output*. In common with ordinary symbols, such "characters" are subject to deletion or even copying, but under no circumstances are they ever printed. On the screen they look gray when displayed, not black like true text characters, and each is in some way associated with a subsidiary level of information underlying the document. Other symbols representing metainformation—and therefore displayed only upon request—include the "hard space" (a blank space whose width is fixed and thus not subject to change when text lines are justified; see below), and explicit "new line" and "new page" signals. The *new line* command is sometimes used, for example, to emphasize a free-standing equation set in the midst of a paragraph. The *new page* command makes it possible among other things to ensure that a heading ushering in a new section will become the first element on a page. The same device might also be invoked to prevent a long paragraph from being split by a page break.

---

[63] Recall that this mark is also associated with a considerable amount of background information (or *metainformation*) summarizing key aspects of the layout of the preceding paragraph.

A "new line" command should *not* be used to increase the space between two ordinary paragraphs. Actually, a full line would be too *much* space to add for this purpose, and the desired effect is in any case better achieved by establishing as one of the *paragraph style* parameters that paragraphs of the type in question are always to be preceded (or followed) by *xx* points of free space. The latter approach has the distinct advantage of enforcing uniform spacing with comparable paragraphs throughout a document, and it requires no special attention whatsoever on the part of the author or typesetter. Moreover, a single change applied in the context of style definition can alter spacing parameters of a host of paragraphs simultaneously.

- As a general rule, judicious application of "styles" will contribute significantly to internal consistency in a printed document, at the same time greatly simplifying the task of typesetting the electronic version.

The amount of space separating certain *characters* in scientific text can also be important. For example, the combination of a *number* and a *unit*, like "5 cm", should be kept intact even if by chance it falls at the end of a line. The pairing might of course be broken if the two parts were linked by a normal blank space. Moreover, the *width* of the space employed should be kept the same everywhere. Both requirements are easily met by utilizing a *fixed* (or *nonbreaking* or *hard*) *space*, typically introduced by a modification of the standard keystroke command for "space", in which the spacebar and some special key are activated simultaneously. Similar considerations apply to the spaces between last names and their associated initials (or abbreviated titles), as in "W. Schmidt" or "Dr. Jones." The width of a fixed space is also immune to changes otherwise mandated by automatic line justification.[64]

- The nonbreaking space plays a particularly important role in the construction of typeset *mathematical expressions* and *formulas*.

In certain instances, for example, such expressions are expected to incorporate (non-breaking) *narrow* spaces, as when you wish to indicate the *multiplication* of two or more quantities (e.g., "*a b*"). A similar narrow space is sometimes called for in the formulation of percentage information, like "6 %". The proper effect is obtained by assigning to an ordinary nonbreaking space an appropriately reduced *type size*.

There are in fact many special considerations involved in the typesetting of mathematical copy, a subject dealt with at some length in Sec. 6.5.6.

Appendix C provides a collection of specific recommendations for authors wishing to prepare highly refined electronic manuscripts for publication (based on suggestions in GREULICH and PLENZ 1997).

---

[64] One of the most important strategies applied in producing fully justified text is systematic expansion or contraction of the spaces between words.

5.4.2 Electronic Editing

Desktop Publishing

The term "desktop publishing" (often shortened to "DTP") first became fashionable in the 1980s.

● Desktop publishing implies preparation at one's own workstation (or desk) of copy suitable—just as it stands—for reproduction and widespread distribution.

Anyone with access to the appropriate tools—a word processor, of course, but also page-layout and graphics software, together with a high-quality printer—can in principle prepare publication-quality documents, either as standard hardcopy or in an electronic state.

Actually, the term desktop *publishing* is misleading, because it is not the literal act of "publishing" that is at issue here, but rather the preparation of material *worthy* of publication, at least from a typographic standpoint. Publishing in the full sense of the word involves much more than an array of technical transformations, as we have shown in Chapters 3 and 4. For this reason, some have suggested that "DTP" might better be interpreted as an acronym for "desktop *processing*", and we are inclined to concur. Occasionally one finds it instead treated as "desktop *printing*", but that, too, is clearly inappropriate.

● Success with a DTP project presupposes not only considerable familiarity with word-processing, graphics-editing, and page-layout software, but also a solid command of the art of *typography* and skill at simulating "camera-ready copy" on a PC screen.

Page makeup at a professional level can quite literally be regarded as an "art" in the aesthetic sense of the word, and it is only possible in a desktop setting if one has access to sophisticated facilities. Important too is the ability to visualize on the basis of a screen image precisely what a corresponding *printed* page will look like, in the process factoring in certain predictable differences between paper output and an electronic representation (deviations from the "WYSIWYG" principle; cf. "Printers" in Sec. 5.2.1).

The transformation of copy (from which all errors have already been eliminated) into a well-designed document requires patience, imagination, and a substantial amount of experience. Obtaining and maintaining the requisite skills in turn entails extensive practice, and on a regular basis. The challenge is rather like that of mastering a musical instrument and sustaining over a period of many years the ability to deliver virtuoso performances on demand.

DTP now plays an important ongoing role in most advertising agencies and in the product-information departments of many corporations—especially where laser-typesetting equipment is available—but relatively few author/scientists engage in it in a serious way. One explanation is that the challenges a publicity specialist faces are fundamentally different from those confronting an author. Advertising brochures tend to be documents of very modest dimension, and design characteristics play a crucial part in their success. With a scholarly book, on the other hand, it is the information content that is dominant, and the volume and complexity of the corresponding information can be quite formidable indeed. Design parameters are significant, but very much secondary. This is one of the reasons we are convinced that the division of labor familiar from traditional book production is likely to persist into the foreseeable future.

Put slightly differently:

● The primary responsibility of the author of a scientific book should be *providing* information, not *processing* it.

Especially with a long document, publication professionals should continue to be the ones who arrange for the *presentation* of material that has been *assembled* by authors, and we see little to be gained from promoting widespread attention to DTP techniques. Sophisticated desktop publishing on the part of scientists will almost surely continue to be limited to unusual situations and exceptional circumstances.

At the same time we would not want to discourage the DTP aspirations of those few authors who happen to have acquired the necessary skills, and who find engaging in such activities to be a source of satisfaction. Certainly no *harm* will flow from presenting an open-minded publisher with professional-quality copy. Moreover, even when preparing reports, research proposals, and other "minor" documents there is reason to strive for attractive results worthy of a professional. We pointed out earlier (cf. Sections 3.2.3 and 3.4.6) that extra effort on the part of authors (and of course the availability of modern writing tools) is why the "camera-ready" contributions in today's "letter journals" might almost be mistaken for traditional print publications, and are a far cry from the haphazard and crude-looking submissions of a decade or so ago.

The author with serious interest in the creative potential inherent in desktop publishing should begin by studying one or more specialized treatises on the subject. Unfortunately, however, most books with the *words* "Desktop Publishing" in their titles are the work of people with backgrounds in computer applications or some phase of public relations. The authors generally are addressing themselves to marketers, or personnel charged with preparing corporate literature, product descriptions, operating instructions, user manuals, and the like. One exception is the book by RUSSEY, BLIEFERT, and VILLAIN (1995), directed specifically at typographic and technical issues facing authors of scientific text, especially researchers contemplating the preparation of camera-ready copy.

A few years ago PLENZ (1995) posed some incisive questions relevant in the present context:

Ten years after introduction of the first MACINTOSH computers, and roughly seven years after DTP began to earn a measure of respectability, various professional periodicals are showing concern for assessing the movement's impact. What professions can be said to have profited as a consequence of the revolutionary developments of the last few years, and which have suffered? Has DTP yet established itself as a part of the "direct line" connecting authors, publishers, and typesetters?

Clearly *authors* class among the winners. As their skill at generating publication-quality copy has increased many have come to experience an increased sense of freedom. *Publishers* can be seen as winners, too, at least to the extent they have been willing to play along.

On the other hand, *typesetting establishments* in the literal sense of the term are now almost a thing of the past. Those few that remain have survived only by evolving into purveyors of file-conversion and scanning services, limiting their scope almost entirely to work at the "pre-press" stage—which itself can be regarded as an offspring of PDF workflow, a new staple of the publishing industry. The *reprographic* sector, in which copy is prepared for offset printing by photographic means, has also witnessed profound change, as bulky and expensive processing equipment has been largely replaced by compact and efficient scanners (including color scanners) representing perhaps one-tenth of the traditional capital investment but—in conjunction with the appropriate software—accomplishing considerably more.

One source to which we are heavily indebted (GREULICH and PLENZ 1997, p. 10) is quite categorical in delineating the appropriate relationship between authors and publishers: "The optimal route [to publication in the future] surely will not be one in which an author's responsibilities extend beyond matters of content to encompass typesetting as well." A brief document entitled *The Manuscript on Diskette* stresses a corollary to this assertion: publishing houses should not press authors to take upon themselves duties formerly reserved for typesetting and page-layout specialists, nor should authors for their part *aspire* to assume too heavy an assignment in this direction. An especially sharp distinction is here drawn between *content* on one hand, which the authors recognize will probably have profited from refinement through word processing, and *layout*. Most of their attention is in fact directed toward layout concerns, and in the course of their presentation they make the cogent observation that programs like PAGEMAKER or QUARKXPRESS are among the most powerful pieces of software in all of data processing. It clearly is not reasonable to expect authors in general to master such specialized and complex tools, or even contemplate doing so.[65]

Strategic Considerations

The advances we have been describing have encouraged authors to become—to a degree—their own typesetters, and thus edged them closer to the printing side of publication. This raises the question of where the publisher's editorial office fits into

---

[65] As always, there are exceptions, of course—and this particular book is a case in point. The very first edition, from 1987, was prepared in its entirety by DTP techniques. We know a number of other scientists who have contributed significantly to the design and layout of their published works, even to the point of preparing single-handedly all the illustrative material for an entire textbook.

the new constellation, or the production department. Will they in the future continue to enjoy their historical status as indispensable intermediaries?

It is quite unlikely that technological advances will make *acquisition specialists* and *editors* redundant, because their efforts have a direct impact on document *content*, a realm in which machines can never assume the upper hand. At the same time it is clear that the tools of the trade, and the nature of the workflow, will necessarily continue to undergo extensive change in the wake of technical developments. The implications are even more profound with respect to the *production staff*, who will find themselves increasingly restricted to organizational and advisory activities (cf. PLENZ 1995).

● One interesting outcome of technological developments in writing and typesetting from a publisher's standpoint has been the *electronic editorial office* (EEO).

The principal new challenge facing such an editorial office is ensuring that authors' manuscripts, submitted in digital form, remain in an "electronic state" throughout processing (cf. Sec. 3.5.4).

● The goal is thus to transform incoming digital copy into publications without the content ever leaving the realm of "bits and bytes".

It is already clear, in a general way, how this is to be accomplished. First of all, the editorial office must *itself* be equipped with appropriate computer-based technology for displaying and electronically modifying virtually any manuscript submitted, subsequently passing it along to others in an optimal form for further processing. Only a few short years ago the "editor of the future"—what SCHNEIDER (1989) has referred to as an "editronicer"—was envisioned as acquiring almost all contributions through long-distance data transfer, subsequently examining the submissions on a monitor, introducing whatever changes seemed appropriate, and forwarding the edited material electronically to a printing facility. There the document would be transferred by laser beam or cathode-ray tube to film—or even directly to printing plates—for immediate publication. This "vision" describes in a crude way what has become everyday reality in many editorial offices, although it is usually not the editor who arranges for the preparation of printing plates. Instead, someone charged with pre-press processing verifies the suitability of an editorial department's POSTSCRIPT or PDF files prior to ordering that printing commence.

● The main obstacle to full exploitation of state-of-the-art technology at every level continues to be lack of *compatibility*, complicated by thorny *organizational issues*.

Three technical systems need to fit well together if the entire process is to function smoothly: those of the author, the printing plant, and the editorial office. Here lies the stumbling block impeding full implementation of a workflow pattern like that sketched above: too much "reformatting" and verification is still required. Nevertheless, everyone assumes that the critical standardization shortcomings will ultimately be resolved.

A more profound set of issues relates to workflow, including ergonomic considerations. There are thus incontrovertible physiological reasons why, for example, it is unrealistic to expect someone to devote hours of uninterrupted editing to a computer screen. The *intellectual* aspect of editing—which entails among other things making a great many comparisons while striving to acquire and maintain a sense of perspective—is compatible only to a limited extent with processing copy at a video terminal.

It is important to stress that so far we have presented only a very abbreviated version of a complete editorial operation. Where, for example, does the work of reviewers fit into the picture, above all in the case of a scholarly journal? And to what extent might freelance personnel have a useful role to play?

● Clearly methodologies of the past will continue to have a significant influence on the publishing world of tomorrow, even in the face of unprecedented technological change.

Both the software and the hardware required to support all aspects of a comprehensive electronic editorial system have existed for some time: starting with composition, continuing through import and conversion (where necessary) of foreign data files, incorporation of typewritten or printed material via OCR (see below), and graphics management and processing, and culminating in output with some type of image processor. An important supporting factor is the potential to transmit instantly not only text but also illustrations and other graphic elements (e.g., chemical structural formulas, cf. Sec. 7.2.5)—in virtually any graphic format—anywhere in the world via the Internet and satellites. Universal adoption of all aspects of the latest methodology is only a matter of time, as shown by the way newspaper editors and advertising agencies have already become prime beneficiaries of the new technology.

Meanwhile, however, the editorial departments of most major journals manage as best they can to bridge the remaining procedural gaps, especially those attributable to software incompatibility, in part by minimizing their dependency on specific software packages employed by their authors—through frequent application of *scanning* technology. The OCR approach makes it practical for an editor to introduce a manuscript available only in printed form directly into a firm's data-processing system where everything can then proceed along a common path toward publication. The chief prerequisite to success with this "work-around" is clean copy, preferably originals prepared on high-quality paper. Quoting PLENZ 1995 once again, from the entry "OCR":

> OCR software undertakes to compare characters a system detects (and stores in bitmap form) with available libraries of alphabetic images. A number of programs of this type boast at least a limited amount of "learning capability", and the best are able to deal even with faxed manuscripts, which are thereby converted into digital text suitable for editing with any standard word-processing system.

Rarely encountered type fonts as well as manuscripts that contain unusual symbols require that the OCR process include a "training phase", during which improperly inter-preted characters are corrected manually, establishing a reservoir of experience that facilitates more automatic operation later.

When *proofreading* a scanned document one should pay especially close attention to specific symbols that are notorious for being easily confused: "l", "I", and "1" can be very difficult to distinguish, for example, as can "O" and "0" (cf. "The Art of Proofreading" in Sec. 3.5.5). *Italic typefaces* generally cause more problems than roman faces (just as they do for the human eye/brain combination!), leading to higher error rates and slower progress. Nevertheless, OCR technology has come a long way in a few short years, and common typefaces are now handled quite routinely. In favorable cases even tables are correctly interpreted and appropriately formatted.[66]

The author proposing to submit for publication only the *printed* version of a manuscript, which will ultimately be scanned, should take pains to provide an exceptionally clean copy based on a very familiar and legible typeface. Random flyspecks on the paper are apt to be interpreted initially as part of the "information content", complicating the procedure unnecessarily. It is usually best, in fact, to supplement a hardcopy manuscript with a digital version of the material, either on a removable medium or sent via the Internet, even if the official *Instructions for Authors* fail to include this as a requirement.

OCR technology is yet another development pointing toward the long forecast arrival of the "paperless office" (cf. "Mouse Techniques" in Sec. 5.3.1). It also demonstrates that electronic document management is not necessarily restricted to outgoing material, but can encompass incoming items as well. Many authors are taking an OCR approach to assembling their personal electronic (full-text) *libraries*, for instance.

By no means underestimate the amount of effort implicit in realizing the elusive "paperless office", however. Scanning and cataloguing a huge backlog of documents is a daunting prospect. Furthermore, the resulting graphic files will occupy a great deal of (digital) storage space, requiring difficult decisions regarding where (and for how long) files of this type are to be preserved. And until the files have been *interpreted* with OCR software it will of course be impossible to access the corresponding information in terms of a convenient "character-string search", so another time-consuming— and error-prone—step lies ahead. Finally, there are complex issues to be confronted regarding the best approach to "document management".

Despite enormous improvement in scanning and OCR technology, we remain skeptical that this avenue represents the long-term solution to problems of electronic publishing. Ordinary scanning simply entails too much manual labor to be attractive. Moreover, the careful scanning of a sheaf of pages takes considerable time, and "time is …" (you know where *that* leads!). In situations involving a combination of text and high-resolution graphics, scan times can be on the order of *minutes* per page. The

---

[66] Error rates with the best OCR programs approximate 1% or less—but that still could lead to as many as 20 or so errors per page! Some professionals contend that manual "typing" still represents a more efficient route to text capture, at least in complex cases, and there is considerable incentive for software developers to continue improving OCR results. One can only hope that the software *prices* will not increase exponentially as the error rate declines further.

probability that text and graphics will need to be dealt with separately also sets limits on the probable level of enthusiasm with respect to scanning efficiency.

We suspect computer processing of material combining text and illustrations will in fact move off in a rather different direction. The IPHOTO utility software that in 2002 became a part of the Apple IMAC computer package (a MACINTOSH hardware line first introduced in 1998) suggests one plausible scenario. What every amateur photographer can now accomplish with a *digital camera* aimed toward palm trees lining a tropical beach (an image conjured up by the icon used to represent IPHOTO in the "dock" on a MAC's desktop)—namely uncomplicated capture of "electronic photos" subject to later computer editing—has the potential to play a valuable and exciting role in the professional world of the scientist as well. Digitized images lend themselves to straightforward incorporation into word-processor documents, for example, which opens a number of promising doors. Indeed, we believe the digital camera will in the very near future become one of the scientific author's standard tools, offering significant advantages over a scanner, not the least being speed.[67]

It is easy to picture an EEO someday being based on a device that, thanks to automatic document feed, captures crisp, digital images of each of the pages constituting a submitted manuscript, subsequently detects and interprets all the text present, and then stores the collected information—text *and* graphics—in digital form for editorial processing. How soon such a system will become widespread is obviously impossible at this juncture to say. The answer will surely depend in part on the extent to which authors, editors, and printing establishments continue to be plagued by compatibility problems.

It is also impossible to assess the likelihood that we will in fact someday see a truly significant decline in consumption of paper products. One thing we can say with confidence, however: it is no longer essential that entire filing cabinets be dedicated to sundry paper documents, or bookshelves to loose-leaf binders. That said, when the time comes to *examine* closely a particular document you are likely still to find yourself printing a hardcopy version, even if you expect to discard it as soon as you finish.

---

[67] Image capture with a digital camera is achieved via a vast array of CCD sensors (currently in the range of 4–5 million!). CCD is an acronym for "Charge-Coupled Device", a light-sensitive semiconductor element. Activating a battery of CCDs is reminiscent of what transpires in the human eye at nerve-ending clusters associated with the retina. CCDs have rapidly become fundamental building-blocks for a wide range of photosensory systems.

5.5 General Formatting Guidelines

5.5.1 Text

Fonts and Units of Measure in Typography

We turn now to a collection of observations related to the *presentation* of scientific text. Most of the suggestions are based on relevant norms, especially DIN guidelines.[68]

Modern word-processing systems provide the user with a wide choice of typefaces, including *proportional type*, both without and with *serifs* (Fig. 5–1 a, b), as well as *nonproportional* (or *monospaced*) *type*, also referred to as *typewriter script* (Fig. 5–1 c).

^a HIM abciixm   ^b HIM abciixm

^c HIM abciixm   serif

Fig. 5–1. Type characteristics.— **a** Sample of a proportional typeface *without* serifs ("Helvetica"); **b** a proportional face *with* serifs ("Times"); **c** a monospaced (nonproportional, typewriter) face ("Courier").

● If you propose with your word processor to prepare a *letter-quality* document, select a "normal" typeface producing upright (straight, roman) characters.

An elegant or decorative typeface may be tempting from an aesthetic standpoint, but it would almost certainly be inappropriate, especially for a scientific document, where *clarity* is of paramount importance. The preferred *size* is typically that labeled "12-point" (see below). With respect to *style*, never select *italic* type for body text since it tends to be relatively difficult to read; italic letters should be saved for distinguishing *symbols* that represent physical quantities (cf. Sec. 6.1.3), and for *highlighting* in a somewhat subtle way important words or brief passages.

---

[68] We beg the indulgence of readers from other parts of the world with respect to our emphasizing *German* standards, an area of scholarship in which one of us was for many years intimately engaged. The DIN (Deutsches Institut für Normung) standards are the ones with which we are most familiar, and it seems unlikely in any case that there exists anywhere a set of standards markedly superior to these. Moreover, in almost every case they reflect practices accepted throughout the world. Our most important sources of guidance (where titles have been translated into English) were DIN 1422 *Publications in Science, Technology, Economics, and Administration*, Part I (1983) *Formatting of Manuscripts and Typescripts*, Part II (1984) *Formatting of Final Copy for Reprographic Processes*; DIN 1421 (1983) *Subdivision and Numbering in Text: Sections, Paragraphs, Lists*; DIN 1338 (1996) *Presentation of Formulas and Formula Typesetting*, especially Supplementary Sheets 2 (1996) and 3 (1980), *Spacing in Formulas* and *Formulas in Typewritten Publications*, together with DIN 5008 (1996) *Composition and Formatting Rules for Word Processing*. We recommend that prolific writers examine the actual documents, which can be obtained directly from the publisher (Beuth, Berlin).

● *Typewritten* text can be simulated by use of the typeface called "Courier", again in 12-point size.

Actually, a (roman) typewriter-like face represents a good choice in many situations, especially for drafts. One reason is psychological: a document that appears to be typewritten is less likely than more "professional-looking" copy to convey prematurely a finished impression. The same is true of a document with a ragged-right margin relative to one in which the text is fully justified.

If more formality is preferred, however, then the best choice is a proportional typeface with serifs (Fig. 5–1 b). Perhaps most common in scientific text (and the face we chose for our book) is "Times" (or a variant like "Times New Roman"), available on virtually every computer. Times combines good *legibility*—the serifs help "direct the eye" along a string of letters—with high *character density*. For example, more than twice as many repetitions of the letter "i" fit in a given amount of space in Times relative to Courier. If you wish your document to look as if it had been printed professionally, Times is an excellent choice. With manuscripts for submission to a publisher, editorial guidelines should always be consulted in case a particular typeface is preferred.

At this point we need to delve a bit more deeply into certain technical concepts and terminology associated with typography in order to facilitate discussion of factors that affect legibility, especially in the context of such things as figure captions, potential size reduction, etc. We do not pretend to have assembled a comprehensive introduction to the study of type, however. Those interested in knowing more about the subject are encouraged to consult BRINGHURST (1996), an outstanding resource. Additional information related to typography as it applies to scientific manuscripts is presented in RUSSEY, BLIEFERT, and VILLAIN (1995).

The traditional unit of measure for height, width, and spacing in the printing trade is the "typographic point" (DIN 16 507). In continental Europe (e.g., France, Germany) the point is closely associated with the name "Didot", and 1 point = 0.3759 mm (often rounded to 0.375 mm).[69] The English system, prevalent especially in the United States, employs a slightly smaller point equal to 0.351 mm. For purposes of unit conversion and type measurement, typesetters make frequent use of a convenient gauge known as a *typometer*. *Paper* sizes in most of the world are described by a system with a metric basis (cf. the second subsection below), but the venerable "inch" (1 inch = 2.54 cm) still holds sway for this purpose in the United States.

In different places and at different times, the "point" has been represented by various symbols, including p, dp (for "Didot point" as distinct from the Anglo-American or "pica" point), P, Pt., and Pkt. The symbol " ″ " is employed to represent the inch (as in "8 1/2″ × 11″ paper").

---

[69] The "point system" of typographic measurement was devised by the 18th century French printer Ambroise DIDOT, who defined the point as 1/72 of the French "royal inch".

Since 1 January 1978, archaic non-metric units have technically been obsolete throughout Europe, so "10-point (Didot) type" is now more properly referred to as "3.76-mm type". Age-old traditions die slowly, however, even in the conscientious, systematically advanced European world—where the printing profession after all enjoys a history much longer than those of DIN and ISO combined! That said, a clear trend toward metric typographic measurements is apparent, reflected by the fact that rulers and grids with most word-processing and page-layout software can easily be set to conform to centimeter/millimeter standards.

● Software for desktop publishing largely originates in the United States. For this reason, "point" measurements in PC programs usually refer to the American (*pica*) point, not the somewhat larger European (Didot) point.

Several distinct approaches have been taken to describing the *size* of type. According to DIN guidelines, for example, type size is established by measuring the height of a capital letter like "H" (cf. the distance labeled c in Fig. 5–2), the implication being that larger symbols seldom occur in traditional typefaces. This of course fails to take into account the effect of *descenders*, like the tails on the "g" and the "y". More often the result of this particular measurement is instead referred to as the less fundamentally significant "cap height".

Fig. 5–2. Terminology and defined quantities important in typography.— h Line height, equivalent to the type size, a ascender length, x x-height, d descender length, c cap height, s line spacing, l leading.

An important objective in establishing a formal measure of type is the ability to express vertical space demands associated with text *passages*, not individual letters. A true "type size" should thus take into account spatial requirements imposed by an entire *family* of type, not a specific subset of symbols. This in turn necessitates envisioning a nominal "type-body" element that includes provision for both descenders and *ascenders* (e.g., the vertical protrusion on an "h"). Put slightly differently, "the *overall size* of a type family is established by measuring the maximum vertical extension exhibited by a complete set of letters" (RUSSEY, BLIEFERT, and VILLAIN 1995, p. 60). A "type size" so defined is thus equivalent to the distance h in Fig. 5–2. It lends itself to straightforward, direct measurement only if one has available a sample of one of the few letters characterized by both ascenders *and* descenders (e.g., *f* and Q).[70]

Publishers and printers have long communicated basic spatial parameters applicable to printed text by expressions like "10/12" (read as "10 on 12 point"), where the first number refers to the type size *h*, while the second is somewhat larger and takes into account blank space always deliberately provided above a line of type, called *leading*.[71] In other words, the second number is a measure of overall *line spacing* (*s* in Fig. 5–2).

● One might say the "type size" represents a (type) *line height* in the strictly literal sense of a minimum permissible separation between two adjacent baselines to ensure that every character in a particular type family will fit, complete with all ascenders and descenders.

By "baseline" is meant the hypothetical line upon which ordinary letters like "H" or "m" appear to rest. Any descenders present will of course fall below the baseline. In contrast to the line *height* so defined, line *spacing* is a measure of the overall physical distance from one baseline to the next in a particular printed passage.

Given our limited goals here it will suffice for the moment to add that the output of a typical ("pica format"[72]) typewriter of a decade or two ago resembled *12-point type*, with a normal capital letter being roughly 2.5 mm high. Photoreduction of typed text by 20% (see Sec. 7.2.2) resulted in approximately 2-mm capitals, comparable to those used in the body text of most books (including this one). We refrain from pursuing further complexities of printing jargon as they relate to type-size and line-height descriptions—including, for example, various other uses of "points" and more obscure traditional measures such as the "cicero".

Specific Type Fonts and Document Formats

For purposes of written communication, all the members of a particular culture share a common set of letters (or similar characters) that they routinely employ as linguistic "building blocks". In the case of our culture this set is of course the familiar *Latin alphabet*. Nevertheless, individuals *within* a culture tend to develop unique *handwriting profiles* which, although based on the accepted standard characters, display a myriad of subtle variations. If text is generated by mechanical or electronic means, however, there will be available only a limited (but possibly very large!) number of discrete stylistic variations on the basic symbols. The latter are known as *typefaces* or *fonts*,

---

[70] For this reason, special tables and typometers have been devised for estimating a true type size from the dimensions of any capital letter that happens to be at hand for inspection.

[71] The terminology is a legacy of the fact that in preparing typeset text, printers once inserted lead "slugs" into their printing forms between lines of characters assembled from stocks of (movable) type.

[72] The *pica* is an English-system unit equivalent to 12 points. Some typewriters were designed to produce smaller so-called "elite" characters appropriate for text with 12-point *line spacing*.

and each will have been "cut" so that the constituent elements conform to a particular design concept (cf. "Keyboards" in Sec. 5.2.1).[73]

Every existing font is attributable ultimately to a specific designer or design team. Font design is justifiably regarded as an art form, one that has blossomed continuously (if sparingly) over the several centuries since the invention of movable type. Individual fonts are assigned unique names in the interest of efficient and unambiguous identification. Fonts commonly encountered in word processing include Times, Courier, Helvetica, and Arial.[74]

● Familiarity with the distinguishing features of a variety of fonts, as well as with the characters of which they are composed, is useful in many ways, especially when it becomes necessary to exchange text files with others.

Complex, systematic "genealogies" of typefaces have been proposed, and various characteristics have been singled out for scrutiny in the assignment of fonts to special descriptive design categories, such as *gothic* and *modern*. Font evolution constitutes an important chapter in the cultural history of the West, and the interested reader is encouraged to consult specialized works on the subject.[75]

Making a selection from the "Font" menu of your word processor is analogous to a printer a century or so ago reaching into a cabinet to withdraw a stockpile of characters representing a specific type variant. A more contemporary analogy—from about the mid-1970s, perhaps—would be a secretary installing a particular type ball in a state-of-the-art electric typewriter. As little as two or three decades ago the notion that writers might someday have ready access to a wide variety of high-quality type fonts for their everyday use would have been dismissed as absurd.

The reason an experienced typesetter seeks to accumulate an assortment of fonts is that the resulting flexibility makes it easier to address a broad spectrum of aesthetic and practical demands. Consider, for example, the font "Times", which was designed for a very specific purpose: to provide *The Times of London* newspaper with a set of relatively slim letters capable of nestling snugly together so that a large amount of (highly legible!) text might be packed into the limited space available on that important publication's pages: a logical goal given a commitment to communicate as much information as possible notwithstanding financial constraints. The fact that the basic letter shapes in Times are augmented with serifs (Fig. 5–1) is also consistent with the

---

[73] The expression "cut" persists as a reminder of the fact that wooden type was once hand-carved—even though there is obviously no "knife" involved today in realizing electronically the fruits of a designer's imagination.

[74] In many cases the names applied are in fact trademarks, use of which presupposes one has permission to do so. Thus, "Helvetica", "Times", and "Palatino" are registered trademarks of Linotype Inc. Other important holders of font trademarks include the International Typeface Corporation and AGFA Compugraphic.

[75] Apart from BRINGHURST (1996), WHITE (1988) offers a good introduction to type in the context of graphic design, and LAWSON (1990) is a very readable source of a wealth of information on the development and singular features of typefaces.

mission of a newspaper, since serifs help prevent the reader's eye from going astray within a block of narrowly-spaced lines of type.

The recent surge in data-file exchange—especially on an international basis—has led to calls for increased standardization of type fonts, although the chief consequence so far apparent has been nearly universal acceptance of the ASCII-defined code equivalencies, as described in Sec. 5.3.3. One PC font that has become almost indispensable to scientists and engineers, and is now regarded as more or less "standard"—although it is at the same time rather unusual—bears the name "Symbol". It consists of a character set in which the various lowercase and uppercase Latin letters have been replaced by corresponding letters from the Greek alphabet (cf. Table 6–5 in Sec. 6.5.5), these being accompanied by a generous selection of symbols useful in mathematical expressions; e.g.,

$$\partial \; \nabla \; \cong \; \equiv \; \Leftrightarrow \; \rightarrow \; \cup \; \supset \; \varnothing \; \in \; \vee \; \angle \; \infty \; \div \; \times \; \otimes \; \oplus \; \aleph \; \wp$$

- Many fonts are devoted to special characters associated with specific fields and disciplines, available either on disk or by downloading from an appropriate Web site.

The inventory of PC fonts, derived from a multitude of sources (in some cases distibuted *gratis*!), is staggering. It embraces alphabets ranging from the elegant and formal to the bizarre and quirky, as well as symbols of every imaginable sort, decorative elements valuable in such special applications as borders, cartoon characters, … the list goes on and on. One of the most widely distributed "pictorial" fonts is "Zapf Dingbats",[76] producing figures like

If a manuscript you are preparing will be subject to processing elsewhere (e.g., on an editor's computer, for example, or with a publisher's typesetting equipment) you unfortunately must reckon with the possibility that special symbols you have painstakingly sought and captured from an exotic font could turn out in the end to be badly corrupted or even obliterated—yet another incentive for early detailed discussions with the editorial office.[77] Bear in mind, too, that an output device can only reproduce symbols from fonts to which it has access. Also, the way symbols are treated can vary somewhat from printer to printer. You should in any case never assume that a piece of digital copy will necessarily be rendered in precisely the same way by printers from

---

[76] "Dingbat" is a technical term for an abstract symbol traditionally available in the printing trade. "Zapf Dingbats" plays the same role with MACINTOSH computers that "Wingdings" does under WINDOWS. Hermann ZAPF, a font designer and typographer born in 1918 in Nuremberg, Germany, was apparently a "Meistersinger" (cf. Richard WAGNER's opera) of calligraphy. From 1977 to 1987 he held the position "professor of typographic computer programs" at the Rochester Institute of Technology in New York. A collection of his work is housed at the library in Wolfenbüttel, where one of the towering figures in German intellectual history, Gotthold Ephraim LESSING, was once active. *Mens* (Lat.: spirit, soul, understanding) and $\tau\upsilon\pi\upsilon\xi$ [Gr.: (printing) type] are indeed closely related!

[77] A symbol occasionally disappears as a consequence of a paragraph being reformatted, for example, especially if font substitution is involved.

two different manufacturers, and both versions are likely to differ markedly from what you see displayed on your computer screen.

Manuscript Style and Markup

The next subject we need to consider is the form that a manuscript as a whole should take in light of various processes to which it may be subject. We begin with line spacing.

Among the problems created by copy with lines set too close together is a lack of sufficient space for editorial markings, a consideration that applies to *all* parts of a manuscript (see also Sec. 3.4.1). A manuscript intended for typesetting should be *double-spaced*, so that every pair of adjacent text lines is separated by a gap with the dimensions identical to those of a word-bearing line (the sum of a line height and leading, corresponding to the line spacing s in Fig. 5–2). With a typewriter (see Fig. 5–3) this would result from four "clicks" of the platen ("single-spacing" would be equivalent to two such "clicks").

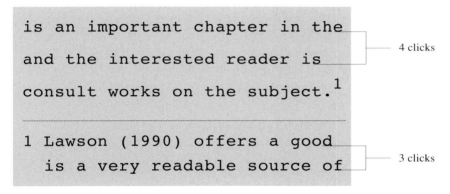

is an important chapter in the

and the interested reader is

consult works on the subject.[1] — 4 clicks

1 Lawson (1990) offers a good

is a very readable source of — 3 clicks

Fig. 5–3. Excerpt from a typed page (reproduced at full scale) containing double-spaced text as well as a footnote.

Double spacing is one of the standard spacing options with every word-processor, and shortcuts exist as well for "line-and-a half" and single spacing. Switching from one spacing option to another requires only a click of the mouse (or issuance of a simple menu or keystroke command), and *custom* line spacings open the way to far greater flexibility. Note that line-space settings apply to entire *paragraphs*, not random sets of lines.

In the case of 12-point type, *single spacing* actually means that, because of leading, adjacent base lines will be separated by a *minimum* (see below) of about 14 points, with approximately 21 points and 28 points being the analogous separations for line-and-a-half and double spacing, respectively.

Earlier (Sec. 3.4.6) we indicated that line-and-a-half spacing is usually requested with "camera-ready copy", sufficient for the purpose because there will be no editing.

With 12-point type, text of this sort is still clearly legible even after a typical photo-reduction to 70% of the original size (which represents approximately the difference between the "A4" and "A5" formats in the most widely employed system for specifying paper dimensions[78]).

DIN 1422-1 (1983) recommends uniform line-and-a-half spacing for manuscripts in general, although closer spacing is advantageous for such special elements as footnotes and figure captions, since it causes them to stand out. "Overlap" problems that might result from the presence of subscripts and superscripts are sometimes prevented by selectively increasing the separation between affected pairs of lines. Word-processing and page-layout programs are thus able to exercise either of two types of line-space control: precise, *fixed* spacing (e.g., *exactly* 16 points), or defined *minimum* spacing (e.g., *at least* 16 points). In the latter case, specific lines are automatically assigned extra space if they contain characters (including subscripts and superscripts) that threaten to come in contact with an adjacent line. This type of adjustment unfortunately confers upon the printed page as a whole a disquieting, untidy appearance, however, so fixed spacing generous enough to prevent overlap is generally preferable (but not *so* generous as to waste paper!).

If a publisher is planning to work directly with your digital files it is important that detailed understandings be reached as early as possible (and adhered to strictly!) regarding all aspects of format. Manuscripts for publication in a journal are subject to strict editorial guidelines regarding line spacing, type size, and the like. Often these are posted on the journal's Web site, from which they can be downloaded.

● Text should normally be supplied on letter-size sheets of paper.

In the United States, manuscript paper almost without exception measures 8 1/2″ × 11″ (216 mm × 279 mm), but in the rest of the world the "A4" format (see the preceding footnote) is considered standard, with the dimensions 210 mm × 297 mm—nearly 2 cm longer than American paper, but a bit narrower. The difference is important to bear in mind if you send documents to colleagues in Europe, for example, since the copying devices at their disposal will be designed to correspond to local standards. Similarly, should you wish to print on *American* equipment files originating in Europe or elsewhere, be sure you use appropriate software to adjust the output format prior to printing so that it conforms to US specifications (8 1/2″ × 11″). Otherwise, one or more lines may be lost at the end of each page.

---

[78] The "A" system of paper sizes originated as a DIN standard, but it has since been recommended by ISO for international usage. A sheet of paper of a particular "A" size (e.g., "A4") is reduced to the next smaller standard size (in this case "A5") by cutting at the line joining the halfway points of the long sides. The two identical fragments that result reproduce exactly the shape of the original, but with only half the area. Repetition of the procedure gives two pieces (here "A6", comparable in size to a postcard) that again preserve the initial shape . The series commences with a rectangular sheet (defined as "A0") whose dimensions are 841 mm × 1189 mm, corresponding to an area of 1 m^2. For any "A" format paper, the ratio of width to height is $1/\sqrt{2}$.

- Text should never be allowed to extend closer to the bottom of the page than about 30 mm, and a margin of at least 25 mm should be reserved on the left.

If you anticipate that a particular document might eventually be *bound* in some way (or collected in one or more loose-leaf notebooks), the left margin should be increased to 35–40 mm. In principle, a *right* margin of ca. 10 mm might suffice, but recall the suggestion (in Sec. 4.3.4) that a 50–60 mm right margin is a convenient place for inserting editorial markings. Margins can be specified to a tolerance of 0.1 mm with word-processing software.

Except for the first page, every page of a multi-page manuscript should bear a prominent *page number* to facilitate reference to specific passages (cf. "Finished Copy" in Sec. 1.4.2). Even with a two-page document (e.g., a letter) there would be no harm in including a number on the second page. It is often recommended that page numbers be centered above the text, separated from the latter by the equivalent of two or three lines, although numbering at the bottom of the page may be preferable in some cases.

Title pages are always left unnumbered, and the same usually applies to *dedication* and *acknowledgment* pages, all elements whose formality would be disrupted by page numbers. Numbers are traditionally omitted also from the first page of a *chapter*— again, largely for aesthetic reasons—although the corresponding pages would of course be taken into account in the overall numbering scheme (*pagination*).

Word-processing programs support automatic introduction of page numbers in the context of a *header* (*running head*) added to the top of each page (or a "footer" at the bottom), as mentioned in the discussion of "Finished Copy" in Sec. 1.4.2. This is a very handy feature well worth exploiting, particularly with a long document. You can also arrange for a header or footer to incorporate the *date* on which the document was prepared (and even the time of day!), along with such other useful information as the author's name, a title (usually abbreviated)—either of the work as a whole or of the current major subsection—and/or an identifier like a report number.

As a document approaches what you perceive to be its "final form" you should begin to direct attention toward ensuring that the right margin will not look *too* ragged.

- *Hyphenating* words otherwise too long to fit at the end of a line is a good way to avoid line truncations more extreme than the length of a syllable; i.e., a few letters.

Bear in mind, however, that if a manuscript will be subject to further electronic processing (e.g., typesetting), any such hyphenation must be confined to *soft hyphens*, which other software will be able to recognize as optional. Reasonably satisfactory results are usually achieved by allowing a word-processing system to hyphenate a piece of text on its own (cf. Sec. 5.3.2), although it will occasionally prove necessary for you to introduce at least some (soft) hyphens manually, especially with unusual words (foreign words, for example) that the program does not know how to hyphenate. Sometimes an *attempt* may have been made to hyphenate such a word, but the result could be incorrect (even ludicrous!) because of the program's reliance on a faulty algorithm rather than a reliable dictionary.

In the case of *justified* text insufficient hyphenation can produce unsightly gaps between the words in certain lines because too much spacing adjustment has accompanied the attempt to fill those line. This is in fact the most prevalent symptom of a need for additional manual hyphenation. Word separations that are supposed to be "fixed", as between a number and an associated unit (see the comments near the end of Sec. 5.4.1), must of course not be *allowed* to change as a consequence of justification, a problem one avoids by assigning *nonbreaking* spaces to all such cases.

Several possibilities exist for introducing *emphasis* into printed text. We first focus attention on a manuscript that ultimately will be published.

To a typesetter, "emphasis" means selective use for highlighting purposes of any of several eye-catching special effects: a special typeface or type-style (such as italics), white text set against a dark background, or underlining, for example.[79] Essentially every desktop or portable computer today offers the possibility of employing *color* with both text symbols and the background, a feature that can prove valuable for on-screen work even if you lack access to a color printer. An editor usually assumes responsibility for determining the extent to which highlighting will be utilized in a published work, as well as the particular approach to be taken. In the early stages of manuscript preparation highlighting with color can facilitate communication among multiple authors, and it in fact occurs automatically if one takes advantage of the "Track Changes" or "Insert Comment" functions in WORD.

● Introduction of highlight requests into a manuscript is an important part of the "markup process" (see Sec. 4.3.5).

When, how, and by whom various markings will be applied is one more important issue for negotiation between author and publishing house, especially when an electronic manuscript is involved.

With a "classical" typewritten manuscript (or *typescript*), markup instructions were always introduced into an editorial copy of the document by hand, by either the author or an editor, for later implementation by a professional typesetter. To the extent that this practice is still observed, the following generalizations apply:[80]

● Usually the preferred technique for highlighting a piece of text is setting it in *italics*. Passages to be *italicized* are traditionally indicated by underlining—by hand, not from a keyboard.

A more dramatic form of highlighting entails use of **boldface** type.

● Boldface treatment is commonly specified with a *wavy* underline.

---

[79] Typesetters, particularly in Germany, once adopted the interesting strategy of adding extra space between *letters* ("spaced letters") for highlighting purposes.

[80] The (American) guidelines presented here are somewhat different from procedures in Europe, where italics are typically indicated by a wavy underline, boldface by a double underline, and small caps by a dotted underline.

The latter expedient is generally reserved for text requiring *exceptional* emphasis, such as an unnumbered section heading, or a warning of some type. Again, the appropriate marks are introduced by hand, preferably with a colored pencil or pen.

Personal names and *trademarks* are sometimes distinguished from surrounding text by setting them in "small caps", a style utilizing for lower-case characters diminutive versions of ordinary capital letters (as in the words "EINSTEIN" and "PAGEMAKER"). Apart from the purposes indicated, however, small caps are rarely seen today in printed works. The traditional signal for small caps is a *double underline*.

For the author or editor with access to a word processor, all these type-styles (and others as well!) can in principle be incorporated into a manuscript directly. As a result, manual mark-up with pencils is now the exception at many publishing houses [cf. "Marking (Highlighting)" in Sec. 5.3.1]. Nevertheless, some publishers even today instruct their authors categorically to *refrain* from employing alternative type-styles in manuscripts they submit.

In a desktop publishing context, these and other "typographic flourishes" are provided by tools integral to the typesetting PC. Be sure to verify, however, that the desktop printer at hand is actually capable of achieving the desired effect in a satisfactory way, although this rarely proves a problem.

Headings, Paragraphs, Equations, and Lists

With reports and other documents that are not to be typeset, it is up to the author to establish the way in which *headings* and other special elements will be treated (Fig. 5–4). A decision is also required about whether or not the convention of starting each major section (chapter) on a new page will be observed.

● All headings should either be left-justified (i.e., set against the left margin) or evenly distributed about the central axis of the body text (*centered*).

In the days when manuscripts were still prepared with a typewriter, centering a piece of text required that one first carefully examine the page to locate the (horizontal) midpoint of the text and then count (and divide by two) the number of (monospaced) characters constituting the heading in question in order to ascertain where typing should commence, an awkward process at best. This is actually one of the reasons why centered headings fell out of favor. With a word processor, centering is of course very easy, but tastes have changed such that, especially with scientific text, many readers now perceive such headings as somehow "forced" looking, and unnatural.

● In the event you *do* decide to center your headings, then *all* of them must be so treated, and the same placement rule should be observed with the headings and captions associated with *figures* and *tables* (and the figures and tables themselves!).

The above admonitions reflect the way "style consistency" has traditionally been interpreted in the printing trade generally. Failure to observe the rules related to figures, tables, figure captions, and table headings thus constitutes a formal style violation.

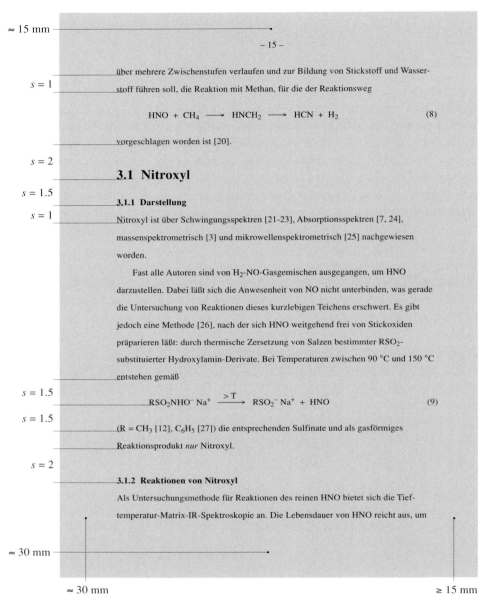

Fig. 5–4. Sample page from a double-spaced manuscript (photoreduced to 50% of the original scale) with suggestions regarding line spacing and margins.– *s* Line spacing (in "lines").

One sometimes encounters a document in which headings have been set entirely in capital letters, but this practice is frowned upon, not only for aesthetic reasons but also in the interest of maximum legibility.

A particular heading's hierarchical status is manifested by the type-style and size assigned to it, *not* by its placement (left-justified, centered, or right-justified, for example). We have already noted in Sec. 5.3.3 the role that distinctive type-styles can play in headings at various levels. The writer with a word processor generally has at his or her disposal far more typefaces, type-styles, and type sizes than there could possibly be any reason to call upon. A conservative approach is in any case preferable when setting type standards for headings.

Headings at the lowest level should be set in type the same size as running text (body text), but distinguished by the use of boldface or italics. Headings at the next higher level should be more prominent (at least 20...30% larger than body text), a trend that is then maintained throughout the hierarchy. The same general typeface (e.g., "Times") is often selected for both headings and running text, but use of a complementary font for headings is also permissible. Bear in mind, however, that here—as in many typographical situations—"less is often more".

With respect to the vertical *positioning* of headings:

● Leave 1.5...2 times as much blank space *above* a heading as *below* it.

In other words, according to this rule of thumb, a header followed by *one* free text line should be preceded by 1.5...2 free lines. In the event that *no* extra space is present below a header, it should still be separated from text (or another heading!) above by approximately one-half the equivalent of a text line.

● The *body* of a document acquires structure through its being divided into *paragraphs*, an observation that has implications with respect to both content and appearance.

There are two standard ways of providing visual evidence of a paragraph break. In the days when most manuscripts were prepared with typewriters, the preferred practice was to insert a *blank line* after each paragraph, analogous to adding extra *leading* in typeset copy. Ordinary blank lines should *not* be introduced between paragraphs in an electronic manuscript, however. Instead, appropriate "space after" and "space before" parameters should be established as a part of defining paragraph *styles* (see Sec. 5.3.3). The precise placement in this regard of a *particular* paragraph can of course be adjusted if special circumstances should dictate raising or lowering it relative to the norm.

Another commonly employed device for indicating the start of a new paragraph (rarely seen when paragraphs are separated by extra space) is indentation.[81]

● From a technical standpoint, indentation means shifting the origin of a line somewhat to the right from the left margin—or the end of a line to the *left* relative to the right margin (or both!).

Thus, precision demands that one distinguish between various *types* of indentation.

"Indentation of a paragraph" is usually understood to mean moving the paragraph's *first line* a short distance away from the *left* margin (i.e. from the "left border line" of

---

[81] Some suggest that first-line indentation should always be *omitted* with the first paragraph following a header, although no strong consensus has developed on the point.

the body text), perhaps by no more than the equivalent of three characters (ca. 5 mm with 12-point type). For text set in a single column one might choose to be a bit more generous, possibly increasing the displacement to the equivalent of five characters.

With a word processor, a single instance of paragraph indentation is most easily accomplished by taking advantage of a special marker included for that purpose on the "text ruler" displayed at the top of the screen (which always applies specifically to the paragraph that contains the cursor). Alternatively, a suitable numerical value (together with the applicable unit!) can be entered through a "Paragraph Format" dialogue box. Some prefer to press the "tab" (tabulator) key as a way of indenting, but this again has the disadvantage of requiring a conscious extra step each time a new paragraph is created. By far the best approach is to specify systematic indentation as a part of *paragraph-style* definition.

As a rule, section and chapter *headings* are *not* indented, since they are already clearly apparent by virtue of their spacing as well as unique type-size and/or type-style characteristics.

● In the *absence* of indentation, it is important that *extra leading* be added between paragraphs.

If *neither* type of signal (indentation or extra space) were to be given it would become much more difficult for the reader to recognize at a glance the end of one paragraph and the beginning of the next, and quite impossible in the event that the last line of the former happened to extend all the way to the right margin.[82]

Sometimes a special paragraph is singled out by indenting it in its entirety (i.e., by displacement of *every* line, possibly even on both the left *and* the right). *Negative* indentation is another technique occasionally exploited, such that the first few characters of a line or paragraph are deliberately relocated into the left margin. In an extension of this strategy, brief notes—called *marginalia*—are sometimes set *entirely* in the margin, where they are used to confer additional structure on a document, especially to help orient the casual reader. If you wish to allow body-text to appear alongside such a marginal note in a document prepared by DTP methods the note will need to be consigned to a specially-placed *text box* or *frame* (without visible borders). Both word-processing and page-layout programs provide tools for that purpose.

● Another way of causing a text fragment to become a distinctive entity is to set in front of it a special eye-catching symbol, and then uniformly indent each text line. (This particular paragraph is a case in point.)

Many books—obviously including this one—employ as the corresponding symbol a large black dot known as a *bullet* (● — or a smaller one like •). Appropriately and sparingly utilized, this device is very effective for capturing a reader's attention. In a manuscript that will ultimately be typeset, a more readily accessible symbol (perhaps

---

[82] The "leading" solution has the minor shortcoming that it fails to solve this particular problem with a paragraph that ends at the bottom of a *page*.

a * or a +) can be substituted, with instructions that this eventually be replaced by a bullet. A more subtle alternative to the bullet is an extended dash of the type ordinarily used to indicate a pause—a so-called "em-dash" (—), a symbol similar to but substantially longer than a hyphen.

If several such text elements are presented together, the collection as a whole is referred to as an *enumeration*. In this case individual elements are more commonly preceded by one of the less intrusive distinguishing marks, for example an asterisk.

● Enumerations composed entirely of single words or very brief phrases, all carefully aligned vertically, are known as *lists*.

A list could be interpreted as a rudimentary form of a *table* (see Chap. 8). From a word-processing standpoint, each list entry (including its associated symbol) constitutes an independent paragraph.

If the elements making up an enumeration are so lengthy that they extend over several lines, continuation lines should usually not be indented, since otherwise a considerable amount of space is wasted, and some readers find the practice distracting.

This is perhaps a good place for us to restate an important general typesetting maxim we have alluded to before, albeit in somewhat different terms:

● Every "special feature" in a document that has the effect of altering the appearance and/or placement of text on a page has the potential for contributing to a disconcertingly "busy" layout.

In other words, the various techniques we have here been describing should all be employed sparingly.

Sometimes the elements in a list are *literally* "enumerated" (numbered), not simply presented. The attention-gathering function is then assumed by a numeral, which could be either arabic (1, 2, 3 …) or roman (I, II, III, …; i, ii, iii, …). Occasionally letters are called upon for the task (A, B, C, …; or a, b, c, …). Arabic numerals in this context are generally followed by periods (implying they should be read as "first", "second", …), whereas a lower-case letter is traditionally accompanied by the "close parentheses" symbol (cf. DIN 1421, 1983); e.g.,

1.		a)
2.	or	b)
3.		c)

It is *not* considered appropriate to use both a period *and* a "close parentheses" mark. Your word processor almost certainly makes the "numbered" option above available automatically should you choose to use it.

If there is no obvious logic behind interpreting the elements of a list as an ordered sequence it is better to utilize a more "neutral" symbol, like the bullet. Among other things this prevents a reference to a particular (numbered) list element from being confused with a reference to a (numbered) heading.

In scientific text a need often arises to establish a clear visual distinction between text and one or more *formulas* or *equations*.

● Formulas and equations should be separated from running text both above and below by *extra* space, equivalent perhaps to one-half the width of a text line.

Individual components in a *block of equations* can be distributed according to the usual line-spacing guidelines, provided each consumes no more space than a normal text line, in which case only the first and last elements would be affected by the "half-line-extra" rule. This generalization is obviously not applicable to complex equations characterized themselves by multiple (vertical) levels, since these always consume more space than a line of text (e.g., equations containing fractions, roots, or summations). Examples of such equations are presented in Sec. 6.5.5 (cf. Fig. 6–3), which also provides quite a bit of information regarding typesetting conventions applicable to mathematical expressions.

● Free-standing formulas and equations should always be indented, and to a significantly greater extent than the first line of a paragraph.

Footnotes

The presence of a footnote is indicated in running text by a *reference mark* at the spot most relevant to that note, whereas the note itself—as the name suggests—is set at the foot of the corresponding page. In the days before word processing, footnote management was for authors a source of much discomfiture (cf. Sec. 3.4.5). Proper placement of such notes was also a labor-intensive exercise, for both typist and typesetter. As a result, many journal and book publishers eventually discontinued their use, replacing them with *endnotes*. When footnotes as such *are* to appear in a publication, the note text with a conventional manuscript should be inserted immediately after the line containing the footnote reference (starting on a fresh line). Alternatively it might be placed directly *below* the corresponding paragraph. Footnote text supplied in this form is always preceded by a repeat of the reference mark, and separated from body text both above and below by horizontal lines that extend across the entire page. In other words, a typescript never actually contains any true *foot*notes. The corresponding information is instead integrated into the manuscript in a more convenient way (from the standpoint of processing), and its proper placement as a footnote is left to a typesetter.

In a document prepared with a word processor, footnotes need pose no real problems at all; everything about them is efficiently attended to by an automatic *footnote function*. Thus, the projected note is first "anchored" in the text with the reference mark of choice, after which its content is typed into what becomes a special "footnote section" of the overall text file. Final placement of the note in the actual document is arranged immediately prior to printing—very discretely—by the program itself. Issuing an "Insert Footnote" command has the effect of opening a dialogue box for specifying the preferred form of the note reference, and also for confirming where the note should be placed (i.e., as a footnote or an endnote). The cursor then jumps automatically to the place for typing the note's content: a separate "footnote window" in "normal" view, but the bottom of the affected page in "page layout view", where the correct reference symbol is already waiting. If notes are to be sequentially numbered, the program does yeoman service by assuming all the related bookkeeping responsibilities.

DIN 5008 (1996) provides the following guidelines regarding footnotes for a document like a report or a dissertation:

> Footnote text appears at the bottom of the page containing the corresponding footnote reference [we would add: set in type 70...80% the size of that employed for body text]. Footnotes are separated from body text above by a footnote line equivalent in length to 10 underline strokes with a typewriter and commencing at the left margin. Line spacing should be the same as that in the body text [but correspondingly narrower if smaller type is used], and the footnote itself is preceded by the appropriate reference symbol.

This is precisely what results in WORD when the "Insert at bottom of the page" option is selected. Proper note placement is assured even if a footnote reference mark occurs very near the end of a page—because page breaks are automatically adjusted to produce an optimal layout. If, because of its length, part of a footnote must necessarily be carried over to the following page, the resulting continuation text is separated from body text above by a different type of line: this time one that extends across the *entire* text block.

If you plan to submit a manuscript for publication, specific guidance should be sought from the publisher regarding the proper treatment of footnotes.

● Necessary punctuation always *precedes* a footnote reference mark set in running text (with the exception of the "em-dash" used to signal a pause).[83]

Footnotes can be labeled either with numbers or with special symbols, such as the commonly employed asterisk (*). Numbers have the advantage of providing consistency throughout a manuscript, even with pages containing multiple notes (where multiple symbols would otherwise be required), and they also facilitate preparation of a separate, unambiguous *footnote manuscript*. The numbers themselves appear in running text as superscripts expressed in rather small type. In scientific manuscripts it is often recommended that such numbers be accompanied by the "close parentheses" mark— e.g., "text[1)]", "text[3), 14)]" or "text[4)–12)]" to ensure clear differentiation with respect to exponents as well as superscripts present for other reasons (cf. Sec. 3.4.5).[84] It is also important to make certain that that no confusion will develop between footnote references and literature references. Problems in the latter regard are best avoided by employing two very different notations, like "text[1)]" vs. "text [1]", for example).

---

[83] The same rule also applies to the (usually numerical) reference marks often utilized in scientific text for citing literature sources (*citation numbers*; cf. Sec. 9.3.2).

[84] The "close parentheses" mark does not appear with the note itself.

The word processor also attends to correct spacing of reference symbols preceding footnote text; e.g.:

^{1} Footnote text.
^{39} Footnote text.

Even if one chooses to structure a document as a whole in term of endnotes, footnotes may be impossible to avoid in conjunction with tables.

● Reference marks for table footnotes typically take the form of superscript *letters*, again accompanied by the "close parentheses" sign.

The same reference symbol—for example a superscript "a"—is repeated immediately below the table (here usually *without* a "close parentheses" mark), followed by that note's text. Letters are generally employed in this case partly to rule out confusion with note references from the main body of the text, but also because tables tend already to contain a great many numbers.

● Never incorporate reference to a footnote or a literature source into a *heading*.

The awkwardness that would attend violation of this rule is obvious if you consider the potential effect on a *table of contents*, or the display of an "outline view" with a word processor. The easiest way to avoid an apparent need for a literature citation within a heading is to begin the section that follows with a sentence or two explicitly designed to summon the essential reference.

5.5.2 *Preparation of Final Copy*

At least one *extra* copy of an original manuscript should always be retained in a safe place as a form of insurance. Excellent duplicates of conventional manuscripts can be obtained with modern copiers: *provided* the equipment is well maintained (with prompt removal of smudges on the windows, timely replacement of toner cartridges, etc.). If you find the first few pages produced to be unsatisfactory, rather than wasting any more time switch right away to a different copier, or arrange for a service call to be placed.

Even in this era of heightened environmental consciousness, insist on being provided with high-quality paper (xerographic grade, bright-white, with at least a "20-pound" weight designation, equivalent to ca. 74 g/m^2) for final copy.[85] Gray, recycled paper obviously has the disadvantage of being less appealing visually, but more important, it is harder on a reader's eyes (including those of a typesetter!). Moreover, additional copies may need to be prepared based on the gray document, and they would almost certainly be poor, which could in turn contribute to a high error rate if some of these

---

[85] The "pound rating" for paper, almost unique to the United States, refers to the total mass of 500 sheets. More common in the rest of the world is specification in terms of "surface weight", with the unit g/m^2.

pages were for some reason scanned. It should go without saying that a dissertation printed on "gray" paper would almost certainly be rejected!

- *High-speed* copiers offer the great convenience of automatic feed for rapid processing of long documents and collation of multiple copies, both valuable time-saving features.

Nevertheless, certain precautions are necessary with automatic feed.

- Every page of the original must be in good form; otherwise, pages are likely to get stuck in the feed mechanism, potentially damaging both the manuscript and the copier. Pages containing taped or pasted corrections must always be processed manually.

Many modern copiers offer the possibility of *size reduction*. With typed text or printed documents based on 12-point type, reproduction at 70% of the original scale results in letters comparable in size to those typically encountered in publications (i.e., with capitals ca. 2 mm high)

From an economic standpoint it is impractical to prepare more than about 100 copies with an office copier. Larger jobs should be contracted out to an offset printer. Professional printing establishments and well-equipped copy shops are in a position also to offer custom bindings of various types. This is advantageous from the protection point of view, and it will greatly improve a document's appearance. Common options include *spiral binding*, *saddle stitching* (essentially a stapling process), and *perfect* (glued) binding, none of which is feasible in a typical office environment.

Publishers normally request two or three printed copies of all submitted manuscripts. In the case of a digital manuscript, be sure you prepare at least two *electronic duplicates* (or *backups*) for archival purposes, preserved on diskette or some other removable medium and carefully stored in safe, isolated locations. An editorial office typically requests one electronic copy together with at least one hardcopy print (which must of course represent the very latest version, identical to the digital file!).

Special considerations related to the treatment of mathematical expressions are discussed in Sec. 6.5.

6 Formulas

6.1 Quantities

6.1.1 Quantities and Dimensions

In the previous edition of the book, material presented in this particular chapter was distributed across *two* chapters entitled, respectively, "Quantities, Units, and Numbers" and "Equations and Formulas". The new arrangement and title, focusing attention on the single broad concept "formulas", establishes a greater degree of parallelism with the two subsequent chapters: "Figures" and "Tables". This set of three chapters emphasizes technical aspects of writing that relate directly to the various "special parts" of a scientific manuscript. Toward the end of the present chapter we have added a brief new section on MATHTYPE and MATHML. The commitment to precise forms of expression implicit in our earlier titles has in no sense been abandoned, however. Indeed, we are still persuaded that only the person who truly understands what is *meant* by "quantities", for example, will be in any position to work with them correctly. Our strong convictions in this regard have shaped the first four sections of the chapter, which are devoted to quantities, units, and numbers, along with a brief excursion into units peculiar to chemistry.

Besides having rolled quantities, units, and numbers under the single heading "formulas" we have also restricted somewhat the scope of our treatment, concentrating on expressions and formulas of the *mathematical* and *physical* type that can be rendered in print (including in a word-processing environment) on the basis of a limited number of standardized symbols. The special universe of "chemical (structural) formulas" is in fact more pictorial than typographical,[1] and is addressed briefly in the context of *figures* (in Sec. 7.2.5 "Structural Formulas in Chemistry") rather than here.

The natural sciences are concerned not only with *qualitative* information, the "how" of things, but also with *quantitative* considerations in the sense of "how much". From one standpoint this complicates matters, but it is also enriching. The resulting attempt to establish the extent to which nature's various manifestations are in fact quantifiable or measurable, an extremely productive endeavor at least since the early years of the 19th century, has conferred special significance on the notion of a *dimension*. Characterizing nature in its entirety requires measurements in many discrete dimensions, where we are here employing the term much more broadly than might be assumed based on the three *spatial* dimensions with which everyone is familiar.

---

[1] This point is explored in depth in a chapter in a German-language treatise on technical terminology, and terminology in general, from an academic standpoint (EBEL 1998).

To "extend over" (or "extend across", or "embrace", or "encompass"),[2] to "persist", to "be heavy", or "be countable"—each of these distinct attributes is associated with its own unique dimension. Let us use "L" (for "length") as a symbol for the notion of "extending over (or across)" in the most general sense. The attribute in question can then be interpreted as manifesting itself within the "dimension" **L**, a dimension useful for examining not only the extent of *lines* but also *surfaces* or *spaces*, at least when raised to the second or third power (i.e., $\mathbf{L}^2$, $\mathbf{L}^3$).

Before we are able to carry out an actual measurement in terms of such a dimension, however, we must first come to terms with the basic concept of a *physical quantity* (or *measurable quantity*), often referred to simply as a *quantity*. The idea of extension, or "being long", is at the heart of the "quantity", represented by the symbol l, for which the corresponding dimension is *also* called "length".[3] To the physicist, a measurement of a height, a depth, a thickness—indeed, a "distance" of any conceivable kind—would classify as a "length" (including, for example, the interatomic distances known as chemical-bond lengths, or wavelengths of every kind: from ocean waves, through sound or radar waves, to X-rays). In practice, quantification of a *particular* length (or distance) entails determining how many replications of an agreed upon "standard" length (or extent) are required to reproduce the length (or extent) of interest in the context of the dimension "length" (**L**). This *standard length*, whatever its magnitude, might in turn be adopted as a fundamental *unit* for use permanently with the quantity length.

● Every concrete example of a "quantity" can be shown to be the product of a *number* and a *unit*.

The particular product associated with a specific situation is referred to as the relevant *value* (or *measured value*) of that quantity. If we take Q to be the symbol for a quantity in the most general sense, then what we have effectively asserted reduces to an equation of the following form:

$$Q = \{Q\} \cdot [Q] \tag{6–1}$$

where $\{Q\}$ stands for the quantity's *numerical* value and $[Q]$ for the corresponding unit.

Equation (6–1) might be restated rather concisely in words as

> "The value of a quantity equals some numerical value times an appropriate unit."

It took a long time to achieve worldwide consensus on the minimum number of *independent* quantities required for describing nature in its entirety. The challenge was

---

[2] There seems to be no single satisfactory word in English to conjure up unambiguously and precisely what is meant here—a concept that should become clearer as you read further (cf. the German verb *sich erstrecken*).

[3] The choice of l as a quantity symbol in this example—and **L** as the symbol for the related dimension—was by no means arbitrary on our part, but rather a case of our adhering to the officially sanctioned national and international practice.

equivalent to answering the question: "How many independent *units* must be defined to ensure that all conceivable natural phenomena will be measurable?" An authoritative answer was provided for the first time in 1960 by the Conférence Générale des Poids et Mesures (CGPM) in the course of establishing a broad, international approach to measurement in general, which culminated in the *Système International d'Unités* (SI; "International System of Units"). The answer is actually quite astonishing:

● It has been reliably established that, at least insofar as can currently be discerned, *seven* fundamental quantities are sufficient for the quantification of all of nature.

This minimal assembly of essential quantities is referred to collectively as the set of official *base quantities*.

The authorized *designations* of the seven SI base quantities are presented along with those of their officially sanctioned units (the *base units*) in Table 6–1 (cf. DIN 1301-1, 1993, *Einheiten: Einheitennamen, Einheitenzeichen*, "Units: Unit Names, Unit Symbols").

Table 6–1. Base quantities and units in the SI.

Base quantity (dimension)	Quantity symbol	Dimension symbol	Corresponding base unit	Unit symbol
length	l	**L**	meter	m
mass	m	**M**	kilogram	kg
time	t	**T**	second	s
electric current	I	**I**	ampere	A
thermodynamic temperature	T	**Θ**	kelvin	K
amount of substance	n	**N**	mole	mol
luminous intensity	I, I_V	**J**	candela	cd

In addition to *names* for the base quantities and their units, a need was recognized for official "abbreviated forms" that would facilitate the concise expression of complex concepts. Thus, Table 6–1 also lists the internationally accepted base-quantity and base-unit *symbols*, together with symbols for the corresponding *dimensions*.

As the table suggests, the various symbols are subject to specific modes of *written expression* with respect to type style.

● Quantity symbols are always to be set in *italic* type, dimension symbols in *bold sans-serif* type, and unit symbols in *plain, upright roman* type.

There are admittedly circumstances under which it is not possible to adhere strictly to these guidelines, but for important reasons the attempt should always be made.

DIN 1338 Beiblatt 1 (1996) *Formelschreibweise und Formelsatz: Form der Schrift-zeichen* ("Supplement 1, Formula Notation and Formula Typography: Character Form") provides the following elaboration:

> Roman ("Antiqua") fonts from Groups I through IV as defined in DIN 16 518 (Fonts with Serifs) are regarded as suitable for use in the typesetting of formulas. Only in special cases should sans-serif linear-roman (grotesque) fonts from Group VI be employed in this context.

The widespread notion that a sans-serif font of the so-called grotesque type is ideal for formulas is thus officially disavowed. This is not surprising, since *serifs* contribute heavily to making a symbol especially legible,[4] and legibility should be a primary consideration, particularly with respect to formulas.

The reason quantity symbols are always to be set in *italics* when they appear in publications is that the practice assists markedly in interpretation on the part of the reader. This is especially true with respect to equations, but it applies as well to scientific text in general.

On the other hand, one can forego if necessary the prescribed *boldface* treatment of sans-serif *dimension* symbols provided all accompanying body text and other formula symbols are set in type *with* serifs (see "Fonts and Units of Measure in Typography" in Sec. 5.5.1). Restricting non-serif type exclusively to dimension symbols is sufficient to ensure that the latter will stand out for easy recognition.

This is an appropriate spot for a bit of further attention to the concept of a "quantity". The fundamental idea is one very familiar to the mathematician, albeit in somewhat modified form and in conjunction with what is referred to as a "variable". An important characteristic shared by quantities and variables is that, in principle, both are able to assume an infinite number of values, expressed in the case of a quantity (as we have seen) as a combination of a numerical value and a unit. All symbols for variables, just like those for quantities, are to be typeset in italics. The same is true with respect to symbols for *general functions*. Consider, for example, the *f*, the *u*, and the *v* in the expression $f(x) = u[v(x)]$. Note that here the "x" is also italicized, because it represents a *variable*.

As previously suggested by the example of the dimension **L**, dimensions in general are subject to being raised to higher powers; that is, any dimension can legitimately be multiplied by itself ($\mathbf{L}^2 = \mathbf{L} \cdot \mathbf{L}$). Multiplication is also allowed with respect to two

---

[4] A "sans-serif" font is one that lacks the extra character embellishments known as serifs (cf. Fig. 5–1 in Sec. 5.5.1.). Most fonts routinely used in the professional typesetting of text (and also in text prepared in a DTP environment) are characterized by serifs, as in the familiar "Times" family, for example. Fonts of this sort usually assume an "upright" form, but a modest (1/6) slant is occasionally encountered. Separate "italic" versions of a font typically display a significantly more drastic slant (e.g., 1/4). Special *math italic* ("mi") fonts have been designed explicitly for use in the typesetting of mathematical expressions. KOPKA and DALY (2003) in their Appendix A, "The New Font Selection Scheme", provide a valuable overview of computer fonts from the vantage point of the TEX user.

or more *different* dimensions. We will revisit this point, incidentally, in the section that follows, dedicated to *derived quantities*.

Such an "accumulation" of dimensions, too, lends itself to expression in the form of a general equation, somewhat analogous to Equation (6–1). Thus, if "dim Q" is taken to represent the dimension associated with the particular quantity Q, with **D** as a general dimension symbol, then

$$\text{dim } Q = (\mathbf{D}_1)^{n_1} \cdot (\mathbf{D}_2)^{n_2} \cdot \ldots \cdot (\mathbf{D}_7)^{n_7} \tag{6–2}$$

Here $\mathbf{D}_1$ through $\mathbf{D}_7$ are to be regarded as symbols for the seven dimensions introduced in Table 6–1; i.e., **L**, **M**, **T**, etc. The exponents n_i in Equation (6–2) will ordinarily be either small integers (positive or negative) or zero. With respect, for example, to the quantity *force*, F, introduced in Table 6–2 (in Sec. 6.1.2):

$$\text{dim } F = \mathbf{L} \, \mathbf{M} \, \mathbf{T}^{-2} \tag{6–2a}$$

where **L**, **M**, and **T** are symbolic representations of the dimensions length, mass, and time.

Equation (6–2) could also assume the form:

$$\text{dim } Q = \mathbf{D} \cdot \mathbf{D}^{-1} = 1 \tag{6–2b}$$

● In other words, a quantity could be so defined that it had the dimension 1.

Claiming in such a case that "Q is dimensionless" would be an inappropriate over-simplification, and could in fact lead to confusion.

An example of the applicability of Equation (6–2b) is seen in the definition of *refractive index*. This definition entails a division process in which both the numerator and the denominator are quantities with the dimension *velocity*, so the resulting dimension must be 1. Obviously Equation (6–2b) is far from a mere formalism.

● Exponents and logarithmic expressions may *only* contain quantities with the dimension 1, because the process of raising something to a power, or taking a logarithm, is strictly numerical.

Thus, no problems are encountered in the course of taking the logarithm to the base 10 of a *number* like 100, but that would certainly not be true of an attempt to carry out the procedure on something like "100 grams". Similarly, a number like 10 is easily squared, but "10 grams" cannot be raised to a power! Unfortunately, the distinction is too often ignored in the scientific literature. For example, contrary to statements in many textbooks, pH is *not* the negative logarithm to the base 10 of the hydronium-ion concentration, but rather of the *numerical value* of the hydronium-ion concentration; cf. Sec. 6.1.2. Note, however, that a definitive declaration is still lacking here with respect to the appropriate *unit* for a hydronium-ion concentration one wishes to convert to a pH.

An additional point worth emphasizing is the fact that quantities with *different* dimensions cannot be added or subtracted, because that would require combining incompatible entities. This represents an example of *dimensional analysis* (what the International Union of Pure and Applied Chemistry calls "quantity calculus"; IUPAC 1988), a technique with which it is often possible to test the validity of an expression or an equation.

● Expressions to the left and right of an "equals" sign in an equation must always be associated with the same dimensions.

On the other hand it is *not* necessary that the two sides of an equation be expressed in the same *units*, as is apparent from a simple unit-conversion equation (potentially the basis for a *conversion factor*); e.g.:

$$1 \text{ m} = 1000 \text{ mm} \quad \text{or} \quad 1 \text{ min} = 60 \text{ s}$$

Finally,

● Identical dimensions cannot be taken as a sign that two *quantities* are identical.

For example, density and mass concentration are very different things despite the fact that they share the dimension $\mathbf{M} \, \mathbf{L}^{-3}$, and the many quantities characterized by the dimension 1 display a great deal of diversity. Thus, counting frequencies of all sorts of periodic occurrences are fundamentally associated with the common dimension $\mathbf{T}^{-1}$ even though the entities subject to counting may be vibrations (*frequency* in a narrower sense), rotations (*revolutions*), or nuclear decompositions (*radioactivity*). A similar principle applies to units. For example, self-inductance L and permeance L, though very different, are both expressed in the unit henry (H, where $1 \text{ H} = 1 \text{ m kg s}^{-2} \text{ A}^{-2}$), and m^{-1} is a unit useful in the measurement of both absorption coefficients and nuclear quadrupole moments.[5]

● A subtle distinction is often made, especially in thermodynamics, between two types of quantities: *extensive* and *intensive*.

Extensive quantities are ones that are affected by the "magnitude" or "expansiveness" of something, while intensive quantities reflect only "condition" or "state". Considering the SI base quantities (Table 6–1), mass would thus be regarded as an extensive quantity, temperature as intensive, because the former is a function of the size of the sample under investigation while the latter is not. A great many derived quantities are intensive, among them the previously encountered refractive index and density.

Equation (6–1) deserves further consideration with respect to how it can contribute to the efficient *labeling* of quantities. There is often reason to provide in the immediate vicinity of a quantity symbol some indication of units applicable to that quantity, since only when this information is available will numerical values convey any real insight.

---

[5] DIN 1301-1 (1993) includes the observation: "Special unit names or unit combinations can be used to provide distinctions between quantities with identical dimensions: for example, the newton meter (N m) as a unit for moment of force rather than the joule, or the hertz (Hz) for frequency of a periodic occurrence and the becquerel (Bq) for activity with a radioactive substance in place of the reciprocal second."

This frequently becomes an issue with *axis labels* for graphs and *headers* for tables. It is not unusual to encounter the use for this purpose of a unit symbol set in square brackets immediately following the quantity symbol of interest—for example "*m* [g]" as an attempt to express the fact that data nearby are to be interpreted as *mass* values expressed in *grams*. Whatever its origin, this notation is most unfortunate, and clearly improper. Notice the role assigned to the square brackets in Equation (6–1): they embrace the symbol for a *quantity*, Q, not that of a unit! Since every quantity symbol (such as m, for instance, signifying mass)— according to the equation—already *has* a unit inherent in it, appending to such a symbol a unit symbol in square brackets, as in our example, would seem to imply (further) multiplication of the quantity in question *by* that additional unit, a nonsensical operation that would lead to an abomination like "6 g^2"! What is called for instead is a notation that provides the requisite information in the form of a *comment*.

- One straightforward, concise way to label a collection of numerical values in terms of the quantity measured and the units employed is simply to append to the quantity symbol a *comma* followed by the appropriate unit symbol.

Alternatively, the comma might be omitted and the unit symbol set in parentheses, possibly preceded by the word "in". Yet another legitimate approach—well-suited to a table header, for example—is setting the unit symbol *below* the quantity symbol. Or one could make use of the *full name* of the quantity rather than its symbol, eliminating all risk of the message being interpreted as a mathematical expression. These suggestions would lead to such labels as the following:

> Blood glucose level, mmol/L
> Blood glucose level (mmol/L)
> Radiation time (s)

- Still another possibility is to show explicitly a *division* of the quantity by its unit.

From Equation (6–1) it is apparent that if some particular quantity is divided by the corresponding unit, what results must be a pure number, and it is numbers of this sort that are what one actually plots in a graph, or enters into a table. Equation (6–1) is thus the origin of perhaps the best solution, one that has come to be quite common in the literature of late, in part as a result of its having been incorporated into various standards and explicitly approved by influential international organizations (e.g., DIN 461, 1973; BS-4811: 1972; IUPAC International Union of Pure and Applied Chemistry 1988, p. 3). We provide numerous examples of this type of notation in Chapters 7 and 8.

The various options proposed above, when applied to the simple case of a length l measured in mm, would result in the following possibilities for labeling the "length" axis in a graph:

> l, mm l (mm) l (in mm) l/mm

6.1.2 Derived Quantities and Functions

A key principle embodied in the SI is that all legitimate quantities in the context of the natural sciences have their roots in seven *base quantities*. By the same token, all quantities apart from these seven can be *derived* from the latter.

● Quantities that are not themselves base quantities are by reason of their origin called *derived quantities*, and in many cases *values* of derived quantities are expressed in *derived units*.

A selection of some of the more important derived quantities, all of which are associated with derived units, is presented in Table 6–2, based in turn on DIN 1301–1 (cf. also Table 6–1 in Sec. 6.1.1). The table entries are arranged in alphabetical order based on

Table 6–2. Important derived units and other units which have been assigned unique names and symbols.

Quantity[a]	Quantity symbol	Name of derived SI unit	SI unit symbol	Expression in terms of	
				SI base units	other units
activity[b]	A	becquerel	Bq	s^{-1}	
quantity of electricity, electric charge	Q	coulomb	C	$s\,A$	
Celsius temperature	Θ	degree Celsius[c]	°C	K	
capacitance	C	farad	F	$m^{-2}\,kg^{-1}\,s^4\,A^2$	C/V
inductance	L	henry	H	$m^2\,kg\,s^{-2}\,A^{-2}$	Wb/A
frequency	ν, f	hertz	Hz	s^{-1}	
energy, work, quantity of heat	E, W	joule	J	$m^2\,kg\,s^{-2}$	N m, W s
luminous flux	Φ	lumen	lm	cd sr	
illuminance	E	lux	lx	$cd\,sr\,m^{-2}$	lm/m^2
force	F	newton	N	$m\,kg\,s^{-2}$	
resistance	R	ohm	Ω	$m^2\,kg\,s^{-3}\,A^{-2}$	V/A
pressure	p	pascal	Pa	$m^{-1}\,kg\,s^{-2}$	N/m^2
plane angle	α, β, γ	radian	rad		
conductance	G	siemens	S	$m^{-2}\,kg^{-1}\,s^3\,A^3$	A/V
solid angle	Ω	steradian	sr		
dose equivalent	H	sievert	Sv	$m^2\,s^{-2}$	J/kg
magnetic flux density	B	tesla	T	$kg\,s^{-3}\,A^{-1}$	Wb/m^2
electric potential	U	volt	V	$m^2\,kg\,s^{-3}\,A^{-1}$	W/A
power	P	watt	W	$m^2\,kg\,s^{-3}$	J/s
magnetic flux	Φ	weber	Wb	$m^2\,kg\,s^{-2}\,A^{-1}$	V s

a   Arranged in alphabetical order of the corresponding *unit symbols*.
b   Activity of a radioactive substance, also called "radioactivity".
c   Strictly speaking, the degree Celsius is not a derived unit, but we will treat it as such for our present purpose; cf. eq. (6–3) and the related discussion in Sec. 6.1.2.

unit symbols. Other frequently encountered quantities and units, grouped by academic (sub)disciplines to which they relate, are to be found in Appendix B. A compilation in English prepared by IUPAC (the International Union of Pure and Applied Chemistry) has been published as a part of what is called that organization's "Green Book" (IUPAC 1988). The acknowledged international authority in the field of measurement, known as *metrology*, is the CGPM. The publications cited reflect standards adopted by the 11th CGPM, held in 1960, which was also responsible for the SI itself.

The potential list of quantities applicable within the natural sciences is limitless. So far only a few hundred of the most important have received "official" names and symbols (HAEDER and GÄRTNER 1980, DRAZIL 1983, FISCHER and VOGELSANG 1993). DIN 1304-1 (1994) provides *general formula symbols* for about 300 physical quantities in the fields of mechanics, electricity and magnetism, thermodynamics and heat transfer, physical chemistry and molecular physics, light and electromagnetic radiation, atomic and nuclear physics, and acoustics. More symbols pertaining to quantities important to specific disciplines are collected in a series of supplements entitled *Zusätzliche Formelzeichen* ("Additional Formula Symbols").

There is no question but that the list of official names and symbols will grow, and the latter have come to constitute an important subset of scientific *terminology*. The hope is that all internationally sanctioned names—with language-specific modifications as necessary—will eventually be acknowledged universally. A number of disciplines have introduced special terminology on their own, which in many cases has later been subjected to the attention of national standards organizations.

Column five of Table 6–2 is worth examining in some detail. The units listed there clearly reflect mathematical operations that relate the various quantities to their underlying base quantities, in essence verifying the validity of Equation (6–2).

● The mathematical operations capable of leading from base quantities to derived quantities are limited to multiplication and/or division.

Consider for a moment the derived quantity *density* (ρ), one of the "mechanical" quantities listed in Appendix B. The corresponding unit, "kg m^{-3}", makes it clear that density is obtained as a quotient of mass divided by volume, since "kg" is a unit of mass and "m^3" is a unit of volume. Turning now to semantics, an acceptable verbalization of the corresponding concept should be so structured as to be strictly accurate, a criterion fulfilled by "Density is mass divided by volume". If one were instead to assert that "density is mass per unit volume" (or something similar) it could be argued that the formal definition had been abandoned in favor of an *interpretation*. Linguistic nuances of this sort, in conjunction with the often imprecise application of concepts by beginners, clearly has an adverse effect on the learning process. We return to our concerns in this respect in our consideration of another unit in Sec. 6.3.1.

- Quantities subjected to combination and transformation through mathematical operations other than multiplication and division can serve as a source of what are known as *functions*.

An example is the *Gibbs Function*, $G = H - TS$, cited along with various thermodynamic *quantities* in the table of Appendix B. Another is pH, which as noted previously is defined as the negative logarithm of the numerical value of a hydronium-ion concentration.

An unusual "quantity" listed in Table 6–2 that is again *not* a (derived) quantity in the strictest sense, but rather a function, is the Celsius temperature. One might suspect that something is abnormal about this entry from the fact that Celsius temperature is reportedly measured in terms of a unit that (judging from column five) is identical to the base unit assigned to the *thermodynamic temperature*, known as the kelvin. What we actually have here is a function ϑ, measured in °C, where 1 °C (degree Celsius) = 1 K (kelvin), the defining expression for which is

$$T/K = \vartheta/°C + 273.15 \qquad (6-3)$$

(The somewhat uncommon symbol ϑ comes from the Greek alphabet. Because in this case the "official" symbol is not directly accessible with most keyboards it is often replaced by t.)

- All the internationally sanctioned quantity symbols are based on upper- or lower-case letters from either the Latin or the Greek alphabet.

In a very few cases, *two-letter* quantity symbols have been approved. Most involve "dimensionless" quantities related to mass or energy transport, such as the Reynolds number. Like all other quantity symbols, these symbols—and they really *do* represent quantities, not simply "numbers": quantities that happen to be characterized by the dimension "1"—should be set in italic type. To prevent the confusion that might be engendered by the presence of multiple letters (suggestive of multiplication!) it is recommended that all such symbols also routinely be enclosed in parentheses, e.g. (Re).

- Additional specificity can be conferred upon quantity symbols by the use of auxiliary identifiers known as *modifiers*.

Whenever possible, such modifiers are set in type significantly smaller than that employed for the quantity symbols themselves. Modifiers might take the form of letters, numbers, or special symbols, including ′ and ″ (the prime and double prime), † and ‡ (dagger and double dagger), ∧ (caret), * (asterisk), ~ (tilde), and ° (degree sign). Zero (0) and the infinity sign (∞) represent examples of *mathematical* symbols occasionally employed as modifiers. A modifier might appear either to the left or to the right of the parent quantity symbol, in either a subscript or a superscript position. In some cases a modifier is placed directly above or below the parent symbol. Illustrative examples include T_0, R_∞, U^*, m'', $\dot{V}$, and $\bar{u}$. The modifier "0" (used to signify a "zero quantity", a reference quantity, or a standard condition, for example) is often improperly simulated by a lower case letter "o", or—equally unsatisfactory—by a degree sign.

Chemists make extensive use of such modifiers ("indicators"), almost always in approved ways (DIN 1304-1, IUPAC). Thus, in conjunction with the symbol for a chemical element, a "left *super*script" is used to convey a *nucleon number* (or *mass number*: the total number of protons and neutrons in the nucleus of the species of interest), whereas a "left *sub*script" reports the corresponding *proton number* (*atomic number*), as in $^{14}_{6}C$. Above and to the right of an element symbol one sometimes finds a *charge number*, expressed as $n+$ or $n-$, while below and to the right there could be explicit indication of a certain *number of atoms* (of the variety in question) to be regarded as present. A typical example[6] of a completely annotated atomic symbol is $^{32}_{16}S^{2+}_{2}$.

With a word-processing program, subscripts and superscripts are easily created, and they can also be positioned vertically with considerable precision, typically in terms of a preferred number of "points" up or down from a standard position.

Introduction of dagger and double-dagger symbols is also straightforward with a word processor, since these symbols are standard components of most fonts due to their frequent use, along with the asterisk, for footnote labeling (Sec. 3.4.5). Other symbols important in research and teaching can pose a real challenge, however. This can be the case, for example, with the *tilde* (~), the caret (^, also called a "hat")—both considered to be in the category of *diacritical marks*[7]—and other signs that may require placement directly above (or below) a symbol, such as the *overbar* used to designate an arithmetic mean, or the dot used in differential calculus to indicate that a particular quantity is subject to change as a function of time.[8]

---

[6] The example reflects a doubly ionized (cationic) sulfur (S) species consisting of two atoms, both with atomic number 16 (actually a redundancy, since sulfur by *definition* has this atomic number) and mass number 32.

[7] Originally, the term "diacritical mark" was reserved for "symbols of emphasis or accent, designed to be placed either above or below the symbol for a letter", as expressed in a technical definition from the field of lexicography. The most familiar example, apart from the tilde (common in the form "ñ", as in "el niño"—which originated as "el nigno"!), is the *cedilla*, found under the letter "c" in a word like "garçon" as an indication that this particular "c" should be pronounced like an "s" rather than a "k", contrary to what one would expect in French for a "c" followed by "a", "o", or "u". This semantic/typographical technicality is ignored in DIN 5008 (1996), *Schreib- und Gestaltungsregeln für die Text-verarbeitung* ("Rules for Typography and Form in Word Processing").

[8] While word-processing programs offer extensive flexibility with respect to subscripts and superscripts, including vertical adjustment to within as little as a fraction of a point, *diacritical marks* are typically made available only in conjunction with a few specific letters to which they commonly apply (e.g. å, é, ç), and their placement is fixed. Freedom randomly to set a mark anywhere one wishes (as with an overbar that needs to appear in conjunction with an italic letter u: $\bar{u}$) typically requires access to special equation-editing software. An example of a program of this kind is EQUATION EDITOR, often supplied as a word-processor "accessory" (as with WORD and APPLEWORKS, for instance). EQUATION EDITOR is actually a subset of the program MATHTYPE, developed by Design Science, Inc. (See Sec. 6.7). Page-layout programs generally offer no special provisions for constructing formal equations, which must therefore be imported as graphic objects prepared elsewhere. The author with frequent occasion to work with unusual symbols whose placement must be precisely controlled would be well advised to investigate the capabilities (but also the complexity!) of LaTeX; cf. Sec. 6.6.

Subscripts and superscripts to the *right* of a quantity symbol are especially common. These are sometimes referred to as *indices* (singular: *index*), and are used to add specificity of many kinds. Such modifiers sometimes consist of multiple letters (or other characters), which may be organized with the help of diagonal lines or parentheses. The standard DIN 1304-1 (1994) provides in the context of Table 10 a number of examples of accepted notations, including the use of "r" for relative, "id" for ideal, and "nom" to indicate a nominal (or "reference") value.

6.1.3 *More Regarding Symbols and Their Representation*

The letters of the Latin and Greek alphabets, upon which the internationally accepted symbolism for quantities and units is based, are too few in number to offer sufficient unambiguous identification to satisfy all the scientific disciplines. This is in fact one of the important reasons why *italics* are employed in conjunction with quantity symbols.

● The availability of a second type style—italics—to complement the set of normal, upright (*roman*) letters effectively doubles the supply of potential symbols, at the same time providing a way clearly to distinguish between the symbols for quantities and those for units.

For example, "m" has been adopted as the standard symbol for (the *unit*) meter, and "*m*" for (the *quantity*) mass; a "g" indicates "gram", whereas "*g*" is the accepted symbol for the (standard) acceleration due to gravity.[9] In spite of this typographic expedient it has still been found necessary to assign multiple roles to a number of the letters of the alphabet. Thus, "*M*" is the official symbol for the quantity "molar mass", but according to IUPAC it can also stand for the magnetic quantum number, mutual inductance, torque, transition dipole moment, or radiant excitance.

● The potential for ambiguous interpretations, and the fact that there is not always total agreement regarding what symbol is "correct" in a given case, are among the reasons why it is recommended that technical scientific documents always include a *directory of symbols*, preferably near the beginning.

This directory (also referred to as a *list of symbols*) should contain a clear definition for each symbol utilized in the text (with appropriate attention to modifiers), irrespective of how "well-known" or typical for such a work the author considers a particular symbol to be. It is also a good idea, incidentally, in conjunction with every listed symbol to note any relevant unit(s), and to cite (or provide text references to) important defining equations or verbal definitions. As author you will find the symbol list to be useful for verifying that all your symbols transfer correctly from one text-processing environment

---

[9] The symbol g (more rigorously: g_n) actually represents not a unit or quantity but a *natural constant*, with (according to IUPAC) the value $9.80665 \ m \ s^{-2}$. A meticulous biochemist would report having conducted a centrifugation not at "5000 g", but rather at "5000 *g*".

to the next. An example of a list of symbols as it might be prepared by an author is shown in Fig. 6–1.

● It is inappropriate in a symbol list (or in a figure legend, or anywhere else, for that matter!) to use an "equals" sign as a way of linking a symbol to explanatory text.

The sole legitimate function of the mathematical "equals" sign is to signify a mathematical equivalency within the context of an equation. One frequently sees abuse of this rule in scientific text. We have given careful attention to the use of proper notation in Fig. 6–1; notice that the only place an "equals" sign appears in this illustration is in conjunction with the *Nusselt number*, a function the origin of which cannot easily be expressed in words and is therefore conveyed here through a defining equation. Obviously a faulty list of symbols you may somewhere have encountered cannot serve as a persuasive alibi for incorrect handling of concepts and symbols within one of your *own* documents.

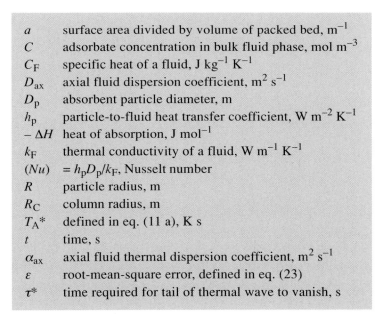

a	surface area divided by volume of packed bed, m^{-1}
C	adsorbate concentration in bulk fluid phase, $mol\ m^{-3}$
C_F	specific heat of a fluid, $J\ kg^{-1}\ K^{-1}$
D_{ax}	axial fluid dispersion coefficient, $m^2\ s^{-1}$
D_p	absorbent particle diameter, m
h_p	particle-to-fluid heat transfer coefficient, $W\ m^{-2}\ K^{-1}$
$-\Delta H$	heat of absorption, $J\ mol^{-1}$
k_F	thermal conductivity of a fluid, $W\ m^{-1}\ K^{-1}$
(Nu)	$= h_p D_p / k_F$, Nusselt number
R	particle radius, m
R_C	column radius, m
$T_A{}^*$	defined in eq. (11 a), K s
t	time, s
α_{ax}	axial fluid thermal dispersion coefficient, $m^2\ s^{-1}$
ε	root-mean-square error, defined in eq. (23)
τ^*	time required for tail of thermal wave to vanish, s

Fig. 6–1. Example of a typical author's list of symbols.

It was from a typographical (and conceptual) point of view that we first introduced in Sec. 6.1.1 the principle of setting in italics all symbols for physical (and mathematical) quantities wherever they appear in text, equations, and formulas. The subject deserves attention from a mechanical standpoint as well, however. Adhering to the rules in the course of typesetting a document, or while preparing text with the aid of a word processor, really poses no fundamental problems—apart from the need for vigilance. On the other hand,

- With copy prepared using a conventional typewriter it is necessary that all quantity symbols be expressed by hand, as a guide for the typesetter.

Careful marking is crucial, because no typesetter can be expected to recognize which symbols in an equation refer to quantities.

- Publishers' guidelines should be strictly observed with respect to the labeling of symbols that need to be set in italics.

For example, some publishers adhere to a convention that characters to be set in italics in formulas and equations should be *underlined* in *green*. It is sometimes permissible to indicate a need for italics within a block of *text* with a wavy underline instead, but this approach is impractical in conjunction with formulas. It is almost impossible, for example, to draw a distinct wavy line under a single letter "*i*" embedded somewhere within an equation. A straight line can be introduced much more easily, and if it is a distinctive color like green (despite the fact that such a line unfortunately will look black on a photocopy!) the resulting strong contrast with the background ensures that a typesetter will see the alert.

- Italicizing symbols is important wherever they happen to occur: not just in body text and equations, but also in figures, tables, headings, and other "special parts" of a manuscript.

Particular attention should be directed to *indices* of all kinds (or what we have earlier referred to as "modifiers"). An index that is in fact both a *variable* and an *integral part* of a quantity symbol—as in C_p, the heat capacity at constant pressure p—must necessarily be set in italics. On the other hand, if the index is merely an annotation, or an abbreviation—as in C_B, for the heat capacity of substance B—then it should appear in ordinary roman type.

An author preparing a book manuscript should establish with the publisher at an early date precisely how material to be set in italic type will be specified, and by whom, and especially the extent to which the task is the author's own responsibility. (Editorial guidelines should be consulted in the case of journal-article manuscripts.) Marking copy for italicization is neither difficult nor especially time-consuming, and it can easily be attended to "on the side" if a green pencil is kept handy during the final reading of a manuscript. Assuming you are comfortable working with a word processor, italics can be introduced as the corresponding characters are "typed" (most conveniently through use of a keyboard shortcut), but you will still need to verify that type-style information is properly interpreted later in the course of further processing.

- As you prepare equations or formulas, remember that anything that would be regarded as a "variable" must be set in italics.

As noted in Sec. 6.1.1, this rule is just as applicable to mathematical variables as it is to physical quantities; even the symbol "*f*" in reference to a *function* in general (not further defined) would be set in italics, as would an "*i*" signifying a running count or an "*n*" to be understood as representing any arbitrary number.

One apparent exception to the rules is the convention of using italics for *natural constants*, like the ideal gas constant *R*, or Planck's constant *h*. The rationale here is that such "constants" actually have essentially all the characteristics of a physical quantity. This would quite obviously be the case prior to reliable experimental determination of a definitive value, but even accepted values necessarily remain subject to refinement.

Approaching the matter of italicization from the opposite direction, we might construct a list of things that should definitely be set in ordinary *roman* type:

- Roman (upright) type is employed not only with the symbols for units but also those that represent specific numbers, "special functions", and defined mathematical operators.

The *special functions* category furnishes such symbols as exp, ln, lg, sin, cos, tan, cot, sinh, etc., as well as Hermitic polynomials $H_n(x)$ and the gamma function $\Gamma(x)$. Symbols for "defined" mathematical operators are ones like Δ, δ, D, and the *differential operator* d (i.e., a differential with respect to the variable x would thus be written as dx!). *Number symbols* that should be set in Roman type include those for the *imaginary* number i and the number that serves as the basis for natural logarithms (e, the Euler or Napier number), as well as the Greek letter π representing the factor that relates the area or circumference of a circle to its radius (or diameter).

Vectors constitute a special class of physical quantities. In a lecture setting these are frequently represented by showing an arrow above the appropriate letter (e.g., $\vec{F}$, $\vec{E}$), but this convention is rarely observed with typeset material.

- In most scientific publications a vector is distinguished by setting it in *boldface italic* type.

Thus, if one is concerned with *directional character*, ***F*** is the symbol for force and ***E*** the symbol for electric field strength. *F* and *E*, on the other hand, refer to the— directionally independent—*magnitudes* of these vectorial quantities: in each case a numerical value times a unit.

One further observation is in order with respect to indices. Occasionally a quantity requires "indexing" so specific that it becomes impractical to rely exclusively upon a simple subscript or superscript notation (see Sections 6.2 and 6.5.3). It has become customary to enclose lengthy "commentary" of this kind in parentheses, and then to set it on the same type line as the quantity symbol itself. Chemists are especially fond of the practice. Thus, between the following two alternatives for "concentration":

$$c_{H_2SO_4} \quad \text{and} \quad c(H_2SO_4)$$

the second is the one usually adopted. This has the advantage of avoiding the need to typeset a "subscript to a subscript" (entailing type of three different sizes set on three different horizontal lines!), which could well result in a symbol that was barely legible.

● It is therefore accepted practice to transform an "index" associated with a quantity symbol into ordinary characters set in parentheses *after* that symbol.

Chemists in certain branches of the discipline (e.g., kinetics and thermodynamics) prefer to represent concentrations in yet another way that also conveniently circumvents the awkward (but "systematic") "$c_{H_2SO_4}$". They attach special significance to *square brackets* such that "$[H_2SO_4]$" is understood to convey the message "concentration of sulfuric acid".

6.1.4 *Quantitative Expressions*

Despite the fact that the natural sciences pride themselves on being "exact", close inspection reveals that many of the expressions and formulations scientists employ routinely in lectures and publications are in fact quite *imprecise*. Consider the following examples from the field of chemistry:

> 3 kg sulfuric acid
> 3 kg sulfuric acid/kg sodium hydroxide
> 10^{12} neutrons/s

The kilogram (kg) can never be anything other than a unit of mass. There is thus no rational basis whatsoever for postulating a special entity called "kilogram sulfuric acid", as *appears* to have happened here. On the other hand, the meaning of a statement like "the mass of a particular sample of sulfuric acid is 3 kg" is perfectly clear, and this statement is easily expressed symbolically in the form "$m(H_2SO_4) = 3$ kg".

● Care should be taken never to mix numbers, mathematical operators, or units together with words, giving what might be mistaken for a mathematical *equation* of some sort.

This is the sin that has been committed in the remaining two examples above. Both contain a mathematical operator: the *slash* (which is a call for division). What could possibly be meant by "divide sulfuric acid by sodium hydroxide", or, in the last example, "divide neutrons by second"? Nonsense! Unfortunately, improprieties of this sort bear part of the responsibility for transforming *stoichiometry* (the "set theory" of chemistry) into something difficult both to teach and to learn![10] We as natural scientists often use our own "language" so carelessly that we are hardly in a position to berate the journalist who frightens the public with a report that the poisonous substance chlorine is a constituent of ordinary table salt!

In the case of the third example above (which may have been encountered in radiochemistry), the one involving neutrons, what one is dealing with is a report of a

---

[10] When it is (truly!) obvious that stoichiometry is in no way involved, a more relaxed mode of expression can be perfectly acceptable. For instance, there is no reason to take exception to an entry like "3 kg sulfuric acid" on an order form.

counting exercise. The fact that *neutrons* are the subject of the count has virtually no effect on the *form* of the expression. One correct formal way of stating the facts would be: "The neutron flux corresponds to 10^{12}/s". Similarly, in clinical chemistry it is wrong to use (either aloud or in writing!) an expression like "5000 leucocytes/mm^3"; much more acceptable is reference to a "leukocyte count of 5000/mm^3" or, better yet, "... of $5.0 \cdot 10^9$/L" (cf. Sec. 6.2.2).

To cite another example, the expression "... a content of 20 mg cantharidine/kg" should be replaced with "... corresponding to a mass-fraction of cantharidine (based on dry mass) of 20 mg/kg". If you regard the latter as too cumbersome, then settle instead for a purely verbal expression, replacing the notation "/kg" in the original with "per kilogram". A particular cell concentration could similarly be reported, for example, as either " $3 \cdot 10^9$ cells per milliliter" or simply "$3 \cdot 10^9$ mL^{-1}". Finally, when labeling the axis of a graph, "Number of nematodes per vessel" is much preferable to the shorter but formally nonsensical "Nematodes/vessel".

Certain disciplines have proven to be especially fertile fields for "pet units" that should be banished. Biochemists, for instance, are notorious for centrifuging samples at "1800 rpm". Why not more properly at (a rotation rate of) "1800 min^{-1}"? The latter formulation spares an uninitiated reader the agony of somehow discerning that "rpm" is not a unit at all, but an *abbreviation* for "revolutions per minute". Molecular geneticists on occasion report nucleic acid sequence lengths in terms of "bp", where 1 bp = 0.34 nm. But bp is again not really a unit: it is an (obscure?) abbreviation for "base pair". Nevertheless, some even resort to building upon this remarkable "unit", inflating it with an SI prefix (cf. Sec. 6.2.3) to "kbp" ("kilobase pair") or even "Mbp" ("megabase pair")—unbelievable! Standard *unit* prefixes have here been grafted directly onto *numbers*, because "bp" is simply an indication of a count, a quantity with the dimension 1. An example of an acceptable notation in this case is "Bacterio-phage FC 174 (NP 5375)", where "NP" is explained as shorthand for "number of pairs".

● Since the natural sciences place great stock in both a high level of *abstraction* and maximum *precision*, they tolerate the ambiguity associated with a *dual notation*.

Depending on the circumstances, either of two approaches can thus be taken to the expression of information (IUPAC 1979) without risk of being contested or forced to compromise. Consider the following two examples:

$$K_c = \prod_i (c_i)^{n_i}$$

$$\alpha(589.3 \text{ nm}, 20 \text{ °C}, 10 \text{ g dm}^{-3} \text{ in water}, 10 \text{ cm}) = 66.47°$$

The first expression above is highly *abstract*, offering a very compressed, mathematical version of the "essence" of an equilibrium constant. The equation itself has been so formulated that it is applicable to an infinite number of distinct equilibrium systems. Equation (6–2) in Sec. 6.1.1 could be described as "abstract" in a similar sense.

By contrast, the expression in the second example is highly *specific* (or *concrete*), conveying very precise information regarding the rotation induced in plane-polarized light by a particular solution of a particular "optically active" compound (here not identified). Based on measurements at the wavelength, temperature, concentration, and path length cited, the rotation was apparently observed to have a value of 66.47°.

Both examples represent perfectly proper equations—but of two very different types!—and each fulfills its purpose. There is no basis with the second, for instance, for objecting that it includes the verbal expression "in water". The "comment" appears exactly where it should, namely enclosed in parentheses *following* the quantity symbol α, so that it can be seen in the role of a modifier (index) or an explanation (see Sec. 6.1.3).

Finally, we feel obliged to repeat our admonition (from Sec. 6.1.3) that an *equals sign* (=) is to be employed only in a *mathematical* context, never with an explanation. Even a report of a melting point in the form "mp = –10 °C" is wrong, because "mp" is an *abbreviation*, not a (quantity) symbol. In other words, this would constitute an *explanation*, not an "equation" in the mathematical sense. The simple, correct way of expressing the observation in question is "mp –10 °C". Alternatively, a quantity "melting temperature" t_m might be appropriated (or devised), opening the way to a very different, but equally correct, formulation that *does* take legitimate advantage of an equals sign: "$t_m = -10 \,°C$".

6.2 SI Units

6.2.1 Base Units and Derived Units

In the preceding section we began our consideration of the SI by discussing dimensions and quantities, with relatively little attention to units. We did stress the fact, however, that each of the seven so-called base quantities has associated with it a unique *base unit*.

● The seven SI base units are the *meter*, the *kilogram*, the *second*, the *ampere*, the *kelvin*, the *mole*, and the *candela*.

These are used in measurements involving the base quantities *length*, *mass*, *time*, *electric current*, *thermodynamic temperature*, *amount of substance*, and *luminous intensity*, respectively. The kilogram (kg) is rather unusual, in that its official name incorporates an SI *prefix* (we examine the subject of prefixes in detail a bit later, in Sec. 6.2.3). The reason the kilogram was selected as the unit of mass rather than the seemingly more fundamental gram (g)—in contrast to the standard in the previously dominant international unit system (the CGS system, for centimeter–gram–second)— is of purely historical significance, and a subject we prefer not to pursue here. We will also dispense with discussing the ways the various base units are defined, and why they are so defined (cf. Sec. 6.3.1, however). For our purposes it is enough to emphasize that, for calibration purposes, there must be clear understanding throughout the world of precisely what each unit actually is (e.g., the precise *length* of a "meter"), as well as the potential for testing in a rigorous way the *accuracy* of that understanding. The reader interested in knowing more about the subject should consult more specialized literature (e.g., IUPAC 1988; the relevant definitions are also provided in DIN 1301-1).

● In addition to definitions and names, each SI base unit also has assigned to it a specific symbol (*unit symbol*), which is intended to be employed in conjunction with numbers (cf. Sec. 6.1.1).

Spellings for official names of the units are permitted to vary somewhat from language to language (e.g., mètre vs. meter), but the symbols take the same form everywhere. As noted, unit symbols usually appear along with—and following—numbers, separated from the latter by a *single space*: "2.4 mg", for example. When expressed verbally, or in writing, if the accompanying numerical information is spelled out in words, complete unit names are also used rather than symbols (e.g., "three meters high", not "three m high"). No period should ever be placed after a unit symbol—except at the end of a sentence, of course!—because these are to be understood literally as *symbols*, not abbreviations. For this reason, alternative "word fragments" should never be used as substitutes for unit symbols or as unit abbreviations (such as "sec" instead of one of the proper forms "second" or "s").

The same unit names are retained when units are raised to powers, as in the *cubic* meter (m^3) or the *square* millimeter (mm^2).

The German standards organization (Deutsches Institut für Normung, DIN) thought it worthwhile to add in its publication DIN 1301-1 (paragraph 7.1) explicit guidelines regarding the grammatical *gender* of units. Thus, this source asserts that

● Words used as the names of SI units are understood in German to be neuter (e.g., *das* Meter).

Despite what appears to be a straightforward pronouncement, 14 "exceptions" are then noted (from "die Sekunde", through "die Tonne", to "der Grad Celsius"), but the venerable "Meter" is not among them. Explicit sanction is thus granted in German dictionaries and grammar treatises to the neuter "das Meter" alongside the everyday masculine "der Meter".

● Unit symbols are capitalized when the corresponding unit name is derived from the name of a person. Otherwise they are written lower-case—with the single exception of the liter, where both l and L are permitted (despite the lack of a "patron").[11]

---

[11] Originally only the lower-case symbol was granted official recognition. Approval of the capital as an alternative was based on the persuasive argument that the letter "l" is too easily confused in print with the numeral "1".

Note that in English, unit *names* that are derived from the names of persons are *not* capitalized!

6.2.2 Derived Units and "Supplementary" Units

We have already mentioned the concept of derived units in the context of derived quantities in Sec. 6.1.2. These are units constructed from base units by some sequence of multiplication and/or division steps. A few have special names, such as "newton" for "kg m s^{-2}" (cf. Table 6–2 and Appendix B).

The practicality of the derived unit concept is evident in its widespread acceptance. The newton, for instance, has been readily embraced by the public generally as a unit of force (as has the "newton meter" for torque, in conjunction with torque wrenches, for example). This can probably be attributed in part to the compactness of the designation. On the other hand, the otherwise fairly rational populace of the United States continues adamantly refusing even to consider abandoning its beloved "English-system" units, like the gallon and the mile (along with the foot, the yard, the inch, etc.), in favor of the metric units adopted essentially everywhere else in the world—this despite peaceful acquiescence long ago on the part of their English and Canadian brethren.

Some would reserve the term "derived unit" for units to which unique names have been assigned, all others (except for the base units) being instead referred to as *compound units*. Examples of the latter are the "nameless" units "C m" for the electric dipole moment and "S m^2 mol^{-1}" for molar conductivity, enunciated "coulomb meter" and "siemens meter squared times mole to the minus one", respectively.[12]

Derived units in the broader sense can also be divided in another way into two categories: "coherent units" and "noncoherent units".

- Derived units composed exclusively of base units, and entailing no numerical factors other than 1, are called *coherent derived units*.

An example would be the density-related unit "mol kg^{-1}". The unit for force, the newton (1 N = 1 kg m s^{-2}), is another coherent derived unit, as is the pascal, a unit of pressure (1 Pa = 1 kg m^{-1} s^{-2}). It is to the SI's credit, incidentally, that numerical factors seldom play a role in the more common derived units.

- Derived units incorporating non-base units, or a combination of base units and *numerical factors* are then *noncoherent derived units*.

One of the latter is the *bar* (1 bar = 10^5 kg m^{-1} s^{-2}), although in this case the numerical factor at least has the virtue of being a power of 10, hence in general harmony with metric principles. A less obvious example of a noncoherent unit is the concentration unit mol/L, so classified because the liter (L), equivalent to 10^{-3} m^3, is *not* an SI base unit.

---

[12] Another way to articulate a unit with the exponent "–1" is "reciprocal second" (s^{-1}), for example, or "reciprocal meter" (m^{-1}).

● Units expressed as combinations of other units are generally written symbolically such that a *space* is left between each pair of constituent unit symbols, although separation by dots of the sort used to indicate multiplication is also allowed.

In other words, one can correctly write either "0.3 N m" or "0.3 N · m", though the latter form is encountered rather infrequently. Once a choice of notation in this respect has been made, that notation should be applied consistently throughout a manuscript, including in tables and figures.

A single space is also left between a unit symbol and its associated numerical value (as already noted above, and in passing near the end of Sec. 5.4.1).

● An alternative to the *negative exponents* that sometimes appear in compound units is insertion of a slash (or "solidus") to indicate *division*. If this option is selected, *groups* of units that follow the slash should be enclosed in parentheses.

Accordingly, both J/(K mol) and J/(K · mol) are appropriate substitutes for J K^{-1} mol^{-1}, but J/ K mol and J/ K · mol are not. IUPAC in its "Green Book" (IUPAC 1988, p. 8) points out that formulations without parentheses actually are legitimate (even if undesirable), but stresses the importance of observing the rule that "multiplication takes precedence over division in the sense that a/bc should be interpreted as $a/(bc)$ rather than $(a/b)c$". We still see here a serious potential for ambiguity (as does IUPAC!), and concur with the recommendation in DIN 1301-1 that parentheses should always be supplied (cf. also Sec. 6.5.5). Also to be avoided are expressions lacking parentheses that contain multiple slashes (i.e., combinations of the type a/b/c). We revisit this "rule" in Sec. 6.5.5.

Written or verbal expressions like "kilograms *per* cubic meter" for kg m^{-3} are no longer considered proper. The units committee of DIN offers two options in this case: either "kilograms divided by cubic meters" or "kilogram meter to the minus three power", both of which underscore the mathematical origin of the compound unit. We alluded to a similar quest for transparency once before in this chapter (in Sec. 6.1.2, in the context of names for quantities).

Table 6–2 and Appendix B include several derived SI units that have acquired special names honoring distinguished scientists of the past. A case in point is the newton (N); others include the watt (W), the volt (V), the pascal (Pa), and the ohm (Ω).[13] As these examples show, the corresponding unit symbols consist of single capital letters (an omega from the *Greek* alphabet in the case of the ohm), or a capital letter followed by a lower-case letter. All other unit symbols for familiar quantities consist of lower-case letters from the Latin alphabet, apart from the "L" for liter mentioned earlier. The symbol "mol", representing the unit associated with the base quantity "amount of substance" (see Sec. 6.3.1), is unique among the symbols for base units in that it consists of three letters, but certain "supplementary" units mentioned below and in Table 6–3 also have three-letter symbols; e.g., "min", "rad", and "bar". The unchallenged record-

---

[13] A valuable and amusing essay on the subject is *Mein Name ist Becquerel* (SCHWENK 1992).

Table 6–3. Supplementary units.

Quantity	Quantity symbol	SI unit	Other units and their names
plane angle	α, β, γ	rad	degree (°), $1° = 1$ rad $\cdot \pi/180$ minute ('), $1' = 1°/60$ second ("), $1'' = 1'/60$
area	A	m^2	hectare (ha), 1 ha = 10^4 m^2 are (a), 1 a = 100 m^2
volume	V	m^3	hectoliter (hL), 1 hL = 10^{-1} m^3 liter (L oder l), 1 L = 10^{-3} m^3 centiliter (cL), 1 cL = 10^{-5} m^3 milliliter (mL), 1 mL = 10^{-6} m^3
time	t	s	day (d), 1 d = 24 h hour (h), 1 h = 60 min minute (min), 1 min = 60 s
pressure	p	Pa	bar (bar), 1 bar = 10^5 Pa = 10^2 kPa $= 10$ N/cm^2
activity (of a radionuclide)	A	Bq	curie (Ci), 1 Ci = $3.7 \cdot 10^{10}$ Bq
exposure to nuclear activity	X	C kg^{-1}	röntgen (R), 1 R = $2.58 \cdot 10^{-4}$ C/kg
dimensionless quantities	n		decibel (dB), $n = 10 \lg(Q_1/Q_2)$, where n is the number of decibels

holder in the letter-count category is "mmHg" ("millimeters of mercury"), used for measuring blood pressure—and occasionally still for reporting atmospheric pressure.[14]

● A few non-systematic units have been accepted (reluctantly) for use within the context of the SI; these we call *supplementary units*.[15]

Most of the "supplementary units" (see Table 6–3) involve the dimensions length and time. Examples include "a" (for are) and "ha" (hectare), both units of area; "L" (liter), a measure of volume; and "min" (minute), "h" (hour), "d" (day), and "a" (year), all measures of time. Despite the unfortunate assignment of a dual role here to the letter "a" (for both "are" and "year"), this particular symbol for year is gradually becoming familiar in expressions involving long periods of time. Archeological works sometimes take advantage of the extended unit "ka" (for "kiloyear") in quantifying *very* large blocks of elapsed time.

Standard DIN 1301-1, already cited several times, deviates from the account we have given in a few particulars. For example, a distinction between "coherent" and "noncoherent" units is not made in the DIN document; indeed, the term "coherent" is not even mentioned. Instead, the DIN standard distinguishes between *three* types of derived SI units. The first is introduced in a Table 2, "Derived SI Units with Special

---

[14] This particular unit is in any case rather strange, especially as it is based on a specific measurement technique, involving a column of mercury, not on that which is to be measured.

[15] The IUPAC "Green Book" refers to them as "units in use together with the SI".

Names and Special Unit Symbols". This table includes 21 units, all defined by relationships such as

$$1 \text{ V} = 1 \frac{\text{J}}{\text{C}} = 1 \frac{\text{m}^2 \cdot \text{kg}}{\text{s}^3 \cdot \text{A}}$$

Notice that no powers of ten or other numerical factors appear, and the same applies to all the other units listed; in other words, all are examples of "coherent units". Sixteen of them bear the names of scientists, a collection to which might arguably be added the "degree Celsius". The remaining units in this particular table are the radian, the steradian, the lumen, and the lux.

Some of the "supplementary units" we present in our Table 6–3 appear in this DIN standard in a different Table 3, entitled "Generally Applicable Units Outside the SI": specifically ones for (plane) *angles* (with symbols °, ′, and ″), *volume* (represented here only by the liter), *time* (min, h, and d), *mass* (including both the gram and the metric ton), and *pressure* (where the bar has survived). The are and the hectare for measuring land areas are consigned to yet *another* table, "Units Outside the SI with Limited Areas of Application", together with the *atomic mass unit* u and the *electron volt* eV, as well as a few special units like the *barn* (b), which atomic physicists employ for measuring atomic cross-sections, the optician's *diopter* (dpt), and the physician's millimeter of mercury (mmHg), mentioned previously.[16]

Several outmoded units of measure that strictly speaking have no real place in the SI thus continue to survive, albeit somewhat tenuously. One of these, the bar, presumably owes its persistent acceptance to the fact that it is a convenient substitute for the clearly obsolete "atmosphere" (atm), one that introduces very little error.

The unit for *atomic mass*,[17] to which passing reference was made above, is defined by the equation

$$1 \text{ u} = 1.660\ 540\ 2 \cdot 10^{-27} \text{ kg}$$

In biochemistry the same unit is instead called the *dalton* (Da), frequently employed in the form "kilodalton", kDa, as for example in an expression like "... a 34-kDa protein".[18] The numerical value when the mass of an atom or molecule is expressed either in daltons or in atomic mass units is equivalent to the numerical value of the *relative particle mass* (see Sec. 6.3.2). The unit "u" is thus technically superfluous,

---

[16] To our surprise, the *decibel* for measuring sound levels seems to have been abandoned altogether.

[17] According to the "Green Book", the correct term is *unified* atomic mass.

[18] Retention of the dalton represents a special concession to biochemists and molecular biologists. This unit has no logical place in the SI, but its elimination would have encountered serious resistance. The dalton is a *counting unit* that treats molecular masses as multiples of the mass of the normal hydrogen atom. A parallel can be drawn here, both formally and in terms of content, not only with the "base pair" ("bp") mentioned earlier, but also with the "K" (for "kilobyte") cherished by information scientists. We would regard it as quite appropriate, in fact, if the information sciences were to be granted *their* own "official" unit, presumably the "byte" (or the "bit", or both), functioning—like the chemist's "u" or "mol"—as a counting unit. Expanding the scope of the SI in this way would emphasize common ground shared by all the "exact" sciences and also pay tribute to what some would call the "premiere science" at this the dawn of the twenty-first century. (We touch upon the point again in Sec. 6.3.1.)

for which reason we have not included it in Table 6–3. Some would argue it is not even a true "unit", but rather a natural constant m_u, the *atomic mass constant*.

The *percent sign* (%) and certain related entities (symbols) could be classified as representing "special units" for expressing "quantities of a relative nature", but their use is not recommended in the context of the SI. Indeed, the corresponding function should really be assumed by numbers alone.

- A better way to view the percent sign is as a *mathematical operator*, to be interpreted as an instruction to "multiply by 0.01" (or 10^{-2}).

In other words, 50% = 0.5. This can be confirmed by "referring" the corresponding 50 to the "100" implicit in the word "cent"; i.e., 50/100 = 0.5. The same principle applies to the occasionally encountered "per-thousand" (or "per mill", ‰), as well as to ppm (*parts per million*), and both ppb and ppt (*parts per billion, parts per trillion*).[19] The relevant conversion factors are as follows:

%	‰	ppm	ppb	ppt
10^{-2}	10^{-3}	10^{-6}	10^{-9}	10^{-12}

In considering the possibility of describing quantities in *relative* terms like these, remember that such comparisons must be between *like* entities. Such "units" could therefore not be utilized for expressing, for example, the overall number of particles present within a given volume. The symbol "‰" is common only in continental Europe. Because not all readers will be familiar with it, its use should be avoided in scientific publications.

Opinions differ regarding the correct typographic treatment of the percent sign. Some interpret "%" as no different from other unit symbols, which means a blank space is required between the sign and its associated number ("15 %"). Others[20] treat this particular symbol as a special case in which, for sake of appearance, the extra space is better omitted. Aesthetic considerations are evidently winning out over formal consistency, since a highly respected German style guide (*Duden Satz- und Korrektur- anweisungen*, p. 49) has tempered its original recommendation, calling now for introduction of only "a small space". We, too, feel more comfortable with the notation "15.5 %" relative to either "15.5%" or "15.5 %" (see Sec. 6.5.6).[21]

---

[19] The names of the two last-cited "operators" presuppose the definitions of "billion" and "trillion" applicable in the *United States* (10^9 and 10^{12}). In much of the world, "billion" refers to a quantity *larger* by a factor of 10^3, and "trillion" actually has various meanings in various places! This dangerous source of potential ambiguity is important to bear in mind.

[20] For example, the *Style Manual* (1978) of the American Institute of Physics, where the observation appears on p. 16 that "some exceptional symbols are not spaced off from the number", the examples cited being %, °, and °C.

[21] The first representation can be achieved with a word processor by "condensing" a fixed blank space by two points. Alternatively, the extra space can be omitted and the last digit before the percent sign instead "expanded" by one or two points. This has the advantage that the expression will never be subject to fragmentation at the end of a line. A custom keyboard shortcut for "Expanded 1 pt" (or perhaps 1.5 pt) can save a great deal of bother in the preparation of a long manuscript.

6.2.3 Prefixes, Decimal Points, and Other Stylistic Matters

In view of the wide application ranges characterizing the dimensions in which we quantify nature—the dimension *length*, for example, encompasses distances from the atomic to the galactic—it would be folly to insist that measurements be restricted to the few units listed in Tables 6–1, 6–2, and 6–3, together with Appendix B. The problem has been solved in a rather ingenious way:

- *Prefixes* are added to extend dramatically the ranges covered by the SI units.

These prefixes serve to expand or contract the units themselves by various factors of 10. That is to say, one is able to increase or decrease a unit's "value" in steps corresponding to multiples of ten thanks to the availability of an extensive set of officially sanctioned prefixes. Like the units themselves, each prefix is associated with a unique *prefix symbol* that can be combined with the relevant unit symbol. Prefix symbols, with but two exceptions, consist of single letters of the Latin alphabet. The exceptions are the ten-fold prefix ("deca"), represented by "da", and the millionth-fold prefix ("micro"), to which has been assigned as a symbol the lower-case Greek letter μ (mu). Symbols for prefixes associated with orders of magnitude of $10^{\pm 6}$ or greater are based on capital letters, all others on lower-case letters (see Table 6–4).[22]

- A prefix symbol must never be allowed to stand alone—as a replacement, for example, for a power of ten.

Use of the prefix symbol "da", while of course permitted, is probably unwise. What is the likelihood, after all, that a typical reader would immediately recognize "1 das" as being synonymous with "10 s"? Several other prefix symbols listed in Table 6–4 are increasingly looked upon with disfavor. Indeed, there is now a clear tendency (cf. DIN 1301-1) to express quantities exclusively with unit clusters whose components reflect "thousand-fold leaps". In other words,

- Routine use of the prefixes hecto (h), deca (da), deci (d), and centi (c) is discouraged.

For example, either "0.320 m" or the equivalent "320 mm" is generally regarded as preferable to "32.0 cm", as is "0.713 L" (or "713 mL") relative to "71.3 cL". In a case like "30 cm", only one "better" alternative is apparent ("0.30 m"), because "300 mm" would imply an unwarranted level of precision.[23] "Recommended" prefix symbols are indicated in boldface type in Table 6–4.

Along with these suggestions, and the comparable recommendations in DIN 1301–1 (1993), one should take note of the further advice that

---

[22] DIN 1301-1 includes the newly adopted prefixes zepto, zetta, yocto, and yotta, extending the overall range of coverage from 10^{-24} to 10^{24}.

[23] This is obviously not to suggest that lengths should not be specified to three-place accuracy; indeed, the Compton wavelength of the proton has been determined and reported to *eight* significant figures. But "300 mm", strictly speaking, is the equivalent of 30.0 cm, *not* 30 cm (cf. Sec. 6.4).

Table 6–4. Unit prefixes. Symbols for the preferred prefixes appear in boldface. The first two and last two entries are relatively recent additions on the part of the CGPM.

Factor by which the unit is multiplied		SI prefix	Symbol
1 000 000 000 000 000 000 000 000	$= 10^{24}$	yotta	**Y**
1 000 000 000 000 000 000 000	$= 10^{21}$	zetta	**Z**
1 000 000 000 000 000 000	$= 10^{18}$	exa	**E**
1 000 000 000 000 000	$= 10^{15}$	peta	**P**
1 000 000 000 000	$= 10^{12}$	tera	**T**
1 000 000 000	$= 10^{9}$	giga	**G**
1 000 000	$= 10^{6}$	mega	**M**
1 000	$= 10^{3}$	kilo	**k**
100	$= 10^{2}$	hecto	h
10	$= 10^{1}$	deca	da
1	$= 10^{0}$		
0,1	$= 10^{-1}$	deci	d
0,01	$= 10^{-2}$	centi	c
0,001	$= 10^{-3}$	milli	**m**
0,000 001	$= 10^{-6}$	micro	**μ**
0,000 000 001	$= 10^{-9}$	nano	**n**
0,000 000 000 001	$= 10^{-12}$	pico	**p**
0,000 000 000 000 001	$= 10^{-15}$	femto	**f**
0,000 000 000 000 000 001	$= 10^{-18}$	atto	**a**
0,000 000 000 000 000 000 001	$= 10^{-21}$	zepto	z
0,000 000 000 000 000 000 000 001	$= 10^{-24}$	yocto	y

● In reporting quantity values, prefixes should be selected such that the cited numerical values fall between 0.1 and 1000.

In other words, "30 mL" is a better choice than " 0.030 L".

● Prefix symbols precede the associated unit symbols, with no intervening blank space; *multiple* prefixes are not allowed.

To avoid possible confusion between the *prefix* m (milli) and the *unit* m (meter), the symbol for meter should always be situated near the end of a composite unit specification (i.e., N m for "newton meter", rather than m N, which—despite the space—might be mistaken for "millinewton".

● When expressing derived units, restrict prefixes to units that appear in the *numerator*.

This recommendation has the interesting consequence in clinical chemistry, to take but one example, that a hemoglobin concentration should no longer be reported as "15.5 g/100 mL", for example—standard practice in the past—but rather as "155 g/L". Incidentally, the restriction here does *not* require that the unit "kg" be excluded from the denominator, because kg is a *base* unit.

6.3 Special Units in Chemistry

6.3.1 "Amount of Substance" and the Mole

The 11th CGPM (see Sec. 6.1.2) incorporated the "countability" of a collection of a vast number of particles as a discrete dimension in the scientific measurement system—for the benefit of chemists. The corresponding quantity is called *amount of substance*, with the *mole* as its unit.

● A mole (unit symbol "mol") is the amount of substance of a system containing as many elementary entities as 0.0120 kg of carbon-12.

The "elementary entities" in question (for terminology, cf. IUPAC 1988) must always be specified. They might be atoms, ions, molecules, or any other "building blocks"—chemists prefer the term "particles"—from nature, including such non-material "entities" as photons. Indeed, the amount-of-substance concept is utilized routinely with respect to photochemical excitation phenomena and the like. In principle one could also regard *galaxies* as "systems" in the sense of the definition, quantifying them in terms of moles, with individual stars constituting the "elementary entities".

The above definition for the mole—the only unit definition we consider in this book—is closely related to the "atomic mass" concept. The atomic "hypothesis" thus becomes a pillar in the international system of standards, casting off at long last its aura of the hypothetical, for without the existence of elementary particles, "countability" in this context would have no meaning. Atoms must be accepted as objects just like screws, envelopes, and countless other things.

● The actual number of entities in a mole—of anything—is the *Avogadro number*, $6.022 \cdot 10^{23}$ (sometimes referred to in Europe as the "Loschmidt number").

This number is often expressed in conjunction with the unit "mole" in such a way as to produce a *natural constant*, the Avogadro constant N_A:

$$N_A = 6.022 \cdot 10^{23} \text{ mol}^{-1}$$

In view of the enormity of the number 10^{23}, "countability" in the strictest sense of the word remains a rather abstract notion when applied to the amount of substance. In order to achieve the underlying goal of a "count", chemists rely on a combination of exact mass measurements for individual particles (obtained with the help of a mass spectrometer, for example) and mass *comparisons* involving equivalent amounts of different materials. If one knew that a screw weighed 1.0 g and a comparable nut 0.50 g, there would be no need to conduct an actual count to establish with certainty that 1 kg of screws contained the same number of "entities" as 0.5 kg of nuts, and that the number in each case must be 1000 (assuming that all the screws and all the nuts were truly identical, and that the mass measurements were highly accurate).

The name "amount of substance" assigned in English to this newly-defined quantity has been subject to considerable criticism, largely because the word "amount" carries with it too much baggage from everyday speech.[24] To simplify phrasing, at least within the confines of chemistry, "amount" alone is accepted as a legitimate abbreviated version of "amount of substance", allowing one to fall back on a formulation like "the amount of glucose present is 1 mol".

Beyond this, a more far-reaching critique asserts that there was no real need to "invent" the concept "amount of substance" in the first place, because its unit, the mole, is nothing more than a counting unit (analogous to the dozen).[25] In our judgement it is pointless to raise such issues now, however, since the SI definition has already been accepted. We see it as more important instead to urge that the authorized vocabulary be employed consistently and correctly. Expressions like "number of moles" (for amount of substance) should be banned forthwith from the literature, because no one would even consider treating the analogous "number of meters" as a synonym for the abstract concept "length"!

6.3.2 Molar Quantities and Mixtures of Substances

In view of the importance of the amount-of-substance concept in chemistry, it should come as no surprise that other concepts have been based upon it. We regard the consequences—which after all flow from one of only seven base quantities—as extremely important, even (or especially!) for a broad scientific readership. For this reason we digress at this point for a brief elaboration.

● Special significance is attached to *extensive quantities* (cf. Sec. 6.1.1) that are based on amount of substance, referred to as *molar* quantities.

The mole is thus the only unit whose name can be transformed into an adjective (disregarding trivial usages like "a three-minute delay" or "a high-voltage source"). Molar quantities are indicated by a subscript "m", and can be characterized generally by:

$$G_m = G/n$$

Here the quantity symbol G might represent mass, volume, heat capacity, or virtually any other extensive quantity of interest with respect to unit amounts of substance, and

---

[24] A similar problem arises in other languages as well; cf. "Stoffmenge" for "amount of substance" in German.

[25] In the information sciences there exist the somewhat similar units *bit* (unit symbol: bit; the term is derived from "binary digit") for the smallest unit of storable information, and *byte* for eight adjacent bits, a unit of computer memory equal to that needed to store a single character. Binary digits here play the same role as particles in the material world, serving as the "atoms" of information science. As has already been noted in Sec. 6.2.2, the bit could reasonably be adopted as an eighth base unit in the SI. In molecular genetics, amount-of-substance information might then be conveyed in terms of bits rather than moles (or base pairs). Information scientists make use of prefix symbols "K" (instead of "k"), "M", etc., so some sort of accommodation would need to be reached here with the SI (cf. Sec. 5.2.1).

n represents a *particular* amount of substance. Units related to molar quantities always contain the factor "mol^{-1}".

● The numerical value of a *molar mass* (the mass of a unit amount of substance for some entity) is identical to what is called the *relative particle mass*.

The quantity molar mass has "g mol^{-1}" as its unit. Note that the old terminology "atomic weight" and "molecular weight" is no longer appropriate, and *definitely* not recommended (IUPAC 1988).[26] The preferred expressions now are *relative atomic mass* and *relative molecular mass*, M_r, respectively.[27]

Additionally it should be noted that certain extensive quantities have also been defined in a relative way with regard to *mass*. These are known as *specific* quantities. An example is the specific heat capacity, *c*, also known as the "mass-based heat capacity" and measured in terms of the unit J/(kg K). But we need to turn our attention now to another important consideration, namely the correct description of *mixtures*.

At first glance the amount-of-substance concept seems applicable only to collections of identical entities (e.g., particles), since different entities cannot be counted together (unless, of course, "apples" and "pears" are both treated as "pieces of fruit"). However, one could certainly determine the various amounts of substance applicable to several different substances in a mixture, and then add all the results together—if there were reason to do so. The amount of substance of *one* of the substances present could then be expressed relative to the sum of *all* the amounts of substance of interest, leading to an *amount-of-substance fraction*, which must of course be a number < 1. The symbol adopted for an amount-of-substance fraction is "κ"; with the more precise formulations "κ_B" or "$\kappa(B)$" in the case of a particular substance B (see appendix B). One-hundred times this number would represent, for each 100 particles in the mixture, the number that are particles of substance B. The same information was traditionally communicated in the past as a "mol-%", or in the case of atoms an "atom-%". The expression "mole fraction" for an amount-of-substance fraction is also now obsolete, and thus should no longer be employed.

Examples of other widespread but nonetheless improper terminology are "wt-%" and "vol-%". It is illogical in any case to combine a mathematical operator (%) with the abbreviated name for a quantity and claim the result to be a unit! Specifications like "% (g/g)" or "% (vol/vol)" are less objectionable—and actually rather efficient because of their brevity—since what has been appended here in parentheses can be regarded as an *explanation*. Optimal (because precise) modes of expression include:

---

[26] These outdated terms never did involve *weights* as such; from a measurement standpoint they referred at most to *masses*, and viewed as pure numbers they were neither the one nor the other.

[27] This preference has in the meantime become part of the standard DIN 1304-1 *Formelzeichen: Allgemeine Formelzeichen* ("Formula Symbols: General Formula Symbols") dated March 1994. M_r is described in this source as the "relative molecular mass of a substance", and the analogous A_r as the "relative atomic mass of a nuclide or an element".

$$\kappa = 0.50 = 0.50 \text{ mol/mol} = 50\%$$
$$\omega = 0.50 = 0.50 \text{ g/g} = 50\%$$
$$\varphi = 0.50 = 0.50 \text{ L/L} = 50\%$$

where for sake of simplicity we have assumed that precisely one-half of an (overall) amount-of-substance (or weight, or volume) is attributable to one component. The quantities symbolized by ω and φ could be called *mass fraction* and *volume fraction*, respectively, in analogy to amount-of-substance fraction.

● Another quantity chemists find useful is the amount-of-substance-based *concentration*, the quotient obtained by dividing the amount (i.e., amount of substance) of some dissolved material by the total *volume* of the corresponding solution.

This quantity has been assigned the symbol c, with the unit "mol L^{-1}". The concentration of a particular component B might be symbolized by c_B or $c(B)$, though more common is the notation [B] (see Sec. 6.1.3). In place of the correct designation "amount-of-substance concentration", many people still refer to "molar concentrations". This practice is especially regrettable—and should be firmly resisted—because the word "molar" now has a very specific (and *different*) meaning (see above). The use of "M" as a substitute for "mol L^{-1}" ("0.1 M HCl") must also be condemned, and the same holds true for the "N" used to signify "normal" (in a characterization based on *equivalent* mass rather than *molecular* mass). It is simply wrong for individual disciplines to invent their own units, or encourage the use of substitute ("ersatz") units! Moreover, the expression "0.1 M HCl" has much in common with the "3 kg sulfuric acid" we emphatically rejected in Sec. 6.1.4. Granted the expression "2 mL of a solution, $c(1/5 \text{ KMnO}_4) = 0.1$ mol/L" is longer and substantially more cumbersome than "2 mL 0.1 N KMnO$_4$", but the first formulation is more generally intelligible, and it is more easily conveyed to students. (We base this assertion on first-hand experience in the training of laboratory technicians; indeed, we fervently hope that the notations M, N, mM, µM, etc. will soon completely disappear from the scene!) There can also be no doubt that precise formulation simplifies the process of carrying out stoichiometric calculations. Even computers become baffled trying to interpret M and N. It is a hopeful sign that labels on commercial reagent bottles are increasingly being revised to conform to the new terminology.[28]

---

[28] In addition to reagent manufacturers, textbook authors (at least in Europe!) began showing an increased willingness to follow the new guidelines, but then along came more trouble in the form of DIN 1304-1 (1994), which claims that the correct unit for amount-of-substance concentration is now "mol m^{-3}". This leads to numerical concentration values 1000-fold greater than ones based on the previously specified unit. where the *liter* was the unit of volume. It is not yet clear to what extent this unfamiliar application of the cubic meter will ultimately be accepted.

6.4 Numbers and Numerical Data

After this lengthy discussion of numerical values and countability, we now must turn our attention to numbers themselves and how they are to be written, and in this context to a few aspects of *mathematical notation* as employed in the natural sciences—too often incorrectly!

● Based on an old rule from the printing trade, the numbers 1 through 12 are always expressed in *words* in running text, while 13 and larger are represented numerically.

Decimal numbers and fractions are of course almost always written in terms of numerals, as in "2.5 liters" or "$1\frac{1}{2}$-fold".

● It is also customary to avoid placing a numeral at the beginning or end of a sentence.

The rule specifying that the first twelve whole numbers should be spelled out rather than represented by numerals need not be observed absolutely. Thus, numerals in running text stand out relative to the surrounding words, and this is a characteristic worth taking advantage of in the case of numbers that have special significance, or in the event that other numbers are already present elsewhere in the sentence (e.g., "Overall, 14 elements were studied, 9 nonmetallic and 5 metallic").

Three forms are acceptable for expressing fractions, as shown in the following examples:

$$\frac{3}{19} \qquad 3/19 \qquad {}^3/_{19}$$

Notations like $5\frac{1}{4}$, 5 1/4, or 5 $^1/_4$—all of which actually mean "5 + 1/4"—should be employed only when there is absolutely no risk of ambiguity.

● Numerals are always to be used in formulations that include a unit symbol or a percent sign (%). In other words, write "2%" or "2.0 g", but "two percent" or "two grams".

● Numbers that contain more than four digits to the left or right of the decimal point should be presented, starting from the decimal point, in *three-digit groups* (*triads*), defined by spaces.

Examples consistent with this rule (cf. Rule 7.6 in DIN 5008) include "9950" and "1.6606", but "10 520" and "1.660 565". In the case of tables, another rule is even more important:

● Comparable numbers in a column are to be aligned vertically based on decimal points (cf. Sec. 8.4.3).

The examples above might thus appear in a table as

9 950	1.660 6
10 520	1.660 565

Word processors generally contain special provisions for decimal alignment of numbers—particularly useful in the construction of tables—in the form of a "decimal tab" function to complement the more familiar left, right, and center tab options.

- In typewritten copy, the space mandated between triads must be introduced with the space bar, but in typeset material and word-processor copy a narrower space is preferred, one the width of a letter "i".

A narrow space can be created with a word processor by assigning a smaller font size to a normal space (10.5 points instead of 12, for example). There exists another possibility as well, however: don't introduce a blank space at all; instead, *expand* the spacing for the character to the *left* of the site in question, using the "character spacing" parameter on the font-format menu. The same techniques can be employed to produce the recommended small gap between a percent sign and the number preceding it, as already mentioned (Sec. 6.2.2, footnote 21). The second approach offers the benefit with respect to creating triads, by the way, that since no "official" space is present, there is no risk of fragmentation if a number happens to fall at the end of a tight line. If you anticipate taking advantage of one of these expedients frequently, you will almost certainly find it worthwhile to create for the purpose a special keystroke combination (cf. "AutoText Entries" in Sec. 5.3.3).

Word-processing programs, like professional typesetting facilities, treat "special" spaces like those just described in a special way, such that they are immune to modification if the program attempts to justify the surrounding block of text. Incidentally: under no circumstances should number triads be defined instead by *punctuation* (commas or periods), as is sometimes done in the business world.

- In German publications (and certain others, including French), decimal "points" are actually *commas* rather than the periods employed in most of the world.

As a matter of fact, IUPAC permits use of either a period *or* a comma to indicate a decimal point (IUPAC 1988, p. 73), consistent with the practice of other organizations as well (IUPAP, ISO). So far it has been impossible to unite the German and English cultures around a single, common symbol. One of the stumbling blocks is the fact that commercial interests in Germany are increasingly turning to periods with respect to triads in monetary expressions, and there has been considerable reluctance to drive a wedge between the scientific and business communities. The situation is complicated further by the tendency among English-speaking authors to introduce *commas* between number triads! "To drive on the right or on the left, that is the question."

International agreement in this whole area would be especially welcomed by editors at multinational scientific publishing houses, who now are frequently required to make extensive modifications in tables and figures, for example, with publications released in multiple languages.[29]

---

[29] Ironically, it is apparently the "trivial" nature of the issue that has so far stood in the way of uniform treatment. Nevertheless, it would obviously be nice to know for sure if one's overdue tax obligation
→

In any case, IUPAC has taken a clear position. Only the *narrow space* is sanctioned for separating triads (as in our examples above). The bitter conflict has thus died down where scientific organizations have the upper hand.

● Numbers should never be expressed to more digits than can be justified by the *precision* of the available information.

If "0.2 m" is what you wish to communicate, then that is how it should be expressed, *not* in the form "200 mm", because the former is indicative of only one significant figure (a zero to the left of a decimal point is ignored in this context), whereas the latter implies *three*! Appropriate choice of a unit prefix, or adopting a suitable "power-of-ten" notation always makes it possible to solve an apparent problem in this regard.

● Careful selection of prefixes and/or powers of ten makes a valuable contribution to the efficient presentation of numbers and quantitative experimental data.

An effort should always be made to exploit the various techniques available in such a way as to achieve the most compact formulation practical. Thus, "1.245 mm" is preferable to either "1.245 $\cdot 10^{-3}$ m" or "1.245 $\cdot 10^3$ µm". Unit prefixes were in fact specifically created to facilitate flexibility of expression.

● Powers of ten can be coupled with associated numerical values using the standard indicators for multiplication: either a *multiplication point* (a "floating dot", ·) or a *multiplication cross* (i.e., "×"—which is *not* the same as a letter "x"; cf. Sec. 6.5.4).

● Especially when undertaking unit conversions it is important to keep in mind with quantitative data the appropriate level of precision, in part to forestall a possible accusation that you are making false claims.

Thus, it would be nonsense to restate "6 inches" as "152.4 mm" simply because the latter set of numbers appeared on a calculator screen. A more reasonable metric value in this case would be "15 cm"—even better, "0.15 m" (cf. Sec. 6.2.3)—assuming you know (or strongly suspect) that the precision of the data is on the order of ± 1/4 inch.

● A *confidence interval* applicable to the *mean* of several experimental values is indicated by appending to the mean value a "±" sign followed by a second number representing the "margin of error" (usually rather small relative to the mean itself).

● Should you wish to specify a *unit* for data presented in this way, all the numerical information—including the "±" sign—should be enclosed in parentheses immediately preceding the appropriate unit symbol.

---

were, for example, \$1.027 or \$1,027(.00). Is it a matter of a single greenback, or might a disaster be looming? So far as we know, ambiguity in this regard has not yet been the cause of any *real* catastrophe, perhaps in part because scientists and other intellectuals are accustomed to *thinking* about what they read, and generally do not accept numbers simply at face value. Nevertheless, the lack of accord could in itself be considered a catastrophe! And reflect, for example, on the crucial paperwork supporting the flight controllers to whom you will entrust your fate on your next transcontinental journey. Must we await the crash of a misdirected jumbo jet to focus serious attention on this vexing problem?

An example of a proper expression would thus be (24 ± 0.3) mm, indicating a range from 23.7 mm (minimum value) to 24.3 mm (maximum value). The formulation "24 ± 0.3 mm" is unsatisfactory because it *appears* that unit information for the mean itself has been omitted [DIN 1333, *Zahlenangaben* ("Numerical Data")]. Regrettably, formulations of the latter type are far from uncommon in the scientific and technical literature. Their use should nonetheless be strongly discouraged. Frequent appearance renders a mistake no less wrong!

In the spirit of the standard DIN 1319-3 (1996), in a quantitative data specification of the form

$$y = \bar{x}_E \pm u$$

$\bar{x}_E$ is the mean value of a set of data points, perhaps corrected for a known systematic error. The number following the "±" sign is the *margin of error u*, defined as what results from dividing the product of the *empirical standard deviation s* and a so-called *t*-value (see below) by the square root of the number of data points contributing to the mean. It is the range "*y*" defined in this way that should be reported as the outcome of an overall experiment. The value assigned to *t* (commonly referred to as the "*t* distribution according to STUDENT"[30]) is generally taken from standard statistical tables, and is itself a function of both the number of experimental data points contributing to the result and the *confidence level* you choose to adopt. Usually, results are communicated in terms of a confidence level of 95%. What this actually means is that if the same experiment (involving the same number of trials!) were to be conducted 100 times, approximately 95 of the experiments would be expected to yield a range-value *y* that captures the *true* mean (something that itself is not really measurable). One of course *hopes* (and in a sense *assumes*) that the present experiment is one of those 95! If you choose to "increase your level of confidence"—to say 99%—the range you would need to accept would be significantly broader. A value of 95% is considered a good compromise.

● Should you for some reason decide to utilize in a report a measure of certainty *other* than the 95%-confidence level, or simply work with the empirical standard deviation *s* directly, then this needs to be noted explicitly in conjunction with the experimental data.

The proper place for information of this sort is often a table footnote. More extensive treatment of your statistical analyses could be made a part of the report's "Experimental" section, or discussed under "Results".

The reader is encouraged to seek guidance in statistical matters from appropriate textbooks (e.g., NETER, KUTNER, NACHTSHEIM, and WASSERMAN 1996; MOORE and

[30] "STUDENT" is a pseudonym for William S. GOSSETT, the originator of this now familiar methodology. His terms of employment in industry at the time were such that he was not permitted to publish his proposal under his own name. A delightful account of the historical circumstances appears in a chapter entitled "That Dear Mr. Gossett" in *The Lady Tasting Tea: How Statistics Revolutionized Science in the Twentieth Century*, by David SALSBURG (2001).

McCabe 2003; Ramsey and Schafer 2002). Access to a reliable *statistics program* makes it possible to delegate some of the statistical evaluation associated with one's work to a computer.[31] A preliminary statistical analysis performed at an early stage of a project can be extremely valuable, since it provides an indication of the reliability of the data accumulated to date, and might help you in deciding whether or not to expand the data pool, or alter the scope of the investigation.

Ranges reported directly are often indicated by a hyphen (e.g., "800 - 1000 bar"), but use instead of the longer "en-dash" is preferable ("800–1000 bar"). Nevertheless, especially in scientific text this type of notation can be a source of ambiguity, since the same symbol is also employed as a minus sign (cf. Sec. 6.5.4). For this reason an alternative notation for ranges has been suggested:

> 800...1000 bar

Another possibility would of course be simply to include the word "to" (or perhaps "through").

A range symbolized as above (with three dots; an *ellipsis*) is treated rather like an expression in parentheses, which is to say any associated unit is written only once.[32] More and more authors are taking advantage of this usage, an encouraging development which has the potential for eliminating a great deal of confusion, especially in tables. The approach is illustrated in ISO 31-11 (1992), as in the example "11-7.20...11-7.25".

It is interesting to note that the "three-dot symbol" is also frequently employed as a range symbol in the context of *databases*: in conjunction with a search for information falling within specific bounds. The declared target range is generally interpreted as *including* the values flanking the dots (an "inclusive range"), just as one would ordinarily interpret a request for results covering "Monday to (or through) Friday" as referring to five days, not three. Thus, a search based on the hours "12:30...17:30" would recover all time-stamped data attributed to the period 12:30:00 PM through 5:30:59 PM. Interestingly, LaTeX typesetting software (cf. Sec. 6.6) offers the option of situating such a set of three dots either on the text baseline or elevated to a more "central" position.

Despite the advantages we have attributed to the ellipsis notation, tradition holds that the dash should be retained in denoting ranges involving years or page numbers.

---

[31] Note that we specifically recommend a *statistics* program—such as Minitab (Minitab, Inc.) or SPSS (SPSS, Inc.). Statisticians have long expressed serious reservations, for example, about the reliability of the statistical functions included in spreadsheet software.

[32] As applied here, the "three-dot" symbol could be regarded as an *exclusion sign*, since it is representative of all the *omitted* "values" between the stated boundaries. This implies that the expression in its entirety must *include* the boundary values. In this sense it is quite analogous to the ellipsis mark (from the Greek ελλειπσιο, "deficit", omission) often found in text (especially in conjunction with quotations), indicating that something has knowingly been left out. Spreadsheet programs frequently use a "two-dot" symbol to convey the same meaning, as in "=SUM (E4..E20)" (cf. Sec. 8.5.1). The three-dot range notation ("...") is appearing with increasing frequency in the literature. We encourage this development, although we are quite aware that the symbol originally had a somewhat different meaning, namely a pragmatic indication in the sense of "and so forth, to", signifying an *omission* that needs to be rectified

$\rightarrow$

6.5 *Working with Formulas and Equations*

6.5.1 *Combining Text with Equations*

In Sec. 5.5.1 under the subheading "Headings, Paragraphs, Equations, and Lists" we briefly discussed line spacing and indentation with respect to equations (cf. Fig. 5–4). Figure 6–3a in Sec. 6.5.5 provides an example of an equation as it might be produced with a typewriter. How should equations in fact be prepared and incorporated into text?

- Mathematical, physical, and even chemical equations are always isolated from surrounding text and made *free-standing*.

The only exceptions to this rule are brief statements like $x = 1$ or $a = 180°$, which one can assume will fit comfortably in the body of a text passage. Lengthy *expressions*— including *terms* that might be relevant in equations—may also need to be introduced as free-standing lines.

- Expressions based on mathematical symbols are isolated and made free-standing whenever it would be inappropriate to permit their division at the end of a line, or if they threaten to disturb the formatting of running text.

This injunction is concerned not only with the *length* of an expression, but also its height. A single subscript or superscript (cf. Sec. 6.5.3) in an expression can ordinarily be accommodated in double-spaced copy (either typewritten or typeset) from the standpoint of vertical space requirements. On the other hand, *fractional* notation (with the possible exception of simple fractions like $\frac{1}{2}$) is not well suited to running text, except when based on a *slash* ("/"; e.g., "$x^2/35$"). With respect to length, roughly six characters should be considered a critical upper limit for incorporating an expression directly into text. Thus, even an expression as brief as $[-K_{elim} (t - t_0)]^k$ would ordinarily need to be consigned to a line of its own because it offers no reasonable opportunity for "syllable division".

- "Isolation" of an equation or expression means not only assigning it to a separate *line*, but also *indenting* it by a systematic (consistent) amount.

Insofar as possible, equations are presented between paragraphs.

- Individual equations should be *numbered* so they can easily be referred to from within the text, and wherever practical, multiple equations should be collected and displayed in *blocks*.

---

in a particular way, as in $I = 1, ..., n$. We regard this slight change in meaning as insignificant compared to the risk of confusing a range with subtraction, especially in a setting like a table. The risk is actually increased by the fact that the hyphen key functions in such a way as to initiate a *real* subtraction in the calculating mode of a wide variety of programs!

Equation numbers (*formula numbers*) are usually set in parentheses on a line with the equations themselves, flush with the right margin.

In point of fact, rules like these, as important as they are with respect to publications, actually have little bearing on manuscript preparation. Indeed, many publishers even today require that an author's formulas be confined to a separate *formula manuscript*. The advent of electronic manuscripts has not really made a great difference in this regard. Thus, matters related to placement and isolation of equations can largely be left in the hands of a publisher, although it is important for you to ascertain from the publisher's editorial office exactly what *is* expected on your part. If your handwriting is reasonably legible you may even be permitted to submit hand-written equations arranged on separate sheets of paper.

● The demands imposed on an author are obviously much greater with copy that will be subjected to *direct reproduction* (i.e., a *camera-ready* manuscript).

If for some reason a formula (expression, equation, etc.) must be directly integrated into a text passage—as in the course of deriving a mathematical relationship, for example—the text should simply be interrupted at an appropriate point, the formula(s) inserted in free-standing-form, and the flow of text then resumed, starting once again at the left margin. This procedure can be repeated as often as necessary (see Sec. 3.4.2). With a word processor, space for such an equation (or group of equations) should be set aside with the aid of the "new line" command, *not* by introducing formulas as separate paragraphs. In this way the formulas remain integral parts of the surrounding text.

On the other hand, it is generally advantageous to define a unique "paragraph style" for application to free-standing equations, a style that ensures proper spacing of the resulting formula block with respect to text above and below. In the early stages of manuscript preparation you may want to consider using simple "place-holders" to indicate where equations and expressions will later appear. The latter can then be constructed in a leisurely way when time permits.

Everything we have discussed here with respect to mathematical equations and expressions is in principle equally applicable to *chemical reaction equations* and individual *chemical formulas*. (For suggestions on the *numbering* of chemical formulas see Sec. 3.4.2).

● A chemical reaction equation can be divided (if necessary) immediately following the reaction arrow. The arrow itself need not then be repeated on the subsequent continuation line.

Figure 6–2 offers an example of how such material is typically handled. If individual chemical species in this example had been assigned *formula numbers* within the body of the text, these would have appeared in the figure *under* the respective formulas, in boldface type. Sometimes, a block of formulas is labeled as a whole with a number of its own (a *scheme* number), typically set against the right margin.

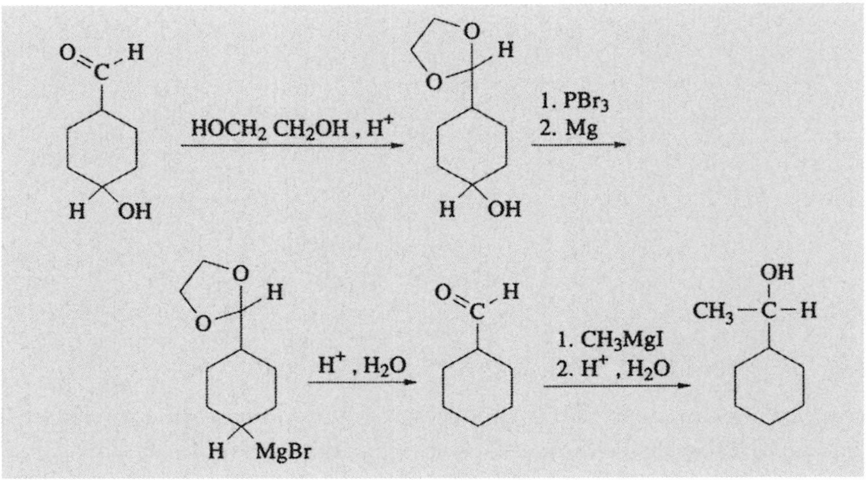

Fig. 6–2. Depiction of a chemical reaction, illustrating the proper treatment when an equation must be divided.

6.5.2 *"Stacked" Expressions and "Fragmented" Formulas*

An expression (or equation) is described as "stacked" if its display requires more vertical space than is available in the equivalent of one line of ordinary type (including subscripts and superscripts that might be present; see the section that follows entitled "Indices"). "Fragmented" equations, on the other hand, are ones whose *length* exceeds the space resources of a standard line.

● Stacked equations must be carefully organized, especially with respect to the most important symbols or characters present.

This concern has to do in particular with proper vertical placement of the equals sign ("="), major operation symbols (e.g., plus and minus signs, "+" and "−"), as well as dominant horizontal *fraction lines*, all of which should coincide with an imaginary horizontal line bisecting the heart of the expression, known as the *formula axis*. There is of course every likelihood that additional signs of the same types, especially operation symbols, will be present elsewhere in the expression as well, as in the numerator or the denominator of a fraction, for example, but these obviously are *not* expected to conform to the formula axis. The significance of this rule will become apparent upon examination of the examples in Fig. 6–3 (Sec. 6.5.5).

● If a long mathematical expression must be divided for presentation, this is best accomplished immediately before a plus or minus sign, but preferably not one embedded in a set of parentheses (or other "fences"; see Sec. 6.5.5). This plus or minus itself then becomes the first symbol in the continuation line, so indented that it falls to the *right* of the most recent equals sign; e.g.,

$$L = E - U = 0.5\, l_1{}^2 f_1{}^2 (m_1 + m_2) + 0.5\, m_2\, l_2{}^2 f_2{}^2$$
$$+ (m_1 + m_2)\, g\, l_1 \cos \delta_1 + m_2\, g\, l_1 \cos \delta_2$$

- An equation can also be broken to the left of an equals sign, which would again constitute the first symbol of the continuation line. A new equals sign of this sort should be adjusted so that it falls directly below any equals sign in the previous line.

Situations subject to this treatment are especially common in the context of mathematical derivations; e.g.,

$$x = a_1 e_1 + a_2 e_2$$
$$= (x_{11} + x_{12}) e_1 + (y_{21} + y_{22}) e_2$$
$$= r(\cos \varphi_1 + i \sin \varphi_2)$$

Fortunately, word-processing packages often provide special facilities and/or auxiliary software ("equation editors") to assist in the preparation of mathematical expressions, with powerful tools for creating and properly arranging at least the most familiar components. An example is the program *called* EQUATION EDITOR, mentioned earlier and distributed in conjunction with WORD (see also Sec. 6.7). Specially engineered software of this type makes it relatively easy, for example, to construct a fraction— whether simple or extremely complex—working directly from the screen in a "what-you-see-is-what-you-get" ("WYSIWYG") environment. All that is normally required is a few selective clicks of the mouse. The expression that results will automatically be properly aligned both internally and with respect to surrounding text. Content for such a fraction is introduced character by character into a pair of pre-established "boxes", one above the other (for the numerator and the denominator, respectively), all in a special window that appears when WORD's "Insert Object: Microsoft Equation" command is issued and a fraction requested. Boxes for the fraction will be seen already to be separated by a straight, horizontal line of the correct "weight" (line width) whose length will vary as a function of the nature and extent of the numerator and denominator content. One or both of the latter could itself assume the form of a fraction (which the program will again properly space), or possibly an integral, a sum, or even a matrix.

An alternative (albeit less "visual") way of creating mathematical expressions incorporating all the preferred typographic characteristics (as regards spacing and symbol style and form, for example) involves mastery of the formal computer language TEX (also available in a more user-friendly version known as LATEX), whose descriptive commands are translated by one's computer as if by magic into a nearly perfect set of display instructions for print purposes (see Sec. 6.6 for a brief introduction to LATEX).

- The scientific author with a frequent need for computer-based mathematical copy, including both stacked and fragmented expressions, is strongly encouraged to seek access to auxiliary software designed specifically for formula construction.

In the sections that follow we turn to a somewhat more detailed look at some of the principles upon which the proper display of mathematical expressions depends, including at least a few details that are almost sure to be new to the novice author—or even the author with a considerable amount of experience!

6.5.3 Indices

In the days when manuscripts were prepared with a typewriter, subscripts and superscripts were necessarily positioned by careful rotation of the *platen* (a rubber roller supporting the paper), producing results like

$$K_a^3 \qquad x_a^3 \qquad SO_4^{2-}$$

Unfortunately, such expressions would be described by a typesetter as primitive at best. Even a relatively simple construction like $e^{-x}2$ presents serious problems: should the "x" in this case perhaps be displaced vertically by *two* "clicks" in order better to accommodate its subscript "2"?

If you do not have at your disposal a good computer-based "equation generator" (along with the skill to use it effectively!), your best recourse is probably to prepare equations required for a manuscript by hand—as separate "figures"—ideally with the aid of drafting templates, but freehand if necessary. These might be pasted exactly where they belong in the manuscript, or kept separate, leaving the publisher with the problem of typesetting them in a professional way. Equations can also be prepared as (digital) *graphic objects*, which can be "pasted electronically" into a word-processing file.

- Sophisticated mathematical typesetting entails the availability of type of various sizes for proper preparation of indices and other features that may be present at various hierarchical levels.

For example, a symbol serving as an index would ideally be substantially smaller (typically by about 30 %) than the primary symbol to which it is attached, and an "index to an index" (for a "three-level expression") smaller still (perhaps only *half* the size of the primary symbol).

The complication of "double indices" can sometimes be circumvented. For example, the exponential function above might also have been expressed in the form $\exp(-x^2)$. Recall, too, the option described in Sec. 6.1.3 of setting an index directly on the text baseline, in parentheses following the primary symbol to which it applies.

- If subscripts and superscripts on a given symbol tend to overlap, try shifting the superscript to the right.

Thus, the subscript is introduced first, *followed* by the superscript. For example, the formulation m_i^0 satisfactorily represents the chemical potential of substance i in the standard state $(^0)$. Superscripts playing the role of *exponents* are often moved to the right deliberately, as in $c_B{}^2$, the square of the amount-of-substance concentration of substance B, to underscore the fact that the entire *expression* is squared. If you think there could be any risk of ambiguity, an entity to be raised to a power should be enclosed in parentheses, with the exponent on the outside: $(c_i)^{n_i}$.

The latest edition of DIN 1338 (August, 1996) no longer recommends simple horizontal displacement. The relevant passage translates as follows:

> If a primary symbol requires not only an index [i.e., a right subscript] but also a right superscript (e.g., an exponent), then the latter should be set directly above the first index character, or else the subscripted expression should be placed in parentheses, to be followed by the exponent. Otherwise the formulation may be unclear; moreover, there is a danger that the exponent to the primary symbol could be misinterpreted as referring to the *index*.

In other words, v_{max} and $(v_{max})^2$ are both acceptable, but v_{max}^2 is not.

Despite this generalized advice, indication of an electric charge ascribed to a chemical species should *always* be displaced to the right if failure to do so would put it directly above a *stoichiometric index* (DIN 32640, 1986; cf. also IUPAC 1990); e.g.,

$$NH_4^+, NO_3^-$$

Computer-based word processing has made it substantially easier to deal with such situations, since it allows characters not only to be set easily at various vertical levels, but also to be expressed in type of various sizes. In other words, individual elements of a formula can be more or less "custom tailored" on the basis of the situation: properly scaled, and raised or lowered by a precisely fixed amount specified in (typographic) points.[33]

The experience of trying to resolve subtle problems of this nature provides considerable insight into the surprisingly high level of metainformation associated with even a simple formula, and also illustrates the sense in which scientific text is almost invariably more troublesome to deal with than "running text" for a typical novel.

As has already been noted, formulas and equations are often manipulated in a word-processing environment in the form of *graphic objects*, as a result of which much of the original meaning vanishes from the concepts "subscript" and "superscript". Graphic elements are positioned much as they would be in a document prepared by hand: selectively, taking into account both formal "rules" and aesthetic factors. This provides the typesetter with considerable flexibility, but at the expense of much invested time, and the required proficiency comes only with a great deal of practice, which means that a purely graphic solution is not always optimal.

6.5.4 *Frequently Encountered Special Symbols*

We consider special symbols only briefly, and almost exclusively from a technical standpoint. For example, questions like the fundamental meaning to a mathematician

---

[33] A typical vertical displacement for a subscript or superscript with 12-point body text is 3 points.

of the *concept* of "identity" will be ignored, with the focus instead on what an identity *symbol* looks like. Much more extensive information along these lines—at least from a mathematical perspective—is provided in the authoritative standard DIN 1302 (1994) *Allgemeine mathematische Zeichen und Begriffe* ("General Mathematical Symbols and Concepts").[34]

There was a time when symbol typography was the concern only of typesetters, not authors, since the former bore full responsibility for achieving professional-looking results. Now that authors, too, engage to an increasing extent in the typesetting of equations, and are sometimes even *required* to do so, it clearly makes sense to expose a broader audience to the most important rules and principles. Space limitations force us to restrict our coverage to those mathematical symbols most often encountered. More comprehensive treatment of "special symbols" in general should be sought else-where.[35]

Basic mathematical symbols of an alphabetic nature (e.g. "exp", "sin", "lim", etc.), can of course be created with the simplest of word processors, but for others one is forced to resort to special symbol collections assembled for this express purpose. The most readily available type font containing a substantial selection of symbols useful for mathematical equations bears the *name* "Symbol", a font for which some word processors provide convenient keyboard shortcuts.

● The symbol for equality, the *equals sign* "=", is one of the few "special symbols" with its own ASCI code, and it is both accessible from every keyboard and repre-sented in nearly every alphabetic font.

---

[34] It is remarkably difficult to formulate a precise definition of what we mean by a "special symbol". Indeed, many symbol collections refrain from even attempting such a definition. A respected encyclopedia, for example, identifies a special symbol only as "one that is neither a letter nor a number." DIN 5007, *Ordnen von Schriftzeichenfolgen* ("Arranging Groups of Symbols"), includes the following characterization, which is broad to say the least: "[Special symbols are] type symbols that serve alongside letters and numbers (basic type symbols) in the presentation of text; e.g., punctuation marks, symbols for words, unit symbols (DIN 1301-1), counting and computing symbols, and the blank space". This interpretation, according to which preparation even of a birthday card would entail the use of special symbols, certainly is not in accord with the way publishing professionals understand the term. More useful in the trade is a formulation like that from the *Lexikon der Satzherstellung* ("Typesetting Lexicon", DORRA and WALK 1990), which translates into English roughly as follows: "**Special symbols**: Unique symbols that find application in conjunction with conventional type symbols, and for which special type forms (usually commercially available) may be required. Special symbols in this sense include mathematical, biological, meteorological, physical, and chemical symbols, as well as pictograms, musical notes, and representations of chess pieces".

[35] Apart from mathematical symbols, symbols are also used—often in abundance!—in a number of scientific and technical disciplines, including chemical engineering, electronics, and information science. The interested reader will find more on this subject in standards such as DIN 28 004-3 (1988) *Fliessbilder verfahrenstechnischer Anlagen: Graphische Symbole* ("Flow sheets and diagrams for processing plants: Graphic symbols") or ISO 3511 (several sections, 1977 and later) *Functions and instrumentation: Symbolic representation*; DIN 40 719-6 *Schaltungsunterlagen: Regeln und graphische Symbole für Funktionspläne* ("Graphic representations in electrical engineering; Rules and graphic symbols for function charts"); and DIN 66 001 (1983) *Informationsverarbeitung: Sinnbilder und ihre Anwendung* ("Information processing: graphic symbols and their application").

The equals sign is one of a number of so-called *relational symbols*, also known as *equivalency operators*, and is the most common relational symbol in equations. (It is no accident that such formulations are *designated* as *equa*tions!) The equals sign signifies equality in the sense of identity: agreement with respect to all characteristics. No additional symbol is thus required for a separate concept "identity". This is the reason why the symbol "≡", once employed to convey the notion "identically equal", is no longer recommended. (This symbol does retain a role in number theory, however, where it serves to indicate *congruency*.) On the other hand, there does exist a need to symbolize the relationship "is by *definition* equal":

● When a quantity is introduced with an explicit understanding that it is to be regarded as having precisely the same meaning as some other quantity, the corresponding relationship is symbolized with " $=_{\text{def}}$ ".

Two accepted alternative formulations are " $\underset{\text{def}}{=}$ " and ":=" (DIN 1302, 1994), and ISO 31–11 recommends yet a third: " $\overset{\text{def}}{=}$ ".

● The notion of *inequality* can be symbolized by ≠.

DIN 1302 contains the recommendation that a symbol with a *perpendicular* crossbar rather than the slanted one be employed (while observing that "≠" can also be used "when necessary for typesetting reasons").[36] Our impression is that this proposal has found little support, so we no longer encourage its implementation. The International Union of Pure and Applied Physics, consistent with ISO 31–11 *Mathematical signs and symbols for use in the physical sciences and technology* (1992), permits the use of either. In a word-processing context, one sometimes sees the relationship "unequal" expressed by "<>".

With respect to symbols for "at most" and "at least", DIN 1302 (1994) provides the following guidance:

● The symbol for "less than or equal to" is "≤", while that for "greater than or equal to" is "≥".

Here again the DIN standard says that if it proves typographically necessary, the symbols ⩽ and ⩾ can be substituted, whereas ISO 31–11 also recognizes ≦ and ≧.

"Plus-or-minus" is covered in most PC type fonts by the special symbol "±", in some cases accompanied by its inverse, "minus/plus" ("∓").

● Apart from the relationship symbols already cited, there also exist a number of other symbols that convey information of a similar type:

≈	"is approximately equal to"	≘	"corresponds to"
≪	"is small relative to"	≫	"is large relative to"

---

[36] Problems in this respect will almost certainly confront the "home typesetter", since we have failed to locate the "preferred" symbol in any of the more than 25 symbol sets we have examined!

DIN 1302 (1994) considers these not to be mathematical symbols "in the true sense", however, since their use signifies a judgement, clarification, or interpretation on the part of the author. They are thus assigned to a separate category called *pragmatic symbols*.

Despite misconceptions to the contrary, it is not correct to attribute to the symbol "~" (the "tilde") the meaning "is approximately equal to"; a tilde instead conveys the message "is proportional to" (or "similar" as the word is used in elementary geometry). Special care should also be exercised with the symbol " ≃ ", present in many font sets, which should be employed in a mathematical context only in the sense of "is asymptotically the same as".

We next consider various *connecting symbols*.

● The most frequently encountered among these are the *plus* and *minus* signs.

Every keyboard grants direct access to the plus sign (+). Most people assume the hyphen key produces a minus sign, but that is not the case. A hyphen (-) is inappropriate because it is too short. One should instead employ an "en-dash" (–), which is almost always *available* in a word-processor setting, but accessible only in special (non-obvious) ways, typically involving some combination of keys, or through one of the program's menus. The *en-dash* was given that name in recognition of the fact that it is roughly as long as the letter "n" is wide. In another important application, the same symbol is frequently incorporated into *range* expressions, as in "800–1000 bar".[37] Potential confusion regarding whether in a particular case the symbol should be interpreted as a sign for *subtraction* or as a "from-to" signal led to the suggestion that range limits instead be linked by the "three-dot" *ellipsis*, "…" (cf. the discussion near the end of Sec. 6.4).[38]

In a complex mathematical expression it is advisable that each minus sign be both preceded and followed by one or perhaps even two blank spaces to ensure clarity and also to emphasize the underlying logical structure. Professionally typeset copy always incorporates such spaces—and for both minus *and* plus signs.

Another extremely important "linking element" with a role to play between numbers and other "mathematical structures" is the symbol for *multiplication*.

● Under most circumstances, a "floating dot" is the preferred multiplication symbol.

The corresponding dot is most often borrowed from the special-purpose font "Symbol". As a matter of fact, however, *most* computer type fonts actually are outfitted with a multiplication dot, but these can only be accessed as "special symbols" because

---

[37] There is yet another type of dash that sees frequent use: an even longer "em-dash", for signifying a break or a pause in text—introduced as here with no adjacent blank spaces. A *pair* of em-dashes flanking a word or phrase has much the same effect as a set of parentheses.

[38] "…" ("dot, dot, dot", the ellipsis) is classified in DIN 1302 (1994) as a fifth "pragmatic symbol". Some fonts include a special character with this form so that the symbol can be manipulated as a single entity. WORD in fact automatically substitutes this character if a sequence of three periods is entered from the keyboard. No extra space should be added either before or after an ellipsis.

of the absence of a corresponding key on the standard keyboard.[39] In many cases no ambiguity results if the dot is simply omitted. That is to say,

● When two quantity symbols appear together, the reader *assumes* that a multiplication is intended.

In other words, "*RT*", just as it stands, would be interpreted as "the ideal gas constant is to be multiplied by the thermodynamic temperature". (In professionally typeset material the corresponding symbols would in fact be slightly separated, but by a very narrow space; see Sec. 6.5.6.)

There exists a second accepted symbol for multiplication as well: the *slanted cross* or *multiplication cross* (×). The "multiply" key on a calculator is often labeled with a *third* symbol: the *asterisk* ("∗"), but neither this nor the cross should be incorporated into a mathematical formula except under very special circumstances. For example, *a* × *b* *does* properly signify the vectorial product of the *vectors* *a* and *b*. In the English-speaking world a multiplication cross is frequently seen in one other scientific context: linking the elements that produce exponential notation (e.g., 3×10^4). DIN 1338 (1996), however, specifies that a dot should instead be used for this purpose, as in $1.32 \cdot 10^6$. Finally, the multiplication cross is occasionally called upon in *format specifications* (e.g., 9 m × 12 m; cf. Rule 7.3 in DIN 5008, 1996). Like its sibling the dot, the multiplication cross is conveniently obtained from the font "Symbol", and it should never be simulated by a letter "x"!

The opposite of multiplication is of course *division*.

● The standard *division symbol* in science is the slash (/), although in complex expressions and displayed equations division is frequently indicated instead by fractional notation, where a horizontal line separates the dividend (numerator) from the divisor (denominator).

Division, just like multiplication, can thus be signaled in alternative ways. The arithmetic division sign ("÷"), familiar mainly from calculators, should *not* be utilized in conjunction with mathematical formulas. DIN 1302 (1994) also cautions against use of a colon (":") to represent division, granting it only a restricted role in *proportions* and "everyday" arithmetic.

6.5.5 Additional Rules for Writing Formulas

● If a denominator (e.g., an expression following a slash) must contain multiple symbols (in a multiplicative relationship), the entire group should be enclosed in parentheses as a way of making clear how far the division instruction extends:

---

[39] Alternatively, a dot of the proper type can be kept on reserve in the form of an "AutoText entry", almost certainly the smallest one in your entire repertoire. Another possibility would be to generate in advance a whole *line* of the dots in some convenient place, from which one could easily be "dragged" to wherever it might later be required, a strategy that facilitates the construction of a considerable amount of mathematical notation in very short order!

$$1/(2\pi) \quad mol/(L\ s) \quad \exp[E_a/(RT)]$$

This rule is too often ignored in scientific text, especially in the context of complex units (like the second example above)—based presumably on the questionable assumption that every reader will know what was intended.

Note also that

● It is never permissible for two or more slashes to appear in a single unstructured expression.

If one attempts to interpret an expression of the type "a/b/c", for example, it quickly becomes apparent that either of the following might have been intended:

$$a/(b/c) = ac/b \quad \text{or} \quad (a/b)/c = a/(bc)$$

One way in many cases of avoiding altogether both quotient notation *and* the slash is to introduce negative exponents; e.g.,

$$ab/c = abc^{-1}$$

Parentheses and other "fence" symbols play an extremely important role in mathematical expressions and equations by introducing additional *structure*, making clear especially which parts of an expression are subject to the effects of various operators.

● Sometimes *nesting* of fences is required—"fences within fences"—in which case the appropriate order for "installing" the fences is (ordinary) *parentheses*, surrounded by *square brackets*, and finally *braces* (also called "curly brackets").

Braces are thus entrusted with the farthest reach, and the innermost fences are parentheses.[40]

$$\{[(\quad)]\}$$

It is particularly important that you remember to use parentheses around multiple elements following a slash in a complex unit symbol.

With respect to *size*,

● Fences should always be erected at least as high as the expressions they enclose.

The same principle applies to symbols with operands, such as those for *roots*, *summations*, *products*, and *integrals*. A word processor makes it quite easy to create the corresponding symbols in any size required:

$$\iiint \quad \Sigma\Sigma\Sigma \quad \Pi\Pi\Pi \quad \text{etc.}$$

We refrain from considering the various symbols associated with specialized branches of mathematics, such as set theory, or symbols for arcane operators like those that fascinate quantum chemists. Use of a number of the more common symbols is illustrated in Fig. 6–3.

---

[40] This sequence is consistent with recommendations of the International Union of Pure and Applied Chemistry (IUPAC).

● If you cannot obtain direct computer access to an indispensable symbol, the best course generally is to introduce it into your manuscript instead by hand.

Carefully executed, this expedient should always be acceptable with a typescript or other manuscript that will in any case later be typeset. Each such symbol should be meticulously prepared using a high-quality felt-tip pen with a sharp point or, better yet, a professional drafting pen. The first time a given hand-drawn symbol appears it should be explicitly identified in the margin, preferably through an explanation enclosed in double parentheses, a convention frequently adopted for conveying instructions or information to an editor or a typesetter; e.g.:

$((Theta))$

Among the most common "special symbols" in scientific text are uppercase and lower-case letters of the *Greek alphabet*.[41] These have been reproduced for reference purposes in Table 6–5. A convenient source of Greek letters for word-processing purposes is the computer font "Symbol". When this character set is utilized in conjunction with "Times", consideration should be given from an appearance standpoint to reducing its point size slightly (e.g., 11 points rather than 12).

a

$$C^* = \lim_{i \to \infty} \left| \frac{c_i}{c_{i+1}} \right| = \left[x_{n-i} \cdot (1 - a) \cdot k_n^2 \cdot \prod_{j=1}^{i} k_j \right]$$

b

$$x = \sum_{i=1}^{n} x_i^2 \qquad g(x) = \frac{e^x}{(x+1)(x-1)} \qquad a = \cfrac{1}{1 + \cfrac{1}{1 + \cfrac{1}{1+x}}}$$

$$x = a + bx^2 + cx^3 + dx^4$$

$$\frac{1}{z} = \frac{1}{a} + \frac{1}{b} \qquad \sqrt{2a} - \sqrt{\sqrt{a-1} + 1} + 1$$

$$\frac{dc(A)}{dt} = -k \cdot c(A) \qquad \frac{\Delta y}{\Delta x} = \lim_{\Delta x \to 0} \frac{h(x + \Delta x) - h(x)}{\Delta x}$$

Fig. 6–3. Typical examples of mathematical formulas prepared for a manuscript— **a** using a typewriter; **b** with the EQUATION EDITOR software supplied with WORD.

---

[41] The Greek *alphabet*—alpha, beta ... (!)—once an unchallenged medium for written communication in the Hellenistic world, would seem from this treatment to have been reduced to a collection of special-purpose symbols. On this point we feel obligated to beg the indulgence of our Greek colleagues. The situation might be generously viewed in the following light: Very often when we find ourselves stymied we seek recourse among the ancient Greeks, who in a very real sense invented the art of thinking.

Table 6–5. The Greek alphabet as rendered in the computer font "Symbol".

Name	Sign	Name	Sign
alpha	A, α	nu	N, ν
beta	B, β	xi	Ξ, ξ
gamma	Γ, γ	omicron	O, o
delta	Δ, δ	pi	Π, π
epsilon	E, ε	rho	P, ρ
zeta	Z, ζ	sigma	Σ, σ, ς
eta	H, η	tau	T, τ
theta	Θ, ϑ (θ)	upsilon	Y, υ
iota	I, ι	phi	Φ, φ (ϕ)
kappa	K, κ	chi	X, χ
lambda	Λ, λ	psi	Ψ, ψ
mu	M, μ	omega	Ω, ω

According to DIN 1338 (1996), the preferred forms for the lower-case Greek letters theta and phi are "ϑ" and "φ", not "θ" and "ϕ".

It is perhaps worth reiterating that special symbols incorporated into an electronic manuscript have a disconcerting habit of disappearing (or becoming corrupted) during subsequent processing. Be prepared to supply the publisher with a copy of any unusual symbol font you may have exploited.

6.5.6 Spacing

The standard DIN 5008 (1996) contains a number of guidelines related to proper use of the *spacebar* for positioning characters in typewritten copy. Several times both in this and the preceding chapter we have referred to the important role played by spaces in formulas of all types. We summarize here the major situations requiring introduction of *blank spaces* (cf. "Special Considerations Applicable to Electronic Manuscripts that Require Typesetting" in Sec. 5.4.1):

– *before* and *after* an equals, plus, or minus sign;
– *before* the symbol for a general or special function (e.g., f, tan), but *after* a function symbol only if there are no fences around the operand (e.g., parentheses);
– *before* and *after* integral signs and symbols for summation or product formation;
– *before*, but *not* after, an operator, such as the differential operator d;
– *between* a numerical value and the corresponding unit;
– *between* an integer and an associated fraction (e.g., $3\,^1/_2$);
– *between* a number and a percent sign (e.g., 0.2 %);
– *between* elements of a complex unit symbol.

A *narrow space* is preferred between a quantity and its multiplier (e.g., $4x$) and between pairs of quantity symbols in a multiplicative relationship (e.g., $a\,b\,c$).

More sophisticated rules apply to the preparation of typeset copy. Thus, spaces of various widths are called for in different situations, such as in conjunction with function symbols (e.g., sin, ln), collections of digits, quantity symbols, and miscellaneous mathematical symbols. These subtle adjustments are all directed toward providing *structure*, which in turn helps the reader interpret an expression.

As a general rule, those parts of an expression most closely related in a mathematical sense should also be set in closest proximity to one another. For example, according to DIN 1338 [Beiblatt 2, *Formelschreibweise und Formelsatz: Ausschluss in Formeln* (Supplement 2 "Written Expression of Formulas, and Formula Typesetting: Spaces in Formulas") (1983, 1996)], the amount of space to be provided between elements subject to multiplication can be expressed as three *space modules*, where a "space module" is defined as equal to the relevant type size in points divided by ten: in other words, 1.2 points in the case of 12-point type (cf. "Fonts and Units of Measure in Typography" in Sec. 5.5.1). Space recommendations are then listed for other situations, again as numbers of modules:

– between the symbols for a function or operator and its operand(s): 2
– between factors of a product: 3
– before and after a plus or minus sign: 5
– before and after an equals sign: 6

The more recent supplement, dated April 1996, includes illustrative examples, along with the following summary observation (translated from the German original):

> If one arranges mathematical expressions of various types in an order reflecting progressively weaker computational links between the components, the resulting sequence (shown below) can serve as a general spacing guide, progressing from the narrowest spaces to the widest:
>
> – Separation between the symbols for functions and operators and the corresponding operands;
> – Separation between elements of a product or a quotient;
> – Separation between elements of a sum or a difference;
> – Separation between the two sides of an equation.

It is also noted here that a "space module" can be equated in the metric system with roughly 0.25 mm. The author with frequent occasion to prepare formal representations of mathematical expressions is encouraged to secure a personal copy of the standard and its supplements, and to become intimately familiar with the many conventions that are the basis for "high-quality" mathematical typesetting.

● With most word processors there exists an interesting option (in the form of a *font format* parameter) of "expanding" the default character spacing in specific instances by a prescribed number of "points". For example:

X–Y	X – Y	X – Y	X – Y
Default spacing	single spaces inserted with the spacebar	"expanded" by 5 pt	"expanded" by 7 pt

Thus, if a formula has been entered from the keyboard in the usual way, parts of the whole can subsequently be selected and "tweaked" after the fact. For example, "*abc*" is transformed easily into "*a b c*" by "spacing expansions" of 2 points for the "*a*" and the "*b*". Frequently recurring expressions of this sort, after proper formatting, can be designated as "AutoText entries" for convenient insertion whenever called for (cf. "AutoText Entries" in Sec. 5.3.3).

● Professional-quality presentation of formulas and equations is truly an art. Unfortunately, much technical material distributed in print—especially when desktop publishing is involved—would class as aesthetically impoverished.

Paraphrasing the respected source *Duden Satz- und Korrekturanweisungen* ("Duden Typesetting and Proofreading Recommendations"):

● Good equation typesetting is distinguished by the contribution it makes to organization already inherent in a mathematical expression. "Effective organization" in turn depends upon close attention to spacing, which should range from the narrow to the negligible in the various parts of an expression.

The passage continues: "Especially with phototypesetting, spacing should never be imposed in a uniform and systematic way, because rational structure and visual impact must take precedence."

6.6 Programmed Typesetting of Formulas

6.6.1 LaTeX as a Formula Generator

In the purely "text" environment of a typical word processor there are normally no provisions whatsoever for sophisticated work with mathematical formulas, equations, or expressions. Nevertheless, exceptions exist: programs that in fact handle complex mathematical notation amazingly well with no sacrifice in the text capabilities one expects from a state-of-the-art word processor. Indeed, in some ways these programs excel in the latter role! We take this prospect as our starting point for the present section.

The best-known and most important programs in the category in question are TeX and LaTeX. Donald KNUTH,[42] the "inventor" of the progenitor, TeX, was himself a typesetter, a craftsman fully conversant with typesetting standards and conventions—especially as they apply to mathematical formulations—and one who had given the situation a considerable amount of systematic thought. The program (or programming

---

[42] According to his home page (www-cs-faculty.Stanford.edu/~knuth/), KNUTH is currently Professor Emeritus of the Art of Computer Programming at Stanford University

language) he developed eventually rose to the status of a "quasi-standard" for publishing activities sponsored by the American Mathematical Society. TEX is in fact intimately concerned with the whole notion of "standards", notably the rules of the game that become second nature to a first-rate classical typesetter.[43]

The scientist/author most likely to thrive on the challenge of working with TEX will be a person already possessing at least a little background in computer programming. This obviously does not describe every potential user, so there was ample incentive for a further stage of development: LATEX, a macro package *based* on TEX that might be described as a "user-friendly" (better: "user-friendlier"!) version of the parent system, one better adapted to the needs of the average scientist interested in preparing text rich in mathematical formulations.[44] Both programs—TEX and LATEX— have spawned enthusiastic national "fan clubs" throughout the world.[45] They also are representative of certain "subsidiary routines" now available as "add-ons" with such conventional text-editing software as WORD.

LATEX in particular has attracted a great many loyal converts, especially among the ranks of university-based mathematicians and physicists, and it is to LATEX that we direct the balance of our attention. Thanks to LATEX's intelligent structure, the user quickly learns to take full advantage of its outstanding facilities for formal management of complex scientific text. A document prepared with LATEX has at its core a predefined layout easily carried forward to the stage of camera-ready copy. It is a simple matter during composition to alternate back and forth between the program's basic *text mode* ("paragraph mode"), structured to meet all the demands normally associated with plain text—regardless of the complexity of the document as a whole—and a separate *mathematical mode* conceived specifically for professional-quality typesetting of equations and formulas. Once the keying-in of a particular formula is complete, the user is able to return effortlessly to "text mode" and resume ordinary composition. Mode shifts can occur as often as necessary, facilitating rapid and convenient assembly of intricate documents studded with mathematical gems.[46]

---

[43] A desire to encourage some degree of standardization in style—not normally a high priority in the United States!—almost certainly was a motivating factor at the program's genesis. TEX might in principle be called upon to supply proper form to a Christmas letter or a set of verses, though that was obviously not the reason for its development. By the way: the authors of the present book are far from devotees of *excessive* standardization and stipulations, a direction in which TEX has some tendency to lean. Rules and standards can be a very good thing, but like most people we become restive when others attempt to impose their will with respect to *our* text.

[44] One source (KNAPPEN et al. 1994) sees things a bit differently, describing TEX as a "typesetter" and LATEX as a "book designer".

[45] The parent organization—and the "local user's group" for the United States—is TUG, the (International) Tex User's Group, with headquarters in Portland, Oregon, and an informative homepage at www.tug.org. The United Kingdom user's group is UK TUG, accessed through www.tex.ac.uk.

[46] A third program mode is available to the user as well: a so-called "LR-mode" that actually has much in common with text mode, since both are fundamentally concerned with text. In this third "environment" one is also able to "rework" mathematical expressions, in much the same way text is edited. You might for example choose this route to modifying some of the default features imposed by "mathematical mode", or injecting a few distinctive characteristics of your own.

- LaTeX is a semantically structured programming "language" which, though it supports processing of all types of text, is of course especially valuable in the context of the sciences, where its unique mathematical capabilities are exploited.

What all this implies we attempt to elucidate in the discussion that follows. The main point to stress at the outset is that *mathematical formulation* served as the point of departure for developing TEX and LaTeX, an activity that entails an unusually high level of abstraction. The scheme that then evolved for analyzing and structuring "text" in the broader sense is in fact quite closely related to the methods and philosophy of SGML, the *Standard Generalized Markup Language* (cf. Sec. 5.4.1).

The latest version (in 2004) is LaTeX 2e (released in June 1994).[47] An excellent overview of the language/program as a whole is provided by *The LaTeX Companion* (GOOSSENS et al.1994), which can be viewed as a fitting companion indeed to Leslie Lamport's *LaTeX: A Document Preparation System, User's Guide and Reference Manual* (LAMPORT 1994). A computer scientist, LAMPORT is the person credited with "inventing" LaTeX (in 1985).[48]

The name TEX, which of course closely resembles "text", was presumably intended also to suggest the word "technology", since the system is so strongly directed at scientists and engineers, the very people who are also heavily engaged in dealing professionally with abstractions in general.

Some readers might be wondering at this point if we really needed to devote so much attention in our book to mathematics. We obviously feel that it *was* necessary, because wherever scientists and engineers engage in quantification of their work, or in computations involving large numbers of variables—or simply labor at more precise formulation of the known laws of nature—they turn increasingly for help to applied mathematicians. Expressing oneself effectively in the mathematical arena, and successfully engaging in the meaningful two-way exchange of information, requires mastery at a reasonably high level of the language and symbolism of mathematicians.

One more parenthetical interjection is perhaps in order: We certainly do not want to be misinterpreted as denigrating LaTeX in referring to it as a "formula program"[49] or a "formula typesetting program", and you can be sure we are acutely sensitive to the fact that its capabilities extend far beyond mere mathematical typesetting. We therefore ask the reader's indulgence with respect to the heading selected for this

---

[47] The software distributors outdid themselves, incidentally, from the standpoint of typographic finesse, in their proclamation of "$L^AT_EX\,2_\varepsilon$", an affectation we see no virtue in emulating.

[48] The first two letters in LaTeX apparently have no significance other than to suggest the name of its creator.—Other useful references: SNOW (1992), SEROUL and LEVY (1991), and SYROPOULOS et al. (2003).

[49] We confess to having rarely encountered the expression "formula program", although the term "graphics program" is quite common. Perhaps this is simply because so many more people engage in preparing graphics as compared to mathematical formulas. "Formula-processing program" (analogous to "word-processing program") seems to us somehow a much less appealing alternative. The object after all is not to "process" or "adapt" formulas, but rather to bring them to paper. The most fitting description would perhaps be "formula composition program", which effectively captures the spirit of the time-honored German expression "Formelsatz", but this may sound too formal. Incidentally: KNUTH embarked upon his ascent into the rarified atmosphere of academe as a dedicated "formula composer".

section, in which the reference to a "formula generator" is intended only to call attention to the fact that, to our knowledge, there exists nowhere a vehicle more effective than LaTeX for working with formulas.

- Formulas generated by LaTeX not only adhere to every established technical standard, they also reflect to the fullest the aesthetic qualities inherent in the most satisfying mathematical expressions.

Consider the matter first from a typesetter's point of view. Imagine you are confronted with the two symbols "g" and "2", mundane constituents selected at random from among a host of alphanumeric ensembles. Whether in a particular application these symbols are intended to be linked as "2 g" or "g^2", however, can only be established from the context, and not everyone would be in a position to answer the question even after scanning the content. The two notations, which differ only in arrangement and type style, convey vastly different meanings, and a person charged with managing such symbols in a text environment must of necessity be at least somewhat familiar with their meaning and utilization.

The command-based typesetting system we are about to examine paradoxically puts the *significance* of mathematical formulas ahead of their *appearance*.[50] The user begins by specifying what is needed *intellectually*—how an expression is to be *interpreted*—leaving the program to attend to the details of appropriate presentation. In the case of a complex fraction, for example, it is not a matter of first requesting a horizontal line of (temporarily) indeterminate length, above which you might later draw another—shorter—line, eventually burying both lines within a conglomeration of miscellaneous symbols. Instead one declares, "What I need is a fraction. This particular term[51] [which, because it is to be a fraction, the program already understands as requiring a fraction line] should have as its denominator *another* fraction" [so LaTeX concludes that a second fraction line will be needed as well]. The next step the user must take is to describe in detail the nature of the components that will define these two fractions.

Rather than presenting a set of rules, we propose to illustrate the process through a number of examples. For a more detailed, systematic introduction see LAMPORT (1994).

The typesetting system as a whole—whether TeX or LaTeX—is built around a particular syntax, best expressed as:

```
\Command name [optional argument]
{obligatory argument}
```

The single most important *command* in LaTeX is present even here: it is a symbol—the "backslash" (\)—which looks like a mirror-image twin of the diagonal fraction line or "slash" (solidus) that has its special place on virtually every keyboard.

---

[50] One might therefore describe TeX and related programs as "content oriented", in sharp contrast to a typical program for word processing or graphics that would better be characterized instead as "appearance oriented".

[51] Based on a dictionary definition, a "term" is "a mathematical expression that forms part of a fraction or proportion, is part of a series, or is associated with another by a plus or minus sign".

Other symbols interpreted by LaTeX as commands or instructions include:

$$ \$ \quad \# \quad \% \quad \& \quad \_ \quad \sim \quad \wedge \quad [\quad] \quad \{ \quad \} \quad < \quad > \quad " \quad | $$

Should you ever need to make *literal* use in a document prepared with this system of the percent sign—as a symbol in its own right—the command nature of "%" would first need to be temporarily set aside. This is accomplished with the aid of the universal "backslash" command already introduced. Thus, "\%" has the desired effect: causing a particular percent sign to be printed as a text element.

Square and curly brackets, included in the brief list of command-symbols above, are used to distinguish an "argument" to which some other command applies, and to clarify its extent. A given command string might in fact incorporate multiple sets of square and curly brackets.

Most of the individual commands in LaTeX are "word-like" and mnemonic in character, based on the English language. For example, the commands

```
\frac{d}{n}
\sqrt[n]{radi}
```

signify in the first case that the variables (or quantities) *d* and *n* are to be presented in the form of a *fraction*, *d/n*, and in the second case that extraction of the *n*th root of something (represented here by `radi`, *radicand*) is to be displayed. (Note that the function designation "sqrt" is somewhat misleading, since it *seems* to imply a *square* root; "radi" here is derived from the Latin "radix" or "root", consistent with the fact that a root is involved).

Formulas that are not intended to be free-standing (i.e., ones that will be *integrated* into the text) are referred to as *text formulas*, and everything included in such a formula —a brief expression, with at most a very few characters—is called *formula text*. This type of expression is prepared in LaTeX with a command structure like

```
\begin(math)  formula text  \end(math)
```

Since typing 32 characters in order to encode what might in the end produce only *one* printed symbol is a nuisance, an option is provided for accomplishing the same thing more efficiently: `\[formula text\]`. More concisely still, one can simply type `$formula text$`.[52] Thus, if one were to enter `$x$`—in the middle of a sentence, perhaps—the consequence would be introduction of a letter "x", albeit one typeset correctly for a mathematical environment, namely in italics (i.e., "*x*", as should always be the case with a variable or a quantity symbol).

From the foregoing one would of course infer that in the longer expression above, `\end(math)` causes the LaTeX system to revert to text mode, as do `\]` and the dollar sign in the more compact notations.

---

[52] For "Let *x* be a prime number to which the expression $y > 2x$ applies", the corresponding input would be: `Let $x$ be a primary number to which the expression $y>2x$ applies.`— The *dollar sign* as a command symbol, like many other commands, has a reversible effect, rather like a two-way switch.

A *free-standing* formula or equation is created with one of the slightly different command strings

\begin(displaymath) *formula text* \end(displaymath)

or

\begin(equation) *formula text* \end(equation)

LATEX even tailors certain aspects of its work to take directly into account whether a particular mathematical expression will be free-standing or not. For example, limits defining a summation might be caused to appear above and below the summation sign in the usual way, or fall instead to the right of it (see below), where the latter arrangement is better suited to an expression embedded in text:

$$\sum_0^\infty x^2 \quad \text{vs.} \quad \sum_0^\infty x^2$$

Similarly, the command int_{a}^{b} produces for inclusion in running text an *integral sign*, on the right-hand side of which, near the bottom and the top, respectively, the limits *a* and *b* will be displayed. In a free-standing formula one would want instead the more familiar—and taller—standard notational form (with limits *directly* above and below the primary symbol), summoned by the command \int\ limits_{a}^{b}. The two display possibilities are thus

$$\int_a^b \quad \text{and} \quad \int_a^b$$

Superscripts and subscripts—interpreted in the broadest sense (as the examples already cited suggest)—are obtained with the specifications ^{...} and _{...}, respectively. In other words, simply typing

x^{2}   a_{i,j,k}

results in

$$x^2 \qquad a_{i,j,k}$$

No problems arise if one requests that both a superscript *and* a subscript be appended to some symbol. The former will generally be set directly above the latter in a way that cleanly avoids interference. Similarly, an "index to an index" is as easily prepared as requested, and the program again takes care of such details as selecting the appropriate type sizes.

If necessary, commands can be *nested*—with the aid of *fences*. In LATEX it is even permissible to nest fences of a single type, as in {{ }}. For example, suppose you wanted to express extraction of the root of a root. LATEX patiently follows your instructions, and is not at all offended by the fact that the request sounds absurd to ears not accustomed to "mathspeak". Thus,

\sqrt[3]{p+\sqrt{p^{2}+q^{2}}}

produces

$$\sqrt[3]{p+\sqrt{p^2+q^2}}$$

Note particularly the tiny (one-digit) "arguments" occupying the innermost sets of fences in the instruction.

Once one has had a little practice, decoding a LaTeX expression is unlikely to present a serious challenge. The first step is always to look for sets of fences. In the example above you might next notice that a *cube root* is being taken, and that another root symbol is incorporated in the argument for the first one. The latter must refer to a *square* root, since no indication appears to the contrary. Incidentally, this example clearly illustrates the difference between a "mandatory argument" and an "optional argument" (see above); that is, there is no term in square brackets in conjunction with the *square* root, because for this default (simplest) type of root none is required. The inherent logic should bolster your confidence that additional study of LaTeX could well be rewarding.

It will be instructive for us to continue just a bit farther in this vein and watch how the program keeps pace. Already in the last example we achieved a result that would not be easy to duplicate manually. Notice, for example that—as displayed—the "embedded" root symbol is noticeably smaller than the one on the outside, and the top line of each root symbol is of a length appropriate to the contents. In other words, there is clear evidence here of considerable typesetting sophistication. The program has also produced root symbols whose overall sizes and lengths are precisely adapted to the corresponding arguments. There is obviously not a vast array of prefabricated root signs of all sorts hidden somewhere within the program, maintained on the off chance they *might* someday prove useful: each sign is "custom manufactured" for the situation immediately at hand. *How* that occurs is something we can quite properly ignore.

● TEX expressions are clearly not very difficult to interpret given a little practice, and they are just as easy to formulate.

The process of creating common expressions like fractions, for example, closely resembles what we have witnessed with respect to roots. This time we start with the general command "\frac". It is important to remember that the first "mandatory argument" associated with this command is always the numerator. The second must therefore be for the denominator. Might there be a need for a second fraction *within* the fraction? No problem!

```
\frac{\frac{a}{x-y}+\frac{b}{x-y}}{1+\frac{a-b}{a+b}}
```

The resulting set of instructions may be awkward to read, but the remarkable thing is that it in fact all fits on *one line*! What results from this particular string of commands is

$$\frac{\dfrac{a}{x-y} + \dfrac{b}{x-y}}{1 + \dfrac{a-b}{a+b}}$$

If we had the time and patience (and the space!), we would at this point present you with a few especially challenging new expressions or equations and encourage you to try your hand at translating them into TEX instructions—with correct answers provided in an appendix. Unfortunately we were obliged to set aside that temptation, but you can for yourself find many interesting examples in the books by LAMPORT and GOOSSENS, et al. cited earlier. Beyond that you should of course refer to the appropriate program handbooks. Work your way through the examples, experiment a bit on your own, and in a short time you will find yourself producing remarkable results.

By the way: the diacritical marks (*accents*) so common in mathematical expressions also do not add a serious complication:

```
\tilde a  ã      \bar a     ā      \check a  ǎ
\dot a    ȧ      \breve a   ă      \vec a    ā⃗
```

Moreover, accessing a Greek letter is simply a matter of spelling it out: "π" is thus specified by `\pi`, "Π" by `\Pi`. LaTeX also places a rich assortment of other symbols for operators, relationships, and negations at your disposal; e.g.:

```
∇ \nabla        ∩ \times        ⋈ \pm
× \cap          ⊂ \subset       ≢ \not\equiv
± \bowtie       ⊢ \perp
```

As noted previously, almost all the LaTeX commands are mnemonic in character, so that anyone with English as a native language is likely immediately to deduce the significance of most of them (and then commit them readily to memory!).

The fact that LaTeX provides more special symbols for mathematical typesetting than, say, a character font like "Symbol" should by now not come as a surprise. Furthermore, the program assumes nearly every responsibility that would ordinarily be vested in a technically skilled typesetter. For example, it acts quite independently in adjusting the widths of required *blank spaces* in almost every conceivable situation. The rules cited earlier (in Sec. 6.5.6) have thus all been taken into account (i.e., proper spacing around an equals sign, etc.).

One frustrating bit of "simple-minded behavior" on the part of the program is worth mentioning, however: it stubbornly regards any letter entered in mathematical mode as a *quantity symbol*. This has the advantage that one need not make a conscious effort to set such characters in italics, but it also results in the nuisance that *names of functions* like "lim", "sin", "log", etc., frequently end up being italicized. Thus, if the latter were typed directly it would appear in the formula under development as a product of the three quantities l, o, and g. To prevent such foolishness one must precede the alphabetic symbol for a function by the familiar "universal" command symbol, as in "`\log`". The same holds true with respect to *unit symbols*, such as "cm".

For all the examples we have cited, as well as much besides, one could construct *custom* commands—*macros*—making it especially easy to generate complex symbol sequences that might be required repeatedly over the course of a manuscript, analogous to "AutoText" entries in WORD.

Viewed as a program, LaTeX presents a stark contrast to the "What you see is what you get" (WYSIWYG) environment favored in so much of today's publishing software. Some have as a matter of principle for years derided the WYSIWYG approximation (recall that this acronym was mentioned earlier in the book), recasting it sarcastically as WYGIWYS ("What you get is what you see"). Substituting the concept of *logical* design for *visual* design requires a fundamental shift in one's thinking. A fitting motto for LaTeX might be "What you get is what you intend", although we hasten to add that we are *not* encouraging adoption of the unsightly acronym "WYGIWYI".

With LaTeX the only thing one sees initially on the screen as the various parts of a mathematical expression are introduced is a seemingly haphazard array of codes—only a very few of which we have so far described, by the way (there are some 900 recognized codes in all!).[53] What remains concealed is how the material will look after it has been formatted. With a few extra keystrokes in the form of a "Preview" command, however, you can quickly unveil the fruit of your efforts, and thereby verify visually that everything is as it should be.[54] If this proves *not* to be the case, it means some rule has probably been overlooked or violated, and you will need to adapt your input file accordingly, confirming seconds later that the right changes have in fact been made. When the time finally arrives to print your masterpiece, you can therefore be confident everything will be perfect.

6.6.2 *LaTeX and Text*

All the activities overseen by LaTeX are subject to preprogrammed, systematic management, not just equation generation. In effect, one's entire document is treated as if it were a "formula". Not surprisingly, the software is unusually well equipped to supervise automatic numbering of free-standing formulas and equations, and to monitor their spacing relative to surrounding text. Concise commands make it possible to establish "shifting environments" for optimal incorporation of figures and illustrations. There is also a meticulously organized scheme for citing external sources, to take but one more example.

Once you have spent a while thinking within the rigid framework of "categories" with respect to a document there is some tendency to react a bit like the Charlie Chaplin character in *Modern Times* (1936), who automatically began to tighten bolts (with

---

[53] Of these, roughly 300 are basic commands, the rest being *macrocommands*, many with adjustable parameters.

[54] LaTeX in fact provides you with a *true* "print preview", since the printer and screen drivers are technically identical.

an imaginary wrench, of course) as soon as he spotted them off in the distance, simply because that had become such a dominant part of his daily routine at the factory. One could argue that such indoctrination might not be a wholly bad outcome in the case of an author given the many opportunities scientific text presents for "tightening things".

With respect to LaTeX as a document development tool, consider for a moment: does a certain piece of text need to be in italics or boldface, or perhaps in small caps? No sooner said than done! The corresponding LaTeX commands are

> `\textit, \textbf, \textsc`

or, more concisely,

> `\it, \bf, \sc.`

But possibly that was not a particularly impressive example. After all, similar changes can be made at least as easily with any word processor. On the other hand, LaTeX definitely shines when it comes to the management of numbered elements, like section headings and literature citations, or references to figures and tables. The software attends to all of these flawlessly—and invisibly—in such a way that if material is later rearranged, inserted, or deleted, all appropriate changes in numbering are implemented throughout the document: instantly!

On the other hand, rigid standardization occasionally becomes a nuisance. For example, the slightest deviation from a predetermined document layout[55] is a major challenge with LaTeX.

Instead of burrowing further into details, we conclude this introduction to TeX and LaTeX by summarizing a few general characteristics. While it is true that the programs demand a great deal of computer time and storage capacity relative to ordinary word-processing software, they clearly offer a number of offsetting advantages:

– pristine mathematical typesetting;
– straightforward support for automatic cross-referencing (with respect to figures, formulas, tables, and footnotes);
– transparent management of large or complex documents, where chapters stored in individual data files can summoned from one central document store;
– provision for both general instructions and custom commands, again through a central document file;

---

[55] The logical structure of a document—and thus its layout characteristics and parameters—must be defined at the very beginning of an input file. This entails assignment to a *document class*, the most important examples of which are `article` (for journal publications, seminar contributions, and brief reports), `report` for longer report-like documents composed of multiple chapters, including dissertations and theses, `book` for (obviously!) books, and (equally obvious) `letter`. Thus, the command `\documentclass{book}` assigns a particular standard book layout to any manuscript in that class. The author is excused (and essentially *excluded*) from further attention to details regarding layout.

– outstanding support for the preparation of tables of contents, indexes, and reference lists, as well as for maintaining strict hierarchical organization, even across multiple data files;
– a logical structure similar to that inherent in SGML, HTML, and even RTF.

Suffice it to say we regard it as a blessing that LaTeX was "invented". The larger and more complex a document, the more advantages the program offers.

6.7 MathType and MathML

Tex and its variants are the only end-user programs that can boast equal proficiency at the typesetting of text, tables, and mathematical formulas. In the DTP realm—an environment with strong appeal for a semiprofessional clientele—the only other software package that begins to address all three areas at once is FrameMaker, which treats formulas as special elements to be manipulated in a WYSIWYG setting, whereas Tex can be said to put text and formulas on a common level.

At the same time one additional program should be mentioned which in recent years has itself become a singular player in the world of formulas, even though it would not yet be regarded as "standard equipment", and may therefore need to be installed as required. The program is MathType (from Design Science, Inc.), a "stripped down" version of which, under the name Equation Editor, is familiar through its distribution in conjunction with programs like Word, WordPerfect, and AppleWorks. In contrast to Tex or FrameMaker, MathType is intended *exclusively* for preparing mathematical expressions. It offers only the most rudimentary of text-editing facilities, but is unmatched for rapid, uncomplicated preparation of even highly complex formulas. Mathematical expressions are constructed either by way of menus and palettes, with a few clicks of the mouse, or, if one takes the time to learn the appropriate shortcuts, via the keyboard. One can also create powerful macros in MathType (especially for use in Word) to simplify formula conversion, numbering, and referencing. According to the Brockhaus treatise *Der Computer and Informationstechnologie* ("Computer and Information Technology", 2002):

> MathType is exceptional in terms of its intuitive operation (based on the WYSIWYG principle), as well as for the filters it provides for Tex, and its smooth cooperation with virtually all word-processing and page-layout software. (MathType formulas can incidentally be treated as OLE[56] objects.) The current versions of MathType [at the time of publication] are 5.0 for Windows and 3.7 for Macintosh. One of the most important recent additions is support for formula output in the "MathML format" for Internet publications.

---

[56] OLE is an acronym for Object Linking and Embedding.

MATHML, the Mathematical Markup Language, was adopted by the World Wide Web Consortium (W3C) in February, 2001, as an official markup language for introducing mathematical content into the Internet. It is closely related to XML (Extensible Markup Language, an Internet variant of SGML), and makes possible high-quality presentation of mathematical formulas with any standard browser. Whereas depicting mathematical structures via HTML requires the availability of *images* (graphic elements, created somehow in advance), MATHML generates sophisticated screen representations from coded information directly. "As a result, expressions of this sort also lend themselves to complex processing dependent upon their intellectual content: using computer-algebra software, for example" (BROCKHAUS 2002; cf. also GOOSSENS and RAHTZ 1999).[57]

---

[57] Systems of the latter sort are exemplified by MATHEMATICA and MAPLE, which are vehicles not only for the *depiction* of equations but also their *solution*. A simplified algebraic system has even been incorporated into FRAMEMAKER page-layout software, permitting actual computations on the basis of formulas once they have been introduced.

7 Figures

7.1 General Considerations

7.1.1 Figures and Figure Numbers

Throughout the book we use the term *figure* in reference to elements in a publication that do *not* consist exclusively of the characters available with an ordinary word-processing system. Other common words or expressions useful in this context are *illustration* and *pictorial matter*, to which can be added *graphics* or simply *picture*.

An interesting distinction can actually be drawn between "figures" and "illustrations". Thus, a "figure" is generally interpreted very broadly (and somewhat philosophically) as an array of symbols which, in contrast to those constituting text, are meant to be interpreted in a *non-sequential* way. "Illustrations" represent a distinct subset of "figures", because they are specifically intended in some sense to depict—through a medium like drawing or photography—*real objects.* A reader thus assumes from the outset that some meaningful similarity must exist between an illustration and the object to which it corresponds. At this level, the broader category "figure" can be seen also to encompass *logical* or *analytical graphics*, "abstractions" (e.g., schemes, diagrams) based not on visible or tangible entities, but rather on principles, concepts, and the like. From the perspective of a scientific publisher, essentially anything in a book that is neither text, nor a formula, nor a table would probably be regarded as a figure.

We have already addressed in Sec. 3.4.3 key points concerning the role of illustrative material in a scientific document, as well as the importance of establishing a clear relationship between each figure present and at least one specific passage within the main body of the text.

- To facilitate convenient mention of figures in text passages, each figure is assigned its own unique *number*. One consequence is that there must then be within the text at least one *reference—by number—*to every figure.

Exceptions to these rules are rarely tolerated in scientific publications, leaving aside the special case of a very small illustration around which text continues to flow, producing an unusually close physical relationship between word and image without at the same time seriously disturbing an otherwise dignified page layout. In any other situation, links between text and illustrative material are established by citing the relevant figure's number. A reference (or citation) of this type acts as a "place-holder" or a surrogate for the figure itself, much like a numerical *literature citation*, which is also in effect a stand-in: for the complete in-text reference that might otherwise be obligatory there (see Sec. 9.3.1). The rules as we have stated them would of course not apply to a document accompanied by only a *single* illustration, since the underlying purpose can in this case be suitably served in a more casual way with a phrase like "as seen in the illustration". The figure will thereby have been explicitly cited, as it should be, so that again the document will contain no illustrative material that has been ignored in the discourse.

● Figure numbers are assigned sequentially, in ascending order, as the corresponding figures are cited in the text. With a lengthy or complex document it is recommended that *double numbers* be used.

An example of what we mean by a "double number" is "3–12", where the "3" identifies the major subdivision of the document in which the figure resides—typically a *chapter* in the case of a book—while the "12" establishes that this particular figure is the twelfth one mentioned *within* that subdivision. Sequence-based numbers of course revert to "1" at the beginning of each major subdivision (e.g., "Fig. 4–1"). A hyphen (better: *en dash*) is employed to separate the two parts of the figure number to avoid possible confusion with *subsection numbers*, in which the traditional separator is a period (or decimal point; cf. "Structure and Form; Decimal Classification" in Sec. 2.2.5).

Apart from localizing figures with respect to a specific section of text, the double-number system also helps keep figure numbers from becoming inconveniently large. Double numbers have the added advantage of reducing the amount of work entailed if *renumbering* should become necessary at some point—for example because of last-minute insertion or removal of an illustration in a nearly completed manuscript. On the other hand, the additional complexity introduced by even more explicit numbers, like "4.5–1", would not be justified. Numerical indicators preceding the dash thus point only to the *uppermost* level of subdivision, resulting in "numbering by chapter".

● A figure number is not considered part of the figure itself, but a component of the accompanying *figure caption*.

It is in fact unwise to incorporate a figure number—or any sort of identification, for that matter—directly into a figure. Much better is to reserve space for that purpose *beside* or *beneath* the figure, where whatever is necessary can be presented in ordinary type (cf. Fig. 7–1). A figure number included as an integral part of a graphic element is distracting, and it severely limits the author's flexibility with respect to using the same figure again someday in a different setting (quite apart from problems that arise if renumbering becomes necessary).

It *is* important, however, to ensure that figure *copy* is clearly labeled with whatever number is tentatively assigned to it, and that it remain labeled until such time as the material assumes its permanent place in the document. The required information can be inscribed with a soft pencil—outside the bounds of the figure itself, naturally—either on the front or the back. In the case of a photo, such notations are necessarily confined to the back. Labels on figure copy for a manuscript that will be submitted to a journal should include, besides the figure number, prominent display of the name of

Fig. 7–1. Layout of a typical page containing figures, showing alternative placements for captions.

the *author* (or senior author) and at least an abbreviated version of the paper's *title*. Many journals specify use of adhesive labels for this purpose.

The rare *unnumbered* figure also must be accompanied by some sort of explanatory note: "((MS. 355, top))", for example—to be interpreted "this figure relates to text at the top of p. 355 of the manuscript".[1]

7.1.2 Figure Captions

Figure Identifiers and Titles

● Text designed to accompany a figure is known as the *figure caption*. In most cases it will eventually be printed directly *beneath* the figure itself.[2]

Sometimes a caption is instead set *adjacent* to the corresponding figure in the interest of saving space. This occurs most often with text set as a single column and a figure occupying less than the full width of the page.

● The first element in a caption is an *identifier*, consisting chiefly of the *figure number*.

---

[1] Notice that we have once again suggested using a "double parentheses" signal in conjunction with information directed specifically to an editor or typesetter (cf. Sec. 3.4.3 under the heading "Relating Figures to the Accompanying Text").

[2] Technically, a caption consists only of what we here call the "figure *title*", but it has become customary to use the word somewhat loosely to cover *everything* printed under (or next to) a figure.

As we have already suggested in passing, this identifier typically assumes the form "**Fig. XY.**" (less frequently "**Fig. XY:**"), and is generally set (as here) in *boldface type*. Occasionally the word "Figure" is spelled out in full, although figure *references* almost invariably employ the abbreviated notation.

In principle, the figure identifier alone might suffice as a caption. After all, there is certain to be at least one explicit reference to that figure in the body text, and an interested reader could seek additional information there! Forcing a reader (or browser) to go to so much extra trouble simply to ascertain a figure's significance, however, is blatantly inconsistent with the goal of facilitating an efficient transfer of knowledge: in this case an understanding of the purpose of that particular graphic object within the document.

● A reader should readily grasp the general meaning of an illustration or figure on the basis of its caption alone, without recourse to the text.

This condition will be met only if the content of the figure is captured effectively in words, at least in summary form. A stunning photographic image derived from an electron-microscopy experiment possesses no communicative value whatsoever if the person looking at it is unable to determine what it depicts. The image remains utterly "mute".

● The second indispensable component of a figure caption is therefore a brief verbal description that could be regarded as the figure's *title*.

A title for a figure is in every sense analogous to the title of a *chapter*, or of a *book*, which is of course why it is given that name. The following can be taken as representative examples of figure titles:

> Schematic drawing of a recycle reactor.
> Flow diagram for the manufacture of nylon.
> Seawater corrosion of a carbon-steel surface (10-fold magnification).
> Solubility of casein as a function of pH.
> Photomicrograph of yeast cells undergoing division.
> Perspective view of the zeolite XY matrix (schematic).
> Typical distribution of the aromatics produced during conversion of
>   methanol to gasoline over a Z-catalyst.

Notice that in one of the examples above there is information regarding the technique used in *obtaining* that illustration. In *every* instance, however, there is a concise summary of what the image depicts. The *form* of the messages will certainly be familiar from other sorts of titles. Note especially the fact that none is a complete sentence. That is to say,

> A collimating lens ahead of the slit projects an image of the light
> source on the diaphragm.

would *not* be an appropriate figure title, although it could well constitute part of an associated *legend* (see below).

In the case of a set of slides designed to serve as "figures" in conjunction with a *lecture* one might easily be tempted to incorporate identifying elements within the framework of the graphic images themselves, but even here we suggest you maintain the practice of providing a separate caption at the bottom of each slide. In our judgement, *all* illustrations should be kept physically distinct from their titles, the latter being incorporated into formal captions. Consistency alone could arguably serve as a legitimate rationale for this advice, but you should also bear in mind the possibility that someday you might want to use one of the figures in *another* publication or presentation, possibly even one in a different language. It would be aggravating to discover that this is ruled out by the awkward presence of a label not easily changed.[3]

Legends

Even a casual reader is often grateful for a bit more information about a figure than its title alone conveys. This would be especially likely to apply with a figure that is inherently difficult to decipher. Moreover, as author you might wish to direct the reader's attention explicitly toward some special detail in the illustration, or perhaps you think it important to point out the conditions responsible for some unexpected feature. With a photomicrograph it could be important to identify immediately the technical innovation responsible for producing exceptionally sharp contrast in that particular image. Ignoring such impulses and merely suggesting that the reader "refer to the text for further information" is clearly inconsistent with the principle that the combination of a figure and its caption *alone* should be meaningful.

● Explanatory details can be provided most conveniently through a *legend* appended to the figure title.

Based on the Latin roots of the word (which evoke the notion of "reading", or "a story"), a "legend" associated with a caption would be expected to help one elucidate how the accompanying figure is to be "read" (cf. "Relating Figures to the Accompanying Text" in Sec. 3.4.3).[4] There may, for example, be a need to clarify the significance of *symbols* with which data points from an experiment have been depicted (see "Graphic Presentations Based on a Set of Coordinate Axes" in Sec. 7.2.3), which could in turn entail describing briefly the experimental conditions associated with various sets of data—or *data series*. Similarly, a legend is the logical place to explain otherwise

---

[3] Now graphics are actually prepared and labeled in most cases by computer-based methods, which means that such labels in fact *are* interchangeable, so the argument is less compelling than in the past. Nonetheless, we continue to believe strongly in minimizing the presence of non-graphic elements in figures of all types, as will be explained presently.

[4] In English the word "legend", like "caption" (see the second footnote in this section), is increasingly being treated as if it embraced *everything* appearing underneath a figure, another illustration of the way a language once exceptionally rich in subtle distinctions is gradually yielding to forces encouraging simplification and a reduced (impoverished?) vocabulary.

obscure identifications: numbers used to label data curves, perhaps, or letters designating various parts of a schematic diagram.

There are advantages, incidentally, in drawing a clear distinction between a clarification of this type—a legend—and the actual title of a figure. For this reason,

- Legend material (explanations, identifications) is often collected in the form of a *block* and set below the figure title, or else separated from title text by the combination of a period and a dash (preferably an *em dash*).

Visual delineation of this sort helps a reader process the information with which he or she is confronted. For example, sometimes a figure title alone will suffice for preliminary orientation, in which case it is better if the title does not metamorphose imperceptibly into a host of (for the moment irrelevant) details. Unfortunately, simple considerations like this too often fall by the wayside in the planning stages of a published work, despite the fact that much could potentially be gained so easily— here simply through conscientious insertion of a few dashes. In a long document (e.g., a book) it is common to include a separate "List of Figures", additional incentive for the author to select compact, pregnant figure titles supplemented with informative legends on those pages containing the actual figures.

- Well-chosen legends go far toward eliminating the need for extensive labeling *within* the various figures.

Figures—depictions of all sorts—should in general be kept as free as possible of "foreign" elements, especially text (referred to officially as *inscriptions*). This advice has its foundation not only in aesthetics, but also in sound economics. It is easier (and thus less costly!) to provide identifications like

L	Light source
l	Prism

in an informative caption so that the figure itself need only display at appropriate points the letter "L" and the number "*l*" (italicized; see below, in the second part of Sec. 7.2.3) as opposed to unsightly verbiage, the placement of which can prove quite troublesome, because a good figure seldom offers sufficient blank space for extensive commentary. Admittedly, figure copy can be adapted to accept integral text much more readily today than in the past, thanks to computers.[5] Most PC graphics programs feature powerful editing tools that allow one to insert independent "labels" of any size into a graphic image wherever they might be required. Such labels also remain subject to future modification—linguistic adaptation, perhaps! The effort required is trivial compared to what would have been expected of the "commander" of a "squadron of drafting pens" in years gone by. By the same token, it is far easier now than "in the old days" for a publisher to establish a *library* of (digital) figures with inscriptions in multiple languages.

---

[5] See the third footnote in this chapter.

- If possible, identification marks in a figure should be limited to symbols readily available from standard type fonts.

"Experiment 1" and "Experiment 2", etc. (or simply *1, 2*) can easily be made part of a typeset legend and then elaborated as necessary, but the same may not be true of an assortment of circles (both "open" and "filled"), triangles, squares, etc.[6] If unusual symbols are truly unavoidable in a figure (see "Graphic Presentations Based on a Set of Coordinate Axes" in Sec. 7.2.3) someone may need to enter their counterparts into the "captions manuscript" by hand.

The average reader understandably prefers it when interesting details are available "on the spot": *within* a figure, for example, rather than consigned to a legend. Some concessions in this direction could perhaps be rationalized even on economic grounds— but only from the *readers* perspective ("time is money", etc.). Depending on the nature of the illustration, and taking suitable account of readability, you will undoubtedly be forced to strike compromises with respect to what details are displaced into a legend and what can actually be accommodated in the figure. The nature of the *publication* can play a role as well. Relative to a professional *monograph*, for example, figures in a *textbook* are usually annotated rather extensively.

The following can be regarded as illustrative of well-conceived and well-organized figure captions:

> **Fig. X–Y.** Longitudinal and transverse sections through an X-ray tube.—
> *1* Anode; *2* Cathode; *3* Voltage source; *4* Aperture; *5* X-ray beam.

> **Fig. Z.** Separation of XXX by temperature-programmed GC.
> Column: 5 m methylpolysiloxan
> Temperature: 28…200 °C; 2 °C/min
> Analysis time: 46 min
> Carrier gas: hydrogen
> Detector: FID

There can be no general rule about which is preferable, continuous text as in the first example or a block layout like that of the second. A block layout has the advantage of greater clarity, but at the cost of space. Still, a little space might well be worth sacrificing if an extensive collection of charts must be presented (gas chromatograms, for example), and the reader is expected to make rapid visual comparisons of the effects of various experimental conditions.

Miscellaneous Technical Observations

- When a figure consists of several smaller figures, the latter are traditionally designated by *letter* (a, b, c …) and dealt with in a single common figure caption.

---

[6] Even if it is easy for you to produce such symbols on your computer screen (e.g., with a font like "Zapf dingbats"), you cannot be sure they will *print* properly for the publisher.

Such a caption could take any of the following forms:

> **Fig. Z.** A set of valves arranged **a** in series, and **b** with feedback provisions.
>
> **Fig. Z.** Alternative valve installations.— **a** Series arrangement; **b** in-corporation of a feedback option.
>
> **Fig. Z.** Alternative valve installations.
> **a** Series arrangement,
> **b** incorporation of a feedback option.

● A figure caption should always be terminated with a period.

Rules like this are actually intended to cover entire *groups* of elements, extending in the present case to literature references, footnotes, and table headers—at least in the eyes of most editors.[7] Although this tradition is occasionally ignored, what *is* important is rigorous uniformity *within* a given document.

● With a manuscript not subject to formal typesetting, figure captions should be treated essentially as part of the text, but compacted (i.e., with "line-and-a-half" spacing, for example, if the body text is double spaced).

This has the effect of creating a clear visual distinction between figure captions and running text. For the same reason, figure captions in printed documents are also set in smaller type.

● If a manuscript is to be *typeset* in the traditional way, figure captions (just like body text) should be double-spaced, but kept separate in the form of a *caption manuscript*.

Generous line spacing makes a typesetter's job much easier, and it also facilitates the introduction of last-minute editorial comments or necessary corrections.

Finally, we consider briefly the *placement* of figures in a document.

● Tradition calls for all figures and their captions to share a common orientation; that is, they should all be set flush against the right—or the left—margin, or else consistently centered on the page.

Put another way, either the left edge (alternatively: the right edge) of each illustration and caption should line up precisely with the left (or right) edge of the corresponding *type page*—also called the *text area*—for example, or else the central axis of each

---

[7] A period sends the signal "here the message ends". A terminal period has even been sanctioned in the literature-citation scheme adopted by the *Vancouver Convention*, which is exceedingly sparing in its tolerance of punctuation (see Sec. 9.4.2). Admittedly, however, no "rules" of this sort are universally recognized. Disagreements even persist between various editorial teams at the very publishing house where this book was produced! One "opinion leader" in the trade (HUTH 1998) openly advocates *inconsistent* treatment with respect to seemingly related components within a document: he thus recommends ending all figure captions with periods, but *omitting* periods after table headers (presumably because periods are never used after headers of *other* types).

illustration should be made to coincide with the central axis of the type page. Whatever rule is adopted, it should apply not only to the figures, but all *captions* as well.[8]

Legal Matters: The Citation of Figures

It may come as somewhat of a surprise, but this seems to us the perfect place to embark on a brief legal excursion. Hopefully every publishing scientist is already conscious of the fact that on ethical grounds alone it is inappropriate to make unfettered use of the work of *others* in one's own publications. *Copyright law* obligates every author to ensure, through proper citation (see Chapter 9), that someone else's intellectual property is clearly acknowledged as such. Illustrations or figures fall under the heading "intellectual property" in a very specific way, which must be taken into account scrupulously as you compose your figure captions.

● Whenever you utilize an illustration or figure from some external source, it is assumed you are in possession of *permission* to reproduce it, and a proper notice of source must be incorporated into the corresponding figure title *prior* to any elaboration.

The following represent examples of appropriate acknowledgements within figure captions:

> **Fig. X.** Genealogy of the yeast employed (from MÜLLER [12], courtesy of the author).

> **Fig. Y.** View of a nuclear spin tomograph (company photo, reproduced with the kind permission of X Incorporated, Y City).

If an illustration has not in fact been literally *reproduced* by photomechanical means, but rather adapted to suit your purposes, then a caption acknowledgement in the following form would suffice:

> ... (based on SCHMITT et al. 1985).

In another publication some years ago (EBEL and BLIEFERT 1982, p. 158) we included the observation below with respect to obtaining permission to reproduce illustrative material:

> Figures (illustrations) of a scientific or technical nature presented in conjunction with previously published text can legitimately be incorporated into another printed work provided the sole purpose is to help explain that publication's content. An attempt to expand upon or more fully document some contention does *not* constitute an "explanation" in this sense, however. Because of this fine distinction it is advisable *always* to seek permission prior to reproduction of an illustration.

---

[8] Our insistence on strict uniformity undoubtedly has its origin in a scientific mindset, and may well seem misplaced to the publication designer more concerned with aesthetics.

This legal provision has, over the years, led to the expenditure of a great deal of effort on the part of publishers and authors related to permission requests (including preparation of the appropriate responses, of course), effort that was in many cases unnecessary. Most such permissions have long been granted almost routinely, normally without monetary compensation, but even when this has not been the case it is questionable whether the associated labor was really worthwhile. A number of scientific and technical publishing houses have in recent years joined together in an initiative sponsored by the International Association of Scientific, Technical, and Medical Publishers ("STM") to abolish much of the ritual involved in seeking authorization to reproduce published illustrations and tables. In particular, member publishers are strongly encouraged to permit almost automatically and without charge the reproduction of up to *three* subsidiary elements derived from a given journal article or book chapter, or *five* from an entire book (as well quotation of modest amounts of text).[9]

Apart from this one move toward "liberalization", copyright protection for illustrations continues in full force. Indeed, significant economic interests are in fact at stake. For example, it is understandably illegal to incorporate into one's work without permission a large *number* of figures "borrowed" from a textbook, say, figures in which a great deal of labor and capital was undoubtedly invested. In case of doubt we recommend that you always seek specific advice from your publisher.

Besides the obvious economic considerations there are also moral issues to be taken into account ("moral rights" in the terminology of copyright law). The original publisher of an English-language textbook might be interested in granting another publisher a translation license, for example, but would not necessarily be free to turn over the rights to associated illustrative material. The author who no longer wishes to stand behind some previously published result, or has in the meantime acquired newer or more reliable data, could understandably be unwilling to see "ancient history" retold over and over again in other contexts!

7.2 Line Art

7.2.1 What Is Line Art?

From the standpoint of technical processing, an important distinction must be made between two types of printed illustrations: *line art* and *halftones*. The latter usually involve *photographic* material, but this fails to reach to the heart of the matter. Line art can itself be photographed, but the resulting photo will still class as line art. The

---

[9] In this context STM has made available a *Permission Guidelines* document, updated periodically (last revised March 2003), together with a comprehensive list of signatory publishing houses. These can be obtained from http://www.stm-assoc.org/committees/guidelines.html or by writing to: STM Secretariat, POB 90407, 2509 LK The Hague, The Netherlands.

real difference between the two types of illustration is the fact that in line art only two states need be considered: *black* and *white* (or "color" and "non-color"), whereas in a halftone figure certain areas will represent transitions *between* black and white—in other words, shades of gray. A real halftone does not in fact contain *true* gray shades, but some parts may *look* gray due to the effect of more or less dense assemblages of black *dots* (see Sec. 7.4).

● There are thus two basic categories of figure *copy*: *line drawings* (in which grays are by definition absent) and *continuous-tone art* (copy that may contain grays and/ or shades of color).

In this section we restrict our attention to line art, deferring the subject of halftones and continuous-tone material to Sec. 7.4. We address first the "classical" approach to preparing such copy (manifested in the use of sophisticated manual *drafting* systems), although this has largely been supplanted by computer technology, to which we devote the subsequent Sec. 7.3 ("Drawing With a Computer") where the emphasis is on creating original graphics electronically, on a computer screen.

It may well be that the crisp, shiny line on parchment (*vellum*), owing its existence to India ink laid down skillfully with a drafting pen, will soon be little more than a dim, nostalgic memory. Only one who has had the privilege of seriously engaging in "mechanical drawing" with traditional tools can imagine the intense satisfaction such work can produce. The more modern approach to drawing—just like text preparation with a computer—nonetheless has its roots deep in the past, and it came to be accepted (sometimes grudgingly) as "the only way" very recently indeed (in the years since ca. 1980). A host of "classical" icons and terms survived the transition, perhaps in silent tribute to past achievements. Thus, age-old symbols for *drawing* (the pencil) and *painting* (a brush and a paint pot, supplemented by a spray can) are still familiar sights in virtually all computer-graphics programs, as is the *palette,* a device almost synonymous with the traditional artist.[10] The square and the circle in this context of course still represent precisely what they always have.

● In line art, every element present is either fully invested with color or completely colorless (i.e., white); "shades" of color are by definition inadmissible.

In the physical sciences, line drawings provide suitable illustrations at least 90% of the time. A schematic diagram ("synthetic picture"), for example, lends itself especially well to this medium's stark contrasts because of its inherently abstract character. On the other hand, the more descriptive sciences, such as biology and geology, would find it very difficult to avoid calling at least occasionally upon continuous-tone illustrations—true photographs—in their attempt to convey fundamentally descriptive information, influenced heavily by impressions. Even with a "drawing" it can sometimes be advantageous to attempt to simulate depth and reality through careful selective application of gray tones, resulting in a type of composition (a *continuous-*

---

[10] One sometimes even sees reference to a concept as anomalous as a "palette of line forms".

tone drawing) that begins to resemble a photograph. Such effects are generally achieved using techniques more akin to painting than drawing.

The term "line art" has long been the subject of criticism because the medium itself is capable of encompassing so much more than mere collections of lines. *Regions* of paper can also be subjected to uniform coloration: circular domains, for example. Moreover, the very notion of a "physical" *line*—let alone a *point*—is in some sense fictitious, especially from the mathematicians point of view. It is quite impossible to depict a true "line" on paper: what we call a "printed line" is really a highly elongated, extremely narrow black *surface*. Nevertheless, more apt or more precise terminology has never gained wide acceptance.

- Line art is readily photocopied, photographed, or transformed by photographic means into the printing plates that serve as the basis for offset printing.

This flexibility with respect to reproduction technology constitutes an enormous advantage when set against the severe limitations imposed by continuous-tone originals. The miracle of *xerography*, which has in so many ways revolutionized the average professional's work day, is in essence a mechanized form of "black-and-white painting", and is thus ideal for the reproduction of line drawings. There is no need whatsoever to simulate specific "gray tones" in the copying process since these are never present in line art. At the same time, nearly everyone has experienced the disappointing results typically obtained from an attempt to duplicate a photograph with a standard photocopier. (The *name* associated with the latter thus proves to be rather ironic!) Even here there has been significant progress in recent years, however, as a result of dramatically increased resolution in the associated scanning process (cf. Sections 7.3.1 and 7.4)—driven in part by developments in telephonic communication—as well as fundamental improvements in every aspect of copying technology and equipment. Enhanced resolution has been important in conjunction not only with scanners of various types, but also computer screens, desktop printers, and indeed every sort of device in which quality is measured in terms of "dpi" (dots per inch; cf. Sections 7.3 and 7.4). Images displaying what seems to be photographic character can now be created, reproduced, and transmitted to the other side of the world from almost anywhere, all in a matter of seconds, and the wonders telecommunication has worked with signals transmitted "through the ether" have been matched by enormous strides in all types of reproduction technology. That said, it remains the case that an attempt to photocopy a (strictly black-and-white) halftone illustration from a newspaper—i.e., a picture that has previously been subjected to bitmap transformation (Sec. 7.4)—and pass the result along over a FAX connection typically leads to an "image" at the receiving end that is barely recognizable.

- Unlike continuous-tone material, line art is easily incorporated by an editorial team or even a novice author into a document in a form suitable for photomechanical direct reproduction—or replication by the offset process.

However, if one is preparing a manuscript destined to be processed for publication in the "classical" way there is little point in merging text and illustrations of any type at an early stage. Indeed, the publisher will much prefer to receive a separate *figure manuscript* (cf. "Relating Figures to the Accompanying Text" in Sec. 3.4.3).

● For publication purposes an author is usually expected to submit along with a text manuscript one complete set of original drawings—or, in the modern variant, a comprehensive set of data files containing computer-based instructions for *preparing* the drawings—together with at least one set of copies (or hardcopy printouts).

Journal editors often wish in addition to receive one or two *additional* copies of all figures (as well as text) for sharing with reviewers. As author you should of course also retain for yourself a set of high-quality *backup copies* of your illustrations. Assuming your line art is not hand-drawn, an emergency set of digital files should be maintained in some secure, isolated location, such as your laboratory or professional office.

Original drawings are usually supplied in $8^{1}/2'' \times 11''$ (letter) format (see "Useful Equipment" in Sec. 7.2.2), and they must be carefully protected against soil and creasing. Art work can be conveniently stored in sturdy file folders—perhaps in individual plastic "sleeves"—and packaged for mailing purposes in heavy, stiff envelopes (braced with cardboard inserts). Data files are usually transmitted to a publisher electronically (by e-mail or FTP, cf. Sec. 3.5.2), with backup copies on diskette or CD dispatched separately in padded envelopes designed specifically for that purpose and custom-made to accommodate the standard formats: $3^{1}/2''$ and 12 cm, respectively.[11]

● Line-art copy for submission to a publisher normally consists of either hand-drawn or mechanically produced "originals" (appropriately labeled) accompanied by one or more clean, high-quality copies, either xerographic or photographic.

The section that follows describes briefly a number of important technical aspects of preparing line-art figure copy. A more extensive treatment is available in *Vortragen in Naturwissenschaft, Technik, und Medizin* ("Lecturing in Science, Technology, and Medicine", EBEL and BLIEFERT 1994), where we deemed it less appropriate to limit ourselves to a concise summary. As a prospective author—especially if you are planning a document the size of a book—you will surely want to seek professional advice from the production staff at the publishing house regarding your treatment of figures. In a favorable case you might find you do not need to assume sole responsibility for preparing final figure copy, in particular for incorporating integral inscriptions.

A *lecture* is of course quite another matter. For that purpose you will have little choice but to make all the arrangements yourself, including preparing any illustrative material you wish to use as a supplement to your oral presentation. Irrespective of

---

[11] The $3^{1}/2''$-diskette, long the primary medium for file transfer, appears well on the way to extinction. Recent MACINTOSH computer models no longer even provide this type of drive.

whether you create line art for this purpose by hand or with a computer—ideally with advice from experienced colleagues—anything the audience sees, including *slides*, *transparencies*, and *software presentations*, will be entirely your responsibility.

Finally, before launching into a systematic (albeit limited) introduction to the design and preparation of scientific illustrations, we wish to recommend explicitly one outstanding book devoted exclusively to this topic: *Preparing Scientific Illustrations; A Guide to Better Posters, Presentations, and Publications* (BRISCOE 1996). This author provides an enormous amount of valuable advice—illustrated by many examples, of both the good and the bad—commencing with the earliest planning stage ("How might a specific idea best be conveyed visually?") and culminating in a treasure trove of technical "tricks of the trade". Anyone aspiring to proficiency in this area should by all means study this source carefully. Another volume with extensive coverage, also dedicated to high technical standards, is the product of a committee of the Council of Biology Editors, now the Council of Science Editors (CBE Scientific Illustration Committee 1988). Finally, we would be remiss to neglect mention of a true classic in the field of scientific illustration, currently in its second edition: *The Visual Display of Quantitative Information* by TUFTE (2001).

7.2.2 *The Preparation of Line Drawings*

Useful Equipment

In the past, scientists and engineers—especially in academia—almost always prepared their own line drawings by hand, using special *transparent* (or translucent) *drawing stock* ("parchment paper", vellum) along with high-quality *pencils* for preliminary sketches and *India ink* as a basis for final illustrations. Also required was a plentiful supply of special-purpose *rulers* and *drafting triangles* as well as standard *templates* of various types, including ones designed for the general preparation of smooth *curves*. Additional indispensable supplies included an assortment of professional *pens* with calibrated tips, to create—in conjunction with the appropriate templates—lines and lettering characterized by very specific *line widths* (e.g., 0.35 mm, 0.50 mm, 0.70 mm, and 1.00 mm). One other crucial component was a professional *drawing board* or *drafting table*.

All necessary inscriptions were introduced—laboriously, by hand—using special alphanumeric templates or *adhesive letters*. Considerable practice was required before one could successfully reproduce complete words or mathematical expressions in presentable form, consisting of neat, clean, *properly spaced* letters and symbols. For publication purposes authors were often advised to submit drawings *without* inscriptions, together with a set of explicit instructions, so that inscriptions could be added by professional draftsmen in the publisher's production department. This service was offered not only to assure high-quality work but also to achieve uniformity throughout a publication with respect to the style and character of lettering.

Figure preparation has obviously undergone revolutionary change in recent years. It is now rare to find anyone preparing complex illustrative material with drawing pens and templates. Today almost everything along these lines is computer-based, perhaps with scanner support for turning an author's preliminary sketch into an "electronic template" (cf. Sec. 7.3.1). Powerful graphics software in competent hands provides the raw material for remarkably high-quality output from laser printers, plotters, and imagesetters. (For more information on the techniques of computer-aided drawing see Sec. 7.3.)

Computer programs for creating line art are now so refined that almost anyone can achieve striking results with surprisingly little training and practice, although the finest work—consistent with the expectations of the most demanding publisher—still reflects the talents and skills of a "master craftsman". Inscriptions, which in the past always presented a major challenge,[12] can now be cleanly composed, properly scaled, and precisely placed with a minimum of effort under the direction of a computer governed by the right software.

● If possible, all illustrations for a particular publication should share a common format.

Generally speaking, copy on letter-size sheets is preferred. Exceptions must obviously be made in the case of certain types of spectra and engineering or architectural drawings, which often are so large they may even need to be folded.

● A separate page should be devoted to each illustration.

Large-format or unusually shaped copy (including spectra and chromatograms) can be quite difficult to reproduce. The same applies to instrumental output in strip-chart form, especially on paper with perforated edges. It is usually necessary with such material first to isolate the crucial segments. *Plotter* output not infrequently proves to be too faint to reproduce well. "Recreating" material of this sort by hand to simplify processing is very time-consuming, but more important it entails a serious risk that details—which someday might be recognized as very important—will be overlooked entirely (e.g., anomalies mistakenly interpreted as "random squiggles"). Moreover, the finished product is unlikely to convey that indefinable sense of something "live" or "authentic". Your best recourse if reproduction turns out to be a problem may be to solicit help from the staff at a professional copy shop, entrusting them with the actual copying process.

Rules associated with drawing as it is practiced by true professionals are the product of a long and admirable tradition, and most of our suggestions below with respect to line drawings are derived from this tradition.[13]

---

[12] One who has never experienced it cannot possibly appreciate the formidable obstacle this seemingly trivial "embellishment" represented. Freehand lettering invariably has a distressingly amateurish look no matter how carefully executed, and alignment problems encountered in conjunction with template-based techniques or adhesive letters can be vexing in the extreme, even for seasoned practitioners!

[13] In addition to a number of German standards (which we have elected not to cite here directly; see
→

The Fundamentals

We begin by considering the optimum *format* for illustrations and pictorial elements in general, again with reference to classical drafting techniques. Many of our comments can be easily adapted to apply to computer-aided drawing as well (cf. also Sec. 7.3).

- If camera-ready line art is to be prepared manually, it is best not to work at publication scale, but rather with enlargements.

Drawing at an enlarged scale greatly simplifies one's work, and it offers two other important advantages as well: unavoidable irregularities become much less evident after the artwork has been reduced in size, and the resulting drawings can be used directly—perhaps after incorporation of a few additional labels—as the basis for slides or transparencies to accompany a lecture.

Reduction can be by essentially any desired factor, but all of one's drawings should be made to the same scale, if for no other reason than to facilitate use of the same type sizes in all inscriptions. Often members of a research group agree on a common policy with regard to the scale of drawings. The guidelines that follow refer to reduction in a *single* dimension; that is, *linear scaling*. Thus, "50% reduction" means that the length of each *side* of an illustration is diminished by half, so the *area* is reduced to 25% of what it was originally.

- According to DIN 15 Blatt 1 (1967), 70%, 50%, 35%, and 25% are generally regarded as "standard reduction values". Notice that each "step" in the sequence is smaller than the previous one by a factor of $1/\sqrt{2}$.

Inscriptions for figures should be so prepared that, after reduction, the lettering is comparable in size to that in accompanying text (i.e., with capital letters ca. 2 mm high). Primary text lettering in drawings for publication purposes should therefore be approximately 4...5 mm high in anticipation of a 50% reduction.

- It can be very helpful to prepare complex drawings not as single entities, but in the form of *paste-ups* of partial representations, in particular since this allows independent preparation of all lettering.

The fact that inscriptions (including italic numerals and roman letters for designating special elements) can be prepared separately means one can initially experiment with their placement, establishing an optimal arrangement prior to pasting (and also before adding "connecting lines" that may be needed). Another advantage of this approach is that it permits reuse of parts of previously prepared drawings (see below).

EBEL and BLIEFERT 1998 for more information), several parts of ISO 128 *Technical drawings, principles of presentation* are quite instructive, including Part 1 (2003): *Technical drawings, General principles of presentation: Introduction and index* and Part 20 (1996): *Basic conventions for cuts and sections.* (The older ISO 128–1982 *Technical drawings, general principles of presentation* was withdrawn in 2001.)

The flexibility of modern reproduction techniques should eliminate any concerns you may have with respect to the viability of what we are here suggesting. Every residual trace of your having taken advantage of a "collage-like" technique is easily disposed of photographically.[14]

It is important that the *adhesive* selected for mounting the multiple parts of a figure have all the right properties. For example, it should dry slowly enough to allow repositioning and precise adjustment of a pasted object. Nevertheless, although elements already incorporated must of course be held securely in place, it should still be possible even after a prolonged period of time to remove various items without damage, permitting later changes in the figure.[15] One adhesive providing these characteristics is the elastic material known as ELMER'S CRAFT BOND ACID-FREE NO-WRINKLE RUBBER CEMENT, manufactured by Elmer's Products, Inc.

There is also another way special elements can be added to a pen-and-ink drawing. If, for example, you have frequent occasion to introduce into your work a specific obscure symbol, and no suitable template or adhesive counterpart is available, xerographic or laser-printer techniques can be used to create custom images on self-adhesive transparent film (e.g., RAYVEN REPROFILM from Rayven, Inc.), which can then be trimmed to size and affixed as needed.

Techniques like these are of course considerably less important now than they once were, especially in the ranks of drafting professionals, as a result of advances in *computer-aided design* (CAD) technology.

It was once regarded as mandatory that a *preliminary pencil sketch* be made of every proposed illustration for the purpose of clarifying in advance such issues as:

- Will the illustration as envisioned in fact be effective?
- Does it supplement the text as usefully as it might?
- Is the proposed distribution of individual elements optimal?
- What size relationships would be most appropriate, from both an aesthetic and a content perspective?
- Given the setting anticipated for the drawing, would a different format be more suitable—from the standpoint of width, for example?
- Will all the required elements fit comfortably in the space available?
- Does the result threaten to appear either "overloaded" or barren?
- In the case of a graph, have provisions been made for the right kinds of axis scales (as discussed in the section that follows)?
- What information will it be necessary to include in an associated legend?
- Does the design call for too many inscriptions?

---

[14] There exist film materials whose presence enhances light scattering in the course of the photocopying process, which in turn decreases further the likelihood that edges of individual components will be visible.

[15] Thus, labels prepared originally for an English-language publication or presentation could later be exchanged for their equivalents in another language for purposes of addressing a different audience.

Preparation of preliminary sketches is strongly recommended even when illustrations are created by digital means. The importance of the questions raised above after all has nothing to do with how artwork is prepared physically.

In the next section we turn our attention to an important subset of line drawings: *graphs*, many of which rely on the presence of a *coordinate system*. This will also serve as a framework for introducing a few more considerations applicable to line art in general. Much of what we suggest is derived from DIN 461 (1973) *Graphische Darstellungen in Koordinatensystemen* ("Graphic Representations in Coordinate Systems"), which we would encourage the interested reader to consult as well.

7.2.3 Coordinate Diagrams

Graphic Presentations Based on a Set of Coordinate Axes

Probably more than half the line drawings scientists prepare are depictions of *functional relationships*. A set of numerical values associated with some particular *quantity* (cf. Sec. 6.1) is plotted against related data for another quantity, in any of several standard ways. Individual data points in such a plot are often connected (or approximated) by *curve segments*, resulting in a visual representation (a *graph*) of a *continuous mathematical function*.

● The *curve plot* is the most common type of line drawing scientists prepare. In most cases the actual curves displayed are approximated on the basis of discrete data points derived from one or more sets of experimental results (or *data series*).

Plots of this type could be described as abstract representations of quantifiable relationships, expressed in graphic terms. An illustration that does *not* represent a mathematical statement in this sense is more properly referred to as a *diagram*.

● Curve plots are analog displays, providing information reminiscent of that from a *gauge* on an analytical instrument.

As a matter of fact, the output of such an instrument can itself often be collected in graphic form by causing an electrical signal proportional to the variable under investigation to drive the pen of a strip-chart recorder, thereby creating a permanent record of the magnitude of the signal as a function of a second quantity—usually *time*. Exceedingly common only a few years ago, graphic recorders of this type have now largely disappeared, giving way to computer-driven plotters. By incorporating into the recording system a certain amount of inertia or "sluggishness", inherent random irregularities ("noise") in an output signal can be suppressed, and apparent discontinuities eliminated, leading to smooth curves—*spectra*, for example, or *chromatograms*. Plots acquired automatically in this way very often become the basis for figures in scientific presentations.

What we have been referring to as "curve plots" are in more formal terms *XY coordinate diagrams*, where "XY" is meant to suggest a presumed functional dependence linking two variables. Such a graph or diagram centers around a continuous, usually smooth line, in most cases accompanied by a set of specific data points. If the number of data points is small, and one wishes to avoid implying the existence of a true functional relationship, the individual points are sometimes connected by straight-line segments. Such lines occasionally are extended so that they produce a *polygon*, in which case it would be better to speak simply of a *coordinate diagram*, omitting explicit reference to any "XY".

● The most familiar *coordinate system* for displaying functional relationships between pairs of quantities is the *Cartesian system*.

A Cartesian coordinate system[16] is based on two mutually perpendicular *axes*: a horizontal *abscissa*, usually representing the *independent variable*—called the *determining* (or *influencing*) *factor* in biomedical work—and a vertical *ordinate* associated with the *dependent variable* (biomedical: *response factor*). In what follows we will usually refer to them simply as the "horizontal and vertical axes". We will not consider here a second relatively common set of graphic conventions, the *polar coordinate system*, useful primarily in conjunction with rotational phenomena.

The expression of a mathematical relationship as a continuous series of points (i.e., a line), consisting of paired values for two linked variables set within a coordinate system, is what produces a *curve* of the type described above. Elaborating upon the coordinate-axis framework by the introduction of a background array of horizontal and vertical lines, regularly spaced and parallel to the axes, results in a *coordinate network* or *grid* against which data can be much more readily interpreted. Thus, with the extra lines it is relatively easy visually to assign rough numerical values to specific points along the displayed curve (see also the subsection below entitled "scaling"). Ordinary graph paper in effect provides just such a coordinate grid based on an undefined pair of perpendicular axes.

Any specific graph constructed according to these principles will be composed ultimately of lines of three fundamental types; that is to say,

---

[16] The reader may be curious regarding the origin of the term "Cartesian coordinates". It is derived from the name of the 17th century French philosopher and mathematician René DESCARTES, credited with being the founder of analytical geometry.— We should note that there is no absolute requirement that coordinate axes be strictly perpendicular to each other. Consider, for example, the latitude and longitude lines on a map of the world, used for specifying a geographical location on the basis of a set of "coordinates"—but in the specific context of an object, the earth, that has the audacity to be a sphere! As a result, the coordinates themselves must inevitably reflect the effects of curvature. They might still be represented two-dimensionally in a typically "cartesian" manner—i.e., as mutually perpendicular straight lines—but to do so forces serious distortion on the surface one is attempting to represent, such that with a map of the world the polar ice cap is stretched to the point of looking larger than Africa! Our sense of geography is influenced more than one often realizes by images subject to this sort of distortion.

● A typical graphic representation in terms of Cartesian coordinates—an XY-coordinate diagram—can be reduced essentially to a pair of *axes*, a *coordinate grid*, and one or more *data curves*, together with a few labels.

Technical Considerations

For a graph to be successful as a figure in conjunction with a piece of text, the reader's attention needs to be properly focused with respect to lines in the three categories described, which should be made to stand out to varying degrees commensurate with their relative significance. The level of impact of various lines can be "tuned" by selectively "weighting" them: assigning to different lines different intensities or widths (thicknesses). The reader's primary concern will of course be with the data curves, which means these should be weighted most heavily. Next in importance are the coordinate axes, followed by lines constituting a coordinate grid, which serve only to facilitate reading the graph. DIN 461 (1973) offers the following recommendation with respect to line widths (see Fig. 7–2):

● A background grid, axes, and data curves, respectively, should be characterized by line widths in the ratio 1 : 2 : 4.

Anyone experienced in preparing such diagrams will immediately recognize that the distinctions called for in this recommendation are quite drastic. Using these standards, the resulting data curves are in fact likely to be so intense that they would obscure precision that might legitimately be associated with the data. In light of this problem, others have put forward alternative recommendations, including the more "modest"

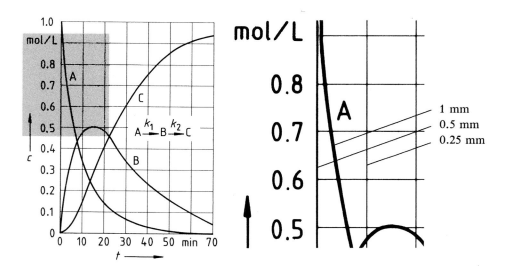

Fig. 7–2. Examples of graphic treatment on the basis of curves set in a coordinate system enhanced with a background grid.— Note the effect of using lines with differing line widths, illustrated here by graphs reflecting two sets of line-width relationships (see the text for details).

line-width relationship $1 : \sqrt{2} : 2$. The example on the left in Fig. 7–2 is based on these modified guidelines, with line widths of 0.35 mm for grid lines, 0.5 mm for axes and major inscriptions, and 0.7 mm for the data curves. On the right is a detail from the same graph, but prepared in accord with the DIN recommendation. We leave it up to the reader to select optimum parameters for a particular situation. Whatever ratio is used, the grid and other auxiliary lines must under no circumstances be allowed to "compete" too effectively with the true source of information in a graph.

Sometimes it is useful to incorporate into a single graph not just *one* curve, but several: typically a *family* of curves derived from multiple (related) sets of data. A reasonable upper limit is four curves per diagram, however, since otherwise the graph is likely to become too complex, especially with curves that intersect.

- Every member of a curve family must in some way be labeled. This might be accomplished either by giving each curve a distinctive appearance, or simply by adding inconspicuous *identifiers*, such as italic *curve numbers* or roman *curve letters*.

The significance of the various identifiers would then be explained in a legend accompanying the corresponding figure caption. DIN 461 (1973) specifically calls for *italic* numbers (an unusual symbol to encounter, by the way!) to minimize the risk that numerical identifiers will be confused with numerical *data*, or with numbers assigned to tick marks along the axes (see below).[17]

- If a single independent variable (e.g., *time*) is to be associated with several dependent variables, all the curves can be constructed with lines of the same *type* so long as there is no significant risk of confusion developing. Otherwise, distinctive characteristics should be associated with the various curves.

Creating distinctive lines is especially important with curves that happen to cross at an acute angle, because the reader will otherwise have trouble determining at a glance which curve leads where. Alternatives to the standard continuous line include *dotted lines* ($\cdots\cdots$), *dot-dash lines* ($-\cdot-\cdot-\cdot-$), and lines based on other symbols or symbol combinations. Fig. 7–8 near the end of this subsection provides an example of multiple line types. In some cases it may be possible to use lines of various *colors*, or lines can be made distinguishable by careful selection of the symbols used to represent the associated *data points*. DIN 461 (1973) recommends that data-point symbols be chosen from among the following:

When symbols like these are employed to represent data points, small gaps are left in the related curves (often *regression lines*) to accommodate point symbols that would otherwise interfere. For aesthetic reasons symbols should be comparable in size to a letter "o" from an inscription. If you wish to indicate the *precision* of reported data

---

[17] *Roman* letters would similarly be clearly distinguishable from any (italic) quantity symbols that might be present.

relative to the ordinate scale, standard deviations *s* can be provided in the form of vertical lines of the appropriate length drawn directly through the various data points (cf. Fig. 7–5b below).

● Occasionally it is not the curves themselves, but rather *areas* they enclose that is of greatest significance. Visual distinctions can be produced in this case with well-chosen background effects, such as *shading, cross-hatching*, or, once again, *color*.

Figures of the latter sort are referred to as *area graphs*. An example of a situation in which this type of presentation might be selected is graphic depiction of the profits realized as a function of crop yields for a series of agricultural crops over a period of several years. A common application from the fields of thermodynamics and materials science is the *phase diagram* (cf. Fig. 7–3). Considerations relevant to the selection of distinguishing patterns and their application are reserved for Sec. 7.2.4.

Sometimes there may be reason to present data points from several sets of measurements on a single graph, but in the *absence* of curves, perhaps because no obvious functional relationships exist, or it seems unwise to imply such a relationship. Presentations in this category are called *scattergrams*, or *XY-scattergrams* if both axes are calibrated.

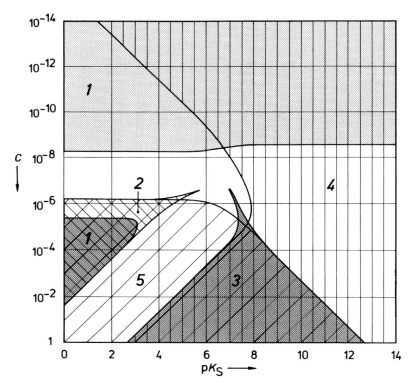

Fig. 7–3. Example of a phase diagram illustrating the effective use of crosshatching and other background patterns.

● Values of an independent variable are generally shown increasing to the right, those of the dependent variable upward.

The standard DIN 461 (1973) offers two options for unambiguous clarification of the relationship between direction and magnitude in a graph. Either the axes are drawn in such a way that they terminate in *arrow points* (directed to the right and the top, respectively; Fig. 7–4a), or directional arrows are incorporated into the corresponding axis *labels* (Fig. 7–4b, c), which are discussed below.

Arrows of some sort are essential with qualitative and semiquantitative graphs, but they are optional in the case of a quantitative graph that includes explicit scales on the axes or coordinate grids, since here the direction of increasing values is obvious from the scale numbers.

This is perhaps a good place to make brief mention of one very special type of "graph". Often during a qualitative investigation initial results or observations are entered by the investigator manually into data forms created for that express purpose, probably with the aid of a computer. Some of the data fields on such a form might require nothing more than a check mark, as in "Situation A (B, C …) was encountered

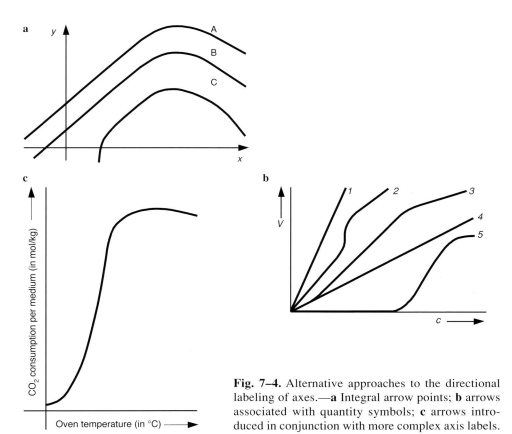

Fig. 7–4. Alternative approaches to the directional labeling of axes.—**a** Integral arrow points; **b** arrows associated with quantity symbols; **c** arrows introduced in conjunction with more complex axis labels.

at time T (at location XX): yes (check mark) or no (leave blank)". Imagine, for example, you are a member of a whale-tracking expedition off Nantucket and you see a breach you want to attribute specifically to "Bessie" or "Fred" or "George" based on the shape of the flukes. A form with an appropriate set of "check-off boxes" would be a very handy thing to have. A great many forms (and tables, cf. Chap. 8) are characterized by a kind of two- or multi-dimensionality, which means in principle you could at your convenience transform the corresponding "snapshots taken on location" into graphic representations. With this possibility in mind, it is helpful if numbers are assigned in advance to the various response options.

Scaling

As indicated above, a quantitative graph requires that *numbers* be provided to calibrate the axes—usually numbers that have associated with them specific *units*, which is to say that indicated values in fact serve to represent *quantities* (we will return to the latter point presently). These calibrated axes thus become *scales* for quantitative interpretation of the data points.

- In scaling a pair of axes it is advantageous to provide short, perpendicular lines called *tick marks* at regular intervals, which are then labeled to reflect the corresponding numerical values.

The interval between tick marks should be so selected that the reader will be able conveniently to estimate values applicable to displayed data points. Numerical labels should be inscribed *below* the horizontal axis and to the *left* of the vertical axis so they will not interfere with the information-bearing part of the graph. It is by no means necessary to label every tick mark, by the way. The reader can surely be trusted to perceive that a mark halfway between "4" and "8" is intended to represent "6".

- Tick marks of this sort (also called *axis divisions* or *scale marks*) should be directed toward the *interior* of the graph; i.e., *upward* from the horizontal axis, and to the *right* from the vertical axis.

This recommendation is largely a matter of tradition, although tick marks directed inward can also be looked upon as "abbreviated" remnants of a grid (cf. Fig. 7–5). In practice, such markings are quite frequently seen pointing in an *outward* direction, presumably to reduce their interference with data points or curves. Tick marks obviously could not be placed inside the type of graph—a mass spectrum, for example—in which the information to be displayed itself consists of lines extending inward from one of the axes. Graphics software often includes provision for both placement options.[18] The COUNCIL OF BIOLOGY EDITORS (CBE) in its carefully developed guideline *Illustrating Science: Standards for Publication* (1988) provides no explicit recommendation

---

[18] On the other hand, some software designed for the preparation of graphs leaves the user no choice but to accept a convention dictated by the programmers. Use instead of an "unstructured" drawing program of course makes it possible for you to design a figure any way you wish.

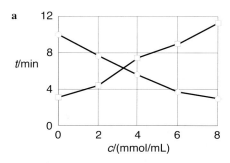
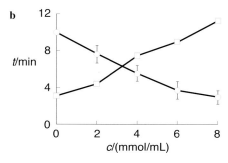

Fig. 7–5. Coordinate systems **a** incorporating a grid and **b** with inwardly-directed tick marks.

on this point. Either approach is considered acceptable. On the other hand, O'Connor (1991, pp. 34, 36) declares categorically that scale markings directed inward constitute poor form.

● Normally, all scale markings in a diagram should be of equal length, although an exception could be made with a graph based on logarithmic scaling, where the beginning/end of each decade might usefully be indicated by a longer mark.

As already suggested, axis markings can in principle be developed into a *coordinate grid* to further assist a reader in the semi-quantitative evaluation of data. In the process it would not be necessary that *every* axis mark be transformed into a grid line, but the very last mark on each axis should definitely be extended, as a way of producing the characteristic "window-effect" one often associates with graphs (see Fig. 7–5a).

● *Zero points* for both the abscissa and the ordinate should be especially clearly labeled as such—and *separately*, even if the two zero points coincide (cf. Fig. 7–5). The *first* and *last* scale marks on each axis also should bear labels despite the fact that others may not.

Sometimes, as in Fig. 7–6, axes in a coordinate system are displayed such that the apparent crossing point does not coincide with the point (0/0). If this is the case it is helpful for labeling purposes to displace the axes slightly (see Fig. 7–6b). Even this measure might not suffice to ensure that "dramatic changes in emission levels" supposedly documented in Fig. 7–6 are exposed for the myth they really are. ("Careful" selection of the coordinate origin in a graph is a ploy sometimes adopted in an attempt to mislead the gullible reader, although obviously not in scientific work!)

● If an axis happens to extend into a region of *negative* numbers, then every scale label indicating a negative value should be preceded by a *minus sign*.

It is standard practice to distribute tick marks at what might be called "normal" intervals; e.g.,

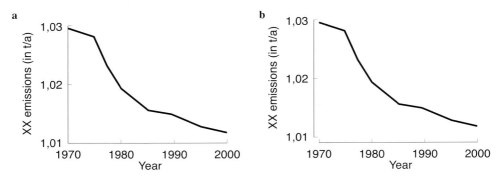

Fig. 7–6. Graphs in which the coordinate origin is shown to be somewhere other than the point (0/0).— **a** Normal placement of the axes; **b** axes offset slightly for purposes of clarity.

	0	5	10	15	20	…

or

	0.0	2.5	5.0	7.5	10.0	…

rather than in a more "quirky" pattern like

	1	4	7	10	13	…

Numerical axis-scale labels are always set in ordinary (roman) type and so oriented as to conform to the orientation of the figure as a whole. In other words, scale numbers for the vertical axis should *not* be rotated to reflect the alignment of that axis. It should not be necessary for the reader to turn a graph around simply to interpret an axis scale.

- When tick marks are extended to produce a grid, scale numbers should always be set along the *left* and *bottom edges* of the graph, even if the apparent point of origin falls within the graphing area (as when both negative and positive values are displayed).

If with a large graph you anticipate that the reader will frequently wish to estimate numerical values, it is helpful to repeat the calibration along the *top* and *right* edges.

In most of the graphs scientists and engineers prepare, the numerical values indicated are in some way associated with *units*, so it is necessary that appropriate *unit symbols* also be present. (By "unit symbols" we mean the standard symbols for simple and derived units introduced in Sec. 6.2.1).

- The usual approach to indicating units is to incorporate the corresponding unit symbols into the same row or column adjacent to the axis used also for numerical values.

Unit symbols belong toward the *right end* of the horizontal axis and the *top* of the vertical axis, in each case between the last two scale numbers (cf. Figures 7–2a and 7–7a). If space proves a problem, the next-to-last number can be omitted.

There exist a number of other common conventions for introducing quantity and unit names or symbols into a graph. Examples are provided in Fig. 7–7.

● It is inappropriate to enclose free-standing unit symbols in parentheses. Remember, incidentally, that such symbols are always to be expressed in *roman* type.

Indicating the applicable unit parenthetically *after* a quantity name or symbol—as in "Time (min)" or "*t* (min)"—is permissible, but under no circumstances should unit symbols be enclosed in square brackets (cf. the comments in Sec. 6.1.1).

● When necessary, a power of ten can also be specified in conjunction with a unit symbol (Fig. 7–8).

An axis entry "10^{-9} N/m" would represent an unambiguous way of indicating that the quantity of interest is here reported in the "unit" 10^{-9} N/m; thus, any number read off the scale must be multiplied by 10^{-9} N/m in order to establish a correct value for the corresponding quantity.

Symbols for *percent* (%), "*perthousand*" (*per mil*, ‰), *parts per million* (ppm), *parts per billion* (ppb), and *parts per trillion* (ppt) are to be treated in essentially the same way as unit symbols, and thus displayed between the last pair of scale numbers.

On the other hand, the *angular* symbols used for expressing degrees (°), minutes (′), and seconds (″) are *not* treated like unit symbols because of the risk of their being overlooked. Instead, these symbols should be applied directly to every affected numerical value along the axis.

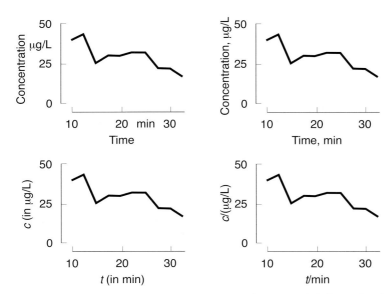

Fig. 7–7. Four approaches to supplying quantity and unit information in conjunction with a graph.

Occasionally it is advantageous to illustrate with a single graph the behavior of more than one quantity (e.g., some combination like concentration, electrical conductivity, and light absorption). In this case separate scales will be required for each quantity, and care must be taken to ensure that each curve in the graph will be associated unmistakably with the correct scale. For example, one might use not only the left (exterior) side of the ordinate for a scale, but also the right (interior) side; or a second scale might be inscribed along the right-hand edge of the graph. Another alternative is to supplement the principal axis with a second axis displaying a different scale (cf. Fig. 7–8).

Axis Labels

The numbers and units discussed above obviously represent one important facet of axis labeling, but in what follows we direct our attention to other information required if a graph is to be "legible": above all identification of the *physical quantities* plotted, which we have so far nearly ignored.

● If possible, the quantity represented by a coordinate axis should be made explicit with the aid of the corresponding *quantity symbol*, set as always in italics.

An example has already been provided in Fig. 7–4 of the way quantity symbols can be introduced in conjunction with axis arrows; that is, the appropriate symbol is placed near the base of an arrow whose purpose is to indicate the direction along an axis in which numerical values increase. In the case of a qualitative diagram, arrows and quantity symbols can be set immediately adjacent to the axes: below a horizontal axis (or the lower boundary of a coordinate grid) or to the left of a vertical axis, in each

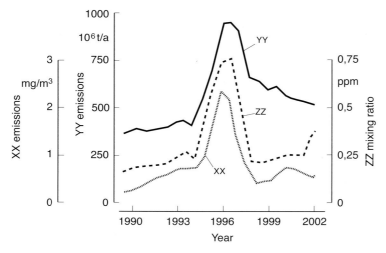

Fig. 7–8. Presentation of values for multiple quantities in a single graph with the aid of supplementary axes.

case near the end. With an axis bearing labeled tick marks, a quantity symbol (and perhaps an arrow) should be displayed not near the end, but toward the middle, and *adjacent* to the scale labels (cf. Fig. 7–7). Such arrows would be dispensed with, however, if one has decided to show the axes themselves as ending in arrow points.

- Another possibility, which is not only permissible but in fact *recommended*, involves *free-standing* axis labels that reflect the *fractional* notation in which a quantity is shown to be divided by its unit.

This would announce that when the axis scales are read to evaluate some point (on a curve, perhaps), what will be obtained is not actually a *quantity*, but rather a pure number (cf. *Quantity* = Value × Unit, Sec. 6.1.1).[19] Examples illustrating this notation are provided in Fig. 7–5 and at the lower right in Fig. 7–7.

There remains yet a third possibility for bringing quantities into the picture.

- If no standard symbol is available for a quantity whose dependency is the subject of a graph, then it is appropriate to present a *definition* of the concept along the axis, preferably in the form of a mathematical expression.

While this expedient is unlikely to become problematic along a horizontal axis, it might on a vertical axis, because the combination of the graph itself together with a (horizontal) label could well require more space than is available.

- If there is no convenient way to avoid a lengthy verbal or mathematical expression in conjunction with the label for a vertical axis, the best solution is to concede, and run the text parallel to the axis, upward from the bottom, even though this means that in order to examine it properly the reader would need to view the text from the right (or rotate the graph 90° clockwise).

Examples of this solution appear in Figures 7–6, 7–7, and 7–8.

A *unit symbol* can easily be incorporated at the end of such an axis label simply by prefacing it with the word "in" (e.g., "*m* in kg" or "Oven temperature in °C"). A long phrase can be subject to the same treatment (e.g., "CO_2 consumption per medium, in mmol/kg"; cf. Fig. 7–6). Nevertheless, we recommend giving serious thought to the alternative of *symbolic* presentation, even if it means "inventing" an appropriate symbol, which you could explain in a legend. Thus, the example just cited could be dealt with

---

[19] Recall Equation (6–1) in Sec. 6.1.1, which—after rearrangement—can be expressed as "the value of a quantity divided by a unit gives a numerical value". The editorial staffs of many highly-regarded journals tend to be a bit stubborn, slow to abandon their unfortunate tradition of setting axis units in square brackets, so we repeat here a piece of advice taken from the IUPAC "Green Book" (IUPAC 1988, p. 3): "In tabulating the numerical values of physical quantities, or labelling [*sic*] the axes of graphs, it is particularly convenient to use the quotient of a physical quantity and a unit in such a form that the values to be tabulated are pure numbers …".— Whenever necessary, powers of ten can be introduced as numerical factors. A tick mark designated "2" on an axis labeled "10^{-3} × NADH-concentration/(mmol L^{-1})" represents an NADH concentration of 2000 mmol L^{-1}; cf. "Working with Units" in Sec. 8.4.2.

using the unorthodox axis label *"VC "*, with an entry in an accompanying legend: *"VC* CO_2-consumption in mmol, based on 1 kg of medium".

● Graphs free of words have the advantage of communicating in a way that is "language-neutral", permitting their utilization in conjunction with documents in a variety of languages.

This consideration should be taken into account with figures of all types, especially halftones, which can prove very difficult to adapt.

Published graphs often contain text in places other than along the axes. Except for concise identifications (e.g., "a", "b" ...), such text should be restricted insofar as possible to areas *outside* the bounds defined by the coordinate axes to avoid compromising the transparency and expressiveness so characteristic of a purely graphic presentation. This applies especially to curves plotted against a coordinate grid. An example of the sort of text we have in mind is that used to single out or characterize some specific data point (e.g., a significant maximum along a curve). Such a point of interest can easily be *linked* to text set *outside* the graph by joining the two with a fine *indicator line* (cf. the right-hand portion of Fig. 7–2). Spectra and chromatograms often contain bands or peaks characteristic of interesting structural elements, or even molecules as a whole, which you may wish to acknowledge with structural formulas (cf. Sec. 7.2.5). Unlike text, these may in fact prove more effective if they are placed *within* the graphing space, provided they can be accommodated without undue "clutter".

More suggestions regarding graphs—together with a host of illustrations—can be found in EBEL and BLIEFERT (2003).

7.2.4 Schematic Drawings and Graphs of Miscellaneous Types

● Scientific text can be enlivened by the presence of *schematic* representations, including *cross-sections* of special apparatus or equipment.

Examples of such illustrations are provided by Fig. 7–9, from which it should be obvious, however, what a challenge preparing such a drawing would present to the novice, whether attempted by hand or with graphics software. It would be beyond our scope to discuss in detail the preparation of drawings like these; indeed, artwork of

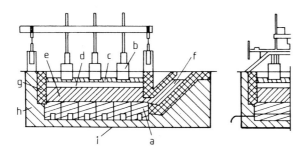

Fig. 7–9. Example of a schematic drawing of an industrial device.

this sort is often the product of a professional illustrator. As an author you would probably be well advised to try to seek permission to reproduce a previously published drawing rather than attempting to create your own.

In some cases—especially if you are writing a book—you may find that if you can put together rough but informative sketches of your ideas, the publisher will be in a position to help you locate professional assistance to transform your thoughts into presentable copy.

● Far less demanding is the preparation of what is known as a *flow chart*, in which the course of some process, or the nature of an organizational structure, is depicted in a highly formalized way.

In this case all that is required, apart from developing a suitable design, is assembly of an appropriate array of horizontal, vertical, and oblique lines together with various squares, rectangles, circles, ellipses, etc. A portion of a typical flow chart—albeit devoid of labels—is shown as Fig. 7–10. Here three basic shapes have been called upon to play their conventional roles:

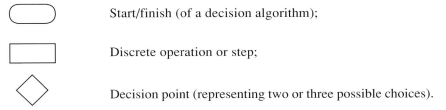

Start/finish (of a decision algorithm);

Discrete operation or step;

Decision point (representing two or three possible choices).

(With respect to simulating data flow and program structure in data processing, see DIN 66 001, 1983.)

A similar approach is effective in depicting production facilities, experimental setups, analytical methods, and the like; cf. DIN 28 004-1 (1988). It is again customary to allow rectangles to represent such things as process steps, basic methods, and components as part of a complex assembly held together by an appropriate network of lines

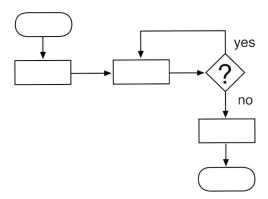

Fig. 7–10. Abbreviated version of a process scheme or decision algorithm, in the form of a flow diagram.

and arrows, and perhaps made visually more interesting through the incorporation of pictorial symbols.

● Popular graphic forms in publications with *statistical* character include *histograms, pie charts,* and *bar graphs.*

A frequency distribution can be portrayed effectively, for example, with a histogram (sometimes called a *step polygon*), in which the independent variable is represented either by points or discrete "platforms", all of equal width, to be joined by oblique or vertical lines, as shown in Fig. 7–11a. If the dependent variable takes the form not of a line but rather free-standing slices (or *collections* of such slices), the result is a *bar graph.* Numerous variants on the latter have been devised, three of which are illustrated in Fig. 7–11b. Individual bars can be displayed in isolation, grouped, or as a continuous

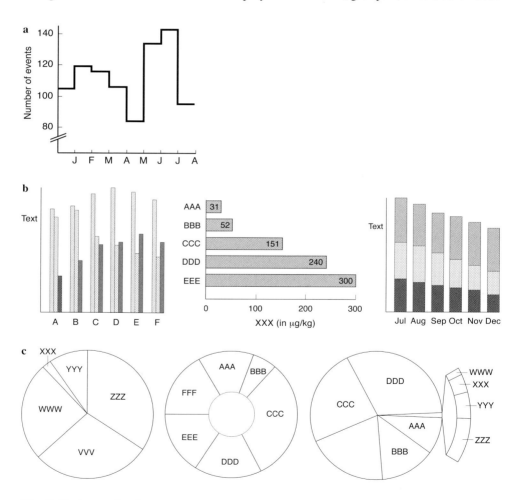

Fig. 7–11. A sample of semi-schematic and schematic representations.— **a** Histogram, **b** bar graphs, **c** pie charts.

array, and they might be arranged either vertically or horizontally. Sometimes the bars are in turn divided into *segments* representing components of some sort. Both positive and negative values can if necessary be displayed (e.g., as in the depiction of deviations from a mean). Bars can also be used to provide information about ranges of validity. The potential variety is limited only by the author's imagination, a fitting note on which to leave this particular subject.

There are few technical complications associated with the preparation of graphics of the type just described. A great many computer programs include tools designed specifically for that purpose, so there is essentially never a need to resort to classical drawing techniques, or to master complex feats of computer gymnastics. Virtually every statistics and spreadsheet package is equipped to generate automatically a wide assortment of graphic representations from previously entered data, perhaps after a bit of preliminary mathematical "massaging". Such data can of course also be output and presented to the reader in tabular form, a topic treated in depth in Chap. 8.

Other conventional display possibilities include the *pie chart* (or *sector diagram*), used for communicating in a highly effective visual way the relative magnitudes of various constituents of a collection of quantities. In this case numerical values are simulated by suitably scaled sectors of a circle, reminiscent of a pie that has been cut into pieces of varying size, as in the first example in Fig. 7–11c. The other two sector diagrams shown in the same figure no longer correspond directly to the "pie analogy", but their interpretation should be obvious.

Graphs such as these can be made more interesting visually (and perhaps easier to fathom) by providing sharper distinctions among the component parts—individual "pieces of the pie", for example—applying to them unique patterns or cross-hatchings, or introducing a three-dimensional effect.

● The first overlay patterns to consider using are ones based on *dots* or *lines*.

Relatively crude patterns of this sort—crosshatchings, for instance—were once applied manually, but it is a time-consuming exercise, and the results are generally disappointing. A better alternative is to take advantage of commercial patterns, distributed in the form of adhesive films that can be cut to any required shape. Today, one almost always resorts instead to computer-based methods, a standard option in virtually every graphics program and typically outfitted with a wealth of distinctive patterns. Representative samples are presented in Fig. 7–12.

As can be seen in Fig. 7–12a, matrices of tiny dots provide a remarkably effective way of conveying the impression of *gray shades* of varying intensity. From a technical standpoint these are still fully compatible with line art. Screening techniques employed in halftone reproduction (see Sec. 7.4) are based on the same principle, namely the fact that the human eye is unable to resolve a fine pattern of dots into discrete components, producing a sense of "grayness" and opening up the possibility of simulating a continuum of shades.

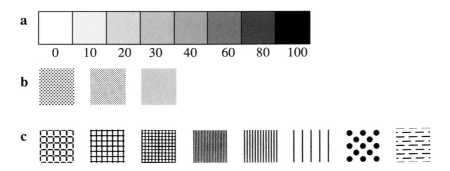

Fig. 7–12. Background patterns of the type commonly available with computer graphics software, useful for labeling two-dimensional elements in line art.— **a** Gray shades simulated by screens with densities ranging from 0% (white) to 100% (black); **b** dot matrices, all characterized by densities of 30%, but based on patterns with 26, 32, and 75 lines per centimeter; **c** miscellaneous line and dot patterns.

● When selecting a dot pattern for use in figure copy, try to avoid ones that are too fine, because these can cause problems in photocopying or offset replication, especially if the image must be reduced in scale.

Depending upon the precise calibration of a copying device, reproduction of a gray pattern sometimes leads to copy that looks white, black, or simply smeared. Photo-reduction complicates things, of course, because it makes the constituent dots shrink, and they also approach one another more closely, with the result that a region which looks fine in an original can become quite unsightly in a copy.

7.2.5 Structural Formulas in Chemistry

A chemist's structural formulas represent another important application of scientific line art. Thanks to the fact that certain structural elements occur repeatedly in countless molecules, special *templates* (*stencils*) were long ago developed to facilitate the process of drawing chemical structures at almost any level of complexity. In recent years the job has become much easier thanks to powerful computer software, including CHEMDRAW (CambridgeSoft Corp.),[20] ISIS/DRAW (MDL Information Systems, a subsidiary of the publisher Elsevier, Inc.), and CHEMWINDOW (BioRad Laboratories), permitting one to prepare with a PC, quite rapidly, publication-quality representations of impressive molecular structures. Once again technology makes it possible to accomplish at your own desk results that a short time ago would have been unthinkable outside the realm of a highly trained professional draftsman.

---

[20] This company in particular, with its home offices at 100 Cambridge Park Drive, Cambridge MA 02140-9802 USA, has over a long period of time established a record of developing exceptionally creative products. Much more information is available from their Website, www.cambridgesoft.com.

We have commented repeatedly on the remarkable extent to which small computers have become an integral part of a typical scientist's professional life, dwarfing the role they initially played (35 years ago?) when they were regarded as little more than "text editors" grafted onto glorified adding machines. The desktop (or laptop) PC is now a familiar feature of most experimental laboratories—or, to put it differently, the scientist's laboratory has to some significant degree migrated to the desktop of his or her personal computer. This is especially apparent in—but by no means restricted to!—the area of graphing and scientific illustration in general. One particular software package that affirms the point in a striking way is CHEMOFFICE (currently distributed—in the fall of 2003—as Version 8.0). An issue of the publication *Scientists @ ChemNews*[21] (vol. 13 no. 3), released shortly before this book went to press, makes the case eloquently. Several articles are devoted to CHEMOFFICE and its various components, including CHEMDRAW, one of which contains the observation "… to call CHEMDRAW just a drawing tool does not do justice to the breadth and depth of its capabilities or its importance to a modern synthetic chemist like myself" (G. MORASKI). The same can be said of its sister program CHEM3D (also from CambridgeSoft), which adds convincing (and mathematically rigorous) three-dimensional character to the "flat" structures CHEMDRAW produces. In this context another author (J. HEFFERNAN) describes CHEM3D as a "visual analysis tool", and "a multi-faceted piece of software" helping the user learn to "appreciate and understand by eye", an ability that now constitutes "a large part of many branches of chemistry".

Some of the more "non-obvious" tasks within the scope of this "scientist-friendly" set of programs include:

– Assignment of the proper name to a structure once it has been "drawn", in strict accord with official IUPAC rules of nomenclature (using the feature "Struct=Name");
– Computation of local energy minima for displayed molecules, based on conformational optimization and the methods of molecular mechanics;
– Prediction of both proton and 13-C NMR spectra;
– Quick calculation of exact mass values for fragments one expects to be prominent in the mass spectrum of an organic compound, through a "mass-fragmentation tool";
– Simulation for electronic-notebook or publication purposes of thin-layer chromatography results, in the form of surprisingly convincing "virtual TLC plates";
– Estimation (again with the help of special "tools") of a wide variety of molecular parameters for any displayed compound, ranging from its critical pressure, temperature, and volume through heat of formation, all the way to the melting and boiling points.

---

[21] This might be described as a "private journal" (officially a "corporate trade magazine") distributed without cost by CambridgeSoft (see the preceding footnote) to a vast audience of chemists. It is currently (2003) in its 13th year of publication. Interestingly, despite its origin, *Scientists @ ChemNews* accepts advertising from a wide assortment of companies.

The above list is by no means comprehensive, but it clearly underscores the fact that far "more than just drawing" is indeed at issue!

We see no need here to treat the subject of structural formulas in greater detail, especially since it is of interest almost exclusively to chemists, most of whom will already regard it as familiar territory, and are already well aware of how structural formulas are expected to look in professional publications [see also DIN 32 641 (1999), *Chemische Formeln* ("Chemical Formulas")]. Details regarding use of the corresponding software are best sought in the associated handbooks. On the other hand, we thought it appropriate to offer our readers the opportunity to admire (in Fig. 7–13) a few modern specimens of the art of structural formula creation. The two examples (chosen at random) come direct from the editorial offices of our publisher.

● For the sake of convenience, structural formulas in books and journal articles are often grouped into *formula blocks* or *schemes*.

This generalization obviously does *not* apply to formulas incorporated into equations (and thus already "grouped" in a different sense).

When submitting a manuscript for publication it is customary to collect all the required structural formulas into a separate *formula manuscript* that will accompany the text (see Sec. 3.4.2). As usual, one must be careful to provide unambiguous reference numbers or an equivalent certain means of identification—a notation like "((F-25))", for example, as shorthand for "Formula 25"—to ensure that every formula will be embedded in the text precisely where it belongs. The author should also retain a full set of high-quality copies of the structures for security purposes.

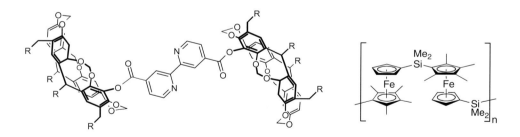

Fig. 7–13. Contemporary structural formulas in chemistry, illustrating the use of 3-dimensional effects. [Reproduced with kind permission from *Modern Cyclophane Chemistry*, a book edited by H. Hopf and R. Gleiter scheduled for publication (2004) by Wiley-VCH.]

7.3 Drawing With a Computer

7.3.1 Overview, and an Introduction to Vector Graphics

Anyone with access to a modern PC, a flexible drawing program, a good printer, and a little experience is in a strong position to prepare impressive line drawings with almost no outside help. Artistic ability is definitely not a prerequisite, although some computer enthusiasts have become adept even at "painting" with a computer, and professionals have been known to produce remarkable works of art with this medium. Occasionally an author will accept the challenge of preparing, independently and in camera-ready form, the illustrations for an entire book—biologically important molecules, for example, or various types of graphs, often enhanced by effective application of color,[22] in the process earning the well-deserved praise and admiration of experts. These are subjects we can unfortunately touch upon only very briefly.[23]

We have seen that in the broad areas of both word processing and document layout the software market is dominated by surprisingly few programs, and much the same can be said when it comes to preparing and modifying line drawings and continuous-tone images. We therefore venture at the end of this chapter (in Sec. 7.5) to provide a concise summary of what seem currently to be the principal options. For now we will continue to restrict our attention to black-and-white line art, since this still meets most of the needs of the average scientist. Suffice it to say, however, that little skill is required to add the exciting dimension of color, a prospect that can prove very seductive.

● Preparing figure copy on a computer screen offers several important advantages relative to the use of classical drawing techniques. No special talent or accessories whatsoever are required for one to

- create straight lines of every description—horizontal, vertical, or oblique, as well as line combinations to represent axes and grids;
- incorporate lines of different types into a drawing, lines which may also display a wide range of widths;
- devise and position creative labels, taking advantage of precisely the right font(s);
- instantly alter the scale of a drawing—repeatedly, if necessary;
- incorporate "prefabricated" graphic elements into a design, custom-tailored to specific layout requirements;
- generate perfect geometrical figures, along with every imaginable type of curve;
- selectively round corners and the junctions formed by intersecting lines in general;
- link specific pairs of points, using either straight or curved line segments;

---

[22] It goes almost without saying that it is no longer acceptable for illustration programs, desktop printers, or creative scientists to be restricted to monochromatic expression.

[23] More information is provided in *Text and Graphics in the Electronic Age* (RUSSEY, BLIEFERT, and VILLAIN 1995), especially Chapter 11.

- establish best-fit curves representing sets of data points;
- modify, clone, or delete individual graphic elements;
- displace or rotate figure components, or perhaps transform them into their mirror images, or adjust them so they suggest three-dimensionality;
- surround a drawing with a more or less elaborate frame;
- embellish selected regions of a figure with cross-hatchings or more imaginative patterns;
- repair perceived blemishes, even ones apparent only though an "electronic magnifying glass";
- selectively modify bitmap arrays obtained by scanning printed copy; or
- plant graphic material exactly where you would like it to appear in running text.

The list above is actually quite random, and it could be extended almost without limit. A few years ago we overheard a colleague remark that one of the few things he had so far not succeeded in doing with his computer was reproducing his own signature, an assignment that can be carried out today with ease simply by scanning and lightly editing a handwritten signature.

Some kinds of illustrative material are better prepared with one type of software than another. A fundamental choice must be made at the outset whether to work with pixel-based representations or assemblages put together instead from mathematically expressed components (*bitmap* vs. *vector graphics*).

Sometimes valuable graphic images originate almost spontaneously as the offshoot of performing a mathematical operation.[24] Thus, most mathematical and statistical software is equipped not only to *process* numerical information but also to *display* it graphically, as for example results of a *regression analysis* for identifying the simple smooth curve that comes closest to matching a set of data points.

Pre-existing illustrative material in hardcopy form is most easily introduced into a computer with a *scanner*, although *digital photography* is rapidly proving to be an attractive alternative. In either case, the resulting electronic images can become the subject of extensive and sophisticated computer modification if you so choose. "Editing" of this sort might have any of several objectives: removal of "smudges" introduced during the transfer process (or even present in the original), for example, or adjustment of color renditions, or a change in image size (and/or resolution) to better meet a specific set of needs. It is important to emphasize that whenever copy is scanned (or captured with a digital camera)—irrespective of whether it originates as a line drawing, an artist's painting, or a conventional photograph—what must transpire is the equivalent of a *screening process*. The coarseness of the hypothetical "screen"— and thus the size of dots that will eventually define one's image (essentially the *resolution*; cf. Sec. 7.4)—is an important parameter to be established in advance. We repeat: *every* image one scans is necessarily obtained as a bitmap (or pixel) graphic.

---

[24] Need we explicitly point out that in a sense *everything* one does with a computer is a mathematical operation? But here we are of course using the word "mathematical" in a much more limited sense.

This is in fact precisely what one is striving for in the case of a continuous-tone original, but it is often an undesirable consequence with a line drawing. If line drawings *are* scanned as a precursor to publication it is important that the process be carried out at exceptionally high resolution (the higher the better; a reasonable minimum might be 1200 dpi), since prints based on the resulting image may otherwise contain lines or curves that are supposed to look smooth but instead have a disconcerting "stepped" appearance.

The latter problem has long been a source of concern for illustrators, especially since the expedient of moving to higher resolution tends to produce bitmap files that are extremely large. Many years ago the first attempts were made to devise special software that would "automatically" transform high-resolution bitmaps into "equivalent" vector graphics. The assignment is a challenging one, however, and such software tends to be not only expensive, but also difficult for the novice to use effectively, so the operation is one best left to professionals. Graphics specialists at publishing houses in fact almost never take advantage of programs of this type, primarily because the images obtained always fall far short of professional standards, and thus demand a great deal of manual "touch-up". One might with justice compare the exercise to asking a computer to take a demanding piece of scholarly text and translate it from one language to another. The translation that emerged *might* prove intelligible, but with the present state of the art it most certainly will not be suitable for direct publication.

Vector graphics for use in a professional setting are almost always prepared manually, entailing a significant investment in terms of both time and labor (although the extent of the challenge obviously varies from project to project). There are thus very good reasons why preparation costs typically fall in the broad range of roughly $20 to $500 per illustration, depending upon complexity. By contrast, the price of professional "fine-tuning" of a digitized photo is unlikely to exceed $50.

Many illustrators (though by no means all!) do, for reasons of convenience, use scanned images as starting points for their vector graphics, but in quite a different way. The scanned pixel graphic is first imported into an appropriate *illustration* program (e.g., ILLUSTRATOR from Adobe or FREEHAND from Macromedia), where it assumes the role of a "background image" or "tracing pattern". A vector-graphic image is then created by hand "on top of" the pattern, with the latter serving as a guide. Programs of this type offer provisions for several discrete "drawing layers", the pixel image residing in a layer of its own at the very "bottom of the stack".

Contours perceived in the imported pattern are "traced" with the aid of an assortment of sophisticated tools, which in turn produce mathematically-defined *vectors* that take the form of straight lines and curves. These lines can be expressed in an infinite number of ways with respect to form and intensity. Moreover, circumscribed *regions* generated in the process can be "filled" as appropriate with gray shades, colors, or more or less complex patterns. Different pictorial elements are commonly consigned to different "layers", the relative transparency of which can be adjusted with respect to layers farther "down in the stack". The "digital artist" is able at any time to view selectively any desired subset of these layers, with or without fills and with or without the underlying pixel graphic. Once all the necessary vector-based elements are present, the background template is erased and the vector layers are combined to produce a single illustration.

The process is an intriguing one, but it can become extremely complex, and in most cases you will probably want to entrust it to someone with the experience necessary to assign optimal line weights, colors, patterns and layer-transparencies intuitively, as well as create, "typeset", and incorporate any needed text elements.

One of the important reasons for encouraging the use of vector-graphic images in publications, incidentally, is the fact that they consume far less storage (and processing!) space than bitmaps. Further incentive comes from a factor fully apparent only at the printing stage: with a laser printer or a laser imagesetter, vector-graphic instructions provide printed images of exceptionally high quality, and at any scale requested.

If you do propose to include figures in a manuscript, before committing yourself to scanning illustrations present in your filing cabinet, or seeking permission to reproduce graphics previously published by others, give serious consideration to the possibility—especially with respect to optimal content, currency, and appropriateness—of treating existing pieces of copy only as *models* from which to prepare altogether new figures. Among other things this will allow you to treat all your graphics consistently from a stylistic standpoint (e.g., line widths, type fonts, and type sizes), and to include precisely those elements most applicable in your current situation.

Also do not overlook the possibility of capturing experimental data from analytical instruments or their associated computers, directly in the form of curves that can then be imported cleanly into a computer-graphic environment. Finally, remember to take full advantage of the opportunity to *archive* all your digital graphics for possible reuse someday in different settings.

Exploiting fully the various techniques we have been describing obviously requires the availability of a printer or plotter capable of producing professional-looking results (including diagonal lines that do not resemble staircases), but this is a luxury affordable today by almost everyone.

In summary, the conclusion is once again inescapable—with regard to imaging technology as in so many aspects of life—that we have in recent years crossed a threshold into a wholly new era.

7.3.2 Miscellaneous Observations

We hope our brief tour has helped a few "newcomers" see a bit more clearly what is possible in the way of illustration as they contemplate preparing their very first extensive—illustrated!—piece of work in print.

The fundamental question for you to address initially is the extent to which graphics can or legitimately should play a role in your particular activity. Is it worth your becoming involved? If so, how difficult will it be? The answer to the latter question is sure to come as a surprise to some:

- So long as you have available a good word-processing system you are already well on the road to preparing at least primitive graphic images.

Word-processing programs like WORD now offer their own special "built-in" graphics facilities, albeit with severe limitations. No additional software whatsoever is therefore required to create simple drawings and place them suitably in a manuscript.

Graphics prepared with other programs can also be "imported" and incorporated into your text documents.[25] A picture so "embedded" can be manipulated in much the same way as a text character: once highlighted in the text window it can be moved to a different spot, adjusted vertically, copied, or of course deleted. Alternatively, one might choose to modify it in some way so that it fits more favorably in its setting (by rescaling, for example).

A much more effective (and flexible) way to combine electronic drawings or illustrations with word-processor text involves the use of special *layout software*, which permits one to create pages worthy of inclusion in a professionally prepared document—or stunning transparencies to accompany a lecture (cf. Sec. 5.2.2).

Brief mention needs to be made of one especially crucial hardware prerequisite: ample *storage space*.

- Illustrations—particularly high-resolution dot-matrix images that include gray shades or multiple colors, and large-scale pixel graphics in general—can consume a considerable amount of memory (RAM) during processing steps, and disk space later for archival purposes. (Such problems are rarely encountered with vector graphics.)

In other words, if you propose to engage in extensive graphic work, be sure to equip yourself with a substantial amount of cost-efficient storage—including RAM. Apart from magnetic devices (e.g., *hard disks*, removables), the most important long-term storage option currently is of the *optical* type, including the CDs and DVDs that in a remarkably short period of time have cornered a market embracing millions (see Sec. 5.2.1).

---

[25] Some attention to *file formats* for graphic images is required, however; see RUSSEY, BLIEFERT, and VILLAIN (1995). Especially if you wish also to make use of your images in the context of the Internet it would be wise for you to learn something about such formats in general, including which ones are appropriate choices for photos. In the case of color photos you will soon discover that the most popular alternative is JPEG (also written JPG), an acronym for Joint Photographic Expert Group, in recognition of the commission that first authorized the corresponding protocol, one which compresses images so they occupy minimal storage space. Another common format is GIF (for Graphics Interchange Format). Web browsers are generally able to deal comfortably with both. A third widely recognized graphic format is TIFF (for Tagged Image File Format).

7.4 Halftones

Much like *xerography* with a dry copier (or printing with a laser printer), modern offset printing—a flatbed process—is capable of depositing on paper only black ink (i.e., ink of one color), thereby establishing a contrast with the white (or absence of color) of the paper itself.[26] For this reason, shades of gray must be *simulated*. This is achieved with arrays of small or larger dots, which can also be more or less closely spaced. Careful adjustment of the dot size and/or density allows one to create in this way an effect almost indistinguishable at a distance from a continuum of gray shades. An image prepared in this way—consisting only of black and white but *seeming* to contain grays—is called a *halftone*. Most halftones originate as (black-and-white) photographs, which differ from halftones in that they are characterized by the presence of a multitude of *true* gray shades. The chief technique for converting a photograph into an image limited strictly to the "colors" black and white is known as *screening* or *rasterizing* (*lat*. rastrum, to rake).

There are two major approaches to creating a rasterized image for printing purposes: one involves "classical" photographic exposure of an *offset film*, using a special camera (a *reprocamera*). The alternative, developed more recently, is based on scanners, computers, and laser imagesetters (cf. UEBEL 1996). In the first of these techniques a fine screen is introduced, prior to exposure, between the camera's objective lens and the photosensitive surface on which the "screened" image is to be formed. The screen itself is known as a *raster* (or *raster film*), and its function is to fragment the image entering through the lens in such a way that it becomes a complex dot pattern. The raster consists of a fine grid of intersecting perpendicular black lines, which in effect behaves like a multitude of tiny windows. Depending upon how much light from the original image passes through a given "window", a smaller or larger dot is generated at the corresponding spot on the light-sensitive surface. The resulting rasterized image, after development, is ultimately transferred to a printing plate.

Rasterizing is an essential processing step, because—as we have seen—the only distinction recognized in offset printing is that between exposed (printing) and unexposed (nonprinting) surface areas. The fact that through rasterizing one is able to simulate so effectively a continuous-tone image is a consequence of the human eye's interpreting a host of closely- and evenly-spaced black *points* against a white background as a *surface*, which looks to be a homogeneous shade of gray.[27]

The second approach to offset printing takes advantage instead of a *gray-scale* (or *color*) *scanner* offering a resolution of at least 300 dpi (dots per inch; 300 dpi is roughly

---

[26] The "ink" in the case of a copier or laser printer is usually referred to as *toner*.

[27] The eye/brain interface reacts to a finely divided mixture of black and white points of light exactly as it would to a balanced shade of gray. Black in effect is the same as "non-light". *Color blending* is achieved in a similar way from the standpoint both of raster technology and the physiology of perception. Thus, a field composed of many closely-spaced yellow and blue points will be perceived by an observer as some shade of green.

equivalent to 120 points per square millimeter). Each light signal registered as a reflection from a tiny portion of the image to be copied, again figuratively a "dot", has attributed to it more than simply black-and-white information: it is in fact labeled with a specific *gray value*. The number of gray shades distinguishable with an efficient scanning device today may be anywhere from 16 to 256, often expressed as a "color depth" between 4 bits ($16 = 2^4$) and 8 bits ($256 = 2^8$). The "electronic image" that results is stored in a computer as a matrix (of dots), and eventually transformed by a high-resolution *laser printer* or *laser imagesetter* into the equivalent of the screened image described earlier—on paper, film, or a photosensitive printing plate ("computer-to-plate technology").[28] Fig. 7–14 contains examples of halftone images produced in these two different ways.

Today, reprocameras are on the verge of obsolescence. It has thus become standard practice for continuous-tone illustration copy to be scanned.

● Careful examination of a halftone print (e.g., with a magnifying glass) clearly reveals the presence of a complex array of dots or points of varying size.

As you consider including graphic material in a publication it is important to remember that

● Rasterizing is required only in the case of continuous-tone copy, *not* line drawings.

a
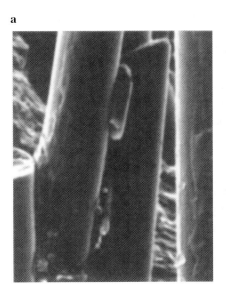

b
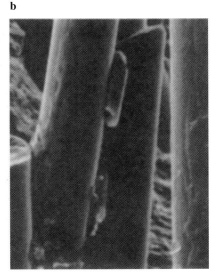

Fig. 7–14. Halftone illustrations prepared **a** by the "classical route", and **b** with a scanner.— Scan resolution 150 dpi, color depth 6 bits, resolution of the laser imagesetter 1693 dpi.

---

[28] Interposed between the computer and the laser-driven imagesetter is another important component: a *raster image processor* (RIP), which arranges for data from the computer to be output at the highest resolution available with the associated printing device (see also " Printers" in Sec. 5.2.1).

Normally the distinction is one that has consequences for an author only if a manuscript is to be subjected to direct reproduction. Thus,

● Continuous-tone images should under no circumstances be pasted into a manuscript that will be reproduced photomechanically.

An illustration of this type can never classify as "camera-ready", unlike text and line drawings, because it must always be screened before reproduction.

So much for technological issues. We will pass over the entire matter of *creating* the images reproduced as halftones (via photography or photomicrography) as lying outside our scope. One reference source that does deal extensively with the preparation, processing, and reproduction of such illustrations in the natural sciences has already been cited: *Illustrating Science: Standards for Publication* by the COUNCIL OF BIOLOGY EDITORS (1988). BRISCOE (1996) also treats the subject in some depth, specifically from the bioscience point of view. Of particular interest to immunologists and molecular biologists is her discussion of the best ways to capture and index electrophoretic blots (gels) for publication or slides. Thus, clear depictions of the striping and band structures revealed through film electrophoresis with polyacrylamide gel are extremely important to those interested in gene technology (and even criminology), and special precautions are required to avoid results that resemble amorphous spots floating along a surface.

In the descriptive sciences it is often necessary to incorporate *color* photographs into publications. Good color reproduction involves the preparation of *four* screened images, which are then combined in a process known as *four-color printing*.[29] The procedure is expensive, so including color illustrations in a journal article sometimes requires that the *author* agree to underwrite at least a portion of the cost. Early discussion of this issue with the publisher is strongly recommended.

The remainder of this section is devoted to highlighting a few miscellaneous matters worth thinking about in conjunction with the inclusion of continuous-tone copy in manuscripts for publication. Toward the end we touch briefly on the situation with halftone images provided in the form of data files.

It is often useful in a photograph to provide the reader with some sense of *scale*, or perhaps to introduce artificially an arrow or an indicator line of some sort. With a dark background the simplest course of action is (or *was*) to paste onto the photographic image, prior to rasterizing, a white slip of paper bearing the required element. A better approach today is probably to leave things entirely in the hands of the publisher (with explicit detailed guidance, of course), who will be in a position to call upon professionals trained in image modification.

● Any sort of editing you wish to specify with respect to continuous-tone copy prior to publication should be explained clearly either on a photocopy of the material or on a transparent overlay.

---

[29] The four colors are ordinarily cyan, magenta, yellow, and black ("CMYK"), commonly utilized in inkjet printers as well.

It would be appropriate to indicate in this way, for example, that only a *portion* of a particular photo is of interest, or to alert the publisher to some essential detail that must remain recognizable. Overlays are attached by taping them over an edge, to the *back* of the photo.[30]

- It is crucial to ensure that photos not become bent. They must also be carefully protected against soil or other damage, but *not* mounted on sheets of paper.

To ward off scratches and the accumulation of dirt we recommend you separate individual photos in a set by covering each with a thin sheet of protective paper, taped—like the overlays described above—to the back, over an edge.

- *Numbers* identifying specific photos can be inscribed on the backs with a soft pencil.

Explicit guidance should also be provided on the back of a photo if there is any chance doubt could arise as to which edge of the picture represents the top.

Authors of scientific manuscripts are often tempted to include photos that have previously been published elsewhere. The legal aspects of this are easily resolved (see "Legal Matters: The Citation of Figures" in Sec. 7.1.2) through correspondence, but that will not lay to rest the technical issues. A halftone illustration you find in a *publication* will already have been screened, and photographing or otherwise copying it and screening a second time is almost certain to result in an image that looks smeared and blurry, and it may even produce interference patterns.[31]

- If you do need to incorporate a previously published photograph into a new publication you should request that the first author (or publisher) provide you with either a print of the original or an original screened image. (This of course assumes you have obtained permission to use the material!)

Finally, how should one proceed in the case of a manuscript that is not scheduled to be typeset, but is already in "finished form", ready for production? "Finished form" here means you are providing data files intended to serve as direct input for a publisher's layout software. Alternatively, you might be supplying the editorial office with a printed copy that is to be scanned and in this way put in digital form. In most cases you will have prepared the digital version yourself, however.

- In a manuscript ready in this sense for production, none of the illustrations (whether halftones or line art) should so far have been incorporated into the text, but sent along instead in the form of separate data files.

---

[30] All this advice is of course relevant only if photos are *acceptable* as figure copy—which can no longer be assumed. We know of publishers that categorically insist on *all* figure copy being submitted in digital form. This is obviously another area in which consultations become important.

[31] This phenomenon, called the *moiré effect*, is a result of the peculiar way that, in certain regions, light rays from a screened original happen to pass through the new screen. The nature of the consequences is strongly dependent upon the precise placement of the screen.

Until a few years ago, manuscripts of conference proceedings, for example, were almost always submitted in "camera-ready" form. Today it is more often the case that those responsible for preparing such a document provide (more or less) print-ready *data files*—typically formatted with text-editing software like WORD.

● Assuming the publisher's production manager approves, illustrative material for manuscripts of this type *will* already have been incorporated into the data files.

In nearly all cases, halftone images present will have been captured by scanning, typically of photographs. It is absolutely essential that the scanning occur at the proper resolution. For technical reasons, most printing establishments rely on input resolutions of at least 250 dpi. Other criteria must of course be met as well, especially when color is involved.

● When working with color copy, and if quality of reproduction is a sensitive issue, you should definitely consider *not* undertaking the scanning yourself, but leaving this step to professionals at the publishing house, or to a printer working under a subcontract.

If only a few copies of a document are to be prepared (as in the case of a dissertation, for example), and you do not need the services of a professional printer, all necessary copies should be prepared at the same time, because "photocopies of copies" cobbled together at a later date will be distinctly inferior. In this circumstance required photographs should be pasted into each copy of the document in the form of original prints.

Specialized information concerning the preparation of illustrations for *lectures* is provided in EBEL and BLIEFERT (1994).

7.5 Overview of Software Useful in Editing Figures, both Line Drawings and Vector Graphics

The summary that follows (Table 7–1) is intended to help orient the author with little or no experience in the area of electronic images. In no sense should the coverage be regarded as exhaustive.

Table 7–1. Software for editing figures.

Program	Distributor (Internet Address)	Category	Typical Field of Application	Operating System	Suitability: Amateur and/or Professional
AUTOCAD	Autodesk (www.autodesk.com)	CAD (computer-aided design) program	Preparation of complex 2- and 3-dimensional drawings	Win, Mac	P
CORELDRAW	Corel (www3.corel.com)	Illustration	Preparation and editing of vector graphics, both artistic and technical	Win, Mac	A, P
CORELPAINTER	Corel (www3.corel.com)	Illustration, editing	Editing of pixel graphics, esp. digital halftone images	Win, Mac	A, P
DESIGNER	Micrografx (acquired by Corel) (www3.corel.com)	Illustration	Preparation and editing of vector graphics, both artistic and technical	Win	A, P
EXCEL	Microsoft (www.microsoft.com/ office/excel)	Spreadsheet program	Diverse calculations (through cells on a work-sheet); also data analysis and simple graphs	Win, Mac	A, P
FREEHAND	Macromedia (www.macromedia.com)	Illustration	Preparation and editing of vector graphics, both artistic and technical	Win, Mac	A, P
IGRAFX FLOWCHARTER	Micrografx (acquired by Corel) (www3.corel.com)	Flow charts	Preparation of flow charts based on symbol patterns	Win	A, P
ILLUSTRATOR	Adobe (www.adobe.com)	Illustration	Preparation and editing of vector graphics, both artistic and technical	Win, Mac	A, P

Table 7–1. (Continued)

Program	Distributor (Internet Address)	Category	Typical Field of Application	Operating System	Suitability: Amateur and/or Professional
LabView	National Instruments (www.ni.com)	Measurement and automation	Virtual instruments; graphic output of measuring devices, process control, etc.	UNIX, Win, Mac	P
Maple	Maple (www.mapleapps.com)	Scientific/mathematical computation	Computer algebra; calculation and display of mathematical functions	UNIX, Win, Mac	P
Mathematica	Wolfram Research (www.wolfram.com)	Scientific/mathematical computation	Computer algebra; calculation and display of mathematical functions	UNIX, Win, Mac	P
Origin	OriginLab (formerly Microcal) (www.originlab.com)	Data analysis	Preparation of charts and graphs based on data series	Win	P
Photoshop	Adobe (www.adobe.com)	Illustration, editing	Editing of pixel graphics, esp. digital halftone images	Win, Mac	A, P
Picture Publisher (now Igrafx Image)	Micrografx (acquired by Corel) (www3.corel.com)	Illustration, editing	Editing of pixel graphics, esp. digital halftone images	Win	A, P
Visio	Microsoft (www.microsoft.com/office/visio)	Flow charts	Preparation of flow charts based on symbol patterns	Win	A,P

8 Tables

8.1 The Logic Behind a Table

In Sec. 3.4.4 we briefly discussed *tables*, and in Sec. 3.4.3 we touched upon the question "A Figure or a Table?" Tables were characterized there as "digital illustrative devices". That is in fact often the case, and in terms of comparing them with curves on a graph the characterization is an apt one. Nevertheless, there are many tables to which this description clearly does not apply. Here we propose to explore the subject somewhat more thoroughly and systematically. What in fact *are* tables, and how does one set them up?

- A table is a concise and highly organized presentation of *verbal* and/or *numerical* information, which might contain graphic elements as well.

In the 4th edition of *Schreiben und Publizieren in den Naturwissenschaften* (EBEL and BLIEFERT 1998), the German counterpart to this book, we dedicated considerable attention to the nature and organization of tables, to the dismay of at least one critic. In the course of this exposition some matters were covered that nearly every reader probably already knows and takes for granted. In what follows, therefore, we have abridged our treatment somewhat, albeit with a sense that in so doing we are depriving our readers, especially those who may still be preparing their first reports, of some useful material.

Too many publications fail to reflect a proper appreciation for the functional and aesthetic standards by which tables should be judged. Colleagues in the publishing business with both editorial and production responsibilities freely acknowledge and deplore the flood of deficient table copy they receive from careless or insensitive authors. There was in fact a time when table preparation was widely regarded as an art entrusted only to the most skilled typesetters. We hope in the ensuing pages to throw some light on what constitutes a first-rate table.[1]

In recent years, computer programs, including spreadsheet packages, have become a welcome source of support and guidance in table preparation, but it remains essential that the user know what to strive for, and how to obtain satisfactory—better: optimal!—results. In another of our books, *Diplom- und Doktorarbeit* ("Theses and Dissertations", 3rd ed.; EBEL and BLIEFERT 2003), we address a wide range of table applications and provide a great many examples.

---

[1] We note in passing that table preparation has been a subject of intense consideration on the part of standards organizations, as illustrated by DIN 2331 (1980) *Begriffssysteme und ihre Darstellung* ("Concept Systems and their Presentation") and DIN 55 301 (1978) *Gestaltung statistischer Tabellen* ("Design of Statistical Tables").

● A table is a device for condensing and systematizing information, where order is established chiefly by careful and strategic distribution of material across two dimensions, embodied in *columns* and *rows*.

● *Columns* are the vertical organizational elements, *rows* the horizontal.[2]

Figure 8–1 can be regarded as the prototype for a table, extracted from a *spreadsheet* environment, in this case the program EXCEL.[3] The scene is dominated by a series of rectangular blocks, referred to as *cells* (or *table cells*), with labels of the form **A1**, **B1**, **A2** … All the "A" blocks (**A1**…**A6**) are in a single *column* labeled "A", whereas the blocks **A1**…**F1**, etc. constitute the *row* labeled "1", containing one block (or cell) for each column.

Table XX. Experimental results A through F obtained at control points 1 through 6.

	A	B	C	D	E	F
1	A1	B1	C1	D1	E1	F1
2	A2	B2	C2	D2	E2	F2
3	A3	B3	C3	D3	E3	F3
4	A4	B4	C4	D4	E4	F4
5	A5	B5	C5	D5	E5	F5
6	A6	B6	C6	D6	E6	F6

Fig. 8–1. Prototype for a 6 × 6 table, containing all the basic elements.

● The upper gray strip serves as a very primitive *table heading* or *lead line* whose content defines the columns below.

Every column has its own entry in the table heading—here shaded gray—in the form of a *column heading* (or simply *head*), also referred to as a *column title* (the term *column descriptor* is occasionally applied as well). Horizontal lines (*rules*) both above and below the heading extend all the way across the table. The first of these establishes an upper border for the table, while the second separates the heading from the *body* of the table. As in Fig. 8–1, the upper rule is sometimes drawn more heavily. A table usually ends with yet another line, below which footnotes may appear.

● The *vertical* gray strip at the left is called in English the "stub", and it is sometimes utilized as a kind of "row identifier".

---

[2] We prefer use of the term "row" to "line", since "line" brings with it too much baggage from the printing world.
[3] Spreadsheet programs first attracted serious attention in the early 1990s. They were enthusiastically embraced not only by the accountants and bookkeepers for whom they were conceived, but also by imaginative scientists who quickly recognized what a valuable contribution they could make to everyday work (cf. Sec. 8.5.1).

The stub plays a less prominent role than the heading, and is often *not* separated from the body of the table by any sort of line (unlike the example in Fig. 8–1). For consistency's sake this element should probably be called the table "flank", as a counterpoint to the "head", but we have never seen that term actually employed.[4]

● The two "identifier strips" embrace what amounts to a coordinate grid, and in fact have much in common with the abscissa and ordinate of a Cartesian coordinate system.

Pursuing this analogy (cf. Sec. 7.2.3), the block shown with a dark gray background in the upper-left corner of our prototype table can be viewed as the *coordinate origin*. In principle it should probably be labeled as shown in Fig. 8–2, which makes the relationship to a coordinate system even more apparent. Tables whose content is limited to numbers and letters (i.e., *alphanumeric* information) are in fact *digital presentations* whose information could also be conveyed by graphs and other diagrams like those we encountered in the *analog* displays in Sec. 7.2.3 and 7.2.4 (histograms, bar charts, etc.). This relationship is the reason why in many software packages charts and tables can be interchanged with "a click of the mouse" (WORD, for example, offers this convenience feature, a subject to which we return in Sec. 8.5.1).

Fig. 8–2. The "coordinate origin" in a two-dimensional table.—The rectangle shown here represents the dark gray cell in the upper-left corner of Fig. 8–1.

In a real table, the labels "*X*" and "*Y*" in the "origin box" of Fig. 8–2 would assume a more descriptive character, like "Quantity" and "Value", "Determining factor" and "Response factor", or "State" and "Year", to suggest but a few rather generic examples.[5] In most cases, however, *nothing* is actually recorded here, perhaps to avoid the nuisance of introducing a diagonal line and arrows. As a result, this "cell" becomes something of a neglected stepchild. Occasionally the space will be utilized to make some reference to the heading, but that hardly contributes to the logical structure of the whole.

The "low profile" associated with the leftmost column (or stub) is to some extent technical in its origin and attributable to the fact that we read and write in a horizontal

---

[4] The "head" of almost anything tends to be more important than its "flank".—In EXCEL and similar applications the gray strips at the top and on the left are in fact not subject to alteration; they function only as formal logical structures. If one wishes to include custom elements equivalent to a *true* table heading these must be consigned to row "1". The same applies with respect to descriptive elements intended to play a similar role in the table "flank", which would of necessity occupy column "A".

[5] "*Y*" in this illustration can be regarded as a "heading for the stub", while the "*X*" serves an analogous function with respect to the column titles (we deliberately avoided here use of the word "heading" with the latter because of the potential for confusion). Both act as "signposts"—an analogy reinforced by the arrows pointing to what transpires on the two sides of the diagonal: toward the "east" and the "south". It seems a pity that signposts of this sort are so seldom in evidence.

way, not vertically. A certain disparity between the coordinates is thus predestined, reminiscent of axis-labeling problems we encountered with graphs (cf. Sec. 7.2). It is an interesting exercise to attempt to picture how a table would look if it were "genuinely" two-dimensional; i.e., with no "discrimination" in favor of the horizontal axis.

The two "identifier strips" in the table of Fig. 8–1 function together to confer upon every "quadrangle" in the assembly an unambiguous alphanumeric descriptor (e.g., **N2**). These descriptors provide one with a sense of orientation (a subject to which we return in Sec. 8.4.3), and in a program like EXCEL they also lend themselves to very effective use in formulating arithmetic operations one would like to see performed. Thus, if you were to attach to some particular cell (**C3**, say) the symbol string "**=C2+A3+B3**", what would in fact appear in that cell is the sum called for by the string expression: a new numerical value which would henceforth be regarded as the content of cell **C3**. Using the mouse to "drag" this same operational string down through several successive lines within column C causes analogous arithmetic operations to be out carried in all the cells affected (e.g., the numerical value corresponding to "**=C3+A4+B4**" would be displayed in cell **C4**, etc.).

● In a spreadsheet context, tables can thus serve as much more than mere repositories of information: they can also be induced to "think".

The one thing still missing from our description of a table is a label to express its overall significance or purpose: a title, perhaps in a form like "Experimental results …". Combining this with a *table number*, such as "**Table XX.**", results in a complete formal *table title* (sometimes called a caption) like the one above the table in Fig. 8–1 above. We return to the title in Sec. 8.4.1.

It is useful to distinguish between tables of two basic types. One could be called "tables of matching entities". Here the content of each cell is an "entity" reflecting two specific characteristics, one established by a column, the other by a row: a concept and a language, for example. In other words, imagine a cell at the junction of a column whose head displayed a *picture* of a bell, and a row with the stub label "English". The cell in question would be expected to contain the *word* "bell". A cell in that same column but a different row, a row labeled "German", should instead display the word "Glocke".

Another concrete example of a table in this general category is the chemist's periodic table. Here each cell presents the name, or the official element symbol (e.g., Ag for silver, argentum), of a chemical element—the element matching the *period* (row) and the *group* (column) corresponding to that cell. This type of table usually provides further information as well, such as an atomic mass for each element.

More common in the natural sciences, however, is a *characteristics table*, in which the functions of axes and cells have in a sense been interchanged. A melting-point table is a case in point. This time the first *column* would be the site of the "entities": names of selected materials that display the characteristic of "melting" (a series of organic compounds, for example—perhaps ones containing the hydroxyl function). In a second column would appear the melting temperatures characteristic of the same compounds. Typically this column would be followed by a third for boiling temperatures, possibly a fourth for densities, etc. With these additions it becomes legitimate for us to speak again in terms of a true second "dimension", but now the title needs to be changed, because "melting points" is too limiting. An appropriate choice might be "Physical Constants of Alcohols". Most of the tables in the *Handbook of Chemistry and Physics* as well as a great many other scientific reference works are of this type: they tabulate *properties*.

8.2 The Significance of a Table

Before proceeding with formal matters in the sections that follow, we wish briefly to pay tribute to tables in general, and perhaps sing their praise a bit. Tables accompany us throughout our lives, and not simply our professional lives. There are tax tables to confront, stock-market tables and tables of weather statistics to skim in the newspaper, "risk tables" that establish our insurance costs—even the calendar on the kitchen cabinet is a table! Tables are simply inescapable. It should come as no surprise that the noun we use to describe this display form—"table"—has been adapted so that it can assume adjectival and verbal roles as well ("tabular", "to tabulate"), consistent with what has happened with the loosely related noun "picture" ("pictorial", "to picture").

Tables are commonplace in reports, articles, and books of all types; indeed, entire *sets* of volumes consist of virtually nothing but tables! Earlier generations of students and scientists were plagued with having to work extensively with *logarithmic* tables, for example, but the needed information would today generally come from the "mind" of a calculator or computer instead, almost invisibly. Actually, *much* of what one formerly would have looked up in printed tables is now more often accessed with a computer.

Tabulation has for some become almost a passion in this computer age. Computers seem to possess a veritable sixth-sense when it comes to arranging and sorting things, and they can deal with organizational issues much more efficiently and effectively than was ever possible with typeset tables: i.e., accumulating facts in columns, shuffling the content of rows and columns, issuing data collections in printed form at the touch of a button, alternating tabular presentations effortlessly between landscape and portrait formats, and much more besides. Almost anyone who has worked extensively with a modern, full-featured (and user-friendly) word processor can attest to how easy it has become to set up even a very complex table on a computer screen (and then print it)—a far cry from the grief that accompanied table preparation in the typewriter era.

An important milestone was the early recognition that a computer's computational prowess can be harnessed to tabulate *new* data from information already available in tabular form, or to transform complex tables into graphic presentations. It takes only a brief exposure to *spreadsheet* and *database* software—both of which have tables at their heart—to become enamored of the capabilities of such programs (see also Sec. 8.5).

It should come as no surprise that much terminology originally derived from the data-processing world has crept into the jargon of publishers and printers. For example, column headings of printed tables are now sometimes referred to in editorial offices as *definitions*, with analogous entries in the stub called *addresses*. The location of a particular compartment or "cell" becomes a *cell address* (cf. Sec. 8.5). For some the (horizontal) table head is a *column-description vector*, and the (vertical) stub a *line-description vector*, and material within a cell, irrespective of its nature (text, numbers, graphics, a function, …) is regarded simply as *data*.

Many authors are extremely fond of embellishing their publications with tables, so much so that editorial offices are forced to publicize a warning: The ratio of tables to text will not be allowed to exceed a certain limit! They say this because otherwise typesetting becomes problematic: layout in terms of orderly columns and pages proves virtually impossible. Breaks are urgently called for at places where nothing can be "broken", for the reason that, in general, there is as much reluctance to perform surgery on a table as on a figure. Hesitancy in this regard obviously does not rule out publication of a work based *exclusively* on tables, for in this case table headings are simply repeated as often as necessary on successive pages to ensure that the material remains intelligible. A good example is the table at the end of Chap. 7.

Quite apart from typesetting concerns, however, it is ludicrous to carry table preparation too far. A table like that in Fig. 8–3 may look impressive at first glance, but printing such a thing is not really worth the trouble. It would perhaps be acceptable in an appendix to a dissertation, but in a journal article this display should certainly be replaced by a simple verbal message:

> Compound **3** was obtained in 79% yield by the reaction of **1** with **2** in tetra-chloromethane. Identical results were achieved at 40 °C and 75 °C. When ethanol was instead used as solvent the yield fell to 24%, again independent of temperature, judging from tests at 20 °C and 75 °C.

● One important task of any table is to increase the *information density* in a document.

Table. Yield of **3** in the reaction of **1** with **2** in several solvents at different temperatures.

Solvent	Temperature	Yield
CCl_4	40 °C	79 %
CCl_4	75 °C	79 %
C_2H_5OH	20 °C	24 %
C_2H_5OH	75 °C	24 %

Fig. 8–3. Example of a pointless table.

What a table obviously *cannot* do is *interpret* data. The reader is forced to draw his or her own conclusions from a tabular presentation. Text has an advantage in this respect, as illustrated in the preceding example, and if text can convey the required message both lucidly and efficiently it is often preferable.

There is also an economic argument to be made against overuse of tables:

● A table is always more difficult and costly to prepare than running text.

This applies to demands imposed upon both author and typesetter. From the author's standpoint the task has been greatly simplified by table-generating tools in word-processing software. Thus, merely expressing a desire for a table with a particular format (e.g., "3 × 4", for example), produces a template ready to accept the relevant data (as implied in Fig. 8–4).

Exp. No.	Δh	Δx	Δy
1	0.35	0.11	0.54
2	0.37	0.09	0.51

Fig. 8–4. Template for a 3 × 4 table as generated by a word-processing program, into which data has subsequently been introduced.

The rows and columns in such a template can easily be adjusted in size, and data entry can occur in any desired sequence. If cell boundaries prove too restrictive, text will automatically be broken into multiple lines, with extra height conferred upon all the cells in the affected row. If there is reason to do so, additional rows and/or columns can be introduced after the fact, wherever required. Specific cells, rows, or columns can be enhanced with distinctive boundary lines of several different types, selective highlighting can be provided by color or shading, and type throughout a table can be subjected to custom treatment (style, size)—as a way of singling out heading or stub text, for example. One could hardly ask for greater flexibility (we return to this point in Sec. 8.4.3).

If for some reason you prefer not to work with a word-processor's table function, tables can instead be created in the "conventional" (manual) way. Even with a manuscript intended for journal publication the editors may sanction such originality—and possibly encourage it! There is one important restriction of which you should be aware, however:

● Only under truly exceptional circumstances should the *spacebar* be used for positioning elements in a table.

A table prepared manually should be handled as a single "paragraph", with line feeds employed to generate rows as required. Make sure that "paragraph marks" and other "invisible" symbols will be displayed on the screen during the construction process so that such things as line feeds and tab signals will be obvious. Next, with the text cursor somewhere in the "table paragraph", "click" on the "ruler" at the top of the screen (see "Keyboards" in Sec. 5.2.1) and create a tab everyplace you wish a column to begin.[6] These should be distributed in such a way that they roughly suit the anticipated cell content, and the table as a whole should fit comfortably into the overall page design. Explicit "cells" are then prepared by pressing the tab key a suitable number of times in each row. The next step is to begin introducing table content, cell by cell.[7] If it appears that more than one line will be required in some particular cell, create what looks like an "additional row" to accommodate the overflow (leaving the corresponding area blank in other cells in that row). Column locations can be revised as necessary by adjusting appropriate tabs with the mouse. To accommodate essential dividing lines (normally three in number: two near the top and one at the bottom; cf. Fig. 8–1 and Sections 8.4.2 and 8.4.3), create empty rows as necessary and use the program's "line-drawing tool" to produce fine, continuous, horizontal lines of the proper length through the (vertical) centers of the new, unoccupied rows. With at least some programs you can "draft" such a table consistent with fairly rigid standards, with respect to line widths for particular "cells", for example.

8.3 The Form of a Table

One of the rewarding consequences of progress in text and data processing is that everyone is now in a position to "typeset" sophisticated tables—in essentially any desired form. With a bit of care the results can be quite professional. Virtually every aspect of such a table can be controlled. Thus, row heights, to take one example, can be specified very precisely in (fractions of) inches, millimeters, or points. There is no need whatsoever to feel restricted by the severe line-spacing limitations once associated with a typewriter (cf. Fig. 8–5a).

But, given a choice, what is the "best" way to lay out a table? What are some of the long-standing guidelines derived from the "classical" publishing industry? In the first place:

● Tables in journal articles and books are almost always set in type 15...20% smaller than that used for running text in the body of a publication.

One advantage of this practice is that it simplifies the process of fitting tables into whatever limited space is available: with small type, columns can accommodate more

---

[6] "Tabs" and "tabulators" bring to mind archaic typewriter (and calculator) technology, but their equivalent in a word-processing or spreadsheet environment constitutes a tremendous advantage when working with tables. The "tabs" in a word-processor present themselves as adjustable markers atop the text ruler. Various *kinds* of tabs (right, left, centered, decimal) are distinguished by characteristic symbols. The equivalent of a set of tabs is also established when a table is created with a custom "table-generating routine", this time to permit customization of the default grid.

[7] The tab key generally causes a text cursor to advance sequentially across the page from cell to cell, while pressing "tab" at the end of a row instead moves the cursor to the first cell of the *next* row.

characters, so it is less often necessary to introduce line breaks. In other words, if the bulk of a manuscript is set in 12-point type with a 16...18-point line spacing, then tables would typically be prepared with 10-point type and a line spacing of 13...14 points. An example is provided in Fig. 8–5b.

Overall table *format* is another important aspect of table "form".

● Thought should be given to table format early in the planning stage of any document.

A table laid out in an awkward way may prove impossible to arrange satisfactorily on the page. The maximum *breadth* of a table is of course limited by the text layout; with letter-size paper in the standard "portrait" (vertical) arrangement, no more than about 6 1/2 inches of breadth will be available. If this is insufficient it will become necessary to work in terms of a "landscape" page orientation, which has the disadvantage of requiring the reader to examine the table from the *right*, or to rotate the page 90° clockwise. (The table at the end of Chap. 7 again serves as an example). Rows will continue to run from left to right, but this time parallel to the long axis of the paper.

a

Table 7. Molar masses M(X) of some natural elements X.

X	Element	M(X) g/mol	X	Element	M(X) in g mol^{-1}
Ag	Silver	107.87	Eu	Europium	151.96
Al	Aluminum	26.98	F	Fluorine	19.00
Ar	Argon	39.96	Fe	Iron	55.85

b

Table 2. Elemental analyses of compunds $CH_3SO_2N(R^1)OR^2$ (R^1, R^2 = H, CH_3).

R^1	R^2	Empirical formula	C calc.	C found	H calc.	H found	N calc.	N found
H	H	CH_5NO_3S	10.8	10.73	4.51	4.52	12.61	12.31
H	CH_3	$C_2H_7NO_3S$	19.20	19.16	5.64	5.71	11.19	11.27
CH_3	H			19.35		5.66		11.20
CH_3	CH_3	$C_3H_9NO_3S$	25.89	25.53	6.52	6.64	10.06	10.36

Fig. 8–5. Examples of tables.—**a** Double-spaced "typewritten" table, appropriate for submission with a manuscript that will be typeset for publication; **b** typeset table as it might appear in a report, a dissertation, or a publication based on direct reproduction of an author's care-fully prepared copy (10p/13p—with 5p extra leading in two places to add structure, cf. Sec. 8.4.3—before photoreduction to 80%).

Fortunately, word-processing software permits one easily to set text—including tables—according to either page arrangement.

A very *narrow* table, with perhaps only two or three slender columns, tends to waste a considerable amount of adjacent space, and the result looks somehow "wrong" from the standpoint of proportions. One solution is to break such a table vertically in the middle and resume data entry from the top (with appropriate repetition of column heads), as illustrated in Figures 8–5a and 8–6.

Table 8. Influence of temperature on the formation of **1** according to eq. (12).

Temperature °C	Yield %	Temperature °C	Yield %
10	5	60	84
20	12	70	83
30	25	80	79
40	51	90	62
50	76	100	32

Fig. 8–6. An example of a "split" table.

- If you find yourself confronted with an exceptionally short, *broad* table running like a strip across the page, try transforming the rows into columns ("inverting the axes") and then splitting them as above.

In principle such a transposition is always possible because of the essentially equivalent status of a table's two dimensions. In which arrangement does a particular table seem easiest to interpret? Which has the more pleasing appearance? Which consumes the least space? Try it out! Keep in mind one relevant observation, however: all other things being equal, it is easier to acquire an overview with a *horizontal* arrangement than a vertical one. To convince yourself of this, type a short sentence, first in the usual way and then again with the words set vertically, as in a column. Which sentence is easier to read? Our eyes are accustomed to reading from left to right, and the horizontal lines bracketing a table heading tend to reinforce the horizontal bias. Once one has acquired a sense of a particular table's structure and function, however, it is actually simpler to *scan* a column than a row—in a search for the largest numerical value, perhaps.

- If your intent is to publish your work in a particular *journal*, take into account that publication's standard page layout as you plan your tables. For example, if a journal features a multicolumn layout, the editor will generally have a preference for long, narrow tables.

In preparing any manuscript for submission to a journal one should carefully consider specific quirks related to the target publication. How many tables are present in a typical article, and how large do they tend to be? How are units expressed in that journal's table headings? How are table footnotes handled—and how much experimental detail do they usually contain? The present section and the one that follows should help make you more sensitive to some of these issues, offering new perspectives you might want to consider—assuming you receive no explicit editorial instructions to the contrary.

8.4 The Components of a Table

8.4.1 Table Title

It would be most convenient at this point if we could present a "typology of tables" as a way of showcasing a finite number of standard options, but to our knowledge no such compilation exists. The alternatives available for dealing with various problems are simply too diverse. Instead, we settle for discussing the various *components* of a table one after another, in conjunction with pertinent design considerations, in an attempt at least to suggest the range of possibilities.

We use the table in Fig. 8–7 as our point of reference, which we will take to be illustrative of a "complete" table. That is to say, it reflects all the characteristics one associates with a typical table, albeit in their simplest imaginable form. We begin with the title.

● What we refer to as the "table title" consists of a *table number* followed by a brief *content summary*, where the latter could in some sense itself warrant being called a title.

Table 1. Comparisons of several basic European words.

Concept	English	German[a]	French	Latin
Parent ♀	mother	Mutter	mère	mater
Parent ♂	father	Vater	père	pater
Sibling ♂	brother	Bruder	frère	frater

[a] In contrast to other European languages, nouns are always capitalized in German.

Fig. 8–7. An example of a complete table.

The table number usually appears in the form "**Table X.**", with a period after the actual number, here represented by "X". This information constitutes a concise *table designator*, and in printed documents it is often set in boldface type (as above). Long documents (this book can serve as an example) may make use of *double numbers* (e.g., "Table 5–3"), just as in the case of figures (see Sec. 7.1.1 for an elaboration).

Actually, tables for a manuscript are treated like figures in a number of ways, such as the requirement that each table be suitably "anchored" in the accompanying text. Almost the sole reason for numbering tables is to simplify making reference to them, and *every* table must be cited by number at least once in the body of the text, as discussed in Sec. 7.1.1 with reference to figures.

Again, just as it is sometimes necessary to supplement a figure caption—which functions essentially as the figure's "title"—with an explanatory *legend*, so it may be appropriate to provide additional clarification following the "name" of a table as well. The result might logically be characterized as a "table legend", but that terminology is not in fact employed.

- An elaboration intended to clarify the role of a table is printed as though it were a part of the "title" in the broadest sense. It should be set *after* the *literal* title, however, separated from the latter by a period and an "em-dash".

We consider it important that—as with figures—any such explanatory material be clearly differentiated visually from the title preceding it. Examples of amplified table titles of this sort are easily located in the professional literature, for which reason we refrain from providing one here. Unfortunately, you will probably discover that relatively little attention is ordinarily directed toward conferring a separate identity upon such supplementary material.

On the other hand, there is one way in which figures and tables truly differ: with a figure, *all* explanatory material must be consigned to the caption and legend, whereas with tables there is the possibility of appending *footnotes* (see Sec. 8.4.4).[8] Nevertheless,

- Table footnotes are intended only for providing details regarding table content.

8.4.2 Table Heading

Simple Table Heads

A *table head* (or *heading*) is the first element of the table proper.

- In order to help distinguish a table heading from the title that precedes it, and from the body of the table that follows, heading text is set between two parallel lines.

These two lines are referred to by editors and printing professionals as *rules*. Just like text, lines can also be subjected to custom treatment within a word-processing

---

[8] Another difference may become apparent as you page through the literature: whereas a figure caption almost always ends with a period, table titles in many publications lack such a terminal period.

environment. In particular, the *width* of a line and its "boldness" can be modified to suit your taste (or that of an editor).

● The various segments into which a table heading is divided announce the content of columns below. These entries must necessarily be concise due to the limited amount of space available.

In a book like this one, a normal text block is wide enough to accommodate roughly 80 to 90 characters (including spaces). A medium-length word contains about 6 or 7 letters, which translates into about 15 words per line. Assuming one allows space equivalent to at least three letters between words, the capacity of a line drops to perhaps ten words.[9] If this will not suffice for a table heading (e.g., because of a large number of columns), wording to define at least some of the columns may need to be spread over two or more lines. You should resist the temptation to resort to the alternative of short, non-standard abbreviations (e.g., "calc." for calculation, our Fig. 8–5b notwithstanding!), and any abbreviation that *is* used should be explained in a table footnote. This rule applies even to abbreviations already introduced in the body of the text, because—again like a figure—every table must be capable of standing on its own.

It is occasionally possible to simplify a table by eliminating related rows or columns. Thus, a row with several entries identical to the ones immediately above might be dispensed with by citing the few exceptions in a footnote. In some cases one can also delete a column if it consists of numbers easily derived from data in a nearby column. Furthermore, in the interest of more efficient use of space, two columns can sometimes be "merged" through use of a heading like

> Yield (in %)
> with/without X

to report results obtained under two sets of conditions (in this case, the presence or absence of an additive).

● Many column headings consist of *symbols* rather than words, especially quantity symbols.

Consider carefully, however, whether the use of symbols will in fact save space, since *data entries* may actually prove to be the limiting factor in column width. In other words,

● The minimum breadth (width) of a column is a function of space requirements associated both with the column heading *and* data entries.

What this of course means is that you must have a clear idea of the proposed content of your table before trying to lay it out. The most logical approach is to let the widest

---

[9] Strictly speaking, the use in tables of smaller type increases row capacity slightly, but our generalization can still be considered roughly applicable.

entry in each column serve as a guide for setting appropriate tabs, or for adapting a "prefabricated grid".

The tab key is a convenient tool for rapidly moving the text cursor forward through a series of several preset positions, the locations of which are determined by the tabs themselves. Once an appropriate set of tabs is in place it is easy to begin entering data for any table column at precisely the right place. Column content is almost always *left justified*,[10] although options usually exist also for centering, right justification, and alignment based on a decimal point (with a *decimal tab*).

● All entries in a given column should be meticulously aligned, with respect both to other entries in that column and to the column head.

With a word processor it is a simple matter, once a table has been constructed, to use the tab settings as a way of fine-tuning column widths, even with data already in place (as, for example, when the need arises to introduce additional information into some cell). Incidentally: most people when working with tables think of the tab function only in conjunction with advancing the cursor from one cell to the next, but it is also possible to make effective use of tabs *within* specific columns or cells. Moving the cursor inside this more limited range may, however, require pressing some auxiliary key in *addition* to the tab key (see your word-processor user's manual for details).

Working with Units

But now we must return to more scientific matters! If a column head reflects *quantities*, then information must always be provided regarding the applicable *units* (cf. Sec. 6.2), in some cases together with an *order of magnitude* (*power of ten*).

● Units are conveniently specified *under* the corresponding quantity symbols, in conjunction with any required order-of-magnitude information.

For example, if the unit symbol "V" were to appear in a column head under a quantity designated "U", a cell in that column containing the number "0.25" would be interpreted as reporting "0.25 V" as that particular cell's U-value. Similarly, a clarification "10^{-10} m" under a quantity symbol "r" would mean that a tabulated numerical value of "1.54" should be understood to signify a radius r of $1.54 \cdot 10^{-10}$ m.

If you are concerned that a "V" standing by itself in such a case would be too inconspicuous, you might choose to precede it with the word "in", or put the symbol in parentheses as though it were an elaboration; e.g.:

Body weight in g	Yield (%)

It is not strictly proper to display such a unit symbol in *square brackets* (cf. the discussion in Sec. 6.1.1). Nevertheless, one does encounter this usage rather frequently,

---

[10] An important exception involving *decimal tabs* is discussed in the section that follows.

especially in chemical, biochemical, and biomedical publications—even in journals produced by the publisher of this book. From a non-standard notation like

$$\Delta\varepsilon_J^{s\in F}(\text{rel.})[\text{eV}]$$

one would probably *guess* correctly what was intended, but it is clearly inconsistent with national and international recommendations. [Even the COUNCIL OF BIOLOGY EDITORS (1994) insists upon the exclusive use of parentheses in this situation.]

Information in a table heading regarding a power of ten (10^x) is a common source of confusion, just as it is in axis labels for figures, because authors do not always make clear whether data are being reported as 10^x-fold values, or if reported numbers are instead to be *multiplied* by 10^x. The notation recommended above is quite unambiguous, and should wipe away all such doubts. Another permissible approach— illustrated again in conjunction with the example used previously—would be to write " $\cdot 10^{10}$ " *after* the quantity symbol r, with the unit "m" once again underneath. This is also open to only one interpretation: as $r\cdot 10^{10} = 1.54$. Even better is to set the power of ten in *front* of the quantity symbol ("$10^{10}r$"), an option that should make it still less likely that someone would read the table incorrectly. We summarize the various ways of dealing with this issue in Fig. 8–8.

r	$r \times 10^{10}$	$10^{10}\,r$	~~r~~	r
10^{-10} m	m	m	~~m $\times\,10^{-10}$~~	10^{-10} m
1.54	1.54	1.54	1.54	1.54
(correct, recommended)	(correct, less satisfactory)	(correct, recommended)	(ambiguous)	(correct, recommended)

Fig. 8–8. Alternative ways of declaring units and powers of ten within a table heading.— In this example the quantity r should be regarded throughout as having the value $1.54\ \cdot 10^{-10}$ m.

With the arrangements in the first four columns of Fig. 8–8, information below the quantity symbol assumes the character of an elaboration: it explains (albeit awkwardly in the case of the fourth column) how numbers in the corresponding column are to be interpreted, and thus what should be written *after* each number. The same elaborations could have instead been set in parentheses.[11]

The last column in Fig. 8–8 illustrates an altogether different approach to quantities and units, taking advantage of the *fractional form* of quantity notation. Merely inserting a dividing line between a quantity symbol and the corresponding unit symbol has produced another accurate statement. Here, the data column should be interpreted as

---

[11] Surely every reader would correctly interpret a column headed "Sales ($)"!

containing nothing but *numerical values*. We have already recommended this strategy for axis labeling in graphs (cf. "Axis Labels" in Sec. 7.2.3), and it is perfectly appropriate here as well. Indeed, the standard BS-4811:1972 treats this as the *only* correct notation for tables, based on the argument that the fundamental purpose of a table ultimately is to organize *numbers*.

Structured Table Headings

It is sometimes helpful to *structure* the entries in a table heading as a way of drawing attention to relationships that link individual columns.

● *Lines* can be used either to *join* or *separate* entries in a table heading, both physically and visually.

Vertical lines in a heading—which are normally continued to the bottom of the table— serve to *separate* columns they fall between. *Horizontal lines*, on the other hand, have a *unifying* effect, essentially creating a table heading *within* a table heading (called a *decked* head). In technical terms the line in this case is a *straddle line*, with a *group designator* (or *spanner head*) above it and two or more *subgroup designators* below.

Examples of structured headings are provided in Fig. 8–9. The heading shown in Fig. 8–9a explains the presence of data based on several different units for the same quantity. In Fig. 8–9b, the use of both vertical and horizontal lines is illustrated.

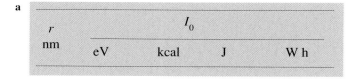

Fig. 8–9. Structured table headings. **a** Use of a "straddle line" between a quantity symbol and symbols distinguishing among alternative units; **b** structure lines both horizontal and vertical.

8.4.3 Table Content

What immediately follows the table header—the actual information to be presented— is referred to as the *table content* or the *table field*. It can be regarded as analogous to a drawer with many compartments.

● Due to the actual or inherent two-dimensionality of a table, each entry has its unique place, which can be described in terms of two coordinates.

These table locations or spaces (cf. Sec. 8.1) might be described in "IT" terms as *cell addresses* or, more directly from the standpoint of layout, simply as *cells* or *fields*. If you imagine that the several columns are labeled sequentially from the left with capital letters (a convention familiar from spreadsheet programs) and the rows from top to bottom by number, then a particular cell, which resembles a "block" in a typical city, could be specified by a descriptor like **B4**, as illustrated in Fig. 8–1 and discussed in Sec. 8.1 (see also Sec. 8.5.1).[12]

Authors sometimes have a tendency to "fence off" a table's cells by literally drawing lines between all the rows and columns. In our judgement such a plethora of lines is more likely to be a source of confusion than clarity; moreover, a table so embellished calls to mind a window covered by "prison-like" bars, hardly contributing to its aesthetic appeal. In all aspects of a manuscript it is best to dispense rigorously with typographic flourishes that fail to make the text more readable. (You need only turn off a "grid lines" command in your "Table" menu.) Extra lines within a table generally fall in this category because they contribute little or nothing to legibility. Occasional horizontal lines separating a table into *segments* can be a valid exception.

Thanks to laser and cathode-ray imagesetters there is for a publisher no longer a *technical* problem associated with introducing vertical lines, though these were once the bane of typesetters. This is one of the reasons why it matters little today from a technical standpoint whether a table (or an underlined text passage) is printed in landscape or portrait format. To the person working with a word-processing or spreadsheet program (and a desktop printer) lines are also no problem, but our advice above still stands.[13]

● The cells in a table are generally distinguished not by lines, but rather through the narrow white strips surrounding them as a consequence of the way design parameters (line spacing, tab settings) are defined.

---

[12] We are unable at this point to resist the temptation of introducing a remarkable analogy, probably unfamiliar to the vast majority of our readers: the street layout established in 1606 (a bit before the Age of Enlightenment) for the city of Mannheim, in southwestern Germany, a pattern that incidentally remains virtually unchanged today. The heart of the city is roughly circular in shape, albeit with a flat "base", across from the center of which the "prince elector's" former palace stands, just east of the Rhine River. Now picture this form overlaid by a grid that has the flat side as its base, with a central perpendicular axis extending out from the palace. This particular grid involves 14 columns and 10 rows. Starting from the left (in a system adopted in its present form in 1811), the columns are numbered 7, 6, ..., 1, 1, 2, ..., 7. The rows, starting from the bottom, are labeled A, B, ..., K ("I" being omitted) in the *left* half and L, M, ..., U in the *right* half. The grid lines are in fact the city's streets. Mannheim is unique in that in most cases these streets have no names, but street *names* are not necessary, because all the *blocks* are clearly identifiable: block C3, for example. Buildings in a block are assigned consecutive numbers, starting at the corner nearest the palace and proceeding around the periphery back to the starting point (counterclockwise in the left half, clockwise in the right), so that a typical address for mail-delivery purposes would be "C3, 19"!

[13] Computer programs also allow the user to add *borders* of several types around individual cells or regions of a table, and cells or groups of cells can be selectively enhanced with background patterns or colors. Used sparingly and under the right circumstances, measures like these can in fact contribute to the clarity of a table (although color is of course helpful only if you have a color printer at your disposal).

The lower boundary of a table is marked by a single horizontal line, below which any required table footnotes would be set.

There is no rule requiring that all the cells in a table contain information. At the opposite extreme, a cell is sometimes called upon to house an unusually large amount of information, in which case one must be careful not to trespass across cell boundaries. Long entries are broken into two or more lines as required, using a manual "line feed" where necessary.

● In a manuscript that will eventually be typeset, table rows should be at least double spaced, although narrower spacing is permissible within individual cells.

An author's spacing suggestions may or may not be respected by the editors when a table is prepared for publication. As a general rule, *uniform* spacing is preferred unless ensuring clarity dictates otherwise (cf. the table in Fig. 8–5b).

Most of what needs to be said about *column* spacing and layout has already been dealt with in the context of table headings. Only one additional consideration requires comment:

● Columns consisting exclusively of *numbers* should in most cases be so arranged that *decimal points* are all strictly aligned.

This alignment, readily achieved with a "decimal tab", is likely to produce "ragged edges" on both the right and the left, and so it *should*, because the goal is to help the reader instantly acquire a clear sense of numerical magnitude, where the decimal point serves as a reference:

```
       12.4        12.4      12.4
        8           8            8
     1120        1120        1120
        0.85        0.85      0.85
```

    correct   incorrect

If numbers are instead either right or left justified (i.e., a sharp edge is maintained along one side), then it is no longer easy to discern which numbers are largest, sufficient grounds for our use of the words "correct" and "incorrect" in the example above.

There is one important exception to this rule, however:

● If no sensible comparisons can in any case be made within a set of numbers, orientation on the basis of a decimal tab is *inappropriate*, because the table in question would be rendered unsightly to no benefit.

An example of a numerical column that should be *left justified* or *centered* is one containing miscellaneous physical data applicable to some chemical system (see Fig. 8–10). Individual numerical values in such a case—for boiling point, critical pressure, etc.—have nothing whatsoever in common: they are even associated with a variety of units, and may well be of different orders of magnitude, so that it would be misleading

Table XY. Selected data for the element fluorine.

boiling point	85.0 K
critical point	52 bar
critical temperature	$- 129 \ ^{\circ}C$
critical volume	$1.74 \cdot 10^{-3} \ m^3/kg$
density at 77.8 K	$1562 \ kg/m^3$

Fig. 8–10. Table containing mutually independent quantities that do not lend themselves to comparison.

to imply comparability. Such numbers should *not* be aligned on the basis of decimal points, and also not right-justified.

According to DIN 55 301 (1978), cells in a numerical table should never be left vacant. If a cell is supposed to reflect a value of zero, then a "0" should be explicitly expressed. This standard permits use of a dash ("–", an "en-dash") to signify an *absence* of data specifically in the case of a statistical table, but the practice is generally discouraged, since the symbol could be mistaken for a minus sign. HUTH (1987, p. 32), on the other hand, recommends that cells indeed be left empty if it is categorically impossible to provide the information called for, because this makes it clear that notes in the corresponding column head or stub entry are also inapplicable with respect to that cell;[14] with cells for which data or observations could in principle exist but happen simply to be unavailable, an omission symbol "…" (*ellipsis*, here signifying "not available") is a proper choice.

Finally, there is the question of alternative means of displaying tabular data altogether, something other than the traditional (but monotonous) row/column pattern.

● Some tables are organized in terms of the left-most column rather than headings at the top.

It often suffices for this purpose simply to group certain related "addresses" in the first column, with or without subheads, perhaps separating the groups markedly in spatial terms (Fig. 8–11). Even individual cells can be provided with structure by clustering certain lines and again interposing subheads. This strategy can sometimes make a table more intelligible, and possibly reduce its width.

Taking advantage of these various suggestions might lead to a complex structure analogous to the one implicit in Fig. 8–12. Examples of a variety of complex table layouts are provided in the appendix to DIN 55 301 (1978).

---

[14] The DIN standard cited above recommends use in this case of the symbol "x" (for "cell closed" or "inapplicable"), a symbol which if necessary would then be explained in a table footnote.

A (Europe)	000	000
A (USA)	000	000
A (Japan)	000	000
B (Europe)	000	000
B (USA)	000	000
B (Japan)	000	000
C (Europe)	000	000
C (USA)	000	000
C (Japan)	000	000

confusing

Europe		
A	000	000
B	000	000
C	000	000
USA		
A	000	000
B	000	000
C	000	000
Japan		
A	000	000
B	000	000
C	000	000

clearer

Fig. 8–11. Tables **a** with, and **b** without internal structure and embedded subheads.

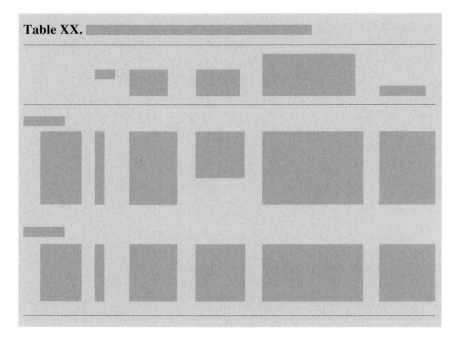

Table XX.

Fig. 8–12. Schematic example of a highly organized table.— The dark gray areas, in which definitions, addresses, and data would appear, can be envisioned as actually being penetrated by pale stripes representing gaps between individual rows.

8.4.4 Table Footnotes

Important details pertaining to a table can be supplied in *table footnotes*.

● Footnotes remain an integral part of a table even though they are printed beneath the line marking the end of the table proper.

The citation mark employed for a table footnote usually takes the form of a lower-case roman letter followed by a "close parenthesis" symbol (cf. Fig. 8–7). The same mark is repeated—usually without the parenthesis—in front of the actual note, immediately below the table's bottom line. Numerals are generally avoided to eliminate any possibility of confusion with data or with literature citations and footnotes associated with the body of the manuscript.

Table-footnote citations can appear almost anywhere within a table, including in column heads and stub "addresses" as well as in conjunction with cell data—but *not* in the title. Footnotes are used for such purposes as identifying uncommon abbreviations or acronyms, or commenting on the content of specific cells.

8.5 Worksheets, Lists, and Databases

8.5.1 Spreadsheets and Worksheets

A *spreadsheet program* is a marvelous tool to have available when a need arises to recalculate numerical data in tabular form on the basis of *formulas*, and in some situations also for the management of text elements. A scientific case in point might involve wishing to express multiple values of some quantity in terms of several sets of units through the use of *conversion factors*, or repeatedly carrying out a particular computation with many sets of data.

The most familiar spreadsheet programs include EXCEL (from Microsoft) as well as LOTUS 1-2-3 (from Lotus Development Corp.) and QUATTROPRO (from Corel). Comprehensive packages like WORKS (from Apple and Microsoft) also offer limited spreadsheet-like functions.[15] Simple applications can even be dealt with in the context of WORD.

A spreadsheet "page" (cf. the fragment in Fig. 8–1), known as a *worksheet*, takes the form of a *gridwork* whose lines divide the sheet into contiguous rectangles known as *cells*. A vertical array of cells constitutes a *column*, a horizontal array a *row*. When the content of a worksheet is printed it is customary to omit the dividing lines (easily accomplished with a straightforward command), but for on-screen work they are indispensable.

---

[15] WORKS provides both spreadsheet and database environments, where computation is at the forefront in the former and data organization in the latter.

● Column and row titles confer upon each cell a unique designation or *address*.

These addresses, like "**B3**" for the cell constituting the intersection of the second column and the third row (cf. Sections 8.1 and 8.4.3), greatly simplify computation. Attaching a *formula* like

"**=C2+C3**"   or   "**=SUM(E4..E20)**"

to cell **C4**, for example, would mean, in the first case, that **C4** would henceforth display the sum of the contents of the two cells immediately above it, or, in the second example, the sum of all entries in column E from row 4 through row 20. Such an expression always begins with an equals sign (=), and the program carries out the specified process automatically. Assuming the cell contents called for in the formulas above are *numbers*, then the requested additions will take fully into account any *signs* ("+" or "−") associated with the data.

Other arithmetic operations—subtraction, multiplication, and division—are of course supported as well, as are a great many *functions*, such as *raising to a power*, computing a *logarithm* with either 10 ($\log_{10}$) or e (ln) as the base, angular functions like sin, cos, and tan, and a variety of statistical functions including *mean, standard deviation*, and *correlation coefficient*. *Regression analysis* can also be performed. "Data" in a spreadsheet context can even consist of text, in which case such operations as "search and replace" and "sort" are available.

These few observations should provide some insight into the power and flexibility inherent in spreadsheet software. Beyond this, worksheet tables can be adapted and formatted in most of the ways one would anticipate based on ordinary table-generation software, including:

– Supplementing the "A/1" notation with true column and row titles, and/or omitting the former when tables are printed;
– Introducing lines at the top and bottom, creating borders for specific cells, and repeating headings as necessary for tables that extend over multiple pages;
– Adding line feeds to cell text;
– Applying special fonts, type styles, or type sizes to specific cells, columns, or rows, and designating the kind of justification to be employed;
– Printing in either portrait or landscape format;
– Pasting-in of text or data copied from other documents;
– Integrating the resulting tables into a "text environment".

In the end, the spreadsheet origin of a table may no longer be at all obvious, but during the preparation phase you would have had available to you all the advantages of program-based mathematical and logical intelligence. Detailed information should be sought in appropriate user's manuals, but we do want to describe at least briefly one important application:

● Spreadsheet programs include extensive provisions for displaying numerical information in *graphic* form.

In scientific work of all types one frequently wishes to examine in a visual way the dependence of one quantity (what the health sciences might refer to as a *determining factor*) upon another (the *response factor*). Creating a pictorial version of a suspected two-way relationship starting with spreadsheet data is simply a matter of designating the two columns whose correlation is to be examined and then requesting a graph of the appropriate kind. This can be of almost any standard type; e.g., a bar graph, area graph, pie chart, scattergram, or line graph (see Sec. 7.2).

In a simple case the purpose might be served nicely with a plot of experiment numbers, spread (as the independent variable) along a horizontal axis, against data drawn from a single spreadsheet column, with the latter being depicted as a set of points, each with a suitable vertical displacement. Successive *pairs* of data points, involving two true variables (from two columns) and interpreted as a *series*, could be similarly plotted, and also connected by straight-line segments. Alternatively, lines could be dispensed with, resulting in a *scattergram*, or a best-fit curve could be computed and displayed graphically in conjunction with the data points. In other words, directly from a spreadsheet program one can easily gain access to any of several data presentations that once would have been achievable only with considerable manual effort on the part of a skilled draftsman.

As you would undoubtedly guess, graphic representations like these can also be selectively (and extensively!) modified and customized by on-screen techniques prior to printing. This extends to providing additional text and graphic elements, including special labels. Admittedly, the flexibility is limited, and it may not satisfy all of a researcher's desires, or every standard established by a journal editor. This is particularly likely to be the case with respect to line widths, arrangement of tick marks along axes, or unit specifications. (All these issues have been touched upon in Sec. 7.2.3).

The ability of modern software to conjure up pictures from numbers is uncanny, and it can be the key to important insights long before the time comes even to consider announcing a set of results to the world in the form of a publication.

8.5.2 *Databases*

Spreadsheet programs (like EXCEL) offer a number of features that can contribute to the management of data. Specific information—symbols, text, numbers—can be located within a worksheet environment using a *search routine*, and "hits" that result can become the basis for new worksheets, for example. These are clearly "database" functions, but in contrast to the situation with true database software their implementation in spreadsheets is rather awkward, and one quickly discovers serious limitations. For example, only sparse provisions exist for coupling the content of multiple worksheets, and it is not possible to create custom data-input forms (with layouts designed to facilitate introduction of cell content), or to tailor reports to meet specific criteria. If you long for the availability of features like the latter you would be well-advised to consider investing in genuine database software.[16]

Domestic activities are useful for illustrating the rudiments of database operation: managing address and telephone lists, for example. Thus, appropriate fields can be created in a special data repository for recording names, home and business addresses, birth dates, telephone numbers, e-mail addresses, etc. Once a set of information has been assembled, the software framework permits you to extract selectively any records present that display certain characteristics you choose to designate, allowing you rapidly to prepare an alphabetized list, say, of all your associates and friends in Australia. From a *calendar* database you could request an overview of conferences and other engagements so far on your schedule for the upcoming month of June.

In your professional capacity you could perhaps profit from establishing a database of experimental observations, or instrument output. Clinical data so compiled, for example, could be examined systematically and rapidly in a quest for previously undoc-umented correlations or anomalies—assuming of course that the right characteristics have been entered for all the relevant cases. Another widely adopted application is maintenance of a *literature database* for one's entire collection of bibliographic information (cf. Sec. 9.2.2).

What are some of the components and features you should expect to find in database software? We begin with a minimalist approach, listing things that are probably self-evident.

● In a typical database, data that are in some way interrelated are *linked* on the basis of more complex units of information known as *records*.

Rather like spreadsheet software, a database program allows the user to define discrete *data fields* customized to suit particular types of data (e.g., text, numbers, or date-and-time records).

● A single record normally consists of data associated with multiple fields.

For every field there must be a formal definition: each is thus assigned a unique *name* and adapted to a specific *data type*, which also presupposes a characteristic format (e.g., numbers always in the form "000.0"). The *size* of a field is also declared in such a definition (e.g., 256 characters). Furthermore, one can establish *conditions* that must be fulfilled by a given piece of data before it will be accepted into a particular field (e.g., "reject if blank", or "ranges excluded").

---

[16] Familiar programs in this category within the PC world include ACCESS (Microsoft), DBASE (dBASE, Inc.), ORACLE (Oracle Corp.), and DB2 (IBM). FILEMAKER (FileMaker, Inc.) and PANORAMA (proVUE Development) are comparable applications that were developed for MACINTOSH users, although FILEMAKER has in the meantime become available for PCs as well.

The combination of a set of field definitions and the corresponding data records is known as a *database table*.

● A *relational database* is one that supports access to several database tables simultaneously, which in turn can be interrelated in complex ways.

One major difference between database programs and other types of software is the extent of the role assigned in the former to *security*. Quite apart from requiring that data be examined for appropriateness at the time of entry, steps can also be take to prevent the accidental *deletion* or *overwriting* of stored information. *Password protection* is generally available as well, meaning that only someone in possession of a valid password can gain access to certain functions. Other security-related measures may include:

– Limiting access to one user at a time in the case of data maintained in a multi-user environment (typically the situation in industry and at major research institutions).
– Ensuring that, in the case of *linked* tables, deletion of a data set from one table leads automatically to deletion of the corresponding information from all related tables as well, an extremely important aspect of maintaining what is called *reference integrity*;
– With a database subject to access via a network (intranet, Internet), arranging for various computers to be able to work with different versions of the data collection simultaneously; one speaks in this context of a *distributed database*.

At some point, comparable data sets in a distributed database will presumably need to be merged. How can one be certain that information introduced by one party will not be overwritten by someone else's obsolete file? Here the principle of *replication* is invoked to assist in automatic data comparison, conducted essentially under the motto "save the good and discard the bad". Examples of database programs with replication capability include ACCESS (from Microsoft) and LOTUSNOTES (from IBM).

Most database programs are sufficiently flexible so that existing field definitions can be altered at any time without undue effort, permitting fields to be added, eliminated, renamed, or revised with respect to type or size. Things become rather more complicated with a relational database, however. Here it is crucial that one consider very carefully, for example, what effects changes in one field will have on *other* fields, perhaps in other tables. Thus, under some circumstances a deletion could mean access would no longer be available to other tables linked to the one being modified.

Generally speaking, a database consists of two parts: a system for managing even very large amounts of data on a hard disk, and a user-interface to give the operator convenient admission to the data.

● The default view of database content in most cases resembles a straightforward table or list.

One of the most important features of a good user interface, however, is a facility for preparing customized *data forms*. These can be of use not only for data input, but also refinement of the way information is output and presented. Whereas a "list view" typically brings to the screen multiple sets of data, but for only a limited number of fields, a "form view" focuses on a single data set, but reveals as many fields as one wishes to see.

● "List" and "Form" views can be interchanged with a click of the mouse.

Forms represent the key to creating timely components for reports or publications with only a few keystrokes—assuming of course that the database is carefully planned and organized at the outset, and that all the truly important aspects of one's information store are incorporated.

● Sophisticated search capability is probably the most essential characteristic.

Issuing a "search" command ordinarily produces a special window with control features familiar from other software ("buttons" for selecting various options, for example) and provisions for specifying appropriate *search criteria*. "Query by example" has become an important byword in this regard. Every record in the database meeting the stated criteria will be quickly identified, which in turn becomes the key to creating unique subsets of records from a master database.[17]

● A set of search results has much in common with a table.

The first piece of information obtained through a search is the number of "hits" in relation to the total number of records in the database. If there appear to be too many, you will probably save yourself considerable time by immediately conducting *another* search, after adding some new criterion, using as your "database" this time only those records singled out in the first pass.

● Adding criteria is the most effective way of *narrowing* a search.

A search can also be conducted such that two or more criteria are taken into account simultaneously. One might, for example, acquire the subset of records fulfilling *all* the cited criteria (criterion 1 *and* criterion 2 *and* ...). Using the terminology of *Boolean algebra* this would correspond to a logical AND connection. If your interest were instead in finding records that fit *either* criterion 1 *or* criterion 2 (an OR connection), the investigation would be pursued in the form of *two* searches, one after the other. Alternatively, two searches can be so arranged that the result is records fulfilling one— and *only* one—of a pair of search criteria, or one might require that all records displaying some particular characteristic be selectively *excluded* in the course of a search.

---

[17] Curiously, search routines with PC databases generally are *not* organized on the basis of the internationally sanctioned *Structured Query Language* (SQL).

Special *operators* are generally available to lend more powerful search support. One might thus be able easily to examine data containing date and time fields, for example, to locate all the records associated with the time window 8 AM to 10 AM on Wednesdays, or probe a collection of numerical data for values of a particular experimental variable in excess of a given predetermined level. One could even ask that *sums* or *mean values* be computed in conjunction with the data from one or more fields.

A carefully planned, targeted search can obviously serve as an excellent basis for a table. Resulting information can subsequently be examined in terms of additional characteristics, or organized in the most convenient way with the aid of a "sort" command. Once created, a set of results can be put in tabular form and printed, or instead incorporated into a report or other document (on screen and/or in print).

Data exchange between two databases, or between a database and some other type of program (e.g., a word processor), usually entails an import/export operation conducted in strictly ASCII terms, although in a few situations special software interfaces support more extensive capabilities (the keyword here is ODC, Open Database Connectivity).

● The emphasis in a database is of course on *content*, not the *form* the content takes.

If information from a database is to be incorporated satisfactorily into a WORD or page-layout document you will want it to exhibit the proper typographic characteristics. In other words, at some point parameters such as typestyle (e.g., italic, boldface, subscript, etc.), justification mode (left, or right, or decimal), and indentation will need to be adjusted—probably manually.

In Sec. 5.3.3 under "Search Operations" we made brief reference to the term "database publishing". Especially because of the influence of the Internet and "Web publishing" it is becoming increasingly important that authors become aware of this concept and what it involves. We cannot delve deeply here into the related topics HTML (Hypertext Markup Language) or XML (Extensible Markup Language) so we must be content with observing that the crucial step is the addition of *format specifications* as information is withdrawn from a database (exported), so that the resulting file—though retaining its ASCII character—will contain not only data but also formatting codes. This obviously requires special programming of the database, a feat relatively easily accomplished with modern database software. A competent programmer well acquainted with the structures of the corresponding HTML or XML codes must simply be told precisely what is to be formatted and how. Even if one has no particular interest in publishing over the Internet, and is committed instead to printed works, database publishing capabilities could still turn out to be relevant, in this case with respect to the introduction of appropriate word-processor or page-layout formatting commands. As a matter of fact, you have probably already had at least limited contact with one very elementary example of database publishing: preparation with a word processor of a set of customized *form letters*. The task entails (with WORD, for example) accessing either an internal or external database (an EXCEL or ACCESS table, perhaps) and incorporating data drawn from a specific field (e.g., names or addresses) into a master WORD document at the site of a special "place-holder". Formatting can be conferred upon the new material in the usual ways, and the resulting personalized documents printed automatically.

More information regarding databases in general is available from SAUER (1997) or MATTHEISSEN (1998), for example, as well as in the user's manuals accompanying specific database software. BOIKO (2001) provides an especially comprehensive introduction to database publishing.

9 Collecting and Citing the Literature

9.1 The Acquisition of Information

9.1.1 Reading and Evaluating the Professional Literature

In order to maintain contact with the work of others and also be in a position adequately to cite the various resources you consult—a solemn obligation of every publishing scientist!—it is essential that you become intimately familiar with the professional literature, and have pertinent reference material nearby when you need it. Reviewing and evaluating the literature on one hand and including relevant references in one's own work on the other are often treated as separate topics, but we have elected to deal with the two in a single chapter. This is actually consistent with certain interesting computer-software applications. Thus, matters of both literature *searching*—especially online searches in external databases—and literature *management*, together with the incorporation of essential acknowledgements (*citations* and *source descriptions*) into new documents as they develop, can be accomplished remarkably effectively with the aid of integrated programs designed specifically to support the scientific author's needs (cf. Sec. 9.2.2). We are therefore extending our treatment beyond simply the *preparation* of publications in the strictest sense, because *literature research* [cf. *fr.* rechercher, search, (re)locate] is such an important aspect of the acquisition of technical insight and creative stimulus.

- It is one's personal reading and analysis of the professional literature that is the source of much of the necessity for providing readers with bibliographic documentation.

Whether a member of a research group, a research director, or even a graduate student learning for the first time to work independently, you face the constant challenge of unearthing on a regular basis from within the vast resources of the published literature every existing report that has a direct bearing on your activities. This applies particularly to journal articles, where new information is normally first made public. *Finding* such articles is a precursor to *examining* them closely and personally analyzing their content—and making sure at the same time that you will be able to locate them again in the future as your work progresses. It would obviously be impossible for us to provide here an exhaustive and systematic account of how one should go about acquiring a grasp of the scientific literature—an undertaking that presupposes a knowledge of the unique structure of the archives and documentation practices of the particular discipline in question. Numerous books have been devoted exclusively to that subject, and from various disciplinary point of view (cf. also "The Organization of a Library" in Sec. 9.1.2); examples include *Information: A Guide for Practicing Chemists, Educators, and Students* (MAIZELL 1998), *Chemical Information: A Practical Guide to Utilization* (WOLMAN 1988), *Chemical Information Management* (WARR and SUHR 1992), *Information Sources in the Physical Sciences* (STERN 2000, a book directed in fact primarily toward physicists), and *Wie finde ich Normen, Patente, Reports?* ("How Do I Locate Standards, Patents, Reports?"; BRESEMANN, ZIMDARS, and SKALSKI 1995).

In every discipline and every country, there are significant new scientific developments almost daily, and researchers and information specialists everywhere are always seeking ever more effective approaches to *information retrieval*. The quest has of course been heavily impacted by recent revolutionary advances in *telecommunication*. Search practices that were universal and taken for granted two decades ago now seem woefully archaic and inefficient, and what makes eminent sense today will undoubtedly be viewed as ludicrous tomorrow. Nonetheless, we wish to offer a few concrete suggestions based on the overall situation as we see it at the moment and from our own personal experience.

We begin by considering the basic challenge of staying abreast (maintaining *current awareness*) of what is already known, as defined by the sum of all the literature resources that have accumulated to date.

● If you truly wish to stay informed of major developments in your discipline you must systematically and on a regular basis peruse at least a significant subset of the important journals devoted to that field, together with select periodicals from neighboring areas, taking careful note especially of new ideas you might yourself be able to use.

It may be you have the good fortune in your professional capacity to be on a "circulation list", such that key journals come automatically to your desk for review as they are issued, the expectation being that you will in turn pass them along (promptly!) to the next person on the list. Perhaps you are able also to indulge in what has become the luxury of maintaining personal subscriptions to some of the most influential serial publications. If you have not already done so you should certainly consider registering your professional interests with at least one of the current-awareness services (*alerting services*) applicable to your discipline. That way you can be assured of receiving notification of much relevant new material without the risk of subjecting yourself to a flood of useless miscellany.[1] Examples include *ISI Discovery Agent* from Thomson–ISI (http://www.isinet.com) and services from the "dot-coms" *Infotrieve*, and *Ingenta*. More restricted services are provided by various commercial journal publishers [e.g., Elsevier (http://www.sciencedirect.com) and Wiley (http://www.interscience.wiley.com)] as well as scientific organizations like the American Chemical Society (http://

---

[1] Some refer to these alerting schemes as *SDI services* (Selective Dissemination of Information), or "subscription search plans". A detailed account of your particular interests is solicited by the host organization and then transformed into an effective search strategy for application on a regular basis to the system's database (which is updated constantly). You then receive lists by mail or e-mail of all the resulting hits.

pubs.acs.org/journals/asap/index.html), with coverage limited largely to their own publications. Any "automated search" has the disadvantage of reinforcing the unfortunate trend toward compartmentalization of knowledge, but this is likely to persist in any case. One important feature of profile-based information services is that you are free to modify the description of your interests whenever you wish—to reflect changes in research emphasis, for example.

There was a time not so very long ago (i.e., as we ourselves were embarking on our careers!) when a serious researcher was expected almost by definition to page conscientiously and meticulously through every issue of at least the most important journals in his or her field—and to *read* a great many of the articles. Engagement at that level has unfortunately become utterly impractical with respect to nearly all the classical scientific disciplines (chemistry as a whole, for instance, or even a major subdiscipline like organic chemistry). The probability has correspondingly been reduced of your unexpectedly brushing against an entirely new line of thought, with the potential to influence decisively the course of your future work.

So how should a researcher in today's world go about confronting the professional literature? One thing is clear: there is little hope of doing so successfully in the course of the daily commute to the office, for example, or in odd free moments here and there.

● An effective encounter with the literature almost always turns out to be a *labor-intensive* experience.

Deriving real insight from a technical document entails not only a serious time commitment, but also patience and concentration. A seat in a crowded bus or plane is hardly an environment conducive to probing complex ideas in depth, or working your way through a web of detail, let alone being truly open to surprises, or having your curiosity unexpectedly piqued. In other words, you must seize the initiative and program a substantial amount of intensive reading time into your regular professional routine.

Successfully processing newly-published research results and their significance, especially if you hope to internalize them, presupposes a certain measure of isolation, and an absence of distractions; it is tempting to say it thrives best in a climate of "constructive leisure". As you proceed, force yourself quite literally to step back occasionally from the printed page and consider the material you are examining in a very critical way. To what extent, for example, has the author of the article at hand availed himself or herself of the most definitive set of methodological tools, given the nature of the problem under investigation? Above all, be prepared to grapple with a reported study's significance from the *author's* point of view before you attempt to assess it from the perspective of your own interests and insight. To paraphrase an observation by Karl JASPERs' regarding the study of a philosophical text (JASPERS 1953, p. 143):

> To read properly you must start with a high level of trust in the work's author, together with a love for the subject on which he or she has chosen to concentrate, adopting for the moment the attitude that everything expressed in the text represents truth. Only after you have allowed yourself to be utterly captivated, indeed engulfed, and have reemerged from the depths of the subject—only then can meaningful criticism commence.

This message is equally applicable to encounters between a natural scientist and the scientific literature. *Reading* is a complex process, the objective of which is to reconstruct a writer's *thoughts* exclusively from a set of symbols on a printed page: truly to *understand* that which has been expressed, and in the intended way. Every successful research scientist is unfortunately compelled to do an enormous amount of reading—and under intense time pressure at that! This requires regular attention to the strengthening of one's personal reading skills, and modification as necessary of long-established habits in the face of new demands. One important consequence of the effort is sure to be increased mastery of the art of speed-reading.

- Effective review of the academic literature requires that you perfect an efficient reading style geared toward distinguishing quickly and accurately between the *relevant* and the *irrelevant*.

Unfortunately, almost the only professional journals today's scientist reads with the intensity JASPERS urges in the passage cited above is the type devoted to overarching informational goals and a mission more akin to the secondary than the primary literature (cf. Sec. 3.1.4): in other words, publications with "review" character, reflecting in their content a very broad disciplinary or even interdisciplinary commitment. Often such a journal has the word "News" in its title, or the label is an especially pithy one succinctly epitomizing the publisher's ambitious scope—like *Nature* or *Science*. Beyond this, a considerable part of your attention will need to be directed to the *primary literature* (journals reporting original research results within more narrow confines), but probably not as an integral part of a program of rigorous, systematic reading.

9.1.2 *Effective Use of a Specialized Library*

"Classical" Resources

Let us assume that, as a natural scientist active in research, or an engineer interested in fundamental ideas, you have already armed yourself by subscribing personally to at least one of the "wide-ranging" publications we have just described, together perhaps with one or two more specialized journals. That is certainly a laudable start, but it is still only the beginning. You will need to address your professional obligations further through weekly pilgrimages to a specialized library at a university or other research institution—each visit lasting a minimum of, say, two hours—where you browse through a wide assortment of additional publications, all the while jotting down complete bibliographic data—possibly along with a few keywords—each time you encounter an article that appears relevant to your research interests. We estimate that the average researcher, in order to maintain reasonable currency, will need to make note of (i.e., *document*) the equivalent of five to ten such publications for every day spent in research. An independent data record should then be established for each documented article. How that might best be accomplished is the subject of Sec. 9.2.

● It is a revealing sign of our times that one can scarcely imagine library work today in the absence of a robust and rapid photocopier.[2]

Articles you perceive to be especially important you will probably want to photocopy immediately in order to have access to them later at your desk. Capturing something in "black and white" at least increases the chance that you will end up carrying new information away with you! If you wish only to ensure being able to locate a potentially interesting source again in the future it may suffice simply to copy the first page, especially since journals are increasingly providing basic bibliographic data at the beginning of every article. In many cases the first page also yields an abstract (or summary) of the paper.

Fortunately, making photocopies in a library is usually both convenient and inexpensive, a development which, as we will see later, has among other things greatly reduced the need for traditional "interlibrary loans". On the other hand, the magazine Eco has lamented in this context an outbreak in recent years of a veritable "collecting fever", ostensibly a symptom of a "neocapitalism of information". It has certainly led to a considerable waste of paper, because much that is copied is never subjected to subsequent evaluation. Another negative outcome may be a psychological factor, with "collectors" coming to *believe* something worthwhile has been accomplished simply because they have acquired a new stack of paper to carry home. It has been suggested that this premature perception of "progress" might actually contribute to a *failure* to examine copied material in a serious way!

We pass over at this point the various legal issues associated with photocopying. Suffice it to say that libraries and other information suppliers are keenly aware of their responsibilities with respect to copyright protection, and they conscientiously attempt to ensure that copying is limited to situations permissible under the "fair-use" standard.

One needn't consult the most recent surveys conducted by library associations to recognize that the threats to our traditional systems of information acquisition and transfer are no longer (in 2003, as we bring these thoughts to paper) perceived as being so grave as once was feared. Technological developments have swept many of the most gloomy predictions aside. How often today, for example, does one take the trouble painstakingly to photocopy an entire article when the same information can be downloaded—perhaps without charge!—from the Internet? Scholars are also increasingly

---

[2] Our younger colleagues will surely be appalled to reflect on the fact that forty or so years ago every trip to the library meant gathering up a pad of paper and a few sharp pencils for a venture that could be counted on to produce a monumental case of "writer's cramp".

recognizing that until a genuine need develops for access to a complete research paper, an abstract can serve the purpose nicely, and abstracts can frequently be acquired quite easily through the Internet, or possibly over the *intranet* connection linking one's office to a library.

At one time there was a burst of enthusiasm for *scanning* articles from the literature. The idea was that one might in this way "borrow" a printed version of an apparently valuable document and subject it readily to the mystical powers of one's shiny new scanner—operated in concert with OCR software, of course—to obtain a *digital* document susceptible to full-text searching. Unfortunately, it soon became abundantly clear that this sort of procedure is both time-consuming and error-prone, and it typically fails altogether to render tables and key illustrative material in usable form.

While the importance of the primary literature can hardly be overstated, one should certainly not overlook secondary sources in one's quest for information. This applies especially with respect to work originating a decade or more ago.

● The older primary literature is often accessed far more easily through *review articles* and *monographs* than in any direct way.

The Organization of a Library

Library collections are usually arranged according to complex, open-ended numerical schemes (in the United States typically either the *Library of Congress system* or the *Dewey decimal system*—with the former becoming increasingly dominant), another topic we will forego pursuing in detail. If you have only limited familiarity with the structure of the specialized library you frequent, by all means ask the staff for a general tour in the interest of more efficient and effective utilization of the available resources.

In a large library it is not unusual for only part of the collection to be in an "open-stack" environment, so that occasionally a particular volume you wish to examine must be requested, and its retrieval may take some time. Books in this category are also frequently excluded from the "circulating collection", which means they must be consulted in the reading room (although photocopying of selected pages may be permitted). If a volume you need *does* belong to the circulating collection, but happens at the moment to be checked out, you can ordinarily reserve it for your personal use upon its return.

● Important *reference works* almost never circulate.

In most libraries recent issues (e.g., those from the current year) of *periodicals* are shelved in a location conveniently accessible to everyone, typically sorted according to discipline. *Bound* volumes of issues from the past few years are also usually within easy reach, whereas older volumes may be relegated to more obscure quarters.

Library professionals will be happy to acquaint you with options that exist for acquiring special materials available only from *other* libraries (including interlibrary loan, fax delivery of select pages, etc.), and of course warn you of costs that may be entailed. Selected aspects of "high-tech" approaches to long-distance information retrieval are dealt with toward the end of this section.

We spoke earlier of a useful supplement to periodic literature searches in the library: subscription to one or more of the current awareness services various information suppliers offer. A librarian may actually be in a position to assist you in pursuing this type of venture, since many libraries now support access to such services—and underwrite all or part of the associated cost. This novel approach to monitoring the literature represents an application of what has been called *push technology*. Thus, rather than your diverting the necessary time and ingenuity to "pulling down" potentially useful information from a large number of sites on the World Wide Web, for example, you arrange for an information service to "push" interesting material to *you* on a regular basis through your personal computer.

A vast array of information resources has also become available in the last few years in the form of CD-ROMs, making it possible to undertake a considerable amount of literature searching in an "off-line" mode.

We obviously cannot hope to provide here a meaningful overview of all the many information sources that now serve the needs of the various scientific disciplines, but we urge you in the strongest possible terms to become intimately acquainted on your own with the reference tools best suited to your particular interests. A useful place to start is the series of *Guides to Information Sources* published by K.G. Saur (Munich; a division of Gale/Thomson Learning). Volumes in this series carry titles of the form *Information Sources in the … Sciences*. Each has been carefully prepared by information specialists and other professionals from the relevant field. Disciplines so far covered (2003) include the earth sciences (WOOD, HARDY, and HARVEY 1989), chemistry (BOTTLE and ROWLAND 1993), and the life sciences (WYATT 1997).

Also of considerable interest, despite being somewhat dated, is the book *Information Sources in Science and Technology* (PARKER and TURLEY 1986). The World Wide Web in general is another important resource. Two highly informative Websites we particularly encourage you to visit are maintained by libraries at the University of Oregon (http://libweb.uoregon.edu/guides) and the University of Leicester in England (http://www.le.ac.uk/li/sources/index.html).

The Library of the Twenty-First Century

Much has transpired in the information world since the previous edition of this book appeared. We have watched the developments with fascination, and of course taken extensive advantage of them ourselves. Though well aware of how rapidly things change, we attempt nonetheless to provide here an update of prospects as we see them.

● The library of the future—on a university campus, for example—will to a great extent be a *virtual library*.

That is to say, it will have increasingly become a service facility providing *advice* and *guidance*, correspondingly less reflective of its original purpose as a *storehouse* for information. Even much of the information tomorrow's library does house will be far from evident. Already today a well-stocked university library includes in its collection hundreds of CD-ROMs, equivalent to the content of thousands of printed books, all available at a moment's notice—but in a form virtually invisible to the prospective reader. The information can be accessed directly only by the robot in command of a "jukebox", for display as requested on the interested patron's monitor. The "turntable" where "your" disk is spinning is likely to be well removed from any part of the library open to the public. A campus-wide or facility-wide *intranet* is employed to establish the required connection, putting you "online" and in touch with what may in fact technically be an "offline" source. The potential is quite remarkable![3]

The most striking aspect of a model network-based information service of this sort is that "everything seems to be tied to everything". In your quest for information you position yourself in front of a screen and then "click" your way merrily from one "Website" to another, from one CD-ROM to the next, from one text passage to a related one, via *links* (also called *hyperlinks*): in-text "nodes" highlighted in a special way on the visual display. The process is a bit reminiscent of "surfing the channels" on a television set. For a scientist, the underlying scheme makes perfect sense, and is in many ways reminiscent of the way literature research has always been conducted. After all, "library work" by its very nature is a matter of moving spontaneously from one source to another, guided by inspiration and intuition. The operation has simply been streamlined. Within the single digital reference source *Ullmann's Encyclopedia of Industrial Chemistry, Fifth Edition on CD-ROM*—one disk corresponding to a print edition comprising 36 volumes of ca. 600 pages each—there are 140 000 (!) of these hyperlinks.

Links in this sense have roughly the same function as the traditional "cross-references" that in printed sources lead one to relevant material in other articles or volumes, but with the profound difference that the incentive to pursue a hyperlink is much greater: the reader can do so instantly, and without stirring from his or her chair. It is important to emphasize that all the links alluded to in the *Ullmann's* case above, together with approximately ten thousand tables and thirty thousand figures—not to mention the surrounding text—exist within the confines of a *single* CD-ROM.[4] Graphics consume by far the most space, with text accounting for at most 15% of the stored data. That the "book of books", the Bible, fits easily on a single CD-ROM in company with a multitude of extracanonical works should be self-evident. One can

---

[3] It is no longer uncommon, incidentally, to find data-processing specialists as key members of a library's staff. These individuals may or may not have had traditional library-school training, but in all probability they will at least be holders of advanced degrees in one of the standard academic disciplines.

[4] One CD-ROM has a storage capacity of approximately 700 megabytes of data, corresponding to perhaps a hundred million words, and media with substantially greater data density are well on the way to becoming commonplace.

only speculate how much theological material has so far been committed to CD-ROM and microfilm in the famed Vatican libraries—although public access to much of it is undoubtedly still restricted.

All the writings and inscriptions that have come down to us from ancient Egypt, such as those found in the temple complex at Karnak, would easily fit in a tiny corner of one of these megastorage devices (assuming transcription were in the form of text rather than graphics). History tells us about a splendid library collection in ancient Alexandria, founded by the first Ptolemy, that eventually grew to encompass more than 700 000 items, including some 40 000 scrolls, representing in a comprehensive way the combined wisdom of Rome, Greece, India, and Egypt. On two of the blackest days in the history of mankind this entire treasure was tragically burned and largely lost forever. Ironically, a handful of CD-ROMs would suffice to keep such a catastrophe within tolerable limits.[5]

But enough of Amenhotep and Pliny and their successors, as well as *Ullmann's*! We must turn our thoughts back to the present and the general.

● Libraries are becoming smaller, but at the same time more effective.

We can expect that the university library of the future will continue to maintain at least a few journal subscriptions, as well as a modest assortment of books—especially basic textbooks—but volumes physically present, printed on paper and bound, will in coming decades constitute only the taproot of what might be characterized as a vigorous "tree of information". Those who use tomorrow's library will be able to orient themselves with ease electronically, and in all likelihood they will check out various resources on their own using the equivalent of barcode readers. Indeed, much of this vision is already reality in many places. If you are one who cherishes the mystical aura emanating from the oak-paneled walls and shelves of the classical library ("and from Erebose arose the spirits of many dead…"), and who mourns the thought of its passing, at least take consolation in the fact that in a facility of the future you will have at your fingertips the knowledge of the entire *world*—and probably be able to take advantage of it many more hours during the week than we have come to expect with conventional libraries.

Library consortia are increasingly engaging in compiling for reference purposes on both a regional and a national basis comprehensive, centralized lists of institutional book and journal holdings. Perhaps the most impressive example is the OCLC *WorldCat* (cf. footnote 12 under "Further Ramifications of the Digital Revolution" in Sec. 3.1.2). Such lists are available for consultation, generally online, at most libraries, and there

---

[5] It is comforting to know that appropriate precautions in this respect are now commonplace. In Germany, for example, there is a sort of national "storehouse of literature", maintained under military supervision in old abandoned mines. Many millions of rolls of film bearing texts from the Middle Ages to the present are here preserved and respected as one of the most sacred and valuable cultural inheritances of the nation. This too is a manifestation of "the literature".

is an expectation that public access to at least a significant part of WorldCat will soon be possible via the Internet.

Remote recourse to the literature—journals in particular—has developed dramatically in recent years. If your local library does not subscribe to some publication of interest, and it is still not available online, chances are good that you can solicit a copy of the relevant portion from some distant library whose collection includes it. A member of the staff there will retrieve what you want and photocopy or scan it for you, and a fax or e-mail copy might well be in your hands within minutes. At most a few days should elapse before the material reaches you by mail in a modern-day substitute for the traditional "interlibrary courier service".

A prerequisite to ordering any article or book is of course being both aware of its existence, and in a position unambiguously to *describe* it. Various *search engines* have been developed to help deal with the first problem: computer programs that in the absence of true bibliographic data attempt to locate material of potential interest on the basis of *keywords*. A basic search of the Internet using a generic search engine like those offered by *Google* or *Yahoo* may produce a few useful results (together with a great deal of chaff), but in scholarly research the careful application of a structured, broad-based electronic identification system of the type increasingly supported by libraries is likely to be far more productive. We mentioned above the OCLC's WorldCat resource, but this is actually only one part of a more comprehensive collection of tools the OCLC bundles under an umbrella known as *FirstSearch*. For example, the latter also supports *title searching* in the periodical literature, and it can even put at your disposal a powerful interlibrary loan mechanism. The FirstSearch route to information is exceptionally user-friendly. Despite its dependence on the Internet it requires no familiarity whatsoever with esoteric URLs or http addresses,[6] descriptors that in themselves provide very little insight into content, and tend moreover to be disconcertingly transient.

● The *World Wide Web* has in a very short period of time firmly established itself as an indispensable information resource—with an impact on research at least as profound as that of the latest generation of analytical instruments.

The Internet is in fact well on the way to being the most important key to locating information sources of every type. That the Web has assumed a central role in the lives of scientists should actually come as no surprise, since it was *developed* by

---

[6] "URL" stands for "Uniform Resource Locator", a specific form of "address" for characterizing information sources available through the Internet; "http", short for "HyperText Transfer Protocol", is an abbreviation familiar to anyone who has worked with the World Wide Web. In conjunction with "://" it acts as a sort of "prefix" to a URL. The "protocol" signified by the "p" in "http" is a set of rules governing communication between Web "servers" and "clients". This is perhaps an appropriate place for us also to mention the often cited "Z39.50 standards", regarded as the definitive guidelines for electronic referencing in general. Their origins and content are maddeningly complex, but for our purposes it will suffice to say that "Z39.50" can be understood as a synonym for ANSI/NISO Z39.50-1995, *Information retrieval (Z39.50) application service definition and protocol specification*.

scientists! Search tools currently employed in a network environment have become incredibly sophisticated, but—paradoxically, perhaps—also increasingly amenable to mastery by the individual researcher. This has in turn lifted some of the burden formerly shouldered by library personnel, freeing them to concentrate more on broader aspects of their primary mission, especially assisting researchers in perfecting *search strategies*.

9.2 Building One's Own Literature Collection

9.2.1 An Author Catalogue

Any reports you envision preparing in the future to describe your personal research efforts, irrespective of whether they will be published, will be expected to include exhaustive formal acknowledgement of all the external sources of information upon which you have relied.

● Only if you conscientiously *document* your reading as you go along will you be in a position to supply these essential acknowledgements.

We use the verb "document" here to cover the collection, organization, and description of written information of all types—especially publications (Sec. 3.1.1), but less formal reports (Sec. 1.2) as well—in such a way as to facilitate subsequent retrieval and unambiguous characterization of the material: essentially what was described earlier in the chapter as "literature management". Our concern is not with major national or international documentation systems or databases, but rather with private collections—an author's (specifically *your*) *personal* literature collection.

Setting up a private literature collection—which in essence means establishing one's own documentation resource—is analogous to an ardent Web surfer identifying a particular series of " favorites" among the myriad addresses that exist on the Web: addresses that have proven worth visiting occasionally for one reason or another. Such a " favorites list" (sometimes called a "bookmark list") requires a certain amount of routine management over time, but that is easily attended to with tools supplied in one's browser. The existence of a list of this sort greatly facilitates access to sites of personal interest. Similarly, the components of a private literature collection could be said to represent one's "favorites" among the countless documents extant within a particular discipline: cherished "addresses" to portals that open onto a highly personal world of information.

Looked at another way, one assembles and maintains such a collection as a kind of summary of developments in some specialty, and as a place to turn for background information, especially when the time comes to prepare one's own documents. Equally important, your personal collection will provide you with the data you need for identifying the various sources upon which you depend and have therefore obligated yourself properly to acknowledge.

● The form and content of a person's source documentation typically reflects the care that he or she devotes to professional work in general.

Documents an active researcher collects and keeps close at hand are in most cases drawn from the *primary literature*: that is to say, articles published in professional journals. *Books* are of course also useful, especially for the purpose of providing an overview, or deepening one's broader understanding. In addition they serve as important repositories of specific pieces of data.

● The act of documentation includes making careful note of all the elements necessary for ensuring that a particular "unit of information" (document) from the professional literature is *retrievable*: assembly of a definitive set of formal *descriptors*, perhaps garnished with a few incisive, well-chosen keywords.

"Documenting" entails recording data related to both place and subject: *localizing* and *identifying* documents, and *characterizing* their content. An important piece of terminology in this context is *bibliography* (from *gr.* βιβλιον, book), a noun most people associate with a list of books and articles an author has consulted, but a word that also encompasses "the history of books and other publications". Perhaps more important for our purposes, "bibliography" can be used to embody "the work of classifying and describing books".[7] *Bibliographic data* constitute the crucial document descriptors which make it possible in principle for anyone who so chooses to gain access to a particular document, at least to the extent that the document can be regarded as a part of the *literature* (*lat.* littera, letter). The selection and expression of such descriptors is an art in itself (see Sec. 9.3).

As we have seen, it has become virtually impossible to encapsulate in a meaningful way how one should in general go about the interrelated tasks of staying abreast of a particular field and keeping a record of relevant source information. Let us simply declare: "to each his own"! Some will adopt as a starting point glancing through recent issues of a journal or two, perhaps over the weekend, in a leisurely way, and saving formal documentation for later. Others proceed in the opposite direction: regularly collecting and cataloguing more or less extensive sets of bibliographic data from some productive source, but putting off looking at the published material until another time, "when the spirit moves". Quite apart from diverse personal traits of this sort there is another factor to take into account in today's world, a consideration almost impossible to ignore which makes it very difficult for us to give dogmatic advice:

---

[7] Curiously enough, a corresponding verb "to bibliograph" is essentially unknown in English, in striking contrast to comparable noun–verb pairs like photography/photograph, autograph/autograph, and even graph/graph, for example.

● Now that comprehensive literature databases for every discipline are available for consultation worldwide and "instantaneously" at any hour via the Web, much of the incentive for personal documentation has faded.

To engage or not to engage in a formal scheme of documentation on one's own—that is here the question. We have no idea how the majority of our colleagues currently feel, or will choose to orient themselves with respect to this particular dilemma, and to our knowledge there has been no systematic study of the issue. Clearly it would be *possible* now for a scientist successfully to conduct research and report the results without ever developing a personal documentation system. Seemingly everything one needs to know can in principle be called up on demand from the Internet. *Browsers*[8] attend to most of the logistics, and by dialing into any of a host of *Internet service providers* (ISPs) one quickly establishes intimate contact with the World Wide Web, able to take instant advantage of information offerings from countless servers scattered across the globe. The individual's role is that of an eager *client* impatient to take advantage of a wealth of available services, most of them free of charge.

And yet… Personally, we find it inconceivable that one would attempt to pursue a scientific research career in the absence of a personal literature collection. For one thing we are not at all enthusiastic about the prospect of repeatedly wasting time crawling around the Internet to pull down from afar information that should already be at hand (if not in the head!). In what follows we will therefore start from the assumption that you, our faithful reader, *will* in the end make an effort to create your own custom-tailored reference system. We take the time-honored (some would probably say "obsolete") *author catalogue* as our point of departure, though we will introduce it in a somewhat cursory way. In fact, many of the principles governing age-old documentation practices remain fully valid today, even if technology has given us new ways to implement them. Certain commercial computer programs, for example, clearly betray structural origins in the "classical" card catalogue.

One fundamental question that must be raised at some point is whether and when it makes sense to prepare and preserve *photocopies* of relevant documents. You will almost certainly not find it possible to assemble at your workplace a comprehensive set of original publications. Even if you did succeed there would still be the problem of devising an *organizational system* that would allow you to find specific items with ease when you need them. The thing we have been referring to all along as a scholar's literature collection is itself in essence just such an organizational system.

● What has traditionally been called a "literature collection" consists not of actual documents, but of a *card file* containing a unique *reference card* for each "collected" document: one document—one card.

---

[8] Wandering around in the Internet is regarded as analogous to "browsing" in an early sense of the word: "feeding or grazing on tender vegetation such as the shoots, leaves, or twigs of shrubs or trees".

The collection we have in mind might therefore more accurately be dubbed a *literature card file*, or (because of the way it is organized) an *author card file*. Files of this type have traditionally been based on standard 3″ × 5″ or 4″ × 6″ cards. Manually recording bibliographic data on everyday file cards could even today be a sensible enterprise for the graduate student concerned primarily with a particular thesis or dissertation topic. A doctoral candidate in the sciences typically reads a total of perhaps 500 documents in this context, 100 to 200 of which may ultimately end up being cited. Creating by hand a file-card collection of that magnitude should not be an overly daunting prospect, and it certainly represents the least "high-tech" (and least costly) approach to document organization.

On the other hand, there are several important benefits to be derived from instead taking advantage of a computer-supported database package (cf. Sec. 8.5.2). Not only is access to individual document records then much swifter and more convenient, but data from the various "cards"—which have been transformed into *digital data records* —can also be incorporated directly into one's dissertation or report as an integral part of preparing the document, with a few selective mouse clicks. It also proves to be extremely easy to isolate and examine the complete subset of records characterized by a particular keyword, for example, and the records will be "re-filed" automatically in their proper places when you are finished with them. Since it is now customary for every member of a research group to have computer access to a common store of information—typically from workstations kept constantly "on-line"—literal file cards and the special boxes designed to house them have begun to seem incredibly old-fashioned and anachronistic, and countless office cabinets dedicated to file-card drawers are seen rusting away in land fills.

In a conventional literature collection, one "card"—a data record—is created (as already noted) for each document of interest. The resulting file is often described as "containing so-and-so many *documents*". Obviously the documents themselves are not actually present within the file; indeed it would be surprising if they would all fit in a single *bookcase*! Nevertheless,

● Modern technology has made it possible for the first time in *fact* to incorporate full-text versions of documents into one's literature file.

The secret is capturing the material in digital form, either by scanning or through downloading from an online database. Dwindling storage space in one's bookshelves can thus be dismissed as a serious problem. We will return to this point later, for the moment turning our attention back to file cards and their properties.

● A single card would traditionally be expected to contain the following information related to a particular document, roughly in the order indicated:

   1. The name(s) of the author(s),
   2. The title of the document,
   3. Additional bibliographic data,

4. Insight into the document's content, and
5. A record of the document's availability,

This might be supplemented by special *localizers* of one's own devising, such as

6. A shelf number, and
7. A document number.

In designing and developing your file, remember that the primary objectives are to ensure that every document you consult in the course of your research can be easily relocated, in a targeted way, and that the file itself includes all the information required for preparing proper bibliographic citations.

● Important prerequisites to deriving maximum benefit from a literature card file are *comprehensiveness* and *accuracy* with respect to the incorporated records.

The *author field* (item 1 in the list above) serves to house not only one of the elements almost indispensable in document characterization, but also the key to organizing the file itself. The usual practice is to arrange all of one's literature cards in alphabetical order according to the principal *name* of the first author. Since for authors from the Western world a person's *last name* is what is most important, last names are entered on literature cards ahead of first names (or initials as the case may be). We recommend that this convention be observed with respect to *all* authors' names, not just those of first authors.

One cannot assume that the first author listed is the person actually responsible for submitting a document for publication. For this reason there is merit in including in your literature file *cross-reference cards* for all coauthors—some of whom may well in fact be principal authors (e.g., the director of the submitting research group). With edited works the author field is used for the name of an *editor*, and with patents that of an *inventor*. In the case of a document that cannot readily be attributed to any individual, the issuing organization, institution, or corporation is credited with authorship. Anonymous documents are filed simply by title.

● If a file contains multiple documents attributed to the same first author, the corresponding cards should be arranged in a sequence analogous to that used with a reference list based on the name–date system; i.e., more or less chronologically (for details see Sec. 9.3.3).

The *title* or *subject title* (our item 2) of a document—a journal article, for example—should always be entered in its *complete* form. Insofar as possible, titles of foreign-language materials should be cited in the original language. (We are not implying here that a New Zealander is expected to be adept at writing in Arabic; on the other hand, nothing should prevent you from transcribing a title expressed in Greek or Cyrillic characters, especially if you are "writing" with a PC.[9]) In the case of a document that

---

[9] If you have never had occasion to explore this capability with your PC, now might be a good time! Clearly one requirement is availability of a font or fonts containing the appropriate characters, but you

→

is a *translation*, the official translated title should be recorded, but the item's *original* title and first publication date should be noted as well (and with a book you might also want to add the name of the translator).

The *additional bibliographic data* field (item 3) should contain all the information one would require to locate and perhaps even obtain a particular document out of the countless millions of documents that have been published: i.e., data that serve to "localize" it. With a journal article, definitive identification of the journal itself, together with a page number and volume number, in principle should suffice. However, documentation specialists insist you also include under the category of routine bibliographic information the name of the author/editor and the complete subject title (cf. Sec. 9.5.1). We have assigned this field the designation "additional bibliographic information" to suggest that not *all* the information present would necessarily class as absolutely essential: a journal article *can* in fact be located simply on the basis of volume and page numbers without knowledge of the author or title. Proceeding from the opposite direction, however, it would be necessary to consult (notoriously fallible!) indexes in any attempt to locate a specific article from the author and title alone. Still, knowledge of the latter is always *useful*, especially when searching an electronic database.

● *Journal titles* are generally to be *abbreviated*, but this should be done only in officially approved ways. The best policy is to employ precisely the abbreviations you would find, for example, in a title list taken from a recent issue of an abstracting journal.

Journal title abbreviations are rigorously standardized, and based generally on an ISO recommendation (ISO 4:1997 *Information and documentation, rules for the abbreviation of title words and titles of publications*). By the way, for convenience sake it is advantageous to keep on hand a complete list of official abbreviations for the various journals you consult frequently, along with the corresponding library call numbers. This can save you a considerable amount of time if you often find it necessary to fill out library loan requests, for example.

● In the case of *books*, important bibliographic data in addition to author(s) and title include an edition number (when applicable), the year of publication, and the location and name of the publishing house.

Beyond this, we recommend that you record a book's *International Standard Book Number* (ISBN). Admittedly this is (not yet) considered part of a "standard" bibliographic entry, and is thus not always included in reference lists, but knowledge of an ISBN can greatly facilitate acquisition of a book, whether through interlibrary loan or purchase—as well as retrieval of book information from an electronic database.

● Another important piece of information with respect to documents of all types is a *page count* (or *page range*).

---

may be surprised to discover that you already *have* one or more of these, without your even realizing it!

For a journal article this means you should include not only the number of the page on which the contribution begins, but also that of the *last* page—in other words a page range. A formal description of a book in a source list almost always reports the total page count for that volume (e.g., "378 pp"), sometimes together with a page *reference* for the specific information of interest. We return to this subject later (in Sec. 9.3) in the context of citations.

The next proposed field, the *content field* (4), should offer insight into the *information* a document presents, beyond what is already obvious from the title. This is the field that typically consumes the most space, so some people refer to it as the *main field*. Even so, space limitations will force you to be concise and selective. Most of what you enter here will of course represent your own subjective impressions; indeed, the field might equally appropriately be entitled "personal observations". On the other hand you should not hesitate here to quote key passages verbatim—transcribing words that in no sense bear your personal stamp.

Be sure that any passage you copy directly from a document (i.e., word-for-word) is set in quotes. Accuracy is extremely important in this regard—just as it would be with numerical information or other concrete data you might wish to cite. The idea is after all to spare yourself the burden of later consulting the same source again to verify details. As noted earlier, an *electronic* literature file can in principle incorporate "full-text" versions of at least some documents.

Your plans related to the "content field" are likely to be the determining factor in your choice of a card size—or in the design of a "standard record" for an electronic card file.

● Extensive detail related to content, as well as lengthy *excerpts* and personal *evaluations*, might better be consigned to numbered and paginated *notebooks*.

This suggestion is especially applicable in the context of the "classical" literature file. If you should decide instead to work from the outset in digital mode, interpret this and the following observations accordingly.

Supplementary literature notebooks should be utilized as efficiently as possible to prevent them from becoming unwieldy. Cards (data records) in a master literature file should be conscientiously annotated in a prominent way to indicate that supplementary information in fact exists, and where it can be found (e.g., "L III, 17–21" for material on pages 17 to 21 of the third in a series of literature notebooks). Each notebook entry should begin with a concise bibliographic identification (or a *document number*; see below) to facilitate locating the corresponding record in the master file. The notebook concept is completely open-ended, and flexible enough to permit inclusion of multiple supplements for a given record, perhaps introduced at different times.

Some who maintain their records on file cards habitually use the cards as a place to paste photocopies of published abstracts. *Keywords* can be listed as well, either drawn directly from the published source or self-assigned. They will be of limited value, however, in most cases serving only to add a bit of extra information about content. The only way such keywords could be used for selective record retrieval is if you took the trouble to construct as well a separate "keywords file", cross-referenced to the master file. This particular shortcoming serves as a powerful argument for adopting digital technology.

The most comprehensive possible record of a document's content is of course the document itself. A set of printed full-text versions of your collection involves either a great deal of photocopying or asking all the various authors for reprints. A much more appealing alternative is somehow securing *digital* files of material you consider important. These considerations lead us quite naturally to the next information fields we describe.

Providing some indication of the state of *availability* (item 5) of a document is important because you may find occasion in the course of your work to consult it again. In the case of a book represented in your personal library, a notation like "office" or "study" would probably suffice. The same applies with respect to an article from a journal to which you subscribe. With a library book, a useful piece of information to record is the *library catalogue number*.

The situation becomes a bit more complicated in the case of journal articles you have in photocopy form. As a first step we suggest you consider creating another information field for what we will call document *shelf numbers* (item 6), a private analogue to the numbers used for pinpointing book locations in a library. These numbers can be useful in monitoring your personal collection not only of obvious items like books or sets of journals, but also compilations (see below) of journal articles. In effect you will be cataloguing the space set aside for your stock of sources. The resulting numbers should be inscribed not only on your file records, but also on the shelved items themselves, as well as on some sort of "signage" to mark your "stacks". The shelf-number field would of course be left blank on any card representing a document not actually a part of your physical collection.

Office-supply stores carry compartmented folders and record boxes that work well for shelf storage of groups of photocopies, although ring binders are a viable alternative. If you decide to use ring binders we suggest you obtain a supply of pre-punched transparent pockets to eliminate the need for punching holes in your documents. These allow you also to deal with loose pages and informal notes in your "supplementary notebooks" (see above).

Any time you remove a catalogued document from its designated shelf location—in the course of manuscript preparation, for example—you should leave behind an indication to that effect to prevent awkward "holes" from developing in the collection.

Finally, we suggest that you also assign a *document number* (item 7) to every item represented in your literature collection, irrespective of whether you actually have a copy of it or not. Enter this number not only in the appropriate field on the corresponding document card, but also on the document itself if you do have a copy.

Document (or *registry*) numbers can be viewed as an independent (and secondary) organizational tool. The numbers are assigned sequentially as documents are catalogued.

You could in principle decide also to *shelve* your collection according to document numbers, which would allow you to dispense with shelf numbers, because the combination of a document number and a generalized storage location would lead you directly to any item you wished to examine. The disadvantage is that the resulting shelf sequence is likely to be inconvenient for browsing.

Since your master card catalogue will be arranged alphabetically by authors, the main purpose of document numbers is to confer upon each document an element of unambiguous identification, not necessarily related to storage location. These numbers are in a sense analogous to the Social Security or passport numbers that help establish the identity of people, although *locating* a person requires access to other information, like an address or telephone number. Just as with Social Security numbers, administering a document-number system entails a certain amount of effort, although with electronic cataloging the burden is minimal.

- Document numbers can be advantageous in other less obvious ways as well—especially for *cross-referencing* purposes, or as temporary substitutes for full bibliographic information in the early stages of manuscript preparation.

A document number also provides a crude indication of when an item entered the collection.

Enough! It is time we now look at some of the ways computer technology can be exploited in the organization and bibliographic documentation of a literature collection.

9.2.2 The Computer-Supported Literature Collection

Management of a literature collection is a task ideally suited to computerization.

- Essentially any computer program offering database functions can serve as a starting point.

It makes no fundamental difference whether such a program is called upon to keep track of an inventory of articles offered for sale by a wholesaler—"widgets" in a range of sizes, styles, colors, and prices obtained from various suppliers and stored in several different warehouses—or the "articles" constituting the literature important to a research scientist. Any general-purpose inventory system can be adapted for use as an electronic author file. This file might in principle be set up in precisely the same way as the conventional card file described in the previous section.

- A standard *database program* puts at the user's disposal an unlimited, unconstrained supply of *data fields* for entering information of almost any type (Sec. 8.5.2).

Various *modes* can be assigned to the fields as required, including a flexible *text* mode and a *numerical* mode that allows field content to be subjected to arithmetic operations. Any field you define can be the object of a full-text search operation in terms of any specified sequence of characters. In a numeric field one can also conduct a search for numerical values within a designated range (e.g., all records characterized by "publication dates" after July, 1998, for example).[10]

● From a structural point of view, a literature "card" can be taken as the model for a *data record* in a database program, with each category of information assigned its own *field*.

The precise form such records take (i.e., the number and nature of the fields, the spatial design of the overall record, etc.) is left entirely up to the user, assuming you elect not to take advantage of commercial *literature management* software (see below), in which case certain parameters are more or less fixed from the outset. Careful thought should be given in advance to the way structural factors might influence the effectiveness with which your ultimate goals will be met.

● Implementing major changes in a database system already in use is at best an awkward and time-consuming chore.

In all likelihood you will want occasionally to print some subset of your data records—for distribution in conjunction with a seminar presentation, perhaps. For this reason the records should be structured in such a way that two, three, or more will fit conveniently on a standard $8\frac{1}{2}'' \times 11''$ sheet. Laying out such a record requires among other things assigning a certain amount of display space to each field, although the limits you set have no effect whatsoever on the true size of the corresponding data field. The latter is a parameter established when the field is first created. Thus, if on some record the content of a particular field exceeds the space provided in the output form, the "overflow" continues to be a part of the stored information (up to the maximum number of characters permitted in the context of the *field definition*), and subject, for example, to searching. Put another way, database programs are so constructed that borders displayed in a data form need not bear any relationship to true field boundaries. More flexibility is provided than might be apparent.

The seven information categories we proposed in the preceding section, augmented perhaps by another for keywords, would constitute a perfectly reasonable set of components for a digital literature database, and therefore a basis for constructing a template to govern the way stored information is displayed. For convenience sake one might label the resulting fields with easily interpreted acronyms, like AUT, TIT, BIB, etc. (for author, title, bibliography, contents, location, shelf number, document number, and keywords), producing a practical set of concise field names. The sequence in which fields are created is irrelevant, so even if "KEY" were created last it could be caused to appear first, at the top of the data template. The most important thing to ensure is

---

[10] This particular type of search could most easily be carried out with a "date" field laid out in the form "yyyy-mm-dd" (y year, m month, d day). Thus, 14 February 1998, or "1998-02-14", would be instantly recognized as *preceding* 10 March 1998, "1998-03-10" even though in a given month the 14th comes *after* the 10th.

that the *database program* itself is so designed—as most are—that all fields are open to searching.

If one so desires, database searches can be conducted on the basis of multiple characteristics (*search terms*), taking advantage of the logical operators AND, OR, and NOT. For example, the instruction

Search for "Author X" AND "since 1995" AND "keyword Y"

would lead to identification of all documents published after 1995 with X as one of the authors, but also an association with the keyword Y. The result should reflect nothing more and nothing less.

Database searches can also be carried out in a stepwise fashion. If, for example, a search of the author field leads to 17 papers by Walker JF, and you consider that to be too many hits, further sifting of this subset of records might reduce the list to only 4 papers by Walker JF published *after* 2001. In other words,

● If the number of hits from an initial search proves excessive, the list can be shortened most easily by searching again using a second—more limiting—criterion.

This sort of strategy is widely employed in the examination of commercial literature and information databases (cf. BARTH 1992), and it is just as applicable to a personal literature file maintained on a PC. Incidentally, most programs allow you to ascertain the *number* of hits prior to a list's being printed or even displayed on the screen.

● With a digital database it is well worth assigning *keywords* to all documents when the records are created.

Keyword characterization does have limitations, however. When the time comes to search the files you may no longer recall how you designated a particular topic—unless of course you have worked on the basis of a standard keyword list derived from some journal or commercial database, or a list of your own devising that you keep on file and update regularly. Relying on keywords recommended for a particular article by the editors of a journal can be problematic, since the terms selected may not have been drawn from a vocabulary list appropriate to your particular interests.[11]

A database program makes it easy to record keywords you assign. It is useful to designate the corresponding field (called KEY, perhaps) as a so-called "lookup" field, to which additions can always be made. An even more flexible approach is to set up a second table (database) for keywords only, and then couple this new table or file with the actual literature file in a relational way. This way there is nothing whatsoever to limit the number or nature of keywords (although their scope will of course be limited in some sense by the literature content). Whether you employ a lookup field or rela-

---

[11] To a librarian or documentation specialist, a master keyword list—"a structured collection of concepts and their (largely everyday) designations employed in a documentation context for indexing, storing, and retrieval purposes" (DIN 1463-1, 1987)—is known as a *thesaurus* (from *gr.* τηεσανροσ, treasure, in this case a "word treasure"). An important consideration in searching by keyword is making sure the search terms have been drawn from the appropriate thesaurus, which may of course be your own creation.

tional coupling your keyword list remains always subject to revision. The search process is facilitated by taking the time to refer in advance to the list of possible terms before composing a search query. In a typical case you might undertake to locate all the documents in your collection associated with the technique "electrophoresis" (a designated keyword), further refined by taking into account the chemical class "nucleotides" in either an "and" or an "or" sense (where "and" would signify that *both* terms must apply for a document to class as a hit whereas "or" indicates that only one need be found to create a match, and it is irrelevant which).

● Electronic databases can be sorted almost instantaneously, and in a multitude of ways, as often as you wish.

Disagreements that once raged about whether document collections should be organized by author or by subject are thus rendered meaningless. With a simple mouse operation the information comprising a database can be arranged with reference to any field you wish to designate, and in either "ascending" or "descending" order (e.g., alphabetically by author or subject in the former mode, or by date in the latter, with the most recent record first).

In some situations it is useful to look in the records for very specific word *fragments*. For example, you might wish to locate all the recorded documents that contain in their titles or summaries the alphabetic string "ionoph", and thus have something to do with *ionophoresis*, *ionophoric* species in general, or *ionophoretograms*, none of which has been designated as a keyword.[12]

Special programs designed explicitly for literature management place at the user's disposal several additional very useful features that involve a certain amount of documentary and bibliographic "intelligence". Thus, they are generally equipped automatically to incorporate select data from external databases, and to forge a powerful link between source data and a word-processing application.

● Such a linkage allows the program to analyze an electronic manuscript containing citations based on *document numbers*, isolate in each case the appropriate bibliographic record, *arrange* all the records in the proper way, extract the relevant bibliographic data, compile it so that it matches the preferred reference style, "append" the resulting list to the evolving manuscript, and finally introduce citations of the *correct* type in place of the document numbers originally present!

In the end you will be left with a fully-referenced document in which everything is consistent with the whims of a particular journal editor, for example, or with one of the "standard" bibliographic styles (as discussed in Sec. 9.4.2). This may, incidentally, entail the selective use of several type styles and appropriate adaptation of certain common expressions (e.g., conversion of the German abbreviation "Hrsg." into the English equivalent "ed.").

---

[12] A word of caution is in order here: not every database program offers the option of searching in terms of character strings *within* words, a feature taken for granted with WORD, for example.

Software of this type is subject to surprisingly few limitations, and any you do encounter may well be attributable at least in part to your *computer*.

● As noted above, special-purpose programs like these facilitate the downloading of selected bibliographic information from an external database into one's personal literature files.

Often such an external database is accessed via the Internet, but major abstracting services like *Chemical Abstracts Service*, *MEDLINE* (Medical Literature Analysis and Retrieval System Online), *BIOSIS* (which includes, among other resources, *Biological Abstracts*), *INSPEC* (a large bibliographic database covering the fields of physics, electronics, etc., and maintained by the Institute of Electrical Engineers, IEE) are increasingly making supplements to their databases available in CD-ROM form, which lend themselves equally well to "mining" with the aid of the special interfaces that accompany literature-management software.

The capabilities, requirements, strengths, and weaknesses of such programs (including ENDNOTE, PROCITE, and REFERENCEMANAGER, all of which are distributed by Thomson ISI ResearchSoft, and VCH BIBLIO from Wiley–VCH) vary widely, and a choice should be made on the basis of your own particular needs and expectations, taking into account advice from colleagues and/or informed specialists.[13]

9.3 Citation Techniques

9.3.1 Citing and Citations

In Part I of the book (e.g., in Sections 1.2 and 2.2.11) we make repeated reference to the crucial role played by citation in scientific work. We turn our attention now to formal and technical issues raised by this need for formal documentation, introducing in the process a certain amount of technical terminology.

The word "citation" can be employed in two quite different ways: it may refer to a *statement* derived from a specific literature source, but it can also represent the *definition* of a source. In the first case one would "cite" an author, using his or her own words, whereas in the second one simply provides a precise description of the origin of some idea or piece of information. The distinction is essentially that between a *quotation* and a *reference*.

● A direct quote should always be set in quotation marks,[14] and the author of the words must be clearly identified.

---

[13] Comparative data for the three Thomson-ISI ResearchSoft packages are available at http://thomsonisiresearchsoft.com/compare/.

[14] There is one important exception to this rule: quotation marks are generally omitted (i.e., *assumed*) with *block quotations*, identifiable by their uniform indentation.

Direct quotes are actually rather rare in scientific documents, in stark contrast to the humanities and legal studies, where the exact words employed in a source often are crucial. In the sciences, attention tends to be directed instead toward what are regarded as *facts* and *data*, which need not be expressed in precisely the words selected by their originator. It is quite sufficient simply to acknowledge that a particular number has, for example, been obtained from some other document, and then produce a detailed description of that source. No purpose whatsoever would be served by setting the number in quotes.

If there ever *should* be need in a paper to establish a link between a literal quotation and a published source it might take a form like[15]

> … We were unable to confirm the conclusion by Meier et al. [23] that "… this reaction takes place only in the presence of traces of heavy metal".

In the example cited, the number in brackets preceding the quotation is obviously not in itself a source description, but a substitute for such a description: a "reference"— more specifically, an "in-text reference". A clarification in the international standard ISO 690-1987 *Documentation—Bibliographic references—Content, form and structure* characterizes such a reference as an abbreviated version of a source citation, introduced into the text parenthetically.

● Source notations (in-text references) typically assume one of two widely accepted forms: *name–date* or *numerical*.

In the first of these approaches, a combination of one or more authors' last names and a year of publication acts as an in-text substitute for a complete source identification, whereas in the other system a comparable role is played by a reference number. (In the event only a very few works are cited, name–date references are sometimes truncated to names alone.) The two sections that follow are devoted to detailed descriptions of these two systems. For the moment we restrict our attention to the question of where and how the corresponding *complete* source information should appear.

● Comprehensive bibliographic data should be provided for the reader either in footnotes or through a free-standing reference list.

Because of the technical problems associated with footnotes (cf. Sec. 3.4.5) it has become customary to follow the second course: compilation of an independent reference list. The list is then generally placed at the end of the document (e.g., a journal article or a report), although in the case of a very lengthy work, like a book, partial reference lists might accompany the individual chapters.

---

[15] The proper English-language placement of punctuation in conjunction with quotation marks is a matter of some controversy. A period, for example, might be set either *before* or *after* a close-quotes symbol. We have elected throughout the book to adhere to the latter convention, except when the period is *part* of the quote.

A reference list is frequently presented without any formal heading, but set in smaller type as a way of distinguishing it visually. One might be tempted to provide a simple title like "References", but the situation becomes more complicated if numbers are used in the text not only as literature citations but also to direct the reader's attention to supplementary information. This would at the very least require the title to be expanded, perhaps to "References and Notes". In the case of a manuscript intended for submission to a journal, guidelines from the publication in question should be consulted to see, for example, if a mix of references and additional commentary is even permitted.

Sometimes even with a list containing *only* bibliographic information an explanatory heading is advantageous as a way of announcing the extent and goals of that particular compilation. An example would be something like "Literature Consulted" for a list in conjunction with a seminar paper, or "English-Language Publications Related to the XY Problem" for an unusually selective list. Especially with a lengthy document there may be reason to provide a list of sources that have no intimate relationship with *specific* passages in the text. Source lists like these are especially common in textbooks, with titles like "Additional Sources" or "Suggested Reading".

DIN 1505-2 (1984) offers a useful stylistic recommendation:

● If in a reference list an entry exceeds a single line, continuation lines should be indented by a few spaces.

Reference lists are expected to be accurate not only with respect to the information they contain, but also regarding the significance of cited sources. Unless otherwise stated, a "source" is understood to be the *original* source, not a document that—like yours—is acknowledging someone *else's* work. Moreover,

● The reader should be able to assume that you, as author of a particular document, have actually *examined* every "source" you include in your reference list.

As trivial as this injunction may sound, it is violated altogether too frequently. One common "justification" for failing personally to consult a reported source is that the referenced information appeared originally in some obscure place, and you know about it only through *secondary* literature (cf. Sec. 3.1.4). Or perhaps the original paper is written in a language with which you are unfamiliar. We are not suggesting you should refrain from providing proper acknowledgements, just that you be *honest* and admit you obtained the information in question "second hand", and cannot therefore testify directly to its accuracy. Perhaps your frankness will give one of your readers the incentive to unearth the original report! Appropriate messages can be conveyed by notations like

> ...; as cited in [3]
> ...; cited by Brown (1989)
> ...; Chem. Abstr. 1966, 64:4953 a

Finally,

● No attempt should be made to "document" *unpublished* results.

The reason for such a categorical statement is that, among other things, a reference list is intended to provide the reader with the tools necessary to verify the accuracy of assertions regarding external sources, and no verification is possible if the information in question has never been made public.

9.3.2 The Numerical System

Although citation by number was alluded to second in the preceding section, it is actually the approach most commonly employed.

● In this system, sources are numbered sequentially in the order of their citation in the text. The assigned number then acts as an abbreviated form of a full bibliographic reference in the text itself, introduced into the narrative as a sort of "place-holder" for "the real thing".

Citation numbers (reference numbers) can be presented in text in any of three ways: as superscripts, in parentheses on the text baseline, or in square brackets; e.g.,

$$\ldots \text{text,}^7 \ldots \qquad \ldots \text{text,}^{7,8} \ldots \qquad \text{text,}^{4,7-10} \ldots$$
$$\ldots \text{text (7), } \ldots \qquad \ldots \text{text (7, 8), } \ldots \qquad \text{text (4,7-10), } \ldots$$
$$\ldots \text{text [7], } \ldots \qquad \ldots \text{text [7, 8], } \ldots \qquad \text{text [4,7-10], } \ldots$$

In the event you do not have convenient access to square-bracket characters with your typewriter or word processor, a slash (solidus) is an acceptable substitute; e.g., / 3 /.

● Superscript citation numbers should be set *after* any punctuation marks present. Parentheses or square brackets on the other hand *precede* punctuation, separated from the actual text by a single space.

The only exception to this rule involves a superscript in conjunction with an "em-dash", signifying a pause. In this case the superscript comes first.

● A citation number should be introduced at the point in the text most closely associated with the published information.

That is to say, you should establish as clear a connection as you can between a bibliographic reference and the word (or series of words) calling for it. It is not necessary, by the way, that citations always be introduced at the end of a sentence; e.g.,

$$\ldots \text{ as has recently}^6 \text{ been shown.}$$
$$\ldots \text{ is apparent from spin-coupling experiments}^7 \text{ as well as } \ldots$$
$$\ldots \text{ was first demonstrated by Smith and Miller}^8 \text{ in the context of } \ldots$$

The last example illustrates that use of the numerical system in no way precludes the possibility of mentioning authors' *names* (which are essential components of the "name–date" citations described in the section that follows).

● Citation numbers in parentheses or brackets on the text line have the advantage that they can easily be annotated as a way of *localizing* the relevant passage in the source document.

It is clearly a service to your readers when you assist them in finding material of interest in a long source document; e.g.,

> … in a recently published review [4, p. 512] …
> … described in the literature (e.g., [2], Sec. 3.4) …

Localizing information could of course be provided instead in conjunction with the complete bibliographic reference: after the formal source description and separated from it by some punctuation mark. DIN 1505-2 recommends a comma for this purpose, typically followed (in an English-language publication) by "p." for "page". The Vancouver Convention (see Sec. 9.4.2) specifies use not of a comma, but rather a colon. Of the two general approaches to providing localization, the one involving information set directly in the text is preferable, especially if a given source is cited more than once, in multiple contexts.

Handling footnotes (or endnotes) is facilitated dramatically by the *automatic footnote management* feature found in most word processors, since this eliminates the need for cumbersome format adjustments. The once typographically awkward introduction of *superscripts* is no longer a matter of concern. Nevertheless, there is a definite trend in recent years toward citation numbers set in parentheses (or brackets) on the text line.

Assigning citation numbers in ascending order as dictated by the text (*sequential numbering*) has one implication that can be a source of problems. Should it become necessary in the course of manuscript development to revise the text in such a way that the sequence of notes changes, all subsequent numbers will need to be altered. With a long document this can—or could, in the past—generate quite a bit of busywork, and the chances are or were great for errors to creep in, with certain citation numbers no longer leading to the correct bibliographic information. For this reason many authors turned to temporary use for citation purposes of *document* or *registry numbers* borrowed from their personal literature files, as described near the end of Sec. 9.2.1, only introducing sequential citation numbers after the manuscript had taken its final form.

● The risk of errors in citation numbers is virtually eliminated by use of an automatic footnote-management system like the one in WORD.

This in itself is a powerful argument for turning to word-processing software, thereby eliminating the need for "provisional numbering": reference *symbols* and reference *text* are firmly linked from the time citations are introduced, and all the bother of maintaining a proper numerical sequence devolves upon software.

Literature citations required in *tables* or *figures* would seem to pose a problem, in that during document development one can never be sure precisely where the element in question will ultimately appear. Too much depends on last-minute page-makeup considerations. But placement is normally the determining factor with respect to how reference numbers are assigned. For this reason it is customary to regard such citations as occurring at the spot where the corresponding table or figure is *referred to* for the first time.

● In long documents, especially books, it is best to assign citation numbers *chapterwise* rather than as a continuous series extending throughout the manuscript.

As a result, the first citation number in every chapter will be "1". This expedient of course greatly reduces the seriousness of problems caused by a need to renumber a set of citations. It also makes it most unlikely you will find yourself dealing with cumbersome three- or four-digit numbers. Numbering by chapter does not, incidentally, rule out the possibility of incorporating references into a comprehensive list at the end of the book, so long as the latter is clearly marked with chapter identifications. In principle an alternative might be to utilize "double citation numbers", analogous to the numbers employed for figures and tables, or (as suggested previously) to append partial reference lists to each chapter. The former solution is extremely awkward, however, and the latter makes it more difficult for a reader to locate a reference of interest.

9.3.3 The Name–Date System

The name–date system (called the "first-element-and-date-method" in ISO 690, 1987) has its roots in the humanities, where "authority" plays an extremely important role (i.e., readers feel it is crucial to know who is responsible for the assertion in question). Adoption of this system is becoming increasingly widespread in science as well, however, especially in the biological sciences.

● A source citation (abbreviated reference) under the name–date system normally consists of one or more author names together with a year of publication.

The citation begins with a name (or names). There is considerable latitude with respect to the style of such a citation, as illustrated by the following examples, the first of which might be described as containing *integrated* abbreviated references:

> Schmidt (1991) and Peterson (1993) report that ...
> Recent in-vitro studies (Smith and Johnson 1998) show that ...
> As is well known (Albert 1995, p. 100), ...
> In fact, the correct value is roughly 15% greater (Klugman et al. 1998).

Ordinarily only a family name is provided unless another author with the same family name is also cited somewhere in the document (requiring that a distinction be made between Smith, J., and Smith, R.S., for instance). As the last example above demonstrates, in the case of multiple authors—certainly more than three, but in many journals more than two—the abbreviation "et al." (*lat.*, et alii, and others) is introduced as a substitute for all names except that of the *first* author. It was once common to employ the phrase "and coworkers" instead, but that is rarely done today in an era when there is a reluctance to emphasize hierarchical status. For the same reason, the name of the *senior scientist* in a collaboration (a thesis adviser, group leader, or chief investigator) may thus appear first, or last, or as part of an alphabetized sequence, indistinguishable from the names of junior colleagues.

The third example above shows one way that an observation in a cited work can be "localized".

"Date" in the name–date system refers to the year of publication. If the same author or set of authors is cited in connection with multiple works published in the same year, lower-case letters are appended to the year, assigned so as to reflect the order of *citation*; e.g.,

> It was previously demonstrated both kinetically (Beckman 1985a) and spectroscopically (Beckman 1985b) that …

With the (sequential) numerical system there is of course never a need to specify an order of citation—that order is apparent from the numbers themselves, and is also reflected in the order of entries in the reference list. Things become more complicated with the name–date system. The organizational principles described below are based on the recommendations of the Commission of Editors of Biochemical Journals, issued under the auspices of the International Union of Biochemistry (IUB) and presented in detail in *Biochemical Journal* 1973: 135. They have actually been around much longer, however, in the guise of the so-called Harvard System.

● Under this system, a bibliographic reference list is arranged in an order partly *alphabetical* and partly *chronological*.

Highest priority with respect to sequencing is conferred upon the name of the first author. If that were the end of it, the list would be put together in the same way as a telephone directory. Complications soon set in, however, because of the problem of coauthorship and multiple authorship in general. If a particular author is represented in a reference list both alone and in conjunction with others, then the publications listed first are the ones attributable to that author only, set in chronological order from earliest to most recent. Next come publications involving *one* additional author, whereby the second authors' names are used to establish an alphabetical order. If names alone are insufficient to establish a proper order unambiguously, publication dates are once again taken into account, with earliest publications first. Contributions by some first author in conjunction with *more than one coauthor* (or in some cases more than *two*; i.e., papers with two or at most three authors)—e.g., "Jones, et al."—are arranged strictly chronologically. Letters "a", "b", "c" … are attached to publication dates as necessary to distinguish entries sharing common names *and* years of publication.

● In a book or paper with citations based on the name–date system, the year of publication is so important that it should always be reported immediately after the author name(s) in the corresponding reference list.

This particular rule is necessary because in many cases the publication date is what distinguishes the reference of interest from others. Unfortunately, some journal guidelines do not permit early date placement, insisting that the year be included as a part of the "imprint section" (see Sec. 9.5.1) near the end of a reference, a nonsensical demand in the context of name–date citation.

● Western-style names are always inverted in the name–date system to permit the sequence-determining family name to appear first.

Proceeding otherwise might make sense in numerical-system references, but alphabetization according to name plays such an important role in the name–date system that name recognition must be given a very high priority.

The rules specify that (largely historical) indications of nobility, like "von" or "de", along with other name *prefixes*, are to be ignored for alphabetization purposes, so they are set *after* the family name, together with the person's initials (e.g., "Hoikstra, C. de"). "Mc", on the other hand, is considered an integral part of a name, to be alphabetized as if it were spelled "Mac". In certain cultures, family names as a matter of course *precede* given names, so that an inversion like that in "Schmidt, John" would be clearly inappropriate. In case of doubt it is best to follow the practice of an appropriate abstracting source.

DIN 5007 (1991) *Ordnen von Schriftzeichenfolgen* ("Arrangement of Sequences of Alphabetic Symbols") recognizes as legitimate two different approaches to alphabetization involving umlauts (e.g., "ä"). For most purposes ("general usage"; e.g., indexes, dictionaries) umlauts are simply ignored. In a *name* list, on the other hand, a letter with an umlaut is treated as if it were followed by an (invisible) "e" (i.e., "ä", "ö", and "ü" are considered equivalent to "ae", "oe", and "ue"). In English-language publications, including *Chemical Abstracts*, an "ä" may actually be *written* as "ae". The author "Bäder, J." thus becomes "Baeder, J." in the English scientific literature.

The following examples illustrate how these various rules contribute to the structure of a properly organized name–date reference list:[16]

> Schmidt, J. (1995)
> Schmidt, W. (1989)
> Schmitt, H.-P. (1994)
> Schmitt, H-P., Hinz, A. (1995)

[16] It is becoming increasingly common to omit the comma that ordinarily precedes a set of "inverted" initials, and the same is true of the periods associated with initials. Indeed, often there would not even be a *space* left between a pair of initials.

Schmitt, H.-P., Kunz, P. (1993)
Schmitt, H.-P., Kunz, P. (1996)
Schmitt, H.-P., Kunz, P., Hinz, A. (1990)
Schmitt, H.-P., Hinz, A., Fischer B. (1996a)
Schmitt, H.-P., Albert, K., Fischer B. (1996b)
Schmitt, H.-P., et al. (1997)
Schmitt, H.-P., et al. (1998)

The order of the entries with dates "1996a" and "1996b" implies that Schmitt's work with Hinz and Fischer has been cited in the text *before* that with Albert and Fischer.

In the case of an edited work, the *editor's* name becomes the "first element" in place of an author's name, accompanied by the notation "(ed.)"; e.g.,

Smith, J. (ed.) (1992)

If no person's name is available for use in a citation, as with a document issued by a government agency, alphabetical placement is determined by the name of the issuing entity.

9.3.4 *A Comparison of the Two Systems*

Clearly there must be advantages associated with both citation systems, otherwise one of the two would have long since been abandoned.

● The main advantage of the numerical system is the *brevity* of citations, but citation according to the name–date system clearly is more *informative*.

Depending on the circumstances, either of the two systems might be perceived as preferable. A journal specializing in review articles is more likely to opt for the numerical system, simply because the presence of a great many name–date citations leads to a significant inflation of the text. On the other hand, the editor of a journal in which contributions tend to reference relatively few other works will probably lean toward the name–date system. The same considerations apply with respect to books.

For the *reader*, name–date citations offer the advantage of making it immediately obvious how recently a particular reference appeared. Often a paper's authorship also allows one to draw certain inferences about the probable significance of that piece of work even before the complete set of bibliographic information is examined in detail. In a sense this restores some of the benefits lost when full references were demoted from footnotes to reference lists. Name–date citations can be more disruptive, however, as the following example demonstrates:

As shown by IR-[8], UV-[9], and NMR-spectroscopic[10] studies, …

As shown by IR- (Arniaud 1992), UV- (Miller 1989), and
NMR-spectroscopic (Rossini 1996) studies, …

One other fundamental disadvantage of the name–date system is the absence of any way to "work backward" from a reference to the passage calling for it. With sources listed alphabetically by author rather than in the order of their citation it becomes exceedingly difficult to locate the citation associated with a particular bibliographic entry, useful as a way of ascertaining its intellectual context. It could at the same time be argued that this drawback is offset somewhat by the ease with which one can determine the number of cited works by a particular author.

Many choose for convenience always to employ name–date citations during the *preparation* of a manuscript as an alternative to the temporary use of document numbers suggested near the end of Sections 9.2.1 and 9.3.2, transforming these into numerical citations later if that proves necessary.

We note in passing that some journals adhere to yet a third reference system. Here one begins by *first* compiling the reference list, organized as it would be under the name–date system (alphabetically/chronologically). The list is then numbered sequentially, and *these* numbers are used for citation purposes. We regard such a "mixed" system—which unfortunately is allowed under both the Vancouver and CBE guidelines (see Sec. 9.4.2)—as (to put it mildly) an "unhappy" choice, since it retains disadvantages associated with both of the more common systems (numerical and name–date) without offering the major benefits of either. Thus the problems entailed in "alphabetization" of a reference list remain, and a specific citation still cannot be found easily in the text. Moreover, introduction of the proper citations themselves must be the *last* step.

Finally, we note one other strategy sometimes adopted: insertion of full reference data directly into the running text. An example of this practice is provided by the widely read "trend reports" (annual reviews) published in the journal *Nachrichten aus der Chemie* in the years prior to 1995. (A change has since been made to numerical citations.) The problems posed by indirect citation are thus eliminated completely, but at the expense of a significant increase in space consumption.

9.4 The Form of a Citation

9.4.1 Standards of Quality

There are a number of criteria that a citation must fulfill if the intended purpose is to be served, and these in turn place severe constraints on citation practice.

● A satisfactory source description consists of a number of *bibliographic components*.

The *choice* of components determines the richness of a source description, and their *sequence* and *assembly* its form.

● A source description must be *intelligible*.

This requirement alone encourages a considerable amount of *standardization* so that readers will not find a reference list confusing with respect to what is meant or where specific information should be sought.

● A source description must be *unambiguous*.

For example, it should be absolutely clear whether a particular source is an *authored* or an *edited* book, and indeed if it *is* a book as opposed to a chapter *in* a book. Similarly: is reference being made to a *contribution* to a conference, or to a volume of *conference proceedings*?

● A source description must be *complete*.

"Cambridge" would not be satisfactory for the place of publication, because one must specify either (UK) or (Massachusetts, alternatively MA) to indicate which of two different cities is meant. As another example, if a book has been released in multiple editions, page-number information is useless in the absence of an identification of the edition.

The question is often raised whether it is necessary to include a *title* in a reference to a journal article. Many would regard a citation lacking an article title as incomplete— and understandably so. The discussion accompanying DIN 1505-2 (1984) includes the following observation:

> Above and beyond unambiguous identification, one should give attention in preparing a citation to information a reader is likely to find necessary or at least interesting (perhaps even information of a negative character), and which might help clarify whether or not that particular source should be consulted directly. This is one reason for always including the subject title of a cited work, an expectation almost always fulfilled today in the humanities and the fields of medicine and engineering, but rarely in the natural sciences.

In our judgement scientists should accept this remark as a formal challenge!

● Source descriptions should be kept as *concise* as possible.

Economic constraints and a desire to prevent publications from becoming unnecessarily bulky are sufficient grounds in themselves for this injunction. The development of internationally sanctioned *journal-title abbreviations* has contributed significantly to the goal of brevity in citations. Additional support comes from the trends toward avoiding both conjunctions and excessive punctuation, and abbreviation of the names of publishing houses.

On the other hand, the quest for brevity must not be allowed to stand in the way of clarity, freedom from ambiguity, or completeness. In the end the most important consideration of all is that every cited document be *identifiable*, to the point that it can easily be ordered through a bookstore, examined in—or borrowed from—a library, requested from an issuing agency or corporation, or accessed through a database.

There is one more general imperative:

● A source description must be *appropriate* to the situation.

It makes no sense to provide a reference to a lengthy source in its entirety when only one particular statement from that work is of interest, because despite the citation a reader would find it virtually impossible to locate the passage in question (assuming the source is not available in digital form). Depending on the circumstances it might be necessary to supply a page number, or even the number of a particular figure on a page. In some cases the relationship between source and text might be such that specification of a relevant chapter will suffice. What one should strive for is the appropriate "depth of citation" for achieving sufficient *localization* (cf. Sec. 9.3.2).

This concern with "appropriateness" implies, incidentally, that source descriptions need not everywhere and for every purpose assume precisely the same form. Indeed, while certain elements of a source description can be regarded as mandatory, others are clearly optional (although they might be required by a particular journal editor!). The various bibliographic elements are in fact often assigned to three categories: *obligatory*, *recommended*, and *supplementary*.

In what follows we will encounter a few additional concepts related directly to libraries and documentation.

● Physically or bibliographically self-contained documents are described as *independent publications*, whereas subunits they may incorporate are *dependent publications*.

The adjective "independent" is meant to be taken quite literally. This status is conferred upon anything that—on a library shelf, for example—is capable of *standing alone*. Independent publications ("bibliographic units that appear separately" in the—translated—words of DIN 1505-2, 1984) comprise in the first place books, patents, standards, and miscellaneous documents issued by agencies or corporate entities. Dissertations, research reports, and similar materials can be included as well provided the concept "publication" is subject to a broad interpretation. A journal would also class as an independent publication: more specifically, a *volume* of a journal, typically consisting of several *issues* that have been *bound* for convenience.

Dependent publications, on the other hand, include such things as articles published *within* journals, or the individual contributions or chapters found *in* books.

● Dependent publications by definition fall within the confines of independent publications.

For purposes of visual differentiation, the titles of independent publications are commonly set in *italics* when they appear in printed works, whereas those of dependent publications would generally be enclosed in quotation marks. This practice is followed in DIN 1505-2 (1984), for example, although the standard itself does not explicitly address the point.

The use of special type in source descriptions is becoming less and less common. This applies not only to italics and boldface, but also to the *small caps* that in high-quality publications were once commonly reserved for authors' names. The development is one that can almost certainly be attributed in large measure to cost-cutting efforts. Without question it represents a loss from a typographic point of view, and it results in diminished *legibility* of references as well. The standard ANSI Z39.29-1977 *American National Standard for bibliographic references* explicitly recommends uniformity in type style "for the sake of simplicity and convenience", though it does not rule out use of special type for calling attention to specific elements.

In this book we have chosen to respect the tradition of italicizing titles of all independent publications, although we have not extended this practice to the examples in Appendix A, since these are intended to reflect official recommendations. We have also adhered to the custom of conferring a bit of dignity on personal family names by setting them, in running text, in small caps.

Dependent publications essentially function as "guests" of "host" independent publications. The concept of "bibliographic level" can be expanded to reveal a somewhat broader spectrum extending from a *collective work*, through a *volume* in a collective work and a *chapter* in a volume, to a *page* in a chapter.

The English language encompasses adjectives for distinguishing in a slightly different way between publications at three hierarchical levels: "collective", "monographic", and "analytic".

- "Appropriate citation" entails characterizing a document in terms of all the applicable hierarchical levels from the highest down to the one of immediate interest.

9.4.2 Standardization in Citation Practice

Background

Unfortunately, there is still no satisfactory consensus regarding how a document "should" be characterized. Authors regularly discover to their dismay that a reference list carefully prepared to meet the specifications of journal A proves to be unacceptable when the paper is instead submitted to journal B. This has long been a source of aggravation, and editors and scientific publishers have been charged with shortsightedness, insensitivity, and even incompetence due to their persistent failure to put this problem behind them.

- There actually are a number of reasons for the lack of uniformity in citation practice.

Much of the problem has its roots in the various scholarly disciplines themselves. Each major domain is associated with its own habits of thought and ways of working, and this is reflected in their approaches to bibliographic documentation and citation. Conventions that are extremely convenient and useful in the area of legal studies, for example, may be totally unsuited to scholarly work in the life sciences. It could arguably do considerable harm to prescribe a *single* bibliographic standard for all disciplines. Further complicating the situation is the diversity and international character of the academic literature.

Another factor of recent origin that cannot be ignored is *information technology* (IT), and the necessity that digital information sources be taken into account. Perhaps this will ultimately prove the key to at last developing an effective set of true bibliographic standards, since the computer world is notorious for its ability to transcend disciplinary and national barriers.

● In order to win broad acceptance, a citation standard must be both comprehensive and universal in its applicability.

Certainly one obstacle to progress has been the number of institutions and interest groups anxious to see their own causes furthered, with no single dominant force. Apart from publishing houses, others with an important stake in the outcome include scientific societies and various "standards committees" operating at both national and international levels. Standards organizations see themselves as the guardians of uniform practice of all kinds, and they have not been hesitant to enter the bibliographic fray. A number of national "citation standards" have already been adopted (e.g., DIN 1505-2, 1984; ANSI Z39.29-1977; BS-5261: Part 1: 1975), albeit with relatively little impact, and sometimes evidence of a disappointing grasp of the underlying issues. Not surprisingly there is already an "international standard" as well, ISO 690-1987, which in fact includes a number of very sensible suggestions. Like all international standards, however, it is intended only to serve as a framework, leaving a wealth of details open to interpretation.

● Some of the most effective initiatives in the direction of uniformity in bibliographic practice have originated in the editorial departments of scientific publishing enterprises.

To anyone familiar with the issues this will not come as a surprise. The publishing industry, after all, bears the brunt of the burden imposed by a lack of consensus in matters of citation. Both "sanitizing" an author's deficient reference list and summarily rejecting the manuscript on grounds of citation shortcomings are genuinely unpleasant prospects. Consistency in and an appreciation for formal matters related to writing and publishing on the part of authors always facilitates the work of an editorial office, allowing it to function more smoothly, and that in turn serves everyone's interest.

● Two important initiatives have produced citation guidelines that have been especially widely discussed and also widely adopted: the *CBE* and *Vancouver Conventions*, both developed in the context of broader attempts to standardize and facilitate the overall preparation of manuscripts for publication

The acronym CBE refers to the Council of Biology Editors, since 2000 known as the Council of Science Editors (CSE), with headquarters in Reston, Virginia.[17] The recommendations from this source have been summarized in the *CBE Style Manual* (5th ed. 1983; 6th ed. 1994). More than 6000 (!) bioscience journals have subscribed to the CBE Convention, which sanctions citations consistent with the name–date system, the (sequential) numerical system, and also the "hybrid system" briefly mentioned earlier (in Sec. 9.3.4) using citation numbers based on prior alphabetic arrangement of a complete set of sources.

The Vancouver Convention

The initiative culminating ultimately in the set of guidelines known as the Vancouver Convention was born as early as 1968, but the name is derived from a gathering of biomedical editors that occurred in January, 1978, in Vancouver, British Columbia. The principles established at that meeting, with some later "fine-tuning", are dealt with in books compiled by one of the participants (HUTH 1987, 1990), who also played a decisive role in development of the American standard ANSI Z39.29-1977 and for a time was chairman of the Council of Biology Editors (CBE).[18] The Vancouver recommendations are consistent with practices implemented with the exhaustive *Index Medicus* database of the National Library of Medicine (NLM), and are obviously therefore reflective of IT considerations. We feel justified in making the Vancouver Convention the centerpiece of our own treatment, especially with respect to Appendix A. In most essential points the Vancouver recommendations are in any case compatible with CBE recommendations and with the ANSI standard.

- A serious attempt was made at Vancouver to design a citation format that was as *language neutral* as possible—and thus attractive for international use—at the same time keeping it "computer friendly". *Punctuation* is employed very sparingly, mainly for separating the various bibliographic elements and groups from one another.

Punctuation marks in a Vancouver citation serve almost exclusively as *logical descriptors*, shedding in the process much of their ordinary meaning. In the terminology of a computer programmer, they are transformed into *delimiting structure codes*. In particular, and contrary to the suggestions in DIN 1505,

- The *period* has been reserved for the task of separating bibliographic groups (and analytical titles from collective titles; see Sec. 9.5.1), with *commas* and *semicolons* used to separate elements from one another *within* a group.

---

[17] The organization was founded in 1957 by joint action of the National Science Foundation and the American Institute of Biological Sciences.

[18] Extensive information related to the Vancouver Convention is also available from the Website of the International Committee of Medical Journal Editors (www.icmje.org) as incorporated in the Committee's *Uniform Requirements for Manuscripts Submitted to Biomedical Journals* (URM).

This in turn means that the period is no longer available as a sign of abbreviation, so abbreviated journal titles, for example, contain no punctuation, as illustrated by the various sample references in column A of Appendix A. Note also in these examples how the *space* can act like a punctuation mark from a structuring standpoint.

● Authors' *initials*, which are set after the corresponding family name and separated from it by a single space, are written *together*, with no spaces and no punctuation.

Vancouver Convention references also utilize in well-defined ways the following additional punctuation marks:

colon, "en-dash", parentheses, and square brackets.

● Whereas *commas* serve to separate within a bibliographic group multiple elements of equivalent stature, non-equivalent elements are separated by *semicolons*.

Notice that the list of allowed punctuation does *not* include the *quotation mark*. The once common practice of setting quotes around *analytical titles* (e.g., the title of a paper from a journal) is therefore no longer permissible if the Vancouver rules are to be followed.

● No mention is made of the use of special type in a reference, leaving publishers a certain amount of latitude in this regard.

It remains to be pointed out that—insofar as possible—connecting words have been strictly avoided, especially the conjunction "and". This in itself disposes of many potential translation problems posed by a reference (e.g., use of *ger.* "und", *fr.* "et", etc.). A reference list will still not be completely language-neutral, however, since the need remains for indicators like "Vol" (*ger.* "Bd") and "ed" ("Hrsg"). Note the absence of periods! Fortunately, modern computer programs can eliminate most of the drudgery associated with such translation (e.g., ENDNOTE; see Sec. 9.2.2).

● One of the important goals of the Vancouver conferees was perfection of reference structures that would be compatible with computer-assisted editing, selection, and organization of bibliographic data in recognition of the growing importance of *literature management software* and *literature databases*.

The Current Outlook

In comparison with the Vancouver guidelines, the recommendations in DIN 1505 seem somewhat narrow, complex, and provincial, and they take insufficiently into account the special circumstances surrounding electronic data processing. It is for these reason we have chosen not to emphasize DIN standards in this particular case. In all likelihood they will in any event be revised in the near future to bring them into closer accord with the American standard ANSI Z39.29. Special attention will undoubtedly be directed as well toward the citation of audiovisual materials and databases, a concern now being considered by a special committee of the National Information Standards Organization (NISO).

Chemists have played a surprisingly minor role in bibliographic developments of the past few years. Reserve on the part of chemistry editors may be attributable in part to the fact that chemistry, as a large discipline, has experienced little pressure to bend to the wishes of other (smaller) disciplines. Another factor may be the exceptionally strong influence wielded by the American Chemical Society with its very widely read and thus heavily cited journals and its sponsorship of the unassailable abstracting publication *Chemical Abstracts*. This huge professional organization has long promulgated its own writing "standards", distributed commercially in book form under the title *The ACS Style Guide: A Manual for Authors and Editors* (DODD 1997). Chemists and chemical journals everywhere have found this particular "style sheet" impossible to ignore (cf. Appendix A).

● Reference formats have attracted uncommon attention on the part of individual scientists in recent years in part because familiarity with the rules allows one to delegate much of the busywork of citation to software.

Reference managing software like VCH BIBLIO, ENDNOTE, and MANUSCRIPT MANAGER (see Sec. 9.2.2) is designed to take account of the Vancouver and CBE recommendations in the course of structuring references. Such programs are equally adept at conforming to unique citation rules adopted by individual periodicals as well, however, so it is not clear whether this represents a force encouraging consensus or diversity ("Thanks to computers we can have it our own way!"). We sincerely hope the former proves to be the case.

In the section that follows we examine more closely the individual *components* of references for various types of documents. In order to establish precisely what, in a particular case, constitutes a "clear, unambiguous, complete, and appropriate" source description (cf. Sec. 9.4.1) one must first become familiar with the elements that need to be considered. The reader who is impatient to proceed without further delay to the "bottom line" may wish to spend a few minutes at this point examining Appendix A.

9.5 Anatomy of a Source Description

9.5.1 General Characteristics

The elements constituting a definitive description of a document fall into several *bibliographic groups*. We begin by addressing these in an abbreviated way (cf. ANSI Z39.29-1977).

● A primary distinction is made between *seven* bibliographic groups.

Author group: The author group consists of one or more names identifying a work's authors, editors, or issuing organizations. In some cases *identifiers* are included, as in the notation "ed" for editor. Organizations in this role (including societies, authorities, companies, etc.) are classified collectively as *corporate sources* in DIN 1505. The author group is omitted in the case of an anonymous publication.

Title group: Titles in any of three bibliographic categories may be present. The title (or *subject title*) of a journal article, a chapter in a book, or some other dependent (i.e., non-independent) publication is referred to as an *analytical title*. Independent publications can be associated with both *monographic titles* (book titles) and *collective titles* (describing collective works, such as a journal or a series of books). Often a distinction is made between a *principal* title and one or more *subtitles*, where the two are separated by a colon. If two or all three types of title apply to a document they are cited in succession, typically in the (ascending) order

Contribution within a volume – volume in a series – series

Edition group: This furnishes identification of a particular edition (through an "edition identifier" in the terminology of DIN 1505-1) together with the names of those involved specifically with its production, such as a translator. A typical entry would be simply "3rd ed" (note again the absence of a period, consistent with Vancouver rules).

Imprint group: This category (called the *publication notice* in DIN 1505-1) includes the place of publication (headquarters of the publisher), publisher's name, publication date (year), volume or issue number (in the case of a journal), and a report number if applicable. The place of publication and name of the publisher—almost always in that order—together constitute the *publisher information*. The publisher's name is presented as concisely as possible, with appendages like "Inc", "Ltd", or "GmbH" omitted. In the event a publisher has multiple offices, only one need be listed. If some "corporate entity" is the publisher, an address should be supplied to facilitate acquisition of the document.

Characteristics group: Provides information regarding the form or perhaps the packaging of a source (e.g., extent, format, number of illustrations), as well as its physical nature (e.g., cassette, diskette, CD-ROM). This is essentially what DIN 1505-1 refers to as the *collation notice*. The page count of a book is often appended to the identification of the publisher, in a form like "300 p" or "300 pp", equivalent to expressing the length of a journal article as a range of the first through the last pages.

Series group: Here one identifies any relationship between a publication (source) and a broader *series*, such as a *volume number* in the case of a book, together with the title of the series itself (although the latter is often supplied as part of the title group, immediately preceding the imprint group).

Supplementary information: A final category is available for such things as extra information that might assist in acquisition of a copy of the document. This would be an appropriate place for an ISBN in the case of a book, a project number in conjunction with a research report, or even a retail price. *Localization* information applicable to an important passage in a document would also appear here, perhaps introduced in a form like

:78 ;   :78–84 ;   p 78 ;   chap 5 ;   (p 523, table 2)

9.5.2 Sources of Various Types

Books and Journals

Our comments here, as in previous pages, are designed to encourage use whenever possible of source descriptions in which the year of publication immediately follows the name of the author(s). This is an especially important consideration in conjunction with name–date citations. The example that follows can be regarded as typical for the formal characterization of a book. It is *adapted* from the first illustration in column B of Appendix A, in that the date has been moved, and a superfluous "Inc" eliminated:

Osler, A.G. (1976). Complement: mechanism and functions.
Englewood Cliffs, NJ: Prentice Hall.

The even more rigorous Vancouver style requires dispensing with parentheses surrounding the year. Thus the first entry in Column A of Appendix A might be rendered as

Eisen HN. 1974. Immunology: an introduction. 5th ed.
New York: Harper and Row; 215.

In the absence of the localizer pointing to p. 215, this reference would be terminated by a period after "Row". The reference above corresponds essentially to the format we have adopted for references in the present book, although in the interest of legibility we normally italicize the titles of independent publications and include a page count (followed by p). In this sense our own source descriptions might tentatively be taken as models. Note that under this slightly modified Vancouver approach only the *first words* are capitalized in English-language titles, apart from proper names, of course. As an example of a source description for a journal article, consider the following:

Bynum WF, Heilbron JL. 1988. Eighteen eighty-eight and all that.
Nature. 331: 27–30.

As already noted in Sec. 9.4.1, journal titles are rarely presented in complete form in a reference list, but are instead abbreviated in "officially sanctioned" ways. (The one-word title here, "Nature", constitutes one of the few exceptions.)

● Use of abbreviated titles is a labor-saving measure with the additional advantage of saving space.

To ensure that journal-title abbreviations will be understandable throughout the world, an exceptionally high degree of standardization is desirable. Unfortunately, even in this elementary matter of communication no true consensus has been achieved. In particular, there are two competing "style guides". *Chemical Abstracts*, *Biological Abstracts*, and *Index Medicus*, three of the largest abstracting services, all adhere to *one* of the major systems (based generally on ISO 4, 1997), and we are unaware of any abstracting media in the field of physics that fail to conform [the Institute of Physics, London; American Institute of Physics (AIP), New York and the abstracting services *Physics Abstracts* and *Physics Briefs* follow the pattern]. One could summarize by observing that natural scientists as a group have committed themselves to sharing a common system.[19]

An extensive list of the abbreviations for journal titles employed in *Chemical Abstracts* is provided in DODD (1997). Curiously, the ISO standard recommends abbreviating city names (e.g., Lond) despite widespread use—and acceptance—among scientists of the unabbreviated designation *Nature* (*London*) for one of the premiere British journals. The International Organization for Standardization has unfortunately failed to do the academic world a favor in its encouragement of even more drastic abbreviation (e.g., "pub" rather than "publ"). Shorter representations are in general more difficult to decipher in an unambiguous way.

The American Institute of Physics (1978) offers the following piece of advice in its *Notes for Authors*:

● If you are uncertain about the correct way to abbreviate a journal title, cite it in full.

We concur, but with the added suggestion that you apprise the editors of your dilemma through a marginal note such as

((abbrev. title ?))

Miscellaneous Documents and Sources

Standards are subject only rarely to citation, but we still find it surprising that none of the sources we have relied upon in developing Appendix A makes any reference whatsoever to standards; only standards themselves touch on the subject. Thus, DIN 1505-2 (1984) treats the matter rather thoroughly (as one might anticipate with a German standard!), suggesting that the *word* "Standard" ("Norm" in German) precede a source description of this type.

● The most important elements for inclusion in the source data for a standard are the abbreviation of the document series or organization responsible for the guidelines (e.g., ANSI, BS, DIN, ISO), the appropriate standard *number*, and the date of issuance. It is helpful in the sense of "recommended data" also to include the

---

[19] The alternative system is that devised by the *World List of Scientific Periodicals*, a publication of the British Union Catalogue of Periodicals.

standard's subject title. The year officially associated with the particular edition in question serves as an appropriate date. Two examples of source descriptions of this type are provided in Appendix A.

Patent descriptions are structured rather like the descriptions of journal articles. A declaration of the issuing country together with a *document type* identification takes the place of a journal title. A *patent number (publication number)* is the equivalent of information leading to a particular article within a journal.

The suggestions that follow are based on DIN 1505-2 (1984). They are applicable not only to patent documents at various stages in the approval process (what is generally regarded as the *patent literature*), but also to copyrights covering *registered designs*. Patents and registered designs—together with trade-marks—constitute forms of legal protection applicable to *property rights* in general. The DIN standard suggests that a source description in this category should begin by making that very fact explicit.

● Such a reference should thus begin with the words "property rights" (or something analogous).

The object is to distinguish clearly between documents of this type and others whose titles might also include letter or number combinations, and thus look similar.

The first defining characteristic after this declaration should be the *country* in which the rights in question were assigned. Increasingly this information is conveyed through an internationally sanctioned code system, analogous in a sense to abbreviated journal titles.

● Standard practice with patent descriptions and other documents related to property rights is to utilize a *two-letter* country code.

These codes are officially defined in ISO 3166-1:1997 *Codes for the representation of names of countries and their subdivisions – Part 1: Country codes*. Examples include "US" for the United States, "GB" for Great Britain, "DE" for Germany (Deutschland; not simply the "D" sometimes seen in other contexts), and "CH" for Switzerland (derived from *Confoederatio Helvetica*). A convenient table listing a great many codes has been posted on the Web by Information Technology Associates at http://www.theodora.com/country_digraphs.html. European patents are distinguished by the letter combination "EP".

Next comes the applicable *protection number* (in the case of patents, a *patent number*). Patent offices issue documents of several different types, so a number alone does not provide sufficient information: a *document type* must also be specified. The latter information, too, is increasingly conveyed in abbreviated form—as a one- or two-character code appended to the protection number, separated from it by a hyphen.

Document-type codes consist of a single capital letter or a capital letter accompanied by a one-digit number. In the case of current United States patent documents, "A1" refers to a patent *application* and "A9" to a *corrected* patent application, for example, whereas "B1" and "B2" refer to *issued* patents, without or with pre-grant publication, respectively. With German patent documents the letter "A" indicates an unexamined application (*Offenlegungsschrift*), "B" an examined application (*Auslegeschrift*), and "C" a true patent document. These and a great many additional designators are now internationally recognized in the context of applicable national traditions. A complete list of such codes, assembled by the *Chemical Abstracts Service*, is available on the Internet at http://www.cas.org/EO/patkind.html.

One final obligatory feature of a patent-document reference is a date, in this case the date of the document's issuance. DIN guidelines recommend that this be cited in the form "year-month-day" following the document number and the code indicating a document type.

The following represents an example of a complete patent description, essentially consistent with DIN recommendations. The patent in question is one based on a set of rights granted earlier (i.e., *priority* plays a role):

> Property rights EP 2013-B1 (1980-08-06). Bayer.
> Pr.: DE 2751782 1977-11-19

The abbreviation "Pr." for "priority" precedes identification of the document in which the claims were originally recognized. This example illustrates one additional feature as well: an *abbreviated* identification of the claimant (the person or organization— here the Bayer Corporation—to whom the patent rights were granted). This is considered a piece of supplementary information, and is not absolutely required. Other supplementary information one might include would be the *inventor*, the *subject title* of the document, and the specific branch of science or technology to which the patent is assigned. Inventors' names, like authors' names with journal articles and other documents, are placed at the beginning of the "title" section, immediately preceding a subject title.

The *subject classification* is yet another piece of information supplied in coded form; e.g.,

> Int. Cl.^{2} C 06B 1/02

A classification would be added at the very end of a patent description. A comprehensive list of international patent classification codes is available from the Website of the World Intellectual Property Organization (WIPO) at http://www.wipo.int/classifications/en/.

We conclude with a recommendation that is not spelled out in any set of guidelines we have seen:

● When the source is a patent document it is extremely helpful to provide the reader with an applicable reference to one of the recognized abstracting publications.

Other types of material for which illustrative examples of source descriptions are provided in Appendix A include journal articles in electronic format, computer programs, CD-ROMs, and World Wide Web home-pages.

Internet sources of all kinds have justifiably been a subject of concern because of their inherently transient nature, unlike traditional print sources that might be inadvertently lost from *one* library collection somewhere, but are unlikely to disappear entirely. Several important initiatives have been developed to tackle the issue, though it is too soon to describe what the definitive solution(s) will look like. Interested readers are encouraged to consult the Web sites of JSTOR (initiated by the Mellon Foundation, and an independent organization since 1995 dedicated to archiving journals, both print and electronic) at http://www.jstor.org/about/ and the Internet Archive (concerned with preservation of Internet resources in general) at http://www.archive.org, together with various links these two sites provide.

As we take leave of you, may we say farewell with a slight modification of a classic maxim?

Quidquid scribis, prudenter scribas, et respice finem.[1]

---

[1] Have we perhaps failed to get across our point? That will of course not do. A rough translation would be:

 No matter what you write, write intelligently, and ponder what could become of it.

The original wording handed down from the ancient Romans was "Quidquid agis, prudenter agas, et respice finem"—with the verb "agere" (to do) rather than "scribere" (to write). For many, this sentence represents the perfect leitmotiv—conceived 2000 years ago—for quality control.

Appendices

Appendix A
Reference Formats

What follows might be regarded as a collection of "model references" for the most important kinds of documents, based on four different sets of guidelines. Juxtaposing them in this way makes it relatively easy to recognize similarities and differences. At the moment there is no such thing as an "ultimate authority" regarding reference format, so it is up to each author to decide upon a course to follow (unless a particular style has been dictated—by a publisher, for example).

The four reference approaches we consider especially important, and upon which this table is based, are:

1. The Vancouver Convention, described by HUTH (1987);
2. The CBE rules, presented in the *CBE Style Manual* (COUNCIL OF BIO-LOGY EDITORS 1983);
3. Recommendations of the American Chemical Society, as described by DODD (1986);[1] and
4. The standard DIN 1505-2 (1984) developed by the Deutsches Institut für Normung.

All four are in some sense dependent upon ISO 690 (1987) *Documentation—Bibliographic references: Content, form and structure*, although they all deviate significantly from its provisions. For further information see Sec. 9.5.

In our table, the four sources are abbreviated Vancouver, CBE, ACS, and DIN. Examples shown are in most cases taken directly from these sources, with only minor modifications (and none that are significant). If citations are to be of the name–date type, you may in some cases wish to move the year forward, placing it immediately after the "first element" (ordinarily the name of an author; cf. Sec. 9.5.2).

The column on the left identifies the various document types, with special features noted in parentheses. Numbers are provided in the far right column to facilitate comparisons.

Not every type of document mentioned in the cited sources is represented here. For example, references to magazines, newspapers, legal statutes, and radio and television broadcasts have been omitted. At the same time, it will be apparent from frequent appearance of the comment "(no example)" that our summary is in some ways more comprehensive than the available sources.

Just as we found it impossible always to cite for all four sets of guidelines examples covering every imaginable situation, the reader will undoubtedly find it necessary on occasion to "invent" reference formats for rarely encountered document types.

---

[1] Essentially no reference-style changes are apparent in the more recent edition (DODD 1997).

Document	Vancouver	A	CBE	B
Book				
with one author[a] *(edition, localization, supplementary information)*	Eisen HN. Immunology: an introduction. 5th ed. New York: Harper and Row; 1974: 215-7.		Osler, A.G. Complement: mechanism and functions. Englewood Cliffs, NJ: Prentice Hall, Inc.; 1976.	
with multiple authors (special-features)	Rowzon KEK, Rees TAL, Mahy BWJ. A dictionary of virology. Oxford: Blackwell; 1981. 230 p.		Eason, G.; Coles, C.W.; Gettingby, G. Statistics for the bio-sciences. West Sussex, England: Ellis Horwood Limited; 1980.	
with editor[a] *(contribution in an edited work)*	Daussert J, Colombani J, editors. Histocompatibility testing. Copenhagen: Munksgaard; 1973: 12-8.		Wood, R.K.S., editor. Active defense mechanisms in plants. New York: Plenum Press; 1982.	
chapter in an edited book[b] *(localization, series title)*	Weinstein L. Invading microorganisms. In: Sademan WA Jr, Sademan W, eds. Pathologic physiology. Philadelphia: WB Saunders; 1974: 457-72.		Kirkpatrick, C.H. Chronic candidiasis. In: Safai, B.; Good, R.A., eds. Immunodermatology. New York: Plenum; 1981: p. 495-514. (Good, R.A. Comprehensive immunology; vol. 7).	
monograph in a series (series title, localization)	Hunninghake GW, Gadek JE, Szapiel SV et al. The human alveolar macrophage. In: Harris CC, ed. Cultured human cells. New York: Academic Press; 1980: 54-6. (Stoner GD, ed; Methods and perspectives in cell biology; vol 1).		(no example)	
volume in a multivolume work (volume and edition information)	Cowie AP, Mackin R. Volume 1: Verbs with prepositions and particles. In: Oxford dictionary of current idiomatic English. London: Oxford University Press; 1975.		(see example 4B)	
multivolume work (editor cited)	(no example)		Colowick, S.P.; Kaplan, N.O. Methods in enzymology. New York: Academic Press; 1955-1963. 6 vol.	
Work with no author/editor	Webster's standard American style manual. Springfield, Massachusetts: Meriam-Webster; 1985. 464 p.		American men and women of science. 15th ed. Jacques Cattell Press, ed. New York: R.R. Bowker Co; 1982. 7 vol.	
Conference proceedings *(contribution in, supplementary information)*	DuPont B. In: White HJ, Smith R, eds. Proceedings of the third annual meeting of the International Society for Experimental Hematology. Houston: International Society for Experimental Hematology; 1974: 44-6.		Giesey, J.P., editor. Microcosm in ecological research. DOE Symposium series 52; 1978 November 8-10; Augusta, GA. 1110 p. Available from: NTIS, Springfield, VA; CONF-781101.	

ACS	C	DIN	D	No.
Stothers, J. B. *Carbon-13 NMR Spectroscopy*; Academic: New York, 1972; Chapter 2.		METZGER, Wolfgang: *Gesetze des Sehens*. 3. Aufl. Frankfurt. Kramer, 1975 (Senckenberg-Buch 53). – ISBN 3-7829-1047-8		1
Littmann, M.; Yeomans, D. K. *Comet Halley: Once in a Life-time*; American Chemical Society: Washington, DC, 1985; p 23.		GRAWFORD, Claude C.; COOLEY, Ethel G.; TRILLINGSHAM, C.C.; STOOPS, Emery: *Biologie der Erkenntnis*. 3. Aufl. Berlin: Parey, 1981. – ISBN 3-489-61084-2		2
Golay, M. J. E. In *Gas Chromatography*; Desty, D. H., Ed.; Butterworths: London, 1958; p 36.		KAEMMERLING, Ekkehard (Hrsg.): *Ikonographie: Theorien – Entwicklung – Probleme*. Köln: DuMont, 1979 (Bildende Kunst als Zeichensystem 1) (DuMont Taschenbücher 83)		3
Geacintov, N. E. In *Polycyclic Hydrocarbons*; Harvey, R. G., Ed.; ACS Symposium Series 283; American Chemical Society: Washington, DC, 1985; pp 12-45.		FRANKE, Herbert W.: Sachliteratur zur Technik. In: RADLER, Rudolf (Hrsg.): *Die deutschsprachige Sachliteratur*. München: Kindler, 1978 (Kindlers Literaturgeschichte der Gegenwart), S. 654-676		4
Jennings, K. R. In: *Mass Spectroscopy*; Johnstone, R. A. W., Senior Reporter; Specialist Periodical Report; The Chemical Society: London, 1977; Vol. 4, Chapter 9.		(see example 3D)		5
(no example)		NEUMÜLLER, Otto-A.: *Römpps Chemie-Lexikon*. Bd. 1. 8. Aufl. Stuttgart: Franckh, 1979		6
(no example)		FRUTIGER, Adrian: *Der Mensch und seine Zeichen / Heiderhoff*, Horst (Bearb.). Bd. 1-3. Echzell: Heiderhoff, 1978-1981		7
(no example)		„Houben-Weyl" *Methoden der Organischen Chemie*. Bd. 13/1. Stuttgart: Thieme, 1970		8
Baisden, P. A. Abstracts of Papers, 188th National Meeting of the Chemical Society, Philadelphia, PA; American Chemical Society: Washington, DC, 1984; NUCL 9.		CID (Veranst.): Chemie, Physik und Anwendungstechnik für grenzflächenaktive Stoffe (4. Int. Kongreß für grenzflächenaktive Stoffe Brüssel 1964). Sect. A, Vol. 1. London: Gordon & Breach, 1967		9[1]

Document	Vancouver	A	CBE	B
Conference proceedings *(indication of a special issue)*	(no example)		(no example)	
Journal *contribution with up to 6 authors*[c] *(with and without title; localization)*	You CH, Lee KY, Chey RY, Menguy R. Electrogastrographic study of nausea patients. Gastroenterology. 1980; 79: 311-4.		Steele, R.D. Ethionic metabolism. J. Nutr. 112: 118-125; 1982.	
(with more than 6 authors or corporate authors)	Brickner PW, Scanlan BC, Conanan B, et al. Homeless persons and health care. Ann Intern Med. 1986; 104: 405-9.		The Committee on Enzymes. Recommended method for the determination of ATPase in blood. Scand. J. Clin. Lab. Invest. 36: 119-125; 1976.	
(pagination by issue)	Seaman WB. The case of the pancreatic pseudocyst. Hosp Pract. 1981; 16 (Sep): 24-25.		Interferon: preparing for wider clinical use. Med. World News 23 (9): 51-54; 1982.	
(accepted, not yet published)	Overstreet JW. Semen analysis. Infertility in the male. Ann Intern Med. [In press].		(no example)	
(no volume numbers)[d]	Nussknacker H, Suite F. Neue Inhalts-stoffe von Juglans regia. Liebigs Ann Chem. 1992: 194-205.		(no example)	
Research report *(supplementary information)*	Ranofsky AL. Surgical operations in short-stay hospitals: United States – 1975. Hyattsville, Maryland: National Center for Health Statistics; 1978; DHEW publication no (PHS) 78-1785.		Zavitkowski, J., editor. The Enterprise, Wisconsin, radiation forest: radioecological studies. Oak Ridge, TN: Energy Research and Development Administration, Technical Information Center; 1977; 211 p. Available from: NTIS, Springfield, VA; TID-26113-P2.	
Government document *(supplementary information)*	National Center for Health Services Research. Health technology assessment reports, 1984. Rockville, Maryland: National Center for Health Services Research; 1985; DHHS publication no (PHS) 85-3373. Available from: National Technical Information Service, Springfield, VA 22161.		World Health Organization, WHO Expert Committee on Specifications for Pharmaceutical Preparations. 28th rep. WHO Tech. Rep. Ser. 681; 1982. 33p.	

ACS	C	DIN	D	No.
(no example)		*Progress in radiology (11. Int. Congress of radiology Rome 1985)*. – Preprints. Teilw. in: *Medica mundi* (1966) Nr. 1		9[2]
Fletcher, T. R.; Rosenfeld, R.N. *J. Am. Chem. Soc.* **1985**, *107*, 2203-2212.		VERKADE, P.: Etudes historiques sur la nomenclature de la chimie organique. Tl. IV; V. In: *Bull. Soc. Chim. France* 1969, S. 3877-3881; 4297-4307		10[1]
(no example)		(no example)		10[2]
Stinson, S. C. *Chem. Eng. News* **1985**, *63* (25), 26.		SCHMIDT, Hans: Aufbruch in Hongkong. In: *Spiegel* 37 (1983-03-14), Nr. 11, S. 172-182		10[3]
Tang, D.; Jankowiak, R.; Small, G. J.; Tiede, D. M. *Chem Phys.*, in press.		(no example)		10[4]
Nussknacker, H.; Suite, F. *Liebigs Ann. Chem.* **1992**, 194-205.		(no example)		10[5]
Stewart, J. M.; Machin, P. A.; Dickinson, C. W.; Ammon, H. L.; Heck, H.; Flack, H. *The X-ray 76 System;* Technical Report TR-446; Computer Science Center, University of Maryland: College Park, MD, 1976.		DUELEN, G.; PRAGER, K.-P.: Mathematische Grundlagen für die Bahnsteuerung von Industrierobotern / Fraunhofer-Institut für Produktionsanlagen und Konstruktionstechnik. Karlsruhe: Kernforschungszentrum Karlsruhe, 1982 (KfK-PFT-E6). – Forschungsbericht. BMFT-Förderprogramm Fertigungstechnik, Projektträger Humanisierung des Arbeitslebens DFVLR-HdA, Identifikation 01-VC 028		11
Interdepartmental Task Force on PCBs. *PCBs and the Environment*; U. S. Government Printing Office: Washington, DC, 1972; COM 72.10419.		(no example)		12

Document	Vancouver	A	CBE	B
Dissertation	Cairns RB. Infrared spectroscopic studies of solid oxygen [Dissertation]. Berkley, California: University of California; 1965. 156 p.		Spangler, R. Characterization of the secretory defect present in glucose intolerant Yucatan ministructure swine. Fort Collins: Colorado State Univ.; 1980. Dissertation.	
Corporate document *(supplementary information)*	(no example)		Eastman Kodak Company. Eastman organic chemicals. Rochester, NY: 1977; Catalog No. 49. 180 p.	
Patent *(inventor, assignee; reference to abstracting journal)*	Larsen CE, Trip R, Johnson CR, inventors; Novoste Corporation, assignee. Methods for procedures related to the electrophysiology of the heart. US patent 5,529,067. 1995 Jun 25.		Harred, J.F.; Knight, A.R.; McIntyre, J.S., inventors; Dow Chemical Co., assignee. Epoxidation process. U. S. Patent 3,654,317. 1972 April 4. 2 p. Int Cl C 07 D 1/08, 1/12.	
CD-ROM *(book)*	(no example)		(no example)	
Standard	(no example)		(no example)	
Journal article in electronic format	Morse SS. Factors in the emergence of infectious diseases. Emerg Infect Dis [serial online] 1995 Jan-Mar [cited 1996 Jun 5]; 1(1):[24 screens]. Available from: URL: http://www.cdc.gov/EID/eid.htm.		Browning T. 1997. Embedded visuals: student design in Web spaces. Kairos: A Journal for Teachers Writing in Webbed Environments 3(1). <http://www.aa.ttu.edu/kairos/2.1/features/browning/index.html>. Accessed 1997 Oct 21.	
Computer program	Hemodynamics III: the ups and downs of hemodynamics [Computer program] 2nd ed. Version 2.2. Orlando (FL): Computerized Educational Systems; 1983.		(no example)	
Personal home page	(no example)		Pellegrino J. 1999 May 12. Homepage. <http:www.english.eku.edu/pelligrino/default.htm>. Accessed 1999 Nov 7.	

[a] *Corporate entities* such as government agencies, companies, and organizations can also act as authors. The corporate entity (e.g., the World Health Organization, WHO) may also be the publisher, and could thus appear in a reference twice; cf. example 12B.

[b] The specific chapter in a book or the contribution in a volume of conference proceedings need not be cited directly; a chapter number may suffice.

ACS	C	DIN	D	No.
Kanter, H. Ph. D. Thesis, University of California at San Fransisco, Dec. 1984.		THIELE, Angelika: *Die Belastung des Stein-bachs durch toxische Metalle*. Münster, Universität, Fachbereich 23, Diss., 1982		13
(no example)		DEGUSSA:Aerosol. Frankfurt, 1969 (RAG-3-8-369 H). – Firmenschrift		14
Norman, L.O. U. S. Patent 4 379 752, 1983.		Schutzrecht EP 2013-B1 (1980-08-06). Bayer. Pr. DE 2751782 1977-11-19		15[1]
Lyle, F.R. U. S. Patent 5 973 257, 1985; *Chem. Abstr.* **1985**, *65*, 2870.		Schutzrecht DE 2733479-A1 (1979-05-15). Henkel. Pr.: DE 2733479 1977-07-25. – Zusatz zu DE 2556376-A1		15[2]
The Merck Index, 12th ed. [CD-ROM]; Chapman & Hall: New York. 1996.		(no example)		16
(no example)		Standard ISO/DIS Draft 1977-05-24		17
		Standard BS 5605: 1978. *British Standard Recommendations for citing publications by bibliographic references*		
Tunon, I.; Martins-Costa, M. T. C.; Millot, C.; Ruiz-Lopez, M. F. *J. Mol. Model.* [Online] 1995, *1*, 196–201.		(no example)		18
BCI Clustering Package, versions 2.5 and 3.0; Barnard Chemical Information: Sheffield, U.K., 1995.		Space Invaders. – Spiel; BASIC-Programm. In: Hewlett-Packard: 2647 A Program Tape. File 4. – Magnetband-Kassette Typ 3M für Grafik-Terminal HP 2647 A		19
ChemCenter Home Page. http//www. chemcenter.org. (accessed Dec 1996).		(no example)		20

[c] The author may also be a *team* of authors; e.g., The Royal Marsden Hospital Bone-Marrow Transplantation Team. – Unattributed articles are introduced with [Anonymous]. Nevertheless, DIN 1505-2 declares that "If no author is listed, some other important responsible party should be cited, such as an editor or a corporate entity." If this also proves impossible, the citation should begin with the title of the document (according to DIN 1505).

[d] The sources make no provision for this case. The two examples in this row have been made up.

Appendix B
Selected Quantities, Units, and Constants

Name of the quantity[a]	Symbol[b]	SI unit[c]	Name of the unit	Other units
Space and Time				
length*	l	m	meter	
breadth, width	b	m	meter	
height, depth	h	m	meter	
radius	r	m	meter	
thickness	d, δ	m	meter	
area, surface	A, S	m^2	square meter	a (are)
				h (hectare)
volume	V	m^3	cubic meter	L, l (liter)[d]
plane angle	$\alpha, \beta, \gamma,$	1, rad	radian	° (degree)
	ϑ, φ			' (minute)
				" (second)
solid angle	ω, Ω	1, sr	steradian	
wavelength	λ	m	meter	
wavenumber	σ	m^{-1}	reciprocal meter	
time*, duration	t	s	second	min (minute)
				h (hour)
				d (day)
				a (year)
frequency, periodic frequency	ν, f	s^{-1}	reciprocal second	Hz (hertz)[e]
relaxation time	τ	s	second	
velocity	u, v, w, c	$m\ s^{-1}$	meter per second	km/h
acceleration	a	$m\ s^{-2}$	meter per second squared	
Classical mechanics				
mass*	m	kg	kilogram	g (gram)
				t (ton)[f]
density	ρ	$kg\ m^{-3}$		g/cm^3

---

[a] SI base quantities are labeled with an asterisk (*).
[b] Recommended by IUPAC.
[c] SI base units and both derived and supplementary units are listed; if necessary, all may be modified with prefixes.

[d] $1\ L = 10^{-3}\ m^3$.
[e] $1\ Hz = 1\ s^{-1}$.
[f] Formerly metric ton; $1\ t = 10^3\ kg$.

Name of the quantity[a]	Symbol[b]	SI unit[c]	Name of the unit	Other units
momentum	$\boldsymbol{p}$	kg m s^{-1}		
angular momentum	$\boldsymbol{L}$	kg m^2 s^{-1}		
force	$\boldsymbol{F}$	N	newton[g]	
torque	$\boldsymbol{M}$	N m	newton meter	
weight	G, W	N	newton	
pressure	p	Pa	pascal	bar (bar)[h]
energy	E, W	J	joule	W h (watt hour)[i]
				eV (electronvolt)[j]
work	W, A	J	joule	
power	P	W	watt	J/s, V A[k]

Molecular physics and thermodynamics

thermodynamic temperature*	T, Θ	K	kelvin	°C (degree Celsius)[l]
Celsius temperature[l]	ϑ, t			°C
number of entities	N			
Avogadro constant[m]	N_A, L			
Boltzmann constant[n]	k			
Planck constant[o]	h			
(molar) gas constant[p]	R			
heat	Q	J	joule	
entropy[q]	S	J K^{-1}	joule per kelvin	
internal energy[q]	U	J	joule	
enthalpy[q]	H	J	joule	
heat capacity[q]	C_p, C_V	J K^{-1}	joule per kelvin	
Helmholtz function,[q] (Helmholtz) free energy, Helmholtz energy	F, A	J	joule	
Gibbs function,[q] (Gibbs) free energy, Gibbs energy	G	J	joule	

Physical chemistry, atomic and nuclear physics

amount of substance*	n	mol	mole	
relative atomic mass	A_r	1		
relative molecular mass	M_r	1		
number of entities	N	1		
atomic mass constant	m_u	kg	kilogram	u (atomic mass unit)[r]
mass (of a substance B)	$m_B, m(B)$	kg	kilogram	g (gram)
molar mass (of a substance B)	$M_B, M(B)$	kg mol^{-1}	kilogram per mole	

[g] $1\ \text{N} = 1\ \text{kg m s}^{-2}$.

[h] $1\ \text{bar} = 10^5\ \text{Pa}$.

[i] $1\ \text{W h} = 3.6 \cdot 10^3\ \text{J}$.

[j] $1\ \text{eV} = 1.602\ 189 \cdot 10^{-19}\ \text{J}$.

[k] $1\ \text{W} = 1\ \text{J/s} = 1\ \text{V A}$.

[l] cf. Eq. (6–3) in Sec. 6.1.2.

[m] $N_A = 6.022\ 136\ 7 \cdot 10^{23}\ \text{mol}^{-1}$.

[n] $k = 1.380\ 658 \cdot 10^{-23}\ \text{J K}^{-1}$.

[o] $h = 6.626\ 075\ 5 \cdot 10^{-34}\ \text{J s}$.

[p] $R = 8.314\ 510\ \text{J mol}^{-1}\ \text{K}^{-1}$.

[q] Molar quantities can be distinguished from the quantities of a system by appending the subscript m; e.g., U_m in J mol^{-1} is the molar internal energy.

[r] $1\ \text{u} = 1.660\ 565\ 5 \cdot 10^{-27}\ \text{kg}$, cf. Sec. 6.2.2.

Name of the quantity[a]	Symbol[b]	SI unit[c]	Name of the unit	Other units
(amount) concentration (of a substance B)	c_B, $c(B)$	$mol\ m^{-3}$	mole per cubic meter	mol/L
amount fraction[s] (of a substance B)	κ_B, $\kappa(B)$	1		
mass fraction (of a substance B)	ω_B, $\omega(B)$	1		%, ‰, ppm, ppb
volume fraction (of a substance B)	φ_B, $\varphi(B)$ ϕ_B, $\phi(B)$	1		%, ‰, ppm, ppb
mass concentration	β	$kg\ m^{-3}$		g/L
molality	b, m	$mol\ kg^{-1}$	mole per kilogram	mmol/kg
volume concentration[t]	σ	1		
molar volume	V_m	$m^3\ mol^{-1}$	cubic meter per mole	L/mol
molar heat capacity	C_m	$J\ mol^{-1}\ K^{-1}$		
molar conductivity	Λ_m	$S\ m^2\ mol^{-1}$		
Faraday constant[u]	F			
activity of a radioactive substance	A	Bq	becquerel	s^{-1} [v]

Electricity, magnetism, light

electric charge	Q	C	coulomb	
electric potential	φ, Φ, V	V	volt	
electric potential difference	U	V	volt	
electric dipole moment	$\boldsymbol{p}$	C m	coulomb meter	
electric current*	I	A	ampere	
electric field strength	$\boldsymbol{E}$	$V\ m^{-1}$	volt per meter	
magnetic field strength	$\boldsymbol{H}$	$A\ m^{-1}$	ampere per meter	
magnetic flux	Φ	Wb	weber	V s [w]
(electrical) resistance, resistance	R	Ω	ohm	
(electrical) conductance, conductance	G	S	siemens	Ω^{-1} [x]
radiant energy	Q, W	J	joule	
luminous intensity*	I	cd	candela	

[s] Formerly: mole fraction.
[t] σ is based on the volume of a mixture, whereas the volume fraction φ relates the volume of a substance to the volumes of the various components prior to mixing.

[u] $F = 9.648\ 530\ 9 \cdot 10^4$ C mol^{-1}.
[v] $1\ Bq = 1\ s^{-1} = 3.703 \cdot 10^{-11}$ Ci; Ci is the symbol for the unit curie.
[w] $1\ Wb = 1\ V\ s$.
[x] $1\ S = 1\ \Omega^{-1}$.

This table was prepared with the aid of DRAZIL (1983) and IUPAC (1988).

Appendix C
The 20 Commandments
of Electronic Manuscripts

Guidelines for authors aspiring to prepare electronic
(digital) manuscripts for publication

1 Always work with a widely accepted word processor like WORD or WORDPERFECT.

2 Provide your publisher at the outset with a test file offering a representative sample of your work, preferably including examples of all the features and special symbols you propose to incorporate—captured in various formats, including a pure ASCII version, a file in the native format of your word processor, and one in RTF (rich-text format).

3 If you plan to submit original graphics in digital form, also include at least one of these in your preliminary sample. The principle of "test it in advance" is especially applicable to graphics!

4 Discuss at length with the publisher the results of this test, and be prepared to modify your procedure and habits as may seem appropriate.

5 Avoid creating files larger than ca. 200 kbyte, subdividing your manuscript into chapters as necessary.

6 Make sure to leave all your text unjustified (i.e., in "ragged right" format). At least in the course of making final revisions arrange to work in "Show ¶" mode so you will have a clear awareness of the structure of your composition.

7 Strictly avoid end-of-line hyphenation, both automatic and manual. If you feel it absolutely necessary in some rare case to divide an expression with hyphens, limit yourself to "optional" or "soft" hyphens.

8 The *only* role of the return key is to terminate a paragraph.

9 Never indent paragraphs manually using the spacebar or tab key. Any indentations must be ones associated with "paragraph styles".

10   Clearly indicate all headings by use of distinctive type sizes and/or styles (ones never employed in the body text). Alternatively, distinguish your headings by special symbols or paragraph styles, agreed upon with the publisher in advance, to make it easy to introduce changes later.

11   Highlight text elements (through special type styles, for example) *only* with advance approval from the publisher.

12   In the event capitalization is selected as a highlighting technique, be sure to use the "All caps" option from the font format menu, *not* the shift key. Again, it is only in this way that it becomes possible to make changes efficiently after the fact.

13   Unless you receive instructions to the contrary, all copy should be double-spaced, with wide margins.

14   Use "nonbreaking" spaces to separate two elements that must always remain in close proximity. If you happen not to have access to a nonbreaking space option, symbolize it with a vertical line instead ("|". ASCII code 124, as in the expression "3|cm").

15   Include true subscripts and superscripts in your digital manuscript *only* if you are sure they will be properly interpreted by software at the publishing house (based on experience with your sample file). If necessary, arrange with the editorial staff for an unambiguous substitute. Note that this injunction also applies to superscripted footnote symbols and citation numbers.

16   Letters of the Greek alphabet and special mathematical symbols should be introduced only with prior authorization from the publisher and *after* your sample file has been evaluated. Again, be prepared to devise an alternative strategy in cooperation with the editorial staff if this is required.

17   Consign all equations and mathematical formulas to a separate "formula manuscript", introducing only appropriate "place-holders" into the text manuscript.

18   Ask the publisher for explicit advice on how footnotes, figure captions, and tables should be dealt with: whether they should be included in the text manuscript or kept separate.

19   Give special attention to the "metainformation" your manuscript contains, including such things as nonbreaking hyphens as well as formats applied to headings, tables, and figure captions. In particular, strive for strict consistency in all your work.

20   Be sure to keep backup copies of everything, and supply the publisher with not only a set of data files, but also an equivalent hardcopy version consistent in every respect with information in the data files.

Appendix D
Conversion Tips

D.1 Conversions Between the MACINTOSH and WINDOWS Worlds

D.1.1 Introduction

Sometimes an attempt to perform what should be a "simple" file conversion runs amok, leading to a useless mishmash of miscellaneous characters or at best a new file in need of very extensive repairs (complete reformatting, for example). Often the problem is not due to a lack of appropriate conversion capabilities (such as missing filters); in many cases the fault can be traced instead to improper use of facilities already at hand. We offer here a few tips to help you avoid some of the most common mistakes.

Essentially all the major word-processing and page-layout software is available in comparable versions for both of the important software platforms (i.e., MACINTOSH and WINDOWS), permitting files—and above all formats—to be passed back and forth between the two fully intact. Font issues may of course interfere, but these arise even between computers sharing a similar operating system. If some required font or type style is unavailable at the receiving end of the transaction, thereby mandating a font substitution, line-break patterns will change, which can wreak havoc with a layout. The solution is straightforward, however: always make sure both computers involved in a transfer share the same set of fonts in all the relevant styles. Once this prerequisite has been met, the only other likely impediment to seamless conversion would be a difference in "Preference" settings (with respect to hyphenation, for example, or letter or word spacings). This of course requires that suitable adjustments be made and the transfer be repeated, but at least it will not be necessary to reformat the entire document.

One other potentially sensitive aspect of a file-transfer operation between nonidentical operating systems has to do with graphics files, which we consider separately from the standpoint of page-layout and word-processing programs.

D.1.2 Layout Programs

So long as all graphics and photos have literally been *incorporated* into a layout file, and the graphic formats are ones both platforms recognize (e.g., EPS, TIF, JPG), no conversion problems should arise. It is still advisable, however, also to pass along copies of the original illustration files, especially EPS files. On the other hand, things become more complicated with illustrations that have not actually been *included* in a layout file, but rather "embedded" in it with the aid of either "publish/subscribe" (MACINTOSH) or OLE (WINDOWS) methodology. Embedded graphics lead to layout files that are much smaller, but the graphic instructions themselves remain buried in files of their own in some folder on the computer's hard disk. If you wish to transfer such a layout file to a different computer, all the original graphic files must be transferred as well, otherwise essential information will be missing and illustrations you expect to see will be replaced by plain gray boxes. Moreover, even when all the appropriate graphic information *is* available, it will still be necessary to reestablish all the defining links, since otherwise the layout program has no idea where to seek the needed instructions. Just as in the case of font problems, by the way, these considerations are equally applicable to an exchange between comparable systems (e.g., file transfer involving two WINDOWS computers).

D.1.3 Word-Processing Programs

Programs in the word-processing category are able to deal *only* with comprehensive data files, which means any associated graphics must necessarily be truly "incorporated". A WORD file containing several bitmap illustrations can easily grow to a magnitude of several mbytes, whereas with graphic information communicated in *embedded* form via page-layout software the master file is unlikely to exceed a few hundred kbytes (not taking into account the obligatory graphics files). Nevertheless, even word-processing files can be subject to "linkage" problems in one sense: namely links involving the *program* responsible for creating an incorporated graphic element in the first place. Thus, when a file is transferred from one platform to another, only if the new system also has the corresponding graphics software installed will it be possible to modify illustrations simply by "double-clicking" on them, which serves as an especially convenient way of calling up the required graphics package.

D.2 File Conversions Involving Two Different Layout Programs

From the outset one must assume that it will not be possible to transfer the complete information content of a page-layout file from one program format to another. The only thing you should anticipate salvaging is *text*, in which at most such minor formatting features as italicization, boldface, and subscripts or superscripts might be preserved. With respect to the way material is arranged on the page, however, or parameters associated with letter and word spacing or hyphenation—to cite but a few crucial characteristics—the various programs differ so profoundly in philosophy, and depend upon such complex data sets, that true "conversion" must be regarded as an unrealistic goal.

Apart from tedious "copy and paste" operations, the simplest approach to transferring the content of various text fields (frames) from one layout program to another, as for example from PAGEMAKER to QUARKXPRESS, is to take advantage of an "exchange format". The format that probably first comes to mind is RTF (rich text format), but if this option happens not to be offered through a "Save as …" command, a good alternative course of action is working by way of one of the WORD filters almost certainly available. Microsoft WORD has come to be regarded as an indispensable piece of software essentially everywhere, and every layout program—without exception— is able to process and export WORD files. In effect, neutral filters like RTF have become virtually superfluous in this context, because with WORD so universally available the latter can almost always be called upon to serve as a "take-off and landing" site.

The suggestion above that "*one* of the WORD filters" be selected was designed to call attention to the fact that various versions of WORD are generally compatible with a particular layout program. Given a choice, the safest procedure is to select a filter adapted to one of the *earlier* rather than more recent WORD versions, since this will increase the likelihood that the layout software available on the other computer (which may itself not represent state-of-the-art!) will interpret the file correctly.

This brings us to perhaps the most important step in DTP file conversion, relevant probably more than 90% of the time: import of a WORD file into a layout program, like QUARKXPRESS, PAGEMAKER, INDESIGN, or FRAMEMAKER. Each of these programs is equipped with WORD filters, but even so the import process sometimes leads to results so poor that one might be tempted to go the long way around via an ASCII file and subsequent reformatting of everything. How can this be?

Let us assume that, in your role as author, you have invested considerable effort in formatting headings, highlighting specific words with italics or boldface, table design, etc. You have even conscientiously assigned "styles" to every paragraph as a way not only of facilitating your own work but also maintaining consistency. Meanwhile, editors at the publishing house have painstakingly created a QUARKXPRESS template with a series of sophisticated styles that will optimize the final appearance of the document. To their dismay, however, they discover that when your file is imported either all formatting disappears, or else, alongside *their* styles in the resulting XPRESS file, your styles suddenly show up as well. Does this mean a massive amount of reformatting is inevitable despite all your efforts? Almost certainly not!

The explanation behind this type of debacle is quite simple: *names* of some of the styles probably need to be revised to create exact matches between the style names in WORD and those in QUARKXPRESS. In practice this means that styles prepared for the layout program are given priority, and style names in WORD are adjusted accordingly.

Securing a perfect match is well worthwhile provided the author has done a reasonably good job of style assignment. It will in fact probably require only a few moments to make all the necessary adjustments thanks to the power of "search and replace" routines. If as author you can ascertain in advance what style names the publisher prefers that you employ, and you take proper advantage of the information, no last-minute modifications whatsoever will be required. When two programs are using comparable styles labeled in precisely the same ways, all of an editor's formatting preferences will be correctly applied as a part of file import: in every paragraph, and to every set of characters designated for display in special type. Only a need for "minor" adjustments with respect to pagination will remain. Thus, it makes absolutely no difference, for example, that you perhaps chose as a font for the style "Heading 1" a 16-point version of Arial, with 12-point Times New Roman as the "Default Paragraph Font". So long as style names are properly matched, the editors' preferences of Franklin Gothic 20 point for "Heading 1", say, and Raleigh 11 point for "Default Paragraph Font" will be scrupulously honored in the course of typesetting.

- Insofar as possible, the principle of ensuring style-name agreement prior to import of a Word file should be observed with all page-layout programs.

Unfortunately, problems will in many cases still occur during import of a file containing tables and formulas. Such elements may even need to be recreated from scratch, or at the very least be reincorporated. At the present time (fall, 2003) only one page-layout program can claim to be an exception in this respect: FRAMEMAKER, version 7.0. This particular program interprets WORD-based tables with absolutely no difficulty, for example. Nothing more should be required after import than perhaps a bit of last-minute "fine-tuning".

FRAMEMAKER is also capable of successfully importing equations prepared with WORD'S equation-editing tool, and such equations can still be edited. The only prerequisite is that formula elements first be transformed into their MATHTYPE counterparts (where MATHTYPE is the professional version of WORD'S EQUATION EDITOR; cf. Sections 6.5.2 and 6.7). Assuming the MATHTYPE program itself is present on the computer running WORD, a WORD-supplied macro can be called upon to accomplish all the necessary changes simultaneously. If an equivalent version of MATHTYPE is also resident on the computer running FRAMEMAKER, editing can be initiated after import simply by "double-clicking" the expression of interest. That is to say, FRAMEMAKER is fully "OLE compliant". This layout program actually supports an equation editor of its own, but the syntax of the latter is accessible for conversion purposes only through the format MIF (Maker Interchange Format), in some ways similar to RTF. The remarkable flexibility associated with FRAMEMAKER means that a publishing house adopting this route to page makeup may find it possible to save considerable effort in the production of books containing large numbers of equations and tables.

One further matter we should touch on in this general context is *filters* included with WORD itself, and how they are employed. These play an important role in the opening of any document created with a program other than WORD. The information that follows seems so far not to be widely appreciated despite its potential importance.

Under the "General" category in WORD'S "Preferences" menu there will be found an inconspicuous checkbox labeled "Confirm conversion at Open". The default setting (e.g., when the program is installed) does *not* have this box checked—and the consequences can be far-reaching. Suppose, for example, you were to attempt to open under these circumstances a document with the file-name extension "txt", signaling that it is an ASCII or "text-only" file. There is a good chance you will be dismayed to find that all letters with umlauts—as well as other symbols with ASCII codes above 128 in the ANSI scheme—have been incorrectly interpreted, and this despite the fact one would assume *especially* with an ASCII file that everything should proceed smoothly. But what has this to do with the "Confirm conversion" checkbox?

Given the opportunity, WORD itself selects the conversion filter to be used when attempting to open a file saved in a non-WORD format. Unfortunately, it does not always choose the *correct* filter. Specifically, WORD apparently cannot distinguish between WINDOWS-ASCII files and DOS-ASCII files. Given the chance, it invariably selects a WINDOWS-ASCII filter. But if the file in question is actually of the *DOS-ASCII* type, many mistakes can result. On the other hand, if you insist on your right to *confirm* the program's choices, you will be given an opportunity to counteract the program's misguided judgement, in the process sparing yourself considerable aggravation. The "bottom line" here is clear: be wary of automatic conversions and claim your prerogative to express an opinion in the choice of a filter.

Literature

Adamski S. 1995. Das Manuskript auf Diskette: Zwischen Euphorie und Enttäuschung. In: Plenz R, editor. 1995 f. *Verlagshandbuch: Leitfaden für die Verlagspraxis*). Hamburg: Input-Verlag.

Adobe Systems Inc. 1986. *Postscript Language Tutorial and Cookbook*. Reading (MA): Addison-Wesley. 256 p.

Adobe Systems Inc. 1999. *Postscript Language Reference Manual*. 3rd ed. Reading (MA): Addison-Wesley. 912 p.

American Institute of Physics. 1978. *Style manual for guidance in the preparation of papers for journals published by the American Institute of Physics and its member societies*. 3rd ed. New York: American Institute of Physics. 56 p.

Barth A. 1992. *Datenbanken in den Naturwissenschaften: Eine Einführung in den Umgang mit Online-Datenbanken* (Bliefert C, Kwiatkowski J, editors. *Datenverarbeitung in den Naturwissenschaften*). Weinheim: VCH. 450 p.

Boiko B. 2001. *Management bible*. New York: Wiley. 816 p.

Bottle RT, Rowland JF, editors. 1993. *Information sources in chemistry* (Series *Guides to information sources*). 4th ed. London: Bowker-Saur. 341 p.

Bradley N. 1996. *Concise SGML companion*. Reading (MA): Addison-Wesley. 336 p.

Bresemann HJ, Zimdars J, Skalski D. 1995. *Wie finde ich Normen, Patente, Reports: Ein Wegweiser zu technisch-naturwissenschaftlicher Spezialliteratur* (Heidtmann F, editor. *Orientierungshilfen*, vol 12). 2nd ed. Berlin: Berlin Verlag Arno Spitz. 283 p.

Bringhurst R. 1996. *The elements of typographic style*. Version 2.5. Vancouver: Hartley & Marks. 254 p.

Briscoe MH. 1996. *Preparing scientific illustrations: a guide to better posters, presentations, and publications*. 2nd ed. New York: Springer. 204 p.

Bugge G, editor. 1929, 1965. *Das Buch der großen Chemiker* (vol 1, 2). Weinheim: Verlag Chemie. 496 p, 559 p.

Butterworth I, editor. 1998. *Impact of electronic publishing on the academic community*. London: Portland Press. 191 p.

Cobb GW. 1998. *Introduction to design and analysis of experiments*. New York: Springer. 795 p.

Council of Biology Editors. 1983. *CBE style manual*. 5th ed. Bethesda (MD): Council of Biology Editors. 324 p.

Council of Biology Editors, Scientific Illustration Committee. 1988. *Illustrating science: standards for publication*. Bethesda, Md: Council of Biology Editors. 296 p.

Council of Biology Editors, Style Manual Committee. 1994. *Scientific style and format: the CBE manual for authors, editors, and publishers*. 6th ed. Cambridge (UK): Cambridge University Press. 704 p.

Daniel HD. 1993. *Guardians of science: fairness and reliability of peer reviews*. Weinheim: VCH. 188 p.

Day RA. 1994. *How to write and publish a scientific paper*. 4th ed. Phoenix (AZ): Oryx Press. 223 p.

Der Brockhaus Computer und Informationstechnologie. 2002. Mannheim: Brockhaus. 1008 p.

Detig C. 1997. *Der LaTeX Wegweiser.* Bonn: International Thompson Publishing. 236 p.

Dodd JS, editor. 1997 (1st ed 1986). *The ACS style guide: a manual for authors and editors.* 2nd ed. Washington (DC): American Chemical Society. 460 p.

Dorra M, Walk H. 1990. *Lexikon der Satzherstellung* (Golpon R, editor. *Lexikon der gesamten grafischen Technik*, vol 2). Itzehoe: Verlag Beruf + Schule. 336 p.

Drazil JV. 1983. *Quantities and units of measurement: a dictionary and handbook.* London: Mansell; Wiesbaden: Brandstätter. 314 p.

Duden-Taschenbuch *Satz- und Korrekturanweisungen* (Duden-Taschenbücher, vol 5). 1986. 5th ed. Mannheim: Bibliographisches Institut. 282 p.

Ebel HF. 1998. Die neuere Fachsprache der Chemie unter besonderer Berücksichtigung der Organischen Chemie. In: Hoffmann L, Kalverkämper H, Wiegand HE, editors. *Fachsprachen: Ein internationales Handbuch zur Fachsprachenforschung und Terminologiewissenschaft*, semivol 1). Berlin: Walter de Gruyter. p 1235-1260.

Ebel HF, Bliefert C. 1982. *Das naturwissenschaftliche Manuskript: Ein Leitfaden für seine Gestaltung und Niederschrift.* Weinheim: Verlag Chemie. 216 p.

Ebel HF, Bliefert C. 1998. *Schreiben und Publizieren in den Naturwissenschaften.* 4th ed. Weinheim: Wiley–VCH. 552 p.

Ebel HF, Bliefert C. 1994. *Vortragen in Naturwissenschaft, Technik und Medizin.* 2nd ed. Weinheim: VCH. 347 p.

Ebel HF, Bliefert C. 2003. *Diplom- und Doktorarbeit: Anleitungen für den naturwissenschaftlich-technischen Nachwuchs.* 3rd ed. Weinheim: Wiley–VCH. 193 p.

Ebel HF, Bliefert C, Russey WE. 1987 (reprint 1990). *The art of scientific writing: from student reports to professional publications in chemistry and related fields.* Weinheim: VCH. 493 p.

Erker G. 2000. Ist die Habilitation für eine Berufung noch notwendig?. *Nachr Chem.* 48:841-843.

Felber H, Budin G. 1989. *Terminologie und Praxis.* Tübingen: Narr. 315 p.

Fischer R, Vogelsang K. 1993. *Größen und Einheiten in Physik und Technik.* 6th ed. Berlin: Verlag Technik.

Funk W, Dammann V, Donnevert G. 1992. *Qualitätssicherung in der Analytischen Chemie.* Weinheim: VCH. 214 p.

Garfield E. 1972. Citation analysis as a tool in journal evaluation. *Science.* 178:471-479.

Goldfarb CF, Prescod P. 2002. *Charles F. Goldfarb's XML handbook.* 4th ed. Upper Saddle River (NJ): Prentice Hall. 1147 p.

Goldfarb CF. 1990. *The SGML handbook.* Rubinsky Y, editor. Oxford: Clarendon. 663 p.

Goossens M, Mittelbach F, Samarin A. 1994. *The LaTeX companion.* Reading (MA): Addison-Wesley. 528 p.

Goossens M, Rahtz S. 1999. *The LaTex Web companion: integrating TeX, HTML, and XML.* Reading (MA): Addison Wesley Longman, 522 p.

Gordon KE. 1993a. *The deluxe transitive vampire: the ultimate handbook of grammar for the innocent, the eager, and the doomed.* New York: Pantheon Books. 175 p.

Gordon KE. 1993b. *The new well-tempered sentence: a punctuation handbook for the innocent, the eager, and the doomed.* Revised and expanded ed. New York: Ticknor & Fields. 148 p.

Gore A. 1992. *Earth in the balance: ecology and the human spirit.* Boston: Houghton Mifflin. 407 p.

Greulich W, Plenz R. 1997. *Das Manuskript auf Diskette.* Hamburg: Input-Verlag. 40 p.

Günther W. 1993. GLP und Akkreditierung. *Nachr Chem Tech Lab.* 41:16-21.

Haeder W, Gärtner E. 1980. *Die gesetzlichen Einheiten in der Technik.* 5th ed. Berlin: Beuth. 268 p.

Heisenberg W. 1973. *Der Teil und das Ganze.* Munich: Deutscher Taschenbuch Verlag. 288 p.

Huth EJ. 1987. *Medical style and format: an international manual for authors, editors, and publishers.* Philadelphia: ISI Press. 356 p.

Huth EJ. 1990. *How to write and publish papers in the medical sciences.* 2nd ed. Baltimore: Williams & Wilkins. 252 p.

Huth EJ. 1998. *Writing and Publishing in Medicine.* 3rd ed. Philadelphia: Lippincott, Williams & Wilkins. 300 p.

IUPAC International Union of Pure and Applied Chemistry. 1979. *Manual of symbols and terminology for physicochemical quantities and units.* Oxford: Pergamon Press. 41 p.

IUPAC International Union of Pure and Applied Chemistry. 1987. *Compendium of chemical terminology – IUPAC recommendations* ("The Golden Book"). Oxford: Blackwell. 456 p.

IUPAC International Union of Pure and Applied Chemistry. 1988. *Quantities, units and symbols in physical chemistry* ("The Green Book"). Oxford: Blackwell. 134 p.

IUPAC International Union of Pure and Applied Chemistry. 1990. *Nomenclature of inorganic chemistry – recommendations 1990* ("The Red Book"). Oxford: Blackwell. 290 p.

Jaspers K. 1953. *Einführung in die Philosophie.* Munich: Piper. 164 p.

Jean G. 1991. *Die Geschichte der Schrift* (Reihe *Abenteuer Geschichte*, Bd 18). Ravensburg: Ravensburger Taschenbücher. 215 p.

Jelliffe R. 1998. *The XML & SGML cookbook: recipes for structured information.* Upper Saddle River (NJ): Prentice Hall. 650 p.

Knappen J, Partl H, Schlegl E, Hyna I. 1994. *LaTex 2e-Kurzbeschreibung.* (ftp://ftp.dante.de/tex-archive/documentation/latex2e-Kurzbeschreibung/)

Knuth DE. 1986. *The TeXbook* Reading (MA): Addison-Wesley. 483 p.

Kopka H, Daly PW. 2003. *A guide to LATEX.* 4th Ed. Reading (MA): Addison-Wesley. 624 p.

Korwitz U. 1995. Angebot und Nutzung elektronischer Informationen aus der Sicht der Zentralen Fachbibliotheken. *4tes Weinheimer Bibliothekartreffen: Referate.* Weinheim: VCH Publishing Group, Private printing.

Lamport L. 1994. *LATEX: a document preparation system.* 2nd ed. Reading (MA): Addison-Wesley. 288 p.

Lamprecht J. 1992. *Biologische Forschung: von der Planung bis zur Publikation.* Hamburg: Parey. 159 p.

Lawson A. 1990. *Anatomy of a typeface.* Boston: Godine. 428 p.

Maizell RE. 1998. *How to find chemical information: a guide for practicing chemists, educators, and students.* New York: Wiley. 515 p.

Maler E, Andaloussi JE. 1996. *Developing SGML DTDs: from text to model to markup.* Upper Saddle River (NJ): Prentice Hall PTR. 532 p.

Marchal B. 2000. *XML by example.* Indianapolis (IN): Que. 505 p.

Matthiessen G. 1998. *Relationale Datenbanken und SQL: Konzepte der Entwicklung und Anwendung.* Bonn: Addison-Wesley. 305 p.

McGilton H, Campione M. 1992. *PostScript by example.* Reading (MA): Addison-Wesley. 640 p.

Megginson D. 1998. *Structuring XML documents.* Upper Saddle River (NJ): Prentice Hall PTR. 420 p.

Moore DS, McCabe GP. 2003. *Introduction to the practice of statistics.* New York: Freeman. 828 p.

Neter J, Kutner MH, Nachtsheim CJ, Wasserman W. 1996 *Applied linear statistical models.* 4th ed. Chicago: Irwin. 720 p.

Niederst J. 2001. *Web design in a nutshell: a desktop quick reference.* Sebastopol (CA): O'Reilly. 618 p.

O'Connor M. 1978. *Editing scientific books and journals: an ELSE–Ciba Foundation guide for editors.* London: Pitman. 218 p.

O'Connor M. 1986. *How to copyedit scientific books and journals.* Philadelphia: ISI Press. 150 p.

O'Connor M. 1991. *Writing successfully in science.* London: Harper Collins Academic. 229 p.

Page G, Campbell R, Meadows J. 1997. *Journal publishing.* Revised ed. New York: Cambridge University Press. 407 p.

Parker CP, Turley RV. 1986. *Information sources in science and technology: a practical guide to traditional and online use.* 2nd ed. London: Butterworth. 328 p.

Partington JR. 1957. *A short history of chemistry.* 3rd ed. New York: St. Martin's. 416 p.

Plenz R, editor. 1995 f. *Verlagshandbuch: Leitfaden für die Verlagspraxis.* Hamburg: Input-Verlag. Looseleaf work.

Plenz R. 1995. DTP: Gewinner und Verlierer In: Plenz R, editor. 1995 f. *Verlagshandbuch: Leitfaden für die Verlagspraxis.* Hamburg: Input-Verlag.

Ramsay FL, Shafer DW. 2002. *The statistical sleuth: a course in methods of data analysis.* 2nd ed. Pacific Grove (CA): Duxbury/Thomson Learning. 742 p.

Reid EE. 1970. *Chemistry through the language barrier; how to scan chemical articles in foreign languages with emphasis on Russian and Japanese.* Baltimore: Johns Hopkins Press. 138 p.

Rheingold H. 2000. *The virtual community: homesteading on the electronic frontier.* Revised ed. Cambridge (MA): MIT Press. 447 p.

Rosenberg AD, Hizer DV. *The resume handbook: how to write outstanding resumes and cover letters for every situation.* 2nd ed. Avon (MA): Adams Media. 176 p.

Roth, K. 1992. Der Fall des Nobelpreisträgers David Baltimore ("Fälschung in der Wissenschaft"). *Nachr Chem Tech Lab.* 40:303-308.

Russey WE, Bliefert C, Villain C. 1995. *Text and graphics in the electronic age: desktop publishing for scientists.* Weinheim: VCH. 359 p.

Salsburg D. 2001. *The lady tasting tea: how statistics revolutionized science in the twentieth century.* New York: W.H. Freeman. 340 p.

Sauer H. 1997. *Relationale Datenbanken: Theorie und Praxis.* 3rd ed. Bonn: Addison-Wesley. 291 p.

Schneider W. 1989. *Deutsch für Kenner: Die neue Stilkunde.* 4th ed. Hamburg: Gruner + Jahr. 400 p.

Schoenfeld R. 1989. *The chemist's English with "say it in English, please!".* 3rd ed. Weinheim: VCH. 195 p.

Schumann L. 1995. *Professioneller Buchsatz mit TeX.* 3rd ed. Munich: Oldenbourg. 366 p.

Schwenk E. 1992. *Mein Name ist Becquerel: Wer den Maßeinheiten die Namen gab.* Frankfurt/Main: Hoechst. 216 p.

Seroul R, Levy S. 1991. *A beginner's book of TeX.* New York: Springer. 282 p.

Snow, W. 1992. *TeX for the beginner.* Reading (MA): Addison-Wesley. 377 p.

Standard ANSI/NISO Z39.14-1997. *Guidelines for abstracts.*

Standard ANSI Z39.16-1979. *American National Standard for the preparation of scientific papers for written and oral presentation.*

Standard ANSI/ISO Z39.18-1995. *American National Standard for the preparation of scientific and technical reports: elements, organization, and design.*

Standard ANSI Z39.29-1977. *American National Standard for bibliographic references.*

Standard ANSI Z39.50-1995. *Information retrieval (Z39.50) application service definition and protocol specification.*

Standard BS-4811: 1972. *Specification for the presentation of research and development reports.*

Standard BS-4812: 1972. *The presentation of theses.*

Standard BS-5261: Part 1: 1975. *Copy preparation and proof corrections: recommendations for preparation of typescript copy for printing.*

Standard DIN 15 Blatt 1. 1967. *Linien in Zeichnungen: Linienarten, Linienbreiten, Anwendung.*

Standard DIN 461. 1973. *Graphische Darstellungen in Koordinatensystemen.*

Standard DIN 1301-1. 1993. *Einheiten: Einheitennamen, Einheitenzeichen.*

Standard DIN 1302. 1994. *Allgemeine mathematische Zeichen und Begriffe.*

Standard DIN 1304-1. 1994. *Formelzeichen: Allgemeine Formelzeichen.*

Standard DIN 1319-3. 1996. *Grundlagen der Meßtechnik: Auswertung von Messungen einer einzelnen Meßgröße, Meßunsicherheit.*

Standard DIN 1333. 1992. *Zahlenangaben.*

Standard DIN 1338. 1996. *Formelschreibweise und Formelsatz.*

Standard DIN 1338 Beiblatt 1. 1996. *Formelschreibweise und Formelsatz: Form der Schriftzeichen.*

Standard DIN 1338 Beiblatt 2. 1996. *Formelschreibweise und Formelsatz: Ausschluß in Formeln.*

Standard DIN 1338 Beiblatt 3. 1980. *Formelschreibweise und Formelsatz: Formeln in maschinenschriftlichen Veröffentlichungen.*

Standard DIN 1421. 1983. *Gliederung und Benummerung in Texten: Abschnitte, Absätze, Aufzählungen.*

Standard DIN 1422-1. 1983. *Veröffentlichungen aus Wissenschaft, Technik, Wirtschaft und Verwaltung: Gestaltung von Manuskripten und Typoskripten.*

Standard DIN 1422-2. 1984. *Veröffentlichungen aus Wissenschaft, Technik, Wirtschaft und Verwaltung: Gestaltung von Reinschriften für reprographische Verfahren.*

Standard DIN 1422-4. 1986. *Veröffentlichungen aus Wissenschaft, Technik, Wirtschaft und Verwaltung: Gestaltung von Forschungsberichten.*

Standard DIN 1426. 1988. *Inhaltsangaben von Dokumenten: Kurzreferate, Literaturberichte.*

Standard DIN 1463-1. 1987. *Erstellung und Weiterentwicklung von Thesauri: Einsprachige Thesauri.*

Standard DIN 1505-1. 1984. *Titelangaben von Dokumenten: Titelaufnahme von Schrifttum.*

Standard DIN 1505-2. 1984. *Titelangaben von Dokumenten: Zitierregeln.*

Standard DIN 2331. 1980. *Begriffssysteme und ihre Darstellung.*

Standard DIN 5007. 1991. *Ordnen von Schriftzeichenfolgen (ABC-Regeln).*

Standard DIN 5008. 1996. *Schreib- und Gestaltungsregeln für die Textverarbeitung.*

Standard DIN 16 507. 1964. *Typographische Maße.*

Standard DIN 28 004-1. 1988. *Fließbilder verfahrenstechnischer Anlagen: Begriffe, Fließbildarten, Informationsgehalt.*

Standard DIN 28 004-3. 1988. *Fliessbilder verfahrenstechnischer Anlagen: Graphische Symbole.*

Standard DIN 32 640. 1986. *Chemische Elemente und einfache anorganische Verbindungen: Namen und Symbole.*

Standard DIN 32 641E 1999. *Chemische Formeln.*

Standard DIN 40 719-6. 1992. *Schaltungsunterlagen: Regeln und graphische Symbole für Funktionspläne.*

Standard DIN 55 301. 1978. *Gestaltung statistischer Tabellen.*

Standard DIN 66 001. 1983. *Informationsverarbeitung: Sinnbilder und ihre Anwendung.*

Standard DIN EN ISO 9000-1. 1994. *Normen zum Qualitätsmanagement und zur Qualitätssicherung – Leitfaden zur Auswahl und Anwendung.*

Standard DIN EN ISO 9001. 1994. *Qualitätsmanagementsysteme – Modell zur Qualitätssicherung/QM-Darlegung in Design/Entwicklung, Produktion, Montage und Wartung.*

Standard DIN ISO 9004-1. 1994. *Qualitätsmanagement und Elemente eines Qualitätssicherungssystems – Leitfaden.*

Standard DIN ISO 9004-2. 1992. *Qualitätsmanagement und Elemente eines Qualitätssicherungssystems – Leitfaden für Dienstleistungen.*

Standard DIN V 8418. 1988. *Benutzerinformation – Hinweise für die Erstellung.*

Standard EN 45 001. 1989. *Allgemeine Kriterien zum Betreiben von Prüflaboratorien.*

Standard ISO 4-1997. *Information and documentation, rules for the abbreviation of title words and titles of publications.*

Standard ISO 31-11:1992. *Quantities and units: mathematical signs and symbols for use in the physical sciences and technology.*

Standard ISO 128-2003. *Technical drawings: general principles of presentation.*

Standard ISO 214-1976. *Documentation: abstracts for publications and documentation.*

Standard ISO 690-1987. *Documentation – bibliographic references: content, form and structure.*

Standard ISO 2108-1992. *Information and documentation: International standard book numbering (ISBN).*

Standard ISO 2145-1978. *Numbering of divisions and subdivisions in written documents.*

Standard ISO 3511-1:1977. *Process measurement control functions and instrumentation – Symbolic representation – Part 1: Basic requirements*

Standard ISO 3511-2:1984. *Process measurement control functions and instrumentation – Symbolic representation – Part 2: Extension of basic requirements*

Standard ISO 3511-3:1984. *Process measurement control functions and instrumentation – Symbolic representation – Part 3: Detailed symbols for instrument interconnection diagrams*

Standard ISO 3511-4:1985. *Industrial process measurement control functions and instrumentation – Symbolic representation – Part 4: Basic symbols for process computer, interface, and shared display/control functions*

Standard ISO 3166-1:1997. *Codes for the representation of names of countries and their subdivisions – part 1: country codes.*

Standard ISO 5966-1982. *Documentation: presentation of scientific and technical reports.*

Standard ISO 7144-1986. *Documentation: presentation of theses and similar documents.*

Standard ISO 8879-1986. *Standard generalized markup language (SGML).*

Standard VDI 4500 Blatt 1. 1995. *Technische Dokumentation – Benutzerinformation.*

Stern D. 2000. *Information sources in the physical sciences.* Westport (CT): Libraries Unlimited. 227 p.

Stix, G. 1995. Publizieren mit Lichtgeschwindigkeit. *Spektrum Wiss.* 3: 34-39.

Syropoulos A, Tsolomitis A, Sofroniou N. 2003. *Digital typography using LaTeX.* New York: Springer. 510 p.

Strunk W Jr, White EB, Angell R. 2000. *The elements of style.* 4th ed. Boston: Allyn and Bacon. 105 p.

The Chicago manual of style. 2003. 15th ed. Chicago: University of Chicago Press. 956 p.

Tufte ER. 2001. *The visual display of quantitative information.* 2nd ed. Cheshire (CT): Graphics Press. 197 p.

Turabian KL. 1996. *A manual for writers of term papers, theses, and dissertations (Chicago guides to writing, editing, and publishing).* 6th ed. Chicago: University of Chicago Press. 308 p.

Uebel JF. 1996. Laserbelichtung von Offsetfilmen. In: Plenz R, editor. 1995 f. *Verlagshandbuch: Leitfaden für die Verlagspraxis).* Hamburg: Input-Verlag.

Warr WA, Suhr C. 1992. *Chemical information management.* Weinheim: VCH. 261 p.

White J. 1988. *Graphic design for the electronic age.* New York: Watson-Guptil. 211 p.

Willard W. 2001. *HTML: a beginner's guide.* Berkeley (CA): Osborne/McGraw-Hill. 569 p.

Wolman Y. 1988. *Chemical information: a practical guide to utilization.* Chichester: Wiley. 292 p.

Wood DN, Hardy JE, Harvey AP, editors. 1989. *Information sources in the earth sciences (Serie Guides to information sources).* 2nd ed. London: Bowker-Saur. 524 p.

Wyatt HV, editor. 1997. *Information sources in the life sciences (Series Guides to information sources).* 4th ed. London: Bowker-Saur. 250 p.

Zimmer DE. 2000. *Die Bibliothek der Zukunft.* Hamburg: Hoffmann & Campe. 331 p.

Index

This index includes entries at two levels only. In many cases, for convenience sake, entry/subentry pairs are provided in permuted form; e.g., as both

outline	*and*	research report	
research report		outline	

You will occasionally encounter "see" and "see also" cross-references as well; e.g.,

Random Access Memory *see* RAM
figure *see also* illustration

The notations "f" and "ff" indicate that the treatment of a subject continues on a subsequent page, or multiple following pages, respectively.—For optimal retrieval results it is suggested that several different approaches be tried.

Notes regarding the production of this book

Text was prepared using Microsoft WORD with various Apple MACINTOSH computer models (IMAC, POWERBOOK, POWERMAC). Laser-printer output was utilized as a hardcopy manuscript. All page layouts were prepared by the authors with Adobe PAGEMAKER. Most graphic elements were created with FREEHAND from Macromedia.

Body text was displayed on the screen and reproduced on paper in 12-point Times with 16-point line spacing, footnotes in 10-point Times with 12-point line spacing, and nearly all the tables in 11-point Times with 13-point line spacing. The width of a standard textblock was 149 mm before reduction to 80% for final printing.

All pages were recorded directly onto thermal printing plates with a TRENDSETTER 3244 digital imagesetter at 2400 dpi resolution.